Drug Discovery in Africa

Kelly Chibale • Mike Davies-Coleman •
Collen Masimirembwa
Editors

Drug Discovery in Africa

Impacts of Genomics, Natural Products, Traditional Medicines, Insights into Medicinal Chemistry, and Technology Platforms in Pursuit of New Drugs

Springer

Editors
Kelly Chibale
Dept. of Chemistry
University of Cape Town
Rondebosch
South Africa

Mike Davies-Coleman
Department of Chemistry
Rhodes University
Grahamstown
South Africa

Collen Masimirembwa
African Institute of Biomedical Science
and Technology
Harare
Zimbabwe

ISBN 978-3-642-28174-7 ISBN 978-3-642-28175-4 (eBook)
DOI 10.1007/978-3-642-28175-4
Springer Heidelberg New York Dordrecht London

Library of Congress Control Number: 2012938653

Printed on acid-free paper

Springer is part of Springer Science+Business Media (www.springer.com)

Professor Johnson Jato. A Tribute

On June 27, 2012, we were deeply saddened to hear of the passing in his sleep of our dear friend and colleague, Johnson Jato, at his home in Yaounde, Cameroon.

In Chap. 2, we discuss some of the significant and novel drug leads discovered from African biodiversity in the US National Cancer Institute (NCI) program. In launching the NCI plant collection program in Cameroon in 1986 through a contract with Missouri Botanical Garden (MBG), Professor Johnson Jato of the University of Yaounde proved to be the key contact for both the NCI and MBG, and prominent among the drug leads discovered were the potential anti-HIV agent, michellamine B, and the antitumor agents, the schweinfurthins.

Dr. James Miller, who at that time was one of the main MBG investigators, now Dean and Vice President for Research at New York Botanical Garden, remarked —"Johnson was tremendously helpful in a country that is often confusing and difficult, and he helped us navigate everything from basic logistics to difficult politics. My most vivid memory of Johnson was that he always did it all in a tremendously cheerful, optimistic way with a great sense of humor. We could not have accomplished what we did without his help." Regarding the michellamine B project and the cultivation of *Ancistrocladus korupensis*, Dr. James Simon, who supervised the project while at Purdue University, noted—"Johnson did seem to be ageless with a super-sized heart of gold and always optimistic (or nearly always anyway). I remember him quite vividly and his leadership in the michellamine B project was instrumental in the ability for all of us to do the work, do it well and navigate it all through the complexities found out there (in Cameroon). He was a generous soul." In the schweinfurthin project, it was the successful scale-up recollections undertaken by Professor Jato in Cameroon in 1998–1999 that permitted the ongoing preclinical studies of schweinfurthin analogues to proceed.

The first interaction between Johnson and the NCI was when he visited Dr. Joe Mayo, Chief of the Biological Testing Branch, in the early 1980s to discuss the possibility of NCI helping with establishing some preclinical testing at the University of Yaounde 1. While no formal collaboration was established, Johnson and Joe became close personal friends, and Joe developed a huge admiration for the level of research Johnson achieved with very limited resources. Their friendship continued over the years, and in 1986, when the NCI established the plant collection contract with MBG, Joe introduced Johnson to Gordon as a possible contact in Cameroon. Johnson and Gordon became firm friends and colleagues and collaborated closely in the various NCI projects in Cameroon. This close relationship extended to all of us, and we kept in contact with Johnson right to the end. We all remember Johnson as an ever-cheerful friend and a resourceful and reliable colleague who approached every project with a determined optimism which enabled him to overcome seemingly insurmountable challenges, and which earned him the respect and admiration of all his colleagues. We can truly say we could not have accomplished what we did without his help and inspiration. Thank you, Johnson. We will miss you!

John Beutler, Gordon Cragg, and David Newman

Foreword

African Drug Discovery: A Window on the World

Africa is a continent of tremendous opportunity: a wealth of natural resources, people talent and energy. However, there is also a great burden of neglected diseases, not only in terms of suffering and lives lost, but also as a loss of economic value. Some of these diseases are shared with the Western world, but many are either especially prevalent to Africa, or even unique to the continent.

This book represents a ray of hope at a difficult time. It is a window on the world of African science, and the perspective on the discovery and progression of new medicines to target key diseases. Bringing together some of the best researchers in Africa with a common goal, to focus on how to develop new medicines in an African context. Drug discovery and development requires being able to pull together people with a wide variety of expertise and background, to form teams with a common goal. It means bringing together the basic science and medical research. It means bridging the gap between science and technology.

But those who invest in new medicines must also have deep pockets, and great patience. Even with the best will in the world, a new pharmaceutical can take a decade to be brought through the development pathway. Traditionally, this research has largely been funded from outside Africa; what are the signs that the political will here is changing? In 2006 in the Khartoum declaration [1], members of the African Union set a target to invest 1% of their gross domestic product in R&D. This target is met by three very different countries—Malawi, Uganda and South Africa—so there is every reason to hope that more will follow if there is the political will [2]. There was also a call for the establishment of centres of excellence in Africa. How many of those will feed into the overall objective of new medicines is not clear yet. However, the establishment of ANDi (the African Network for New Drug and Diagnostic Innovation) [3] is surely an important milestone in this process. The Cairo declaration of the African Ministerial Council on Science and Technology (AMCOST) [4] made two other strong recommendations: support of South-South cooperation in science, technology and innovation and the establishment of a Pan-African Intellectual

Property Organization (PAIPO). This book contains many examples of the former and hopefully sows the seeds for the latter.

At a time when the productivity of the pharmaceutical industry is static, and its financial star is waning, it is pertinent to ask why of all technologies the countries in Africa should prioritise investments in new medicines. One key to success in the identification of new medicines is the proximity to patients, having researchers who understand on a daily basis the devastating effects of the diseases they are tackling. African approaches to neglected disease will not only come from the heart, but will have this insight. Second, the patrimony of the continent is its natural resources, which can be used for the benefit of all Africa's people. We need to work alongside traditional healers to understand these products better [5]. We need to follow the Chinese approach of *dao-xing-ni-shi* (acting in the reversed direction): [6] use observational clinical studies to improve our clinical understanding of how these natural products work, before becoming too reductionist. These natural products have existed side by side with the population for generations, and so the information known about their clinical aspects is critical. The traditional remedies of yesterday have been improved by the skills of our chemists, pharmacologists and physicians into the medicines of today [7]. In a similar way, the traditional remedies of today contain the seeds of the ideas for the medicines for tomorrow. The focus on natural products from Africa is clear in this book; half the chapters are dedicated to some aspect of the work. We need to develop an African way forward for natural products. They key here is that the way these products are moved forward must be seen from an African perspective. We must avoid the tendency to 'cut and paste' the western experience of drug development.

Ultimately, science is about people, and developing a new generation of scientists. Beyond funding, this requires training of the next generation of scientists, and opportunities for scientific exchange between countries [8]. It is my hope that the readers of this book will benefit not only from the scientific content, but from the improved visibility and interaction with the new generation of African science leaders. Ultimately, the chance to change the future of Africa is in their hands. The window on African science is open; we need to all work together to keep it that way.

Tim Wells
Chief Scientific Officer
Medicines for Malaria Venture
Geneva
Switzerland

References

1. EX. CL/Dec.254 (VIII) http://www.africa-union.org/root/au/Documents/Decisions/com/AU6th_ord_Council_Decisions_Jan2006_Khartoum.pdf
2. UNESCO world Science Report

3. http://www.andi-africa.org/
4. http://www.nepadst.org/doclibrary/pdfs/cairo_declaration_2006.pdf
5. Willcox ML, Graz B, Falquet J, Diakite C, Giani S, Diallo D (2011) A "reverse pharmacology" approach for developing an anti-malarial phytomedicine. Malar J 10(Suppl 1):S8
6. Lei SH (1999) From Changshan to a new antimalarial drug. Soc Stud Sci 29:323–358
7. Wells TNC (2011) Natural products as starting points for future anti-malarial therapies: going back to our roots? Malar J 10(Suppl 1):S3
8. Murenzi R (2011) Give the new generation a chance. Nature 474:543

Preface

It is vital for African scientists to enhance the drug discovery capability of the continent to address African health needs. This book highlights the status of early stage drug discovery activities in Africa with a view to capturing strengths, weaknesses and opportunities within this field. The book demonstrates that at present, drug discovery expertise in Africa exists in part, yet has huge potential to build capacity and competency in the relevant areas of drug discovery including target identification, hit discovery, medicinal chemistry, preclinical pharmacology as well as drug metabolism and pharmacokinetic studies.

The following 17 chapters cover diverse topics from target identification and validation, hit identification and hit to lead medicinal chemistry progression, through to drug delivery systems. Some chapters offer a historical overview of Africa's efforts at drug discovery from traditional medicines through to natural product-driven search for hits against infectious and non-communicable diseases. While the sources of the contributions clearly indicate that only a few places in Africa have established the competencies to champion modern drug discovery, network-based initiatives such as the African Network for Drugs and Diagnostics Innovation, ANDI, demonstrate an emerging strategy to build capacity for drug discovery research across Africa. Some chapter contributions have been made by international researchers who have committed their research efforts to finding healthcare solutions for Africa. These colleagues are a vital link for African researchers to the international community.

To the best of our knowledge, this is the first book dedicated to contemporary drug discovery approaches in Africa. Although traditional medicines' research has often characterised research agendas at many African institutions, they have rarely yielded verifiable results with respect to the treatment or control of infectious and non-communicable diseases responsible for the high morbidity and mortality in Africa. Reasons for this lack of participation in drug discovery by African institutions and scientists are varied and include a lack of a common culture of research and innovation, limited government and/or private sector financial support for drug discovery research, and poor access to technological platforms and pharmaceutical industry expertise. Contributions in this book seek to break the myth that it is not possible to do good science in Africa, as they demonstrate emerging world class scientific work by researchers on the African continent. The book therefore hopes to inspire a new generation of African scientists to courageously build on the cases presented in this book, to witness African scientists contributing significantly to the discovery of medicines for diseases that are partly responsible for the stagnation of Africa's social and economic progress.

The drug discovery process is a multidisciplinary undertaking requiring teams composed of biochemists, medicinal chemists, pharmacologists, molecular biologists, computational chemists and many others. In putting together this book, we have tried to bring to bear the contributions of these various experts towards the one objective of drug discovery. In a conventional pharmaceutical industry setting, experts work in integrated teams with a shared project vision and committed resources to meet set targets and deliverables. Chapters presented here, however, reflect scattered efforts by individuals in time and space, involving isolated aspects of the drug discovery process. This suggests that presently little will change in the emergence (or lack thereof) of drugs from Africa. This book, however, aims to highlight identified areas of expertise which we hope can be forged into an effective drug discovery pipeline.

The book covers current sciences and technology for drug discovery: crystallography in discerning ligand–enzyme interactions in the design of angiotensin-converting enzyme (ACE) inhibitors, harvesting the chemical diversity of the plant and marine biodiversity of Africa; novel approaches in target discovery against *Mycobacterium tuberculosis*; application of in silico, in vitro and in vivo Drug Metabolism and Pharmacokinetics (DMPK) in the whole drug discovery value chain; exploration of nanotechnology as a drug delivery vehicle to rescue old drugs by addressing their PK and safety limitations; and repositioning of some drugs for the treatment of infectious diseases. The authors address new ideas emerging on how to increase chances of identifying lead compounds from natural products, given the clear need for a paradigm shift from the traditional approaches that generally resulted in either reports of medicinal plant extracts having activity in a standard microorganism-based assay, or the purification and structural elucidation of natural products, publishing a paper and storing the pure compounds in a laboratory cupboard. The book also takes stock of some important initiatives towards drug discovery across the continent, and highlights the failure of most governments to honour their promise to fund science, technology and innovation.

This book targets African life sciences institutions and their leaders to inspire them to give high priority to science and technology that supports the drug discovery process. It also targets young scientists, and encourages them to see the exciting opportunities in the field of drug discovery and development. Above all, the book targets the policy makers on the need to have budget lines for drug discovery initiatives, because we believe that results from such investments can have positive national and continental implications. The book ultimately aims to demonstrate to the international community the seriousness with which Africa is taking the need to engage in drug discovery research.

As editors of this book, we are grateful to all the leading researchers who have given up their time to write the chapters in this book. They have demonstrated the team spirit that is required for a successful drug discovery campaign. We would also like to thank the scientists who reviewed these chapters, ensuring they meet international standards. We are grateful to Dr. Aloysius T. Nchinda for coordinating the writers and reviewers, completed with exceptional levels of professionalism, and wish to thank Dr. Heather Davies-Coleman for proofreading the chapters and ensuring consistency in the presentations.

Rondebosch, South Africa — Kelly Chibale
Grahamstown, South Africa — Mike Davies-Coleman
Harare, Zimbabwe — Collen Masimirembwa

Contents

Contributors

Ivan Addae-Mensah Department of Chemistry, University of Ghana, Legon, Ghana, a-mensah@ug.edu.gh

Kerstin Andrae-Marobela Department of Biological Sciences, University of Botswana, Gaborone, Botswana; Center for Scientific Research, Indigenous Knowledge and Innovation (CesrIKi), Gaborone, Botswana, marobelak@mopipi.ub.bw

Nyasse Barthélemy Laboratory of Medicinal Chemistry and Pharmacognosy, Faculty of Science, University of Yaoundé I, Yaounde, Cameroon, bnyasse@yahoo.com

John A. Beutler Molecular Targets Laboratory, Center for Cancer Research, Frederick, MD, USA, beutlerj@mail.nih.gov

Gregory L. Blatch Biomedical Biotechnology Research Unit (BioBRU), Department of Biochemistry, Microbiology and Biotechnology, Rhodes University, Grahamstown, South Africa, g.blatch@ru.ac.za

Kelly Chibale Department of Chemistry and Institute of Infectious Disease and Molecular Medicine, University of Cape Town, Rondebosch, South Africa, Kelly.Chibale@uct.ac.za

J. Coates Advanced Materials Division, Mintek, Johannesburg, South Africa

Gordon M. Cragg Molecular Targets Laboratory, Center for Cancer Research, Frederick, MD, USA, newmand@mail.nih.gov

Michael T. Davies-Coleman Department of Chemistry, Rhodes University, Grahamstown, South Africa, m.davies-coleman@ru.ac.za

Ross G. Douglas Division of Medical Biochemistry, Institute of Infectious Disease and Molecular Medicine, University of Cape Town, Observatory, Cape Town, South Africa

Mthandazo Dube Department of Biological Sciences, University of Botswana, Gaborone, Botswana

Adrienne L. Edkins Biomedical Biotechnology Research Unit (BioBRU), Department of Biochemistry, Microbiology and Biotechnology, Rhodes University, Grahamstown, South Africa, a.edkins@ru.ac.za

Timothy G. Geary Institute of Parasitology, McGill University, Ste-Anne-de-Bellevue, QC, Canada, timothy.g.geary@mcgill.ca

Eric M. Guantai Division of Pharmacology, School of Pharmacy, University of Nairobi, Nairobi, Kenya, eguantai@uonbi.ac.ke

Rose Hayeshi Council for Scientific and Industrial Research, Polymers and Composites, Pretoria, South Africa, RHayeshi@csir.co.za

R. Hewer Advanced Materials Division, Mintek, Johannesburg, South Africa, raymondh@mintek.co.za

Lonji Kalombo Council for Scientific and Industrial Research, Polymers and Composites, Pretoria, South Africa

Lebogang Katata Council for Scientific and Industrial Research, Polymers and Composites, Pretoria, South Africa

Sami A. Khalid Faculty of Pharmacy, University of Science & Technology, Khartoum, Sudan, khalidseek@hotmail.com

F. H. Kriel Advanced Materials Division, Mintek, Johannesburg, South Africa

Yolandy Lemmer Council for Scientific and Industrial Research, Polymers and Composites, Pretoria, South Africa

Collen Masimirembwa Department of DMPK/PD and Toxicology, African Institute of Biomedical Science and Technology, Harare, Zimbabwe; Department of Clinical Pharmacology, University of Cape Town, Cape Town, South Africa, collenmasimirembwa@yahoo.com

Paula Melariri Council for Scientific and Industrial Research, Polymers and Composites, Pretoria, South Africa

Valerie Mizrahi MRC/NHLS/UCT Molecular Mycobacteriology Research Unit, DST/NRF Centre of Excellence for Biomedical TB Research, Institute of Infectious Disease and Molecular Medicine, University of Cape Town, Cape Town, South Africa, valerie.mizrahi@uct.ac.za

Masego Mokobela Department of Biological Sciences, University of Botswana, Gaborone, Botswana

Keitseng N. Monyatsi African Regional Industrial Property Organisation (ARIPO), Harare, Zimbabwe

Mbaki Muzila Department of Biological Sciences, University of Botswana, Gaborone, Botswana

Leah Mwai Kenya Medical Research Institute (KEMRI)/Wellcome Trust Collaborative Research Program, Kilifi, Kenya

David J. Newman Natural Products Branch, Developmental Therapeutics Program, Frederick, MD, USA, gmcragg@verizon.net

Barbara N. Ngwenya Okavango Research Institute (ORI), University of Botswana Maun Campus, Maun, Botswana; Center for Scientific Research, Indigenous Knowledge and Innovation (CesrIKi), Gaborone, Botswana

Aku N. Ntumy Department of Biological Sciences, University of Botswana, Gaborone, Botswana

Belle Nyamboli Council for Scientific and Industrial Research, Polymers and Composites, Pretoria, South Africa

Alexis Nzila Departments of Chemistry and Clinical Pharmacology, and Institute of Infectious Disease and Molecular Medicine, University of Cape Town, Cape Town, South Africa, alexisnzila@yahoo.co.uk

John Okombo Kenya Medical Research Institute (KEMRI)/Wellcome Trust Collaborative Research Program, Kilifi, Kenya

Dorcas Osei-Safo Department of Chemistry, University of Ghana, Legon, Ghana

Philippe Rasoanaivo Ecole Supérieure Polytechnique, Université d'Antananarivo, Antananarivo, Madagascar; Institut Malgache de Recherches Appliquées, Antananarivo, Madagascar

Solofoniaina Razafimahefa Institut Malgache de Recherches Appliquées, Antananarivo, Madagascar

Boitumelo Semete Council for Scientific and Industrial Research, Polymers and Composites, Pretoria, South Africa

Bongani Sethebe Department of Biological Sciences, University of Botswana, Gaborone, Botswana

Angelina Sosome Department of Biological Sciences, University of Botswana, Gaborone, Botswana

Edward D. Sturrock Division of Medical Biochemistry, Institute of Infectious Disease and Molecular Medicine, University of Cape Town, Observatory, Cape Town, South Africa, edward.sturrock@uct.ac.za

Suthananda N. Sunassee Department of Chemistry, Rhodes University, Grahamstown, South Africa, snsunassee@gmail.com

Hulda Swai Council for Scientific and Industrial Research, Polymers and Composites, Pretoria, South Africa

Roslyn Thelingwani Department of DMPK/PD and Toxicology, African Institute of Biomedical Science and Technology, Harare, Zimbabwe; Department of Chemistry, University of Cape Town, Cape Town, South Africa

Eliane Ubalijoro Institute of Parasitology, McGill University, Ste-Anne-de-Bellevue, QC, Canada; Institute for the Study of International Development, Peterson Hall, McGill University, Montreal, QC, Canada, eliane.ubalijoro@mcgill.ca

Digby F. Warner MRC/NHLS/UCT Molecular Mycobacteriology Research Unit, DST/NRF Centre of Excellence for Biomedical TB Research, Institute of Infectious Disease and Molecular Medicine, University of Cape Town, Cape Town, South Africa, digby.warner@uct.ac.za

Chapter 1
Overview of Current Drug Discovery Activities in Africa and Their Links to International Efforts to Combat Tropical Infectious Diseases

Barthélemy Nyasse

Abbreviations

AMANET	African Malaria Network
AMCOST	African Ministerial Council on Science and Technology
ANDI	African Network for Drugs and Diagnostics Innovation
API	Active Pharmaceutical Ingredients
APOC	African Programme for Onchocerciasis Control
APRIORI	African Poverty Related Infection Oriented Research Initiative
ASTIII	African Science, Technology and Innovation Indicators Initiative
ATPS	African Technology Policy Studies Network
AU	African Union
CARTA	Consortium for Advanced Research Training in Africa
CMH	Commission on Macroeconomics and Health
COHRED	Council on Health Research for Development
EDCTP	European and Developing Countries Clinical Trials Partnership
EMEA	European Medicines Agency
FDA	Food and Drug Administration
GATBDD	Global Alliance for TB Drug Development
GAVI	Global Alliance for Vaccines and Immunization
GDP	Gross Domestic Product
GHPs	Global Health Partnerships

B. Nyasse (✉)
Laboratory of Medicinal Chemistry and Pharmacognosy, Faculty of Science, University of Yaoundé I, Box 812, Yaounde, Cameroon
e-mail: bnyasse@yahoo.com

K. Chibale et al. (eds.), *Drug Discovery in Africa*,
DOI 10.1007/978-3-642-28175-4_1,

GIBEX	Global Institute for BioExploration—Africa
GSPOA	Global Strategy and Plan of Action on Public Health, Innovation and Intellectual Property
HRP	Special Programme for Research and Training in Human Reproduction
HRSA WHO	Health Research Systems Analysis
HVI	HIV Vaccine Initiative
IAVI	International Aids Vaccine Initiative
ICGEB	International Centre for Genetic Engineering and Biotechnology
IGWG	Intergovernmental Working Group
IP	Intellectual Property
ISHReCA	Initiative to Strengthen Health Research Capacity in Africa
JSPS	Japan Society for the Promotion of Science
MCTA	Malaria Clinical Trials Alliance
MDGs	Millennium Development Goals
MDP	Mectizan Donation Programme
MMV	Medicines for Malaria Venture
MVI	Malaria Vaccine Initiative
NAPRECA	Natural Products Research Network of Eastern and Central Africa
NEPAD	New Partnership for Africa's Development
OECD	Organisation for Economic Co-operation and Development
PDPPPs	Product Development Public-Private Partnerships
PEI	Polio Eradication Programme
PMPA	Pharmaceutical Manufacturing Plan for Africa
RBM	Roll Back Malaria
SACORE	Southern African Consortium for Research Excellence
SADC	Southern African development Community
SAMI	South African Malaria Initiative
SIGN	Safe Injection Global Network
TDR	Special Programme for Research and Training in Tropical Diseases
TRIPS	Trade and Related Aspects of Intellectual Property Rights

1.1 Introduction

The driving force for drug discovery and development by pharmaceutical firms has always been the foreseeable profit from drug sales [1–5]. Since most infectious diseases prevail in Africa and the people living there have poor purchasing power, the market for drugs, diagnostics, and vaccines is unattractive to both pharmaceutical firms and other research institutions [6–8]. As a result, there has been reluctance for local governments to invest in drug research but also for pharmaceutical companies to engage in the development of drugs [1], addressing diseases that mainly affect developing countries in general and Africa [1, 9] in particular. Although there are limited research activities in progress in the disease-endemic countries to discover new effective and cheap drugs, it is not yet possible to fully develop leads and drug candidates, even from natural products despite Africa possessing a competitive advantage resulting from its very rich biodiversity. Poor economies [10–12] and technological capabilities and a lack of human resources and good management in many African countries are the major constraints to progress in research and development for drugs and diagnostics.

The lack of market incentives in Africa to drive health product research and development (R&D), fragile health services infrastructure, and weak regulatory environments are all blamed for the high burden of infectious tropical diseases [1, 9]. The increasing global awareness about this inequity led to promises of the availability of new funding from governments. However, the Abuja Declaration of 2001 that set a target for all African countries of 15% of public spending for health has not been fulfilled. The medium- to long-term focus on meeting other international targets such as the Commission on Macroeconomics and Health (CMH) [13] and the Millennium Development Goals [14] (MDGs) remains difficult to attain; the CMH target of US $34 per capita spending is based on an essential package of health services, while the MDG targets directly seek, among others, significant health improvements. Conservative estimates of what is required to attain the CMH and MDG targets seem to indicate, however, that the financing gap between currently available funds and present requirements is well beyond the reach of many African countries. A total of US $20 to US $70 billion per annum until 2015 (compared to US $10 billion total health aid in 2003) has been identified, well beyond the reach of many of the low-income countries. The inefficient management and organization of science and technology in these countries compounds the problem. In 2007, only South Africa, Uganda, and Malawi invested more than 1% of their gross domestic product (GDP) on science and technology [13, 15]. Without a firm commitment by the majority of African governments to raising the level of R&D funding from its current level of less than 0.3% of GDP to at least 1%, no science policy will be effective in generating and sustaining endogenous research [16]. However, a much needed boost in product development activities for some diseases has occurred, through, for example, public–private partnerships [10, 11, 17] with the entry of global health partnerships (GHPs) and new private foundations onto the international funding scene. These new global institutions have also greatly enhanced awareness of issues around HIV/AIDS, tuberculosis, and malaria. There are partnerships owned by the public sector with private sector participants, for example, the Global Alliance for Vaccines and Immunization (GAVI), Roll Back Malaria (RBM), Stop TB Partnership (Stop TB), Safe Injection Global Network (SIGN), Global Polio Eradication Programme (PEI), the Special Programme for Research and Training in Tropical Diseases (TDR), and the Special Programme for Research Development and Research Training in Human Reproduction (HRP). Partnerships are sometimes principally orchestrated by companies such as in the case of Action TB and can be legally independent such as the International Aids Vaccine Initiative (IAVI), Medicines for Malaria Venture (MMV), and the Global Alliance for TB Drug Development (GATBDD). Examples of large partnerships hosted by a civil society NGO include the Malaria Vaccine

Initiative (MVI), the Mectizan Donation Programme (MDP), and the HIV Vaccine Initiative (HVI).

This chapter seeks to reflect the predominant situation in the majority of African countries with no pretention to fully represent the continent's heterogeneity as far as drug discovery and development is concerned. It is essentially an update of previous studies related to both Africa's contribution to world scientific production [18–25] and to the mapping of health innovation in Africa [26–28]. It is organized around regional initiatives—African institutions involved in drug discovery including international collaborations for drug discovery.

1.2 Regional Initiatives

Over the past years, the concept of an indigenous African institution focused on health product research and innovation has been raised and discussed in Africa [27–31]. The African Union (AU) has adopted a clear policy position '*to pursue, with the support of some partners, the local production of generic medicines on the continent and make full use of flexibilities within the Trade and Related Aspects of Intellectual Property Rights (TRIPS) and Doha Declaration on TRIPS and Public Health*'. To this end, the AU adopted the Pharmaceutical Manufacturing Plan for Africa in 2007 and gave mandate to a technical committee to coordinate the efforts. Several reports [1, 10, 26, 27] have highlighted the need for African governments and international agencies to support the transition of African science from fragmented and isolated activities to more coordinated and integrated R&D efforts across the continent.

However, a determination to move on has been promoted through the Africa's New Partnership for Africa's Development (NEPAD) [29, 31–33] which stipulates that '*Scientific and technological capacity for health cannot, thus, be reduced to equipment, funding and number of health scientists and technicians. It is the configuration of skills, policies, organizations, non-human resources, and overall context to generate, procure and apply scientific knowledge and related technological innovation to identify and solve specific health problems. The capacity is built through interactive processes of creating, mobilizing, using, enhancing or upgrading, and converting skills/expertise, institutions and contexts*'. To achieve this vision, NEPAD promised to provide US $21 billion per annum for R&D in Africa. While this fund is still in the pipeline of promises, the NEPAD Planning and Coordinating Agency has been established from a successful transition of the former Secretariat as the technical body of African Union after achieving its integration into the Union. This shift is to re-energize the implementation of the NEPAD program and hopefully to facilitate the flow of new and significant research funds.

Two new mechanisms have also emerged that aim to put African countries in the driver's seat for access, research, and local medicine production—the Global Strategy and Plan of Action on Public Health, Innovation and Intellectual Property

(GSPOA) [34, 35] and the African Union Pharmaceutical Manufacturing Plan [36]. Together, they form the first comprehensive framework and promise of long-term funding to support countries' strategies for pharmaceutical innovation. The GSPOA is probably the most important initiative to date that enables developing countries to access the drugs they need and to support innovation in these regions. Adopted in 2008 by the World Health Assembly [34], the GSPOA reached an international consensus on the need to provide long-term support and financial mechanisms for needs-driven research and development on '*diseases that disproportionately affect developing countries*'. This initiative is a mechanism to ensure long-term, needs-driven research and development and a funding framework for medicines that affect developing countries. The strategy proposes clear objectives and priorities for promoting innovation, building capacity, improving access, and mobilizing resources. The funding levels total US $149 billion between 2009 and 2015—an average of US $21 billion per year. It has eight core elements: prioritizing research and development needs; promoting research and development; building and improving innovative capacity; transfer of technology; application and management of intellectual property; improving delivery and access; ensuring sustainable financing mechanisms; and establishing monitoring and reporting systems.

1.2.1 Noordwijk Medicines Agenda and Yaoundé Process

In parallel with the Intergovernmental Working Group (IGWG) and GSPOA, the Organisation for Economic Co-operation and Development (OECD) and the Netherlands together brought a group of countries to discuss stimulating innovation and accelerating development and delivery of medicines for neglected and emerging infectious diseases. The approach included medicines, vaccines, and diagnostics needed by developing countries.

The resulting call for action was called the Noordwijk Medicines Agenda [37], a positive step by OECD countries to become more active in providing incentives for drugs research and production to deal with neglected diseases. The action plan recommended, among other things, increased networking and partnerships between research actors.

The Yaoundé Process [38] grew out of a need identified by the Cameroon's Minister of Public Health, supported by some African participants at the Noordwijk meeting, to develop a '*complementary African agenda*' *to look at medicines*' access and long-term socioeconomic development from the perspective of the African countries. This '*complementary agenda*' aimed to develop and strengthen an African vision on health innovation and medicine R&D and production and build north–south partnerships. The Council on Health Research for Development (COHRED) was engaged in 2007 to facilitate this work, named the '*Yaoundé Process*'.

The primary goal of the Yaoundé Process is to strengthen health innovation in Africa and complement the following: (a) the GSPOA on Public Health, Innovation and Intellectual Property by putting its principles into action at regional and country level and (b) the work of NEPAD for harnessing political support of African leaders, harmonizing regulations and processes in public health and health-care systems and for '*shaping and driving a new research and innovation agenda*' [31, 32].

The Noordwijk Medicines Agenda and the Yaoundé Process support African countries, global health players, and the GSPOA in assessing the situation of innovation activities, projects, and programs in Africa today. There is a need to identify African countries' specific needs for technologies and skills to strengthen pharmaceutical R&D, production, and delivery to improve populations' access to medical products; to implement the GSPOA at region and country level; and to inform NEPAD's efforts to harness political support of African leaders, harmonize regulations, and processes in public health and health-care systems and shape a new research and innovation agenda.

Prior to the above initiatives, the African Ministerial Council on Science and Technology (AMCOST) adopted Africa's science and technology plan of action [32] in 2005. This plan focuses on improving the quality of science, technology, and innovation policies in six areas:

- Supporting the African Science, Technology and Innovation Indicators Initiative (ASTIII)
- Improving regional cooperation in science and technology
- Building public understanding of science and technology
- Building a common African strategy for biotechnology
- Building science and technology policy capacity
- Promoting the creation of technology parks.

1.2.2 ANDI as a New Coordination Body

The GSPOA has paved the way for a greater focus on supporting developing countries to participate in the discovery, development, and delivery of the products that African governments need the most. This resolution builds upon other commitments by African governments such as the NEPAD health targets, the Abuja Declaration of March 2006, the Accra Declaration on Health Research adopted in June 2006, and the Algerian Declaration on Research for Health in the African Region adopted in June 2008. Despite this promotion, African countries or institutions have not demonstrated sustainable capacities to move from basic research to discovery of a new chemical entity to registration and commercialization of a single new drug product [39].

ANDI (African Network for Drugs and Diagnostics Innovation) came at a time when various stakeholders were seeking concrete ways to meet these commitments and to promote sustainable product R&D and capacity development in developing countries, especially in Africa [30]. ANDI is a platform to help support African institutions participate in discovering, developing, and manufacturing health products. ANDI held its first successful meeting in Abuja (Nigeria) in 2008. Within the particular context of sub-Saharan Africa, possibilities exist to build up a sustainable African-led R&D innovation by strengthening and utilizing existing capacity and infrastructure, promoting collaborative efforts directed toward sustained delivery of affordable health products including those based on natural products and traditional medicines.

1.2.3 Other Initiatives

Other initiatives exist on the continent that can contribute to improving health research, and drug discovery and development. At the most basic level are initiatives focusing on institutes of higher learning. Efforts are directed at training scientists in particular techniques, or for researching specific diseases of high prevalence in Africa, for example, through the African Programme for Onchocerciasis Control (APOC) and the African Malaria Network (AMANET). While these and several others are reputed for building and strengthening disease-specific research capacity, efforts in building up technical or managerial capacities that are applicable across national research systems have been much less deliberate and sustained. Examples of such initiatives are given in Table 1.1.

In addition to such initiatives, other organizations were created to coordinate efforts in the field of drug discovery and development, for instance, the Asia/Africa Center for Drug Discovery initiated by Meiji Pharmaceutical University (Japan) for the purpose of educating and training internationally active researchers and pharmacists in drug discovery. In the same trend, the Asia and Africa Science Platform Program was established by an independent administrative institution, the Japan Society for the Promotion of Science (JSPS), to establish core research institutions concerned with relevant fields within Asian and African countries, to foster young researchers by building sustainable relations with core institutions from Asia and Africa.

1.3 African Institutions

Drug research is concentrated in African countries where science and technologies receive significant attention such as in South Africa, Tunisia, and Egypt. A report [18] on the state of science and technology in the African continent based on two scientometric indicators (number of research publications and number of patents awarded) indicates that Africa produced 68,945 publications over the period 2000–2004 or 1.8% of the world's publications. In comparison, India produced 2.4% and Latin America 3.5%. More detailed analysis reveals that research in Africa is concentrated in just two countries—South Africa and Egypt which

Table 1.1 Examples of initiatives to address research capacity gaps

Focus	Examples of initiatives
Higher education	Nelson Mandela Institute
Drug discovery and development	African Network for Drugs and Diagnostics Innovation (ANDI)
	AU/NEPAD Panel on Biotechnology
	Global Institute for BioExploration—Africa (GIBEX-Africa)
	International Centre for Genetic Engineering and Biotechnology
	(ICGEB)
	Natural Products Research Network of Eastern and Central Africa (NAPRECA)
	Special Programme for Research and Training in Tropical Diseases (TDR)
Clinical trial capacity, including ethics review committees	African Poverty Related Infection Oriented Research Initiative (APRIORI)
	Council on Health Research for Development (COHRED)
	European Developing Countries Clinical Trial Platform (EDCTP)
	Initiative to Strengthen Health Research Capacity in Africa (ISHRECA)
	Malaria Clinical Trials Alliance (MCTA)
	Special Programme for Research and Training in Tropical Diseases (TDR)
Knowledge management	WHO Health Research Systems Analysis (HRSA)
	WHO International Clinical Trials Registry Platform (ICTRP)
Disease focused	African AIDS Vaccine Programme (AAVP)
	African Programme for Onchocerciasis Control (APOC)
	African Malaria Network Trust (AMANET)
	South African Malaria Initiative (SAMI)
	Special Programme for Research and Training in Tropical Diseases (TDR)
	Mali Malaria Research Centre
Policy development at national and regional level	AU/NEPAD Consolidated Science and Technology Plan of Action
	African Science, Technology and Innovation Policy Initiative—UNESCO collaboration
	African Technology Policy Studies Network (ATPS)
	UN Science and Technology Cluster support to the AU Consolidated Plan of Action
Influencing policy at global level	Global Strategy and Plan of Action on Public Health, Innovation and Intellectual Property (GSPOA)
	IQsensato
	Knowledge Ecology International
	MSF Access Campaign
	OXFAM
	Third World Network

produce approximately 50% of the continent's publications—with the top eight African countries producing more than 80% of the continent's research. Similarly, Africa produces few patents, with 88% from South Africa. South Africa is also considered the leading African country in science and technology [24–28].

According to ANDI landscape mapping of African health innovation [27], capabilities within Africa in health product R&D including commercialization do exist as evidenced by the data gathered in Table 1.2 (additional details are presented in Annexes 1, 2, and 3).

1.3.1 Pharma Firms in Africa

In 1998, more than 97% of R&D activities occurred in developed countries [1]. Africa's capacity for pharmaceutical R&D and local drug production is among the lowest globally. Overall, 37 countries have some pharmaceutical production, and only South Africa has limited primary production of active pharmaceutical ingredient (API) and intermediates [36].

National capacity for local production has increased, with, for example, Egypt and Tunisia producing between 60% and 95% of their national requirements for essential medicines [26, 36]. Significant production activity also occurs in Nigeria with more than 200 pharmaceutical companies, and Ghana and Kenya with approximately 20 and 40 companies, respectively [13, 26, 28, 40]. Other locations such as Uganda and the United Republic of Tanzania have a handful of active manufacturers with even smaller numbers reported elsewhere (such as Cameroon, Côte d'Ivoire, Democratic Republic of the Congo, and Malawi). Despite the existence of these companies, the majority of pharmaceutical products are imported.

With the partial exception of South Africa, production in sub-Saharan Africa is generally limited to final formulations, characterized by non-complex, high-volume essential products, encompassing basic analgesics, simple antibiotics, anti-malarial drugs, and vitamins. The local production of medicines has for instance been identified as an important development objective by the African Union (AU) through its 2007 Pharmaceutical Manufacturing Plan for Africa (PMPA). At the subregional level, fostering pharmaceutical production features as one component in the Southern African Development Community's (SADC) Pharmaceutical Business Plan (2007–2013) and a Regional Pharmaceutical Manufacturing Plan of Action is at an advanced stage of preparation within the East African Community (EAC) [40]. In addition, the domestic pharmaceutical industry has also been earmarked as a priority sector in a number of countries, including Botswana, Cameroon, Ghana, Kenya, and the United Republic of Tanzania.

The disparities in global pharmaceutical research, development, and production are a reflection of underlying global inequities in health and socioeconomic development. However, low levels of pharmaceutical innovation reflect low levels of government commitment to financing health research and development in science

Table 1.2 Examples of capacities in African institutions

R&D stage	Capacity and gaps in health product R&D		Examples of centers/countries with capacity
	Examples of available capacity	Examples of gaps	
Basic exploratory research relevant to health products	• Identification of different protein targets for the development of drugs, diagnostics and vaccines • Use of molecular methods such as regular polymerase chain reaction (PCR), quantitative PCR (Q-PCR) and genomic sequence and analysis using different software, genetic engineering, probe hybridization techniques, biological and molecular cloning, evaluation of immune markers for laboratory diagnosis of infections, serological assays involving the use of rapid tests, ELISA-based evaluations and immunofluorescent assay techniques • Reference laboratories designed for diagnosis as well as clinical and vaccine research • Screening, lymphocyte phenotyping, cytokine activation and production, epitope tracking, HLA typing, etc. • Biotechniques: DNA sequencing flux cytometrics, ELISPOT, ELISA, Western blot, cell cultivation, etc.	• Data management including protection • Ethical challenges • Translation of basic research into innovative products (translational research)	Annex 1

- Virology: viral load calculation, study of genetic resistance to ARV and viral subtype sequencing
- Biological profiles for the follow-up of PVVS, and vaccine and clinical research
- Analysis of data obtained from sequences and subtypes of viral origin, immune polymorphism, the clinical progression of the diseases and the most relevant social determinants
- Epidemiological baseline studies and evaluation of in vivo, in vitro activities of antimalarial compounds
- Ligation-mediated PCR genotyping
- Epidemiology and immunodiagnosis of *Schistosomiasis haematobium*
- Studies on circulating adhesion molecules in schistosomiasis
- Synthetic chemistry
- Computational modeling
- Systems biology
- Biomarkers of *Mycobacterium tuberculosis* and other pathogens
- Functional annotation of reconstructed genome transcripts
- Host-parasite relationships: induction of regulatory immune mechanisms
- Molecular epidemiology and prognostic markers

(continued)

Table 1.2 (continued)

R&D stage	Capacity and gaps in health product R&D		Examples of centers/countries with capacity
	Examples of available capacity	Examples of gaps	
Hit identification and hit to lead identification	• Synthesis and evaluation of compounds as antiparasitic and antineoplastic agents • Centers have the capacity to screen plant medicines as sources of lead compounds for treatment of infectious diseases such as malaria and Buruli ulcer, as well as noncommunicable diseases like diabetes mellitus and cardiovascular and renal disease • In vitro efficacy assays: HIV, TB, protozoan and helminth parasites Cancer and biosafety level 3 (BSL 3) environments available • Synthetic chemistry • Parallel and robotic synthesis capability	• Access to chemical libraries • Throughput of primary assays. High-throughput screening, in silico screens • Analysis and prioritization of hits • Choice of in vitro efficacy assays • Medicinal chemistry, in vivo ADME/PK testing, construction of SAR Proactive coordination and management of projects • Data and project management including protection	• Pharmacognosy departments at Faculty of Pharmacy—Helwan and Cairo Universities • Chemistry of Natural Products, National Research Center—Egypt • University of Cape Town (UCT)—South Africa • Council for Scientific and Industrial Research (CSIR)—South Africa • African Institute of Biomedical Science and Technology (AiBST)—Zimbabwe
Lead optimization	• Computational modeling • Parallel and robotic synthesis capability • Spectroscopic characterization of secondary metabolites • Chromatography	• Lead optimization chemistry and QSAR CaCo2 testing • Cytotoxicity microsome stability, PAMPA • Metabolite identification PK, toxicokinetic animal models, efficacy animal models	• Pharmacognosy departments at Faculty of Pharmacy— Helwan and Cairo Universities • Chemistry of Natural Products, National Research Center—Egypt • University of Cape Town (UCT)—South Africa

		• Detailed medicinal chemistry, ADME, construction of SAR, toxicology • Proactive coordination and management of projects, candidate selection	• Council for Scientific and Industrial Research (CSIR)—South Africa • African Institute of Biomedical Science and Technology (AiBST)—Zimbabwe • Department of Chemistry University of Botswana
Natural products and traditional medicines	• Significant capacity in this area. Centers have the capacity to screen plant medicines as sources of lead compounds for treatment of infectious diseases such as malaria, TB, Buruli ulcer, as well as noncommunicable diseases like diabetes mellitus and cardiovascular and renal disease • Evaluation of plant drugs using medicinal phytochemistry, pharmacology and toxicology. Including isolation of active ingredients, structure elucidation using NMR and MS and chemical profiling of botanical extracts using HPLC MS/MS • Pharmacology, physiology and biochemistry equipment including laboratories • NIPRD have developed natural product-based formulation called Niprisan for the treatment of sickle cell anemia	• Isolation and identification of active components • Proactive coordination and management of projects, aligning extracts to classical drug discovery pathway • Lack of knowledge in IP issues • Toxicology and teratogenicity studies • Data protection Scaling up production and marketing • Quality assurance of drugs: quality control and surveillance • Development of standard documentation	Annex 2

(continued)

Table 1.2 (continued)

R&D stage	Capacity and gaps in health product R&D		Examples of centers/countries with capacity
	Examples of available capacity	Examples of gaps	
	• Experience and facilities to study medicinal plants with antimalarial activity, immunology and HIV/AIDS • IMPM developing anti-HIV tests kits from its spin-off called CAM Diagnostics • Labothera laboratories developed HEPASOR, a hepatoprotective and hepatocurative drug made up of protoberberine extracted from *Enantia chlorantha* (Annonaceae)		
Management of intellectual property	• Mainly administered by Ministries of Industry and Ministries of Justice • Available IPR laws • Various private law firms in West Africa offer IP services • Some African countries, e.g., Cameroon, are members of WIPO and are party to the Paris Convention for the Protection of Industrial Property	• Limited practical capacity in this area • Existing IP laws are generally not appreciated by researchers and the general populace and therefore not enforced • Preparation and filing of patents • Management of intellectual property, licensing principles	Many African countries and regional organizations such as ARIPO in East Africa and OAPI in Central and West Africa
Preclinical toxicology and safety pharmacology	• Basic toxicological studies including drug/chemical-induced toxic injury, drug metabolism and drug-drug and drug-herbal interactions. Reproductive and genetox testing	• General gap in the area of safety pharmacology • Microsome stability, PAMPA, metabolite identification PK, toxicokinetic animal models,	Noguchi Memorial Institute for Medical Research (NMIMR)—Ghana Nigerian Institute of Medical Research (NIMR)—Nigeria

	• NMIMR conducts research into biomonitoring and prevention of poisoning from mycotoxins and toxic heavy metals • NMIMR has the VICAM Afla Test equipment capable of analyzing aflatoxins in foods to meet international standards • NMIMR has the LUMEX mercury analyzer capable of measuring mercury in both biological and environmental samples • NMIMR is also equipped with a spectrofluorometer, clinical chemistry and hematology autoanalyzers, HPLC, analgesimeter and a plethysmometer	binding panels, hERG, mutagenicity and carcinogenicity studies, GLP facilities	National Institute for Research in Public Health (INRSP)—Mali Centre d'Etudes des Ressources Vegetales—Groupe de Recherches Biomedicales (CERVE/GRBM)—Congo Pharmacognosy departments at Faculty of Pharmacy—Helwan and Cairo Universities—Egypt Kenya Medical Research Institute (KEMRI) Council for Scientific and Industrial Research (CSIR)— South Africa African Institute of Biomedical Science and Technology (AiBST)—Zimbabwe
Clinical Studies	• Many African research institutions possess facilities and expertise for phases I and II and III clinical trials according to GCP • Observation wards, resuscitation equipment, medical and paramedical staff etc., available • Many collaborate in multi-center drug, vaccine and diagnostic clinical trials	• Protocols used in these trials are not locally developed although might contain some input from local investigators High-end scientific work is carried out in the West or North, with the African centers supporting by facilitating the collection of biological samples or information	Annex 3
Regulatory expertise	• Most possess drug regulatory agencies in the form of Food and Drugs Board (FDB)	Most rely heavily on prior approval from regulatory agencies in the West and North such as the	•Food and Drugs Board of many African countries • Ministries of Health

(continued)

Table 1.2 (continued)

R&D stage	Capacity and gaps in health product R&D: Examples of available capacity	Capacity and gaps in health product R&D: Examples of gaps	Examples of centers/countries with capacity
	• Levels of expertise and powers vary from country to country	European Medicines Agency (EMEA) and the US Food and Drug Administration (FDA)	
Raw material processing and active pharmaceutical ingredients		Ability to process raw material is limited Most import purified forms to use in manufacturing	Several pharmaceutical companies in North and South Africa and few in sub-Saharan Africa LaGray Chemical Company—Ghana
Formulation, manufacturing including production of finished dosage forms, packaging and labeling	• Numerous local companies • Mainly produce and package analgesics, antimalarials and other essential medicine • KEMRI has fully developed 2 diagnostic kits—HEPCELL kits for detection of hepatitis B virus and KEMCOM kits for detection of HIV in the blood	Formulation capacity Non-GMP	Several non-GMP facilities available Kenya Medical Research Institute (KEMRI) LaGray Chemical Company—Ghana GMP facility GMP facilities available in North and South Africa University of North-West (Potchefstroom)—South Africa
Marketing	Extensive capacity. Host to local branches of many big international pharmaceutical companies from the West and North	Price barriers and competition from Chinese and Indian pharmaceutical companies Inadequate marketing laws to protect indigenous marketing companies	Numerous
Phase IV/ Pharmacovigilance	Spontaneous reporting and other pharmacoepidemiological methods available to systematically collect and analyze adverse events associated with the use of drugs, identify signals or emerging problems and communicate how to minimize or prevent harm	• Parallel structures and lack of empowerment • Difficulty in collecting data remotely. Most initial reports have to be manually transmitted • Four West African countries (Togo, Benin, Nigeria and Ghana) appear to be full members of the WHO Programme for International Drug	Food and Drugs Board (FDB) and the Centre for Tropical Clinical Pharmacology and Therapeutics (CTCPT) of the University of Ghana Medical School (UGMS) collaborate in carrying out pharmacovigilance in Ghana Two parallel structures controlling pharmacovigilance in Cameroon,

Monitoring at the Uppsala Monitoring Centre programme No Central African country is a full member of the WHO Uppsala Monitoring Centre Only two Central African countries are associate members, Cameroon and the Democratic Republic of the Congo Uganda and Tanzania are the only East African countries that are full members of the WHO Uppsala Monitoring Centre Kenya, Ethiopia and Zanzibar are associate members • Translation of PV information into policy	the Directorate of Pharmacy and Drugs (DPM) and the National Drugs Committee University of Cape Town—National Adverse Drug Event Monitoring Centre—South Africa

and technology in most African countries. With the exception of South Africa, which invests 0.9% of its GDP in research and development, R&D intensity in the rest of sub-Saharan Africa is generally less than 0.3% of GDP [16, 41]. Further, the potential for private sector contributions to bridging financing gaps is hampered by the high business risk posed by unstable political environments, poor governance, and weak or absence (of) legal and regulatory frame.

Pharmaceutical innovation in Africa reveals a complex landscape [26] with multiple actors—including governments, pharmaceutical companies, UN agencies and other international organizations, NGOs, public–private partnerships, and civil society organizations. These players are engaging with countries at various levels and sectors, addressing different steps of the pharmaceutical innovation process (Fig. 1.1).

In an attempt to improve the landscape of pharmaceutical sector on the continent, UNIDO, with German funding, has been rendering advisory and capacity-building support since 2006, under a global project that aims at strengthening the local production of essential generic drugs in developing countries. Specific emphasis has been placed on the promotion of small and medium enterprises, business partnerships, investment, and South–South cooperation.

1.4 Partnerships

A host of public-private partnerships (PPPs) have emerged to address the aforementioned challenges [10, 11, 22, 26–28]. Their approach is to stimulate R&D for neglected diseases while minimizing business risk. This R&D is typically done as not-for-profit or no-profit-no-loss by partnerships involving public, multilateral and bilateral agencies, pharmaceutical companies, NGOs, and philanthropies.

These partnerships, especially those developing new medical products based on the needs identified by the most disadvantaged countries—known as product development public-private partnerships (PDPPPs)—are seen as a positive force. They have raised great expectations for expanding the pool of products available for improving the health status of the most deprived populations. However, some prevalent diseases and conditions, for example, trypanosomiasis, schistosomisasis, and filariae, are still truly neglected, partly because of restricted potential markets.

However, new partners such as the Novartis Institute for Tropical Diseases in Singapore are engaged in developing medicines for neglected diseases. The Bill & Melinda Gates Foundation is similarly trying to stimulate the emergence and application of innovative scientific techniques and approaches to neglected diseases and speed up the development of new drugs, diagnostics, and vaccines [42].

As part of its African Institutions Initiative toward strengthening research capacity in Africa, the Wellcome Trust announced the formation of seven new international consortia [43], each led by an African institution. By developing research networks and building a critical mass of sustainable research capacity across Africa, local ability to tackle disease and poverty will be improved.

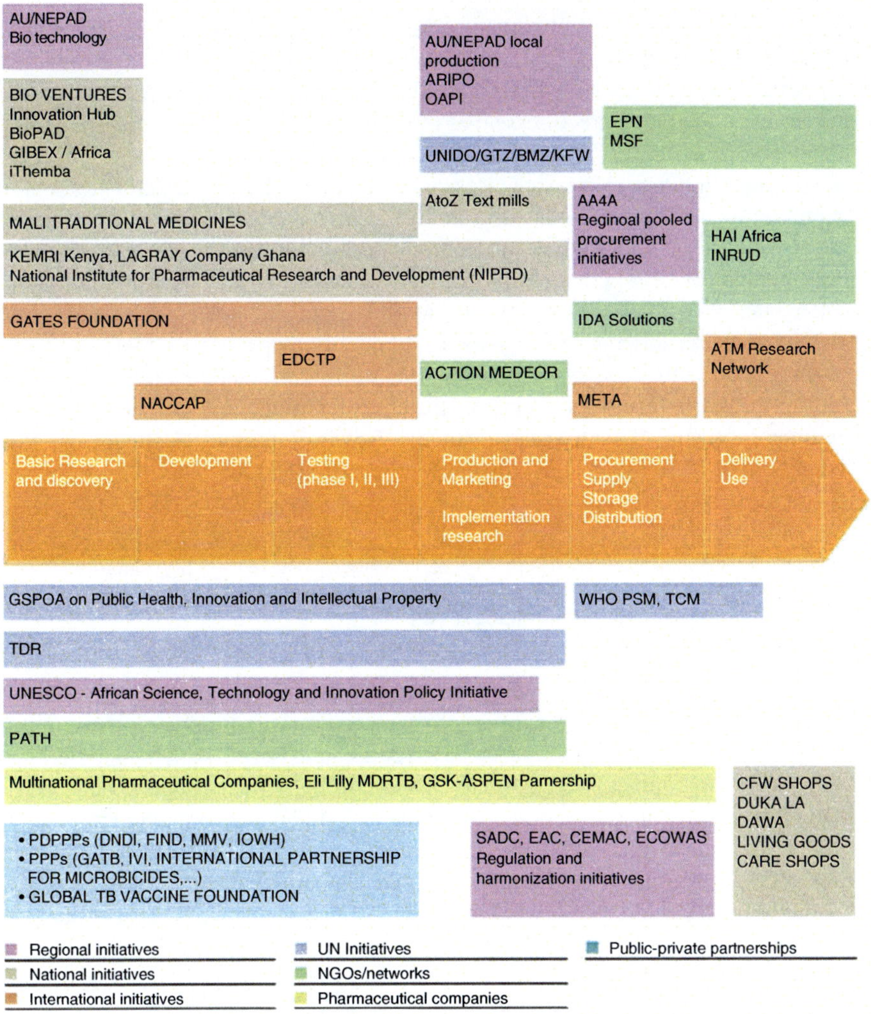

Fig. 1.1 Examples of pharmaceutical innovation initiatives in Africa mapped along the drug development and access pipeline. Extracted from reference [26]—a large number of initiatives are identified across Africa showing the contribution of international agencies, and other health programs effectively shaping the innovation and medicines access policies for the continent, without the transparent or explicit involvement of national governments

An example is the Southern Africa Consortium for Research Excellence (SACORE) which consists of institutions based in Malawi, Zambia, Zimbabwe, Botswana, South Africa and the UK. Postgraduate training is its main focus enabling PhD programs to begin in some of these southern African countries for the first time. The Consortium for Advanced Research Training in Africa (CARTA) consists of institutions in Kenya, South Africa, Tanzania, Uganda, Malawi, Nigeria, Rwanda, the USA, Australia, Switzerland and the UK. CARTA aims ‘to train and retain a critical mass of networked African researchers with complementary research skills, able to work in multidisciplinary environments'. Critical thinking skills, analytical techniques, writing and publication planning abilities, and grant management knowledge are all essential to generate top researchers in Africa who can compete on equal terms with those elsewhere.

The One Medicine Africa-UK Research Capacity Development Partnership Programme for Infectious Diseases in Southern Africa (SACIDS) [43] consists of medical, veterinary, and wildlife academic and research institutions in the Democratic Republic of Congo, Mozambique, Kenya, South Africa, Tanzania, Zambia, and the UK. SACIDS will enable new MSc courses in molecular biology and epidemiology, promote biosafety systems and quality management in laboratories, develop a continuing professional development program, launch a secure internet platform to share resources, enhance distance learning, and provide research apprenticeships for African students to the Royal Veterinary College and the London School of Hygiene and Tropical Medicine in the UK.

1.5 Conclusions and Perspectives

Despite the various challenges that Africa drug discovery and pharmaceutical industry face in their effort to evolve into a research-driven industry, the current availability of scientific talent, cost advantages, and large patient populations provide a unique competitive edge. Given the size of Africa's population, a self-reliant health-care industry is critical to reduce the burden of maintaining its health. Africa is nowadays undergoing a demographic transition and industrialization process. This constitutes a good opportunity to encourage at the same time R&D activities and sustain drug discovery and diagnostic innovation. While the pace of this transformation will be contingent upon deploying resources strategically and effectively negotiating the current and future challenges, it can be catalyzed by a collaborative, resilient effort of the government, academia, and industry.

It has been observed [26–28, 39] that few and isolated product discovery and development activities are ongoing in Africa. To achieve coherent and sustainable product innovation, greater effort is needed to bring groups working in this area together, so they may join forces, share lessons, and explore a more coordinated approach to health R&D and innovation. In the specific area of drugs, our study shows that only a limited number of African countries or institutions have demonstrated the capacity to move from basic research to discovery of a new chemical entity to registration and commercialization of a new drug product. The same applies to natural products-based drugs such as Niprisan, a very promising whole-plant extract with anti-sickling properties, which failed scale-up because of poor business strategy and management.

However, ANDI intends to achieve a strategic mechanism to support relevant, continent-wide activities in a coordinated and structured manner. It should be noted that African institutions do have an untapped potential to expand their work in diagnostics development and vaccine research, especially through genomics. Indeed, several African institutions have already been able to discover and commercialize new diagnostic tools or vaccines. Examples include the following:

- Kits for specific antibody detection for diagnosis and screening of schistosomiasis, fascioliasis, and hydatidosis produced by the Theodor Bilharz Research Institute (TBRI)
- Vaccines and antivenoms from VACSERA (Egypt)
- An HIV immune-enzymatic test by CAM Diagnostics (Cameroon)
- Multidrug resistant tuberculosis line probe assays (MDR-TB LPA) by the University of Stellenbosch (South Africa)
- A rapid diagnosis of the rabies virus and an immunochromatographic strip for the detection of visceral leishmaniasis made by the Institut Pasteur of Tunisia
- A hepatitis B surface antigen screening test (HEPCELL) by KEMRI (Kenya)
- A visually readable portable malaria dipstick test developed at the University of Ghana

Significant capacity also exists in the areas of basic research, lead identification, clinical trials, and marketing. However, the challenge is the lack of sustainable mechanisms to translate findings from basic research into concrete products and to further optimize and commercialize such findings [27, 28, 39]. African hospitals and institutions also have the capacity to undertake clinical trials. However, it is not clear whether most of the clinical centers can yet carry out studies to international good clinical practice (GCP) standards. Major gaps identified were in the areas of lead optimization, preclinical good laboratory practice (GLP) safety assessment, and raw material processing. In the area of manufacturing, several companies exist, but it is not clear whether all these firms can manufacture products to international good manufacturing practice (GMP) standards. In all cases, there is a need for better management and coordination of research and procedures and a need for a greater emphasis in research to support the use of traditional medicines and better management of local knowledge, including intellectual property.

Annexes

Annex 1 Examples of Countries/Institutions with Capacity to Carry Out Basic Exploratory Research Relevant to Heath Products

Country	Institution
Botswana	Department of Chemistry, University of Botswana
Burkina Faso	Institut de recherche pour le développement
Cameroon	Centre de Recherche pour la Santé des Armées (CRESAR)
	Centre International de Référence Chantal Biya, Centre Hospitalier Universitaire (CIRCB/CHU)
	Organisation de Coordination pour la lutte contre les Endémies en Afrique Centrale (OCEAC)
	Laboratoire de Chimie des Substances Naturelles, Département de Chimie Organique, Faculté des Sciences, Université de Yaoundé I
	Centre de Biotechnologie de Nkolbisson, Université de Yaoundé I
	Institut Médical et d'Etudes des Plantes Médicinales
Congo	Centre d'Etudes des Ressources Végétales/Groupe de Recherches Biomédicales (CERVE/GRBM)
Egypt	Department of Pharmaceutical Chemistry, Faculty of Pharmacy, Cairo University
	Genetic Engineering and Biotechnology Research Institute, Minufiya University
	Theodor Bihar Research Institute, Giza
	Zoology Department, Faculty of Science, Cairo University
Ethiopia	Addis Ababa University
	Jimma University
Gabon	Centre de Primatologie, Centre International de Recherches Médicales (CIRMF)
Ghana	Noguchi Memorial Institute for Medical Research (NMIMR)
	Department of Chemistry and Biochemistry, University of Ghana
	Faculty of Pharmacy, Kwame Nkrumah University of Science and Technology
Kenya	Kenya Medical Research Institute (KEMRI)
	Department of Biochemistry, Kenyatta University
	Department of Chemistry and Department of Biochemistry, University of Nairobi
	Faculty of Pharmacy, University of Nairobi
Madagascar	Institut Malgache de Recherches Appliquées (IMRA)
Mali	Malaria Research and Training Centre (MRTC)
	Department of Traditional Medicine of the National Institute for Research in Public Health (INRSP)
Morocco	Laboratory of Immunology, Biochemistry and Molecular Biology
	Faculty of Sciences and Technologies, Cadi Ayyad University, Béni Mellal
Nigeria	Nigerian Institute of Medical Research (NIMR)
	University of Nigeria
	Nsukka, Federal University of Technology Owerri
	University of Benin
	University of Ibadan
	University of Maiduguri
	University of Calabar, University of Jos and others

(continued)

Country	Institution
	Faculty of Pharmacy, Obafemi Awolowo University
	Departments of Chemistry and Biochemistry, Obafemi Awolowo University
South Africa	Council for Scientific and Industrial Research (CSIR)
	South African National Bioinformatics Institute (SANBI)
	University of Cape Town (UCT), many South African universities
Sudan	University of Khartoum, Institut de Endemic Diseases
Tanzania	Department of Chemistry, Dar es Salaam University
Tunisia	Laboratoire des Mycobactéries, Institut Pasteur de Tunis
	Laboratoire d'immuno-Oncologie Moléculaire, Faculté de Médecine de Monastir, Avenue Avicenne
Zambia	The Malaria Institute at Macha
Zimbabwe	African Institute of Biomedical Science and Technology (AiBST)

Annex 2 Examples of Countries/Institutions with Capacity to Carry Out Natural Products Research

Country	Institution
Cameroon	University of Buea
	Institut medical et d'etudes des plantes medicinales (IMPM)
	Laboratory of Medicinal Chemistry and Pharmacognosy at the Faculty of Science, University of Yaoundé I
	Laboratory of Chemistry, University of Dschang
Egypt	Schistosome Biological Supply Center (SBSC), Theodor Bilharz Research Institute
	Department of Pharmacology, Faculty of Pharmacy, University of Alexandria
Ethiopia	The Aklilu Lemma Institute of Pathology, University of Addis Ababa
Ghana	Centre for Scientific Research into Plant Medicine (CSRPM)
	Kwame Nkrumah University of Science and Technology, School of Pharmacy
Kenya	Kenya Medical Research Institute
Madagascar	Institut Malgache de Recherches Appliquées (IMRA)
Mali	Department of Traditional Medicine of the National Institute for Research in Public Health (INRSP)
Morocco	Laboratory of Endocrinian Physiology and Pharmacology
Nigeria	National Institute of Pharmaceutical Research and Development
	University of Ibadan, University of Nigeria Nsukka, University of Lagos
	Federal University of Technology Owerri
	University of Ife, Imo State
South Africa	University of Cape Town (UCT)
	Medical Research Council
	University of Pretoria
	University of Limpopo
	University of Western Cape
	University of Witwatersrand
	University of KwaZulu-Natal
Sudan	University of Khartoum
	National Center for Research

Annex 3 Examples of Countries/Institutions with Capacity to Carry Out Clinical Trials

Country	Institution
Angola	Faculdade de Medicina, Agostinho Neto University, Faculty of Medicine
Benin	Regional Center for Entomological Researches of Cotonou (CREC) Réseau Béninois de Recherche et de Communication sur le SIDA Université Nationale du Benin, Faculte des Sciences de Santé
Burkina Faso	African Malaria Vaccine Testing Network (AMVTN), Muraz Center (MC) African Malaria Vaccine Testing Network (AMVTN) Centre National de Lutte contre le Paludisme (CNLP) Ecole Supérieure des Sciences de la Santé Ouagadougou University Hospital Yalgado Ouedraogo Hospital
Burundi	ANSS—Burundi Programme National de Lutte contre la Lèpre et la Tuberculose (PNLT) University of Burundi, Faculty of Medicine
Cameroon	Ministry of Public Health Organisation de Coordination pour la Lutte contre les Endemies en Afrique Centrale (OCEAC) Centre de Recherche pour la Santé des Armées (CRESAR) Centre International de Référence Chantal Biya, Centre Hospitalier Universitaire (CIRCB/CHU) University of Yaounde, Faculty of Medicine Central African Republic Centre National de Reference des MST et du SIDA National University Hospital
Côte d'Ivoire	OCCGE Institut Pierre Michet Université d'Abidjan-Cocody, UFR Sciences Médicales Université de Bouake, Faculty of Medicine
Democratic Republic of the Congo	Université Catholique de Bukavu, Faculté de Medicine University of Kinshasa, The Faculty of Medicine: Ecole de Santé Publique University of Kisangani, Faculty of Medicine University of Lubumbashi, Faculty of Medicine
Ethiopia	Armauer Hansen Research Institute (AHRI) Ethiopian Health and Nutrition Research Institute, Vaccine Research and Development Task Force Jimma University, Institute of Health Sciences Medical School of Gondar University of Addis Ababa, Department of Medicine; School of Pharmacy
Gabon	International Center for Medical Research (CIRMF) The Albert Schweitzer Hospital, Medical Research Unit Université Omar Bongo, Faculté de Médicine et des Sciences de la Santé (FMSS)

(continued)

Country	Institution
Gambia	Medical Research Council (MRC) Laboratories
	University of The Gambia Medical School
Ghana	Centre for Tropical Clinical Pharmacology and Therapeutics
	Komfo Anokye Teaching Hospital
	Korle Bu Teaching Hospital
	Navrongo Health Research Center
	Noguchi Memorial Institute for Medical Research
	Severe Malaria in African Children (SMAC) site
	Kintampo Health Research Centre
Kenya	International Centre for Insect Physiology and Ecology (ICIPE)
	Kenya Medical Research Institute (KEMRI)
	Kenya AIDS Vaccine Initiative (KAVI)
	Moi University, College of Health Sciences
	The African Medical and Research Foundation (AMREF)
	United States Army Medical Research Unit—Kenya (USAMRU-K)
	University of Nairobi
Madagascar	Institut Pasteur de Madagascar
	Universite d'Antananarivo, Faculté de Médecine
	Université de Madagascar (Mahajanga), Faculté de Médecine
Mali	University of Mali, Malaria Research and Training Center
Morocco	Ibn Rochd Hospital
Mozambique	Centro de Investigacao em Saude Manhica (CISM)
	Universidade Catolica Beira, Medical School
	Universidade Eduardo Mondlane, Faculdade de Medicine
Namibia	University of Namibia, Faculty of Medical and Health Sciences
Nigeria	Ahmadu Bello University Teaching Hospital
	Federal Medical Center
	Nigerian Institute of Medical Research
	Nnamdi Azikiwe University
	Obafemi Awolowo College of Health Sciences
	University of Calabar, University of Calabar Teaching Hospital
	University of Ibadan, College of Medical Sciences
	University of Ilorin Teaching Hospital
	University of Jos, Jos University Teaching Hospital
	University of Nigeria, College of Medicine
	University of Port Harcourt Teaching Hospital
Republic of Congo	Laboratoire National de Santé Publique
Rwanda	National University of Rwanda, School of Public Health
	National University of Rwanda, Faculty of Medicine
	Treatment and Research AIDS Center (TRAC)
Senegal	Research and Development Institute
	University of Dakar (Université Cheikh Anta Diop)
South Africa	Aurum Research Unit, Aurum Health Research
	Global Alliance for TB Drug Development, South Africa c/o Medical Research Council
	Medical Research Council, South Africa (MRC)
	Global Alliance for TB Drug Development
	South Africa c/o Medical Research Council
	Medical University of Southern Africa (MEDUNSA)
	South African HIV Vaccine Action Campaign (SA HIVAC)

(continued)

Country	Institution
	The South African National Institute for Virology (NIV)
	University of Cape Town, Faculty of Medicine
	University of Natal, Faculty of Medicine
	University of Pretoria, Centre for the Study of AIDS
	University of Stellenbosch, Faculty of Health Sciences—School of Medicine
	University of the Free State, Faculty of Health Sciences
	University of Transkei (UNITRA), Faculty of Medicine and Health Sciences
	University of Witwatersrand
Sudan	Juba Teaching Hospital
	Tropical Medicine Research Institute (TMRI)
	University of Khartoum, Institute of Endemic Diseases
Tanzania	African Malaria Vaccine Testing Network
	Hubert Kairuki Memorial University (HKMU), Faculty of Medicine
	Muhimbili University College of Medical Research
	National Institute for Medical Research
	Tanzania Essential Health Interventions Project (TEHIP)
	The African Medical and Research Foundation (AMREF)
	The Ifakara Health Research and Development Centre
	The Kilimanjaro Christian Medical College (KCMC)
	Vignan's International Medical and Technological University, Faculty of Medicine
Uganda	AIDS Information Center
	AMREF Uganda
	Joint Clinical Research Centre (JCRC)
	Makerere Institute of Social Research
	Makerere University, Institute of Public Health
	Mbarara University of Science and Technology
	Med Biotech Laboratory
	Mildmay Palliative Care Center
	National Chemotherapeutics Laboratory (NCTL)
	The AIDS Support Organisation (TASO)
	Uganda AIDS Commission
	Uganda Virus Research Institute
Zambia	Chainama Hills College Hospital
	Tropical Diseases Research Centre
	University of Zambia, School of Medicine
	Uganda Virus Research Institute
Zimbabwe	Batsirai Group
	Biomedical Research and Training Research Institute
	Blair Research Institute, TVBU
	Chitungwiza Hospital
	The Medical Research Council of Zimbabwe
	Training and Research Support Centre (TARSC)
	University of Zimbabwe-Bulawayo College of Health Sciences
	University of Zimbabwe, Faculty of Medicine
	University of Zimbabwe-Bulawayo College of Health Sciences

References

1. Trouiller P, Olliaro P, Torreele E et al (2002) Drug development for neglected diseases: a deficient market and a public-health policy failure. Lancet 359:2188–94
2. Gregson N, Sparrowhawk K, Mauskopf J et al (2005) Pricing medicines: theory and practice, challenges and opportunities. Nat Rev Drug Discov 4:121–130
3. Ullman F, Boutellier R (2008) Drug discovery: from project driven research to innovation studios and process factories. Drug Discov Today 13(11/12):543–550
4. David E, Tramontin T, Zemmel R (2009) Pharmaceutical R&D: the road to positive returns. Nat Rev Drug Discov 8:609–610
5. Ruffolo RR (2006) Why has R&D productivity declined in the pharmaceutical industry? Expert Opin Drug Discov 1:99–102
6. Kong De-Xin, Li X-J, Zhang Hong-Yu (2009) Where is the hope for drug discovery? Let history tell the future. Drug Discov Today 14(3/4):115–119
7. Clark RL, Johnston BL, Mackay SP et al (2010) The drug discovery portal: a resource to enhance drug discovery from academia. Drug Discov Today 15(15/16):679–683
8. Glicksman M (2008) The drug discovery process in academics. Opportunity knocks. BIOForum Europe 6:34–35
9. Nwaka S, Ramirez B, Brun R et al (2009) Advancing drug innovation for neglected diseases – criteria for lead progression. PLoS Negl Trop Dis 3(8):e440. doi:10.1371/journal.pntd.0000440
10. Nwaka S, Ridley RG (2003) Virtual drug discovery and development for neglected diseases through public-private partnerships. Nat Rev Drug Discov 2:919–928
11. Tralau-Stewart CJ, Wyatt CA, Kleyn DE et al (2009) Drug discovery: new models for industry–academic partnerships. Drug Discov Today 14:95–101
12. Hughes B (2008) Pharma pursues novel models for academic collaboration. Nat Rev Drug Discov 7:631–632
13. Commission for Macroeconomics and Health (2001) Investing in health for economic development. http://whqlibdoc.who.int/publications/2001/924154550x.pdf. Accessed 26 Aug 2011
14. Millenium project. Commission by the UN Secretary General and supported by the UN Development Group. http://www.unmillenniumproject.org/. Accessed 29 Aug 2011
15. Nordling L (2011) Upturn for African technological investment. Continental survey reveals that some countries have been exceeding a 2010 investment goal. Nature http://www.nature.com/news/2011/110526/full/news.2011.320.html. Accessed 26 Aug 2011
16. UNESCO Science Report 2010: The status of science around the world. From http://unesdoc.unesco.org/images/0018/001899/189958e.pdf. Accessed 7 Oct 2011
17. Nishtar S (2004) Public – private 'partnerships' in health – a global call to action health research policy and systems, 2:5 doi:10.1186/1478-4505-2-5
18. Pouris A, Pouris A (2009) The state of science and technology in Africa (2000–2004): a scientometric assessment. Scientometrics 79(2):297–309
19. Pouris A (2010) A scientometric assessment of the southern Africa development community: science in the tip of Africa. Scientometrics 85:145–154
20. Tijssen RJW (2007) Africa's contribution to the worldwide research literature: new analytical perspectives, trends, and performance indicators. Scientometrics 71(2):303–327
21. Narvaez-Berthelemot N, Russell JM, Arvanitis R et al (2002) Science in Africa: an overview of mainstream scientific output. Scientometrics 54:229–241
22. Hofman KJ, Kanyengo CW, Rapp BA et al (2009) Mapping the health research landscape in sub-Saharan Africa: a study of trends in biomedical publications. J Med Libr Assoc. doi:doi:10.3163/1536-5050.97.1.007
23. Uthman OA, Uthman MB (2007) Geography of Africa biomedical publications: an analysis of 1996–2005. Int J Health Geogr. doi:10.1186/1476-072X-6-46
24. Adams J, King C, Hook D (2010) Global research report—Africa. Leeds, UK, Thomson Reuters. http://researchanalytics.thomsonreuters.com/m/pdfs/globalresearch report-africa.pdf. Accessed 1 Aug 2010

25. UNECA (United Nations Economic Commission for Africa) ECA/ISTD/07/ – Building Science, Technology and Innovative Systems for Sustainable Development in Africa, January 2007 ISTD Division. http://www.uneca.org/estnet/ecadocuments/builingsandt_innovation.pdf. Accessed 3 Oct 2011
26. Berger M, Murugi J, Buch E et al (2009)Strengthening pharmaceutical innovation in Africa. Council on Health Research for Development (COHRED); New Partnership for Africa's Development (NEPAD). http://www.policycures.org/downloads/COHRED-NEPAD_Strengthening_Pharmaceutical_Innovation_AfricaREPORT.pdf. Accessed 20 Sep 2011
27. Nwaka S, Ilunga TB, Da Silva JS et al (2010) Developing ANDI: a novel approach to health product R&D in Africa. PLoS Med 7(6):e1000293
28. Al-Bader S, Masum H, Simiyu K et al (2010) Science-based health innovation in sub-Saharan Africa. BMC Int Health Hum Rights 10(Suppl 1):S1, http://www.biomedcentral.com/content/pdf/1472-698X-10-S1-S1.pdf. Accessed 26 Aug 2011
29. Juma C, Serageldin I (2009) Freedom to Innovate: Biotechnology in Africa's Development. In: African Union (AU) and New Partnership for Africa's Development (NEPAD); 2007:5,6. http://www.nepadst.org/doclibrary/pdfs/biotech_africarep_2007.pdf. Accessed 22 Aug 2011
30. Volmink J, Dare L (2005) Addressing inequalities in research capacity in Africa. BMJ 331:705–706
31. Kalua FA, Awokedu A, Kamwanja LA, Saka JDK (eds) (2009) Science, technology and innovation for public health in Africa. Monograph, NEPAD Office of Science and Technology, Pretoria, Republic of South Africa
32. NEPAD: Consolidated S&T Action Plan. Pretoria, Republic of South Africa: NEPAD Office of Science and Technology (2006). http://asiandrivers.open.ac.uk/NEPAD_health_final.pdf. Accessed 19 Sep 2011
33. Juma C (2005) Going for growth: science, technology and innovation in Africa. London: The Smith Institute. http://www.smith-institute.org.uk/file/GoingforGrowthScienceTechnologyandInnovationinAfrica.pdf. Accessed 22 Aug 2011
34. World Health Assembly (2008) Global strategy and plan of action on public health, innovation and intellectual property. WHA 61.21. http://apps.who.int/gb/ebwha/pdf_files/A61/A61_R21-en.pdf. Accessed 22 July 2011
35. World Health Assembly (2009) Global strategy and plan of action on public health, innovation and intellectual property. WHA 62.16. http://apps.who.int/gb/ebwha/pdf_files/A62/A62_R16-en.pdf. Accessed 22July 2011
36. First Meeting of the Technical Committee on the Pharmaceutical Manufacturing Plan for Africa October 2007 Report. www.africa-union.org http://www.africa-union.org/root/au/Conferences/2007/october/sa/Pharmaceutical/DOCS/REPORT.doc. Accessed 19 Sep 2011
37. Noordwijk Medicines Agenda, 21 June 2007. http://www.oecd.org/dataoecd/62/11/38845838.pdf. Accessed 22 Aug 2011
38. Yaoundé Process. http://www.cohred.org/wp-content/uploads/2011/05/Yaounde2009_en_leaflet.pdf. Accessed 19 Sep 2011
39. Nyigo VA, Malebo HM (2005) Drug discovery and developments in developing countries: bottlenecks and a way forward. Tanzan Health Res Bull 7:154–158
40. Conference of African Ministers of Industry (CAMI-19) 2011. Regional Pharmaceutical Manufacturing Plan of Action. http://www.unido.org/fileadmin/user_media/UNIDO_Worldwide/Africa_Programme/CAMI/RoundTable_Pharma.pdf. Accessed 20 Oct 2011
41. UNESCO institute for statistics fact sheet: a global perspective on research and development, October 2007, No. 05 http://istic-unesco.org/meetingdoc/A%20global%20perspective%20on%20research%20and%20development%20-%20El-Tayeb.pdf. Accessed 7 Oct 2011
42. Press release Bill and Melinda Gates Pledge $10 Billion in Call for Decade of Vaccines http://www.gatesfoundation.org/press-releases/Pages/decade-of-vaccines-wec-announcement-100129.aspx. Accessed 20 Oct 2011
43. Wellcome trust. News and Features. http://www.wellcome.ac.uk/News/Media-office/Press-releases/2009/WTX055742.htm. Accessed 20 Oct 2011

Chapter 2
The National Cancer Institute and Natural Product-Based Drug Discovery in Africa

John A. Beutler, Gordon M. Cragg, and David J. Newman

Abbreviations

CA1P	Combretastatin A-1 phosphate
CA4P	Combretastatin A-4 phosphate
CBD	United Nations Convention on Biological Diversity
CNARP	Centre National D'Applications des Recherches Pharmaceutiques, Madagascar
CSIR	South African Council for Scientific and Industrial Research
DCTD	Division of Cancer Treatment and Diagnosis
DM1 and DM4	Maytansanoid derivatives conjugated to carrier molecules
DTP	Developmental Therapeutics Program, formerly Cancer Chemotherapy National Service Center (CCNSC)
FDA	US Food and Drug Administration
HIV	Human Immunodeficiency Virus
LOC	NCI Letter of Collection
MaB	Monoclonal Antibody
MDR	Multidrug Resistance
MBG	Missouri Botanical Garden
MOU	NCI Memorandum of Understanding

The opinions expressed in this chapter are those of the authors, not necessarily those of the US government.

J.A. Beutler
Molecular Targets Laboratory, Center for Cancer Research, Building 560-1, NCI-Frederick, Frederick, MD 21702-1201, USA
e-mail: beutlerj@mail.nih.gov

G.M. Cragg (✉) • D.J. Newman
Natural Products Branch, Developmental Therapeutics Program, NCI-Frederick, P.O. Box B. Frederick, MD 21702-1201, USA
e-mail: gmcragg@verizon.net; newmand@mail.nih.gov

K. Chibale et al. (eds.), *Drug Discovery in Africa*,
DOI 10.1007/978-3-642-28175-4_2, © Springer-Verlag Berlin Heidelberg 2012

NCI	US National Cancer Institute
NIDDK	National Institute for Diabetes Digestive and Kidney Diseases, NIH
NIH	US National Institutes of Health
NPB	DTP Natural Products Branch
NPR	Natural Products Repository Frederick, Maryland, USA
OTT	NIH Office of Technology Transfer
SCG	Source-Country Government
SCO	Source-Country Scientific Organization
USDA	US Department of Agriculture

2.1 Background

The United States National Cancer Institute (NCI; http://www.nci.nih.gov) was established in 1937, its mission being "to provide for, foster and aid in coordinating research related to cancer." In 1955, NCI set up the Cancer Chemotherapy National Service Center (CCNSC) to coordinate a national, voluntary cooperative cancer chemotherapy program, involving the procurement of drugs, screening, preclinical studies, and clinical evaluation of new agents. The responsibility for drug discovery and preclinical development at NCI now rests with the Developmental Therapeutics Program (DTP; http://dtp.nci.nih.gov); subsequent clinical development, generally up through phase II human trials, is conducted by its companion program, the Clinical Trials Evaluation Program (CTEP; http://ctep.cancer.gov), both being major components of the Division of Cancer Treatment and Diagnosis (DCTD). Thus, for the past 50 years, NCI has provided resources for the preclinical screening and clinical development of compounds and materials submitted by public and private scientists and institutions worldwide and has played a major role in the discovery and development of many of the available commercial and investigational anticancer agents.

During this period, more than 500,000 chemicals, both synthetic and pure natural products, have been screened for antitumor activity using a variety of screening methods. These have ranged from in vivo studies against murine tumors, through human tumor xenografts in immunodeficient mice, to isolated human tumor cell lines and molecular targets expressed in a variety of formats, with the initial systems changing over time. The success of this effort has depended on close collaboration with organizations worldwide, and international collaboration continues to be an important feature of the NCI programs [1].

2.1.1 Achievements: 1955–1982

While most of the materials initially screened were pure compounds of synthetic origin, the program also recognized that natural products were an excellent source of complex chemical structures with a wide variety of biological activities. The original plant collections from 1960 to 1982 were performed by the US Department of Agriculture (USDA) through an interagency agreement with NCI and involved the random collection of over 35,000 plant samples, mainly from temperate regions. In this period, marine invertebrates were generally collected by academic investigators, mainly funded through grants from the NCI, while microbial samples were obtained from pharmaceutical companies and research institutes, such as the Institute of Microbial Chemistry in Japan, some of which were funded through contracts with the NCI. From 1960 to 1982 [i.e., more than 30 years before the United Nations Convention on Biological Diversity (CBD) to 10 years before], over 180,000 microbial-derived, some 16,000 marine organism-derived, and over 114,000 plant-derived extracts were screened for antitumor activity by the NCI, and as mentioned above, a number of clinically effective chemotherapeutic agents have been developed [2, 3].

2.1.2 Contract Collections (1986–Present): The NCI Letter of Collection

After a lapse in funding from 1982 to 1986, the systematic collection of marine invertebrates and terrestrial plants resumed, coordinated by the DTP Natural Products Branch (NPB; http://dtp.nci.nih.gov/branches/npb/index.html). Marine organism collections originally focused in the Caribbean and Australasia, but, in 1992, were expanded to the Central and Southern Pacific and to the Indian Ocean (off East and Southern Africa) through a contract with the Coral Reef Research Foundation, which is based in Palau in Micronesia. With the renewal of the contract in 2002, collections are now performed worldwide. Terrestrial plant collections have been carried out in over 25 countries in tropical and subtropical regions worldwide through contracts with the Missouri Botanical Garden (MBG) in Africa and Madagascar, the New York Botanical Garden (1986–1996) in Central and South America, the University of Illinois at Chicago in Southeast Asia, and the Morton Arboretum and World Botanical Associates in the United States mainland and territories. Over 60,000 plant samples have been collected, and the permanent repository of over 120,000 extracts is intended to be a resource of potential agents for the treatment all human diseases.

NCI collection contractors are required to obtain all necessary permits, including visas, collecting, shipping, and export permits, from the appropriate source-country agencies or departments. The NCI provides contractors with the NCI Letter of Collection (LOC; http://dtp.nci.nih.gov/branches/npb/agreements.html; http://ttc.nci.nih.gov/forms/) for transmission to the appropriate source-country authorities and scientific organizations (SCOs). The LOC states NCI's willingness to collaborate with local scientists and/or authorities in the discovery and development of novel drugs from organisms (plants, marine invertebrates, microbes) collected in their country and/or

Table 2.1 African countries with which NCI had a letter of collection agreement

Source country	Source-country organization and date of agreement
Gabon	Centre National de la Recherche Scientifique et Technologique (CENAREST), Libreville, 1993
Ghana	University of Ghana, Legon, 1993
Madagascar	Centre National D'Applications des Recherches Pharmaceutiques, Antananarivo, 1990
Tanzania	Traditional Medicine Research Institute, Muhumbili University College of Health Services, University of Dar es Salaam, 1991
Tanzania	Amended agreement signed with The Institute of Traditional Medicine, Muhumbili University College of Health Services, 2009

territorial waters, and, if requested, the NCI will enter into formal agreements based on the LOC with the relevant source-country government agency or organization. LOC agreements have been signed with four African countries (Table 2.1), and it is notable that the first agreement with any country worldwide was signed with CNARP, Antananarivo, in 1990, 2 years prior to the signing of the CBD.

Collections by MBG were performed in Cameroon and the Central African Republic without the finalization of a formal LOC agreement, but the authorities in these two countries were fully aware of the terms of the LOC and granted the necessary permits for MBG collections without requiring a formal agreement. In this respect, the NCI is totally committed to the terms of the LOC irrespective of whether or not a formal agreement has been signed [4]. This absence of formal agreements was not due to any lack of effort on the part of the MBG and/or NCI staff to solicit formal agreements; indeed, in the case of Cameroon, NPB staff interacted extensively with government representatives and scientists, both in Cameroon and during NCI-sponsored visits to NCI and MBG (Table 2.2). The purpose of these visits was to provide opportunities for source-country officials and scientists to observe the NCI drug-discovery facilities and the processes to which their raw materials are subjected, and to discuss collaboration in the drug-discovery process. From 1989 to 2001, 18 officials and scientists from African countries visited NCI one or more times for periods of 1–2 weeks, either to discuss participation in NCI contract collections or direct collaboration in the drug-discovery process (Table 2.2).

2.1.3 Source-Country Collaboration

In carrying out these collections, NCI contractors work closely with qualified organizations in each of the source countries. Botanists and marine biologists from source-country organizations (SCOs) collaborate in field collection activities and taxonomic identification, and their knowledge of local species and conditions is indispensable to the success of the NCI collection operations. SCOs, when necessary and relevant, provide facilities for the preparation, packaging, and shipment of

Table 2.2 Short-term (1–2 weeks) African visitors to the USA sponsored by NCI

Year	Visitor	Institution	Country
1989	Dr. Elimweka Mshiu	U. Dar Es Salaam	Tanzania
	Dr. Johnson Jato	U. Yaounde 1	Cameroon
	Dr. Robodo Andriantsiferana	CNARP[a]	Madagascar
	Dr. Feetham Banyikwa	U. Dar Es Salaam	Tanzania
1990	Dr. Rogasian Mahunnah	U. Dar Es Salaam	Tanzania
1991	Dr. Blandine Akendengue	CENAREST[b]	Gabon
	Dr. Johnson Jato	U. Yaounde 1	Cameroon
1992	Dr. Johnson Jato	U. Yaounde 1	Cameroon
	Dr. Ivan Addae-Mensah	U. Ghana, Legon	Ghana
1993	Dr. Lucien Obame	CENAREST[b]	Gabon
	Dr. Lucienne Nze-Ekekang	CENAREST[b]	Gabon
1994	Dr. J. Rajaonarivony	CNARP[a]	Madagascar
	Dr. Johnson Jato	U. Yaounde 1	Cameroon
	Dr. Thomas Tata	Ministry of the Environment	Cameroon
1995	Dr. J. Rajaonarivony	CNARP[a]	Madagascar
	Mr. J. Edou	Office of the Prime Minister	Cameroon
	Dr. Johnson Jato	U. Yaounde 1	Cameroon
	Dr. T. Mbenkum	Ministry of the Environment	Cameroon
	Dr. G. Chavanduka	U. Zimbabwe	Zimbabwe
	Dr. P. Mashava	U. Zimbabwe	Zimbabwe
	Mr. R. Chadwick	Legal rep., U. Zimbabwe	Zimbabwe
	Dr. W. Phillips	U. Ghana, Legon	Ghana
1997	Dr. P. Mashava	U. Zimbabwe	Zimbabwe
2001	Dr. M. Andriantsoa	CNARP[a]	Madagascar

[a]CNARP: Centre National D'Appliques Recherches Pharmaceutique, Madagascar
[b]CENAREST: Centre National de la Recherches Scientifique et Technologique, Gabon

the samples to the NCI's Natural Products Repository (NPR) in Frederick, Maryland. In a significant number of cases, these interactions have materially aided the procurement of both the initial collection permits and, most importantly, the specific export documentation required by the country of origin.

The collaboration between the SCOs and the NCI collection contractors, in turn, provides support for expanded research activities by source-country biologists. The deposition of a voucher specimen of each species collected in the national herbarium or repository expands source-country documentation of their biota. NCI contractors also provide training opportunities for local personnel through conducting workshops and presentation of lectures, both in-country and at the contractor's US facilities. In the context of plant collections in African countries, during the contract cycle from 1996 to 2001, MBG offered one-month curatorial workshops at their facilities in St. Louis, Missouri in May 1999 and March 2001. Through its contract with MBG, the NCI supported the attendance of seven botanists from Madagascar, Ghana, Tanzania, and Zambia, and participants were instructed in collections management, botanical research methodology, biodiscovery, conservation, and global information systems.

Table 2.3 Long-term (1–12 months) African visiting scientists sponsored under the auspices of the NCI LOC

Year	Visitor	Home institution	Country	US host institution
1990	Dr. Z. Mbwambo	U. Dar Es Salaam	Tanzania	NCI
1993	Mr. C. Mutayabarwa	U. Dar Es Salaam	Tanzania	NCI
1994	Dr. R. Andriamaharavo	U. Antananarivo	Madagascar	NIDDK[a], NIH
	Dr. J. Jato	U. Yaounde 1	Cameroon	NCI
1995	Dr. W. Phillips	U. Ghana, Legon	Ghana	VPISU[b]
1996	Dr. V. Rasimison	CNARP[c]	Madagascar	Washington U., St. Louis
1997	Dr. Sadri Said	U. Dar Es Salaam	Tanzania	U. Oklahoma
	Ms. J. Ropivia	CENAREST[d]	Gabon	UIC[e]
2000	Dr. M. Lamidi	CENAREST[d]	Gabon	U. Mississippi
2002	Dr. R. Andriamaharavo	U. Antananarivo	Madagascar	NIDDK[a], NIH
2003	Dr. Ladislaus Mdee	U. Dar es Salaam	Tanzania	UIC[e]
2004	Dr. Johnson Jato (Fulbright Scholar)	Bamenda U. Sci. and Technology	Cameroon	NCI

[a]*NIDDK* National Institute for Diabetes, Digestive and Kidney Diseases, NIH
[b]*VPISU* Virginia Polytechnic Institute and State University
[c]*CNARP* Centre National D'Appliques Recherches Pharmaceutique, Madagascar
[d]*CENAREST* Centre National de la Recherches Scientifique et Technologique, Gabon
[e]*UIC* University of Illinois at Chicago

In addition, through its LOC and agreements based upon it, the NCI has invited 10 African scientists nominated by SCOs to visit its facilities, or equivalent facilities in other approved US organizations, for 1–12 months to participate in collaborative natural products research involving the screening and bioassay-directed fractionation of extracts (Table 2.3). The LOC also dictates benefit-sharing and use of source-country resources in the event of the licensing and development of a promising drug candidate. Successful licensees are required to negotiate agreements with source-country government (SCG) agencies or SCOs, dictating terms of collaboration and compensation. The terms apply irrespective of whether the potential drug is the actual isolated natural product or a compound structurally based upon the isolate, a synthetic material for which the natural product provided a key development lead, or a method of synthesis or use of any aforementioned isolate, product, or material. The percentage of royalties negotiated as payment varies depending upon how closely the marketed drug relates to the originally isolated product. The first milestone in the licensing agreement is that a signed agreement must be presented to the NIH's Office of Technology Transfer (OTT), the group within NIH that formally licenses all NIH patents, within 1 year of the initial granting of the license.

The original formulation of the NCI policies for collaboration and benefit-sharing embodied in the LOC predated the drafting of the CBD (http://www.biodiv.org/convention/articles.asp) by at least 4 years, and as noted in Sect. 2.1.2, the first agreement was signed with CNARP, Madagascar, in 1990.

Table 2.4 MOU between NCI and African organizations: direct collaborations

Country	Organization and date of MOU
Zimbabwe	Zimbabwe National Traditional Healers Association (ZINATHA), 1994
S. Africa	Council for Scientific and Industrial Research (CSIR), Division of Food, Biological and Chemical Technologies (BIO/CHEMTEK), 1996
S. Africa	Rhodes University, 1998

Table 2.5 Long-term (1–12 months) African visiting scientists sponsored by NCI under the auspices of the MOU

Year	Visitor	Home institution	Country	US host institution
1994	Dr. P. Mashava	U. Zimbabwe	Zimbabwe	NCI
1997	Ms. H. Van Vuuren	CSIR, Pretoria	S. Africa	NCI
1999	Dr. P. Mashava	U. Zimbabwe	Zimbabwe	NCI
1999	Dr. M. Davies-Coleman[a]	Rhodes University	S. Africa	NCI

[a]Visiting scientist supported through NIH Research Fellowship to NCI Laboratory of Drug Discovery, Research and Development

2.1.4 Direct Collaboration with Source-Country Organizations: The NCI Memorandum of Understanding

As mentioned in Sect. 2.1.2, the collections of plants and marine organisms have been carried out in over 25 countries worldwide through contracts with qualified US botanical and marine biological organizations working in close collaboration with qualified SCOs, and all collections are performed subject to the terms of the NCI Letter of Collection. Particularly in the area of plant-related studies, source-country scientists and governments are becoming increasingly committed to performing more of the drug-discovery operations in-country, as opposed to simply exporting raw materials. The NCI has recognized this fact for several years, and contract collections of plants are now being de-emphasized in favor of establishing direct collaborations with qualified organizations in the source countries where the necessary expertise and infrastructure exist.

The NCI has established collaborative agreements [Memoranda of Understanding (MOU); http://dtp.nci.nih.gov/branches/npb/agreements.html; http://ttc.nci.nih.gov/forms/] with over 20 SCOs suitably qualified to perform in-country processing, including three from Southern Africa (Table 2.4).

In establishing these agreements, the NCI undertakes to abide by the same policies of collaboration and compensation as specified in the LOC. Depending on the availability of the necessary resources, NCI also assists the SCOs in establishing their own drug-discovery programs through training in techniques of antitumor screening and natural product isolation. NCI has sponsored long-term visitors from 18 countries worldwide since 1988 for such purposes of collaboration and training, including three from Southern Africa (Table 2.5).

It is anticipated that the discovery of novel anticancer drugs will be performed by the SCO at its own expense, with assistance from the NCI in terms of free secondary in vitro and in vivo testing. All results from such secondary testing are considered the sole intellectual property of the SCO since the NCI regards such testing as a routine service to the scientific community and can be used by the SCO to apply for patents covering promising inventions. The NCI will commit its resources to collaborating with the SCO in the preclinical and clinical development of any SCO-discovered drug which meets the NCI selection criteria and will make a sincere effort to transfer any knowledge, expertise, and technology developed during such collaboration to the SCO, subject to the provision of mutually acceptable guarantees for the protection of intellectual property associated with any patented technology.

2.1.5 NCI Screening Agreement

As noted in Sect. 2.1, the NCI has played a major role in the discovery and development of many of the available commercial and investigational anticancer agents. Organizations or individuals wishing to have pure compounds tested in the NCI drug screening program, such as pharmaceutical and chemical companies or academic research groups worldwide, may submit their compounds for free testing through an online submission process which can be accessed at http://dtp.nci.nih.gov/docs/misc/common_files/submit_compounds.html. A screening agreement (http://dtp.nci.nih.gov/docs/misc/common_files/canagr.html) may be signed with the NCI Division of Cancer Treatment and Diagnosis (DCTD) which includes terms stipulating confidentiality and levels of collaboration in the drug development process. Should a compound show promising anticancer activity in the routine screening operations, the NCI may propose the establishment of a more formal collaboration for further development, such as a Cooperative Research and Development Agreement (CRADA) or a Clinical Trial Agreement (CTA) (http://ttc.nci.nih.gov/forms/).

2.2 African Plant-Derived Anticancer Agents: Recent Developments

Plants have a long history of use in the treatment of cancer. Hartwell [5], in his review of plants used against cancer, lists more than 3,000 plant species that have reportedly been used in the treatment of cancer. In many instances, however, the "cancer" is undefined, or reference is made to conditions such as "hard swellings," abscesses, calluses, corns, warts, polyps, or tumors. Many of these claims for efficacy in the treatment of cancer, however, should be viewed with skepticism because cancer, as a specific disease entity, is likely to be poorly defined in folklore and traditional medicine [6]. This is in contrast to other plant-based therapies used in traditional medicine for the treatment of afflictions such as malaria and cutaneous fungal infection, which are more easily defined and which are prevalent in the regions where traditional medicine systems are used extensively. Of the plant-derived anticancer drugs in clinical use, among the best known are the so-called vinca alkaloids, vinblastine (VLB), and vincristine (VCR), isolated from the Madagascar periwinkle *Catharanthus roseus*. *C. roseus* has been used by various cultures for the treatment of diabetes, and vinblastine and vincristine were first discovered during an investigation of the plant as a source of potential oral hypoglycemic agents. Their discovery, therefore, may be indirectly attributed to the observation of an unrelated medicinal use of the source plant. It is interesting to note that, though the plant was originally endemic to Madagascar, the samples used in the discovery of VLB and VCR were collected in Jamaica and the Philippines. More recent semisynthetic analogues of these agents are vinorelbine (VRLB) and vindesine (VDS). These agents are primarily used in combination with other cancer chemotherapeutic drugs for the treatment of a variety of cancers.

Other important plant-derived, clinically used agents are etoposide and teniposide, which are semisynthetic derivatives of the natural product epipodophyllotoxin; the taxanes, paclitaxel (Taxol®) and docetaxel (Taxotere); topotecan (hycamptamine) and irinotecan (CPT-11) semisynthetically derived from camptothecin; and homoharringtonine. These and other important anticancer agents in clinical and preclinical development have been comprehensively reviewed, and interested readers should consult this review for details [2].

2.2.1 The Combretastatins: Models for Combinatorial Chemistry

The combretastatins were isolated from the South African "bush willow," *Combretum caffrum* (Eckl. & Zeyh.) Kuntze, collected in Southern Africa in the 1970s for the NCI by the United States Department of Agriculture (USDA), working in collaboration with the Botanical Research Institute of South Africa. These collections were part of a random collection program aimed at the discovery of novel anticancer agents. Species of the *Combretum* and *Terminalia* genera, both of which belong to the Combretaceae family, are used in African and Indian traditional medicine for the treatment of a variety of diseases, including hepatitis and malaria. Several *Terminalia* species have reportedly been used in the treatment of "cancer."

The combretastatins are a family of stilbenes which act as anti-angiogenic agents, causing vascular shutdown in tumors which leads to tumor necrosis [7, 8]. Two water-soluble analogues, combretastatin A-4 phosphate (CA4P; fosbretabulin disodium) and combretastatin A-1 phosphate (CA1P; OXi4503) (Fig. 2.1) have advanced into clinical trials under the sponsorship of Oxigene, Inc. To date, there have been 12 clinical trials involving CA4P against a variety of cancers, mostly in combination with other agents such as carboplatin and paclitaxel, as well as

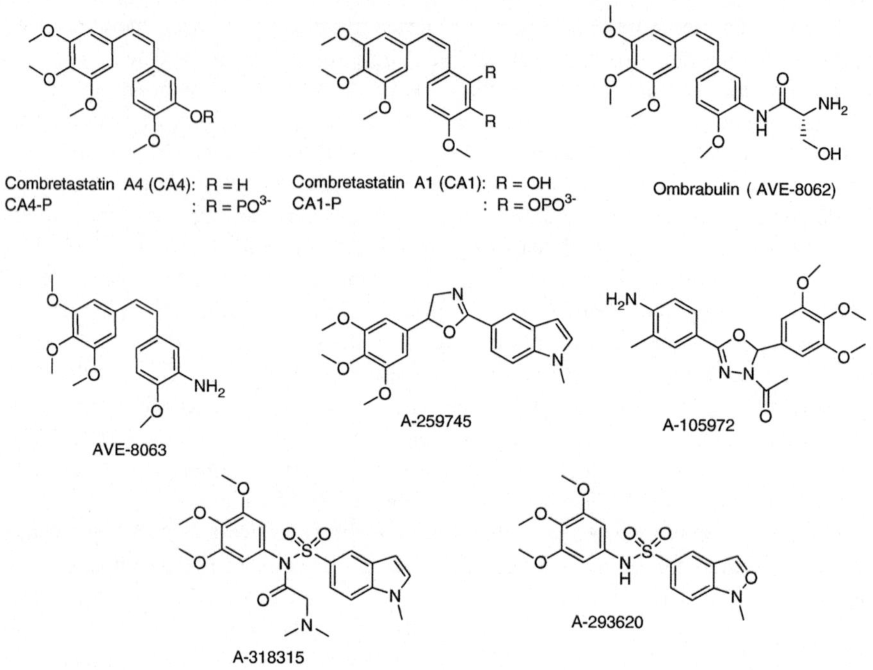

Fig. 2.1 Some combretastatin analogues and mimics

radiation treatment. Currently, two phase II trials are ongoing, one against non-small cell lung cancer (NSCLC) in combination with carboplatin, paclitaxel, and bevacizumab and the other against anaplastic thyroid cancer (ATC) in combination with carboplatin and paclitaxel. The clinical experience with CA4P has been reviewed [9]. Combretastatin A1P, like CA4P, shows excellent activity in preclinical studies, and a phase I clinical trial is under way, evaluating the safety and tolerability in patients with solid tumors, while a phase Ib/II trial is assessing the safety, tolerability, and efficacy, specifically against solid tumors growing in the liver [8]; in addition, patients are currently (June, 2012) being recruited for a phase I trial against Acute Myelogenous Leukemia (AML) and Acute Myelodysplastic Syndromes (AMS) [http://www.clinicaltrials.gov/ct2/results?term=cancer+combretastatins].

CA4P has also shown promising activity against ophthalmological diseases such as myopic macular degeneration (MMD), where all 23 patients in a phase II clinical trial achieved the primary endpoint of the trial, stabilization of vision. Parallel phase I/II clinical trials using CA4P for the treatment of exudative age-related macular degeneration (AMD) are also in progress [8, 10].

A number of combretastatin analogues and mimics are also being developed [8, 11]. The combretastatin chemical class has served as a model for the synthesis of a host of analogues containing the essential trimethoxy aryl moiety linked to substituted aromatic moieties through a variety of two or three atom bridges, including heterocyclic rings and sulfonamides. Some of these are shown in

Fig. 2.1. This is an impressive display of the power of a relatively simple natural product structure to spawn a prolific output of medicinal and combinatorial chemistry. One of these analogues, ombrabulin (AVE8062), is currently in a phase III clinical trial against advanced soft tissue sarcoma [http://www.clinicaltrials.gov/ct2/results?term=cancer+combretastatins].

2.2.2 *Maytansine: Targeting Toxic Natural Products*

A recurring liability of natural products, at least in the area of cancer chemotherapy, is that, although many are generally very potent cytotoxins cell growth inhibitors, they have limited solubility in aqueous solvents and exhibit considerable toxicity, often resulting in a narrow therapeutic index. These factors have led to the demise of a number of pure natural products as promising leads, including the plant-derived agents bruceantin and maytansine. An alternative strategy for the utilization of such agents is to investigate their potential as "warheads" attached to monoclonal antibodies specifically targeting epitopes on the tumor of interest. The promise of this approach to cancer therapy has been the subject of several reviews, and readers are referred to them for further details of developments in this area [12–16].

A promising candidate for such an approach is maytansine [17–19]. Maytansine (Fig. 2.2) was isolated in extremely low yield ($2\times10^{-5}\%$ based on plant dry weight) in the early 1960s from the Ethiopian plant *Maytenus serrata* (Hochst. Ex A. Rich.) Wilczek (Celastraceae family) collected for the NCI as part of a random collection program performed through a collaboration with the US Department of Agriculture (USDA). The novel structure and very potent in vitro activity of maytansine prompted great interest in pursuing further preclinical, and possible clinical, investigation. In order to obtain sufficient quantities, other species of *Maytenus* and related members of the Celastraceae family were surveyed; *Maytenus buchanii* (Loes) R. Wilczek collected in Kenya was shown to produce a seven times higher yield, while *Putterlickia verrucosa* (E. May ex Sonder) Szyszyl collected in South Africa proved to be the richest source, with a yield eight times that from *Maytenus buchanii*. Given the greater ease of large-scale collections of *M. buchanii*, however, a large recollection (about 9,000 kg) was undertaken in Kenya in 1976. Even given the improvement in yields, these remained extremely low, but the extreme potency of maytansine in testing against cancer cell lines permitted the production of sufficient quantities for pursuit of preclinical and clinical development. In appreciation of the collaboration of Kenyan authorities in the recollection process, the NCI donated supplies of the antileukemic drugs, vinblastine and vincristine, and preliminary studies aimed at the cultivation of *M. buchanii* were initiated [private communication, USDA Program Coordinator Dr. Robert Perdue (deceased, July 2011)].

Unfortunately, the very promising activity observed in preclinical animal testing of maytansine failed to translate to significant efficacy in human clinical trials, and it was dropped from further study in the early 1980s. Related compounds, the ansamitocins, were subsequently isolated from a microbial source, the

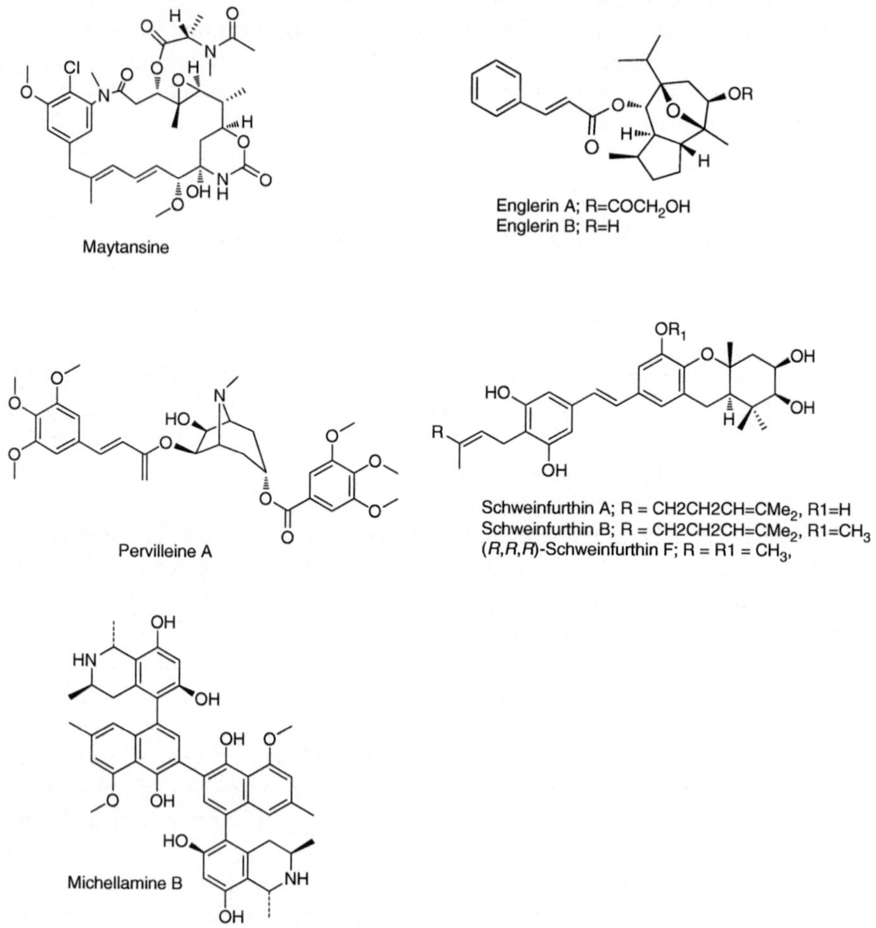

Fig. 2.2 Maytansine, englerins, schweinfurthins, pervilleine A, michellamine B

actinomycete *Actinosynnema pretiosum*, posing the question as to whether the maytansines are actually plant products or are produced through an association between a microbial symbiont and the plant. While this remains a topic of continued interest, the microbial production of the closely related ansamitocins allowed for easier production of larger quantities of this class of compounds, and this factor, together with their extreme potency, stimulated heightened interest in pursuing their development.

Maytansanoid derivatives, DM1 and DM4, can be conjugated through either thioether or disulfide linkages with various monoclonal antibodies (mAbs) targeting a variety of cancers [17, 19]. Conjugates have been prepared with huC242, a mAb directed against the *muc*1 epitope expressed in pancreatric, biliary, colorectal, and gastric cancers. HuC242-DM4 has undergone a phase I clinical trial for the treatment of CanAg-expressing tumors such as carcinomas of the colon and pancreas

[20], and a further phase I trial of this agent is ongoing (April 2011) in patients with inoperable or metastatic colorectal cancer, pancreatic cancer, or other solid tumors (http://www.cancer.gov/drugdictionary/?CdrID=492706). DM4-conjugated anti-Cripto monoclonal antibody, BIIB-015, is in a phase I study in patients with relapsed/refractory solid tumors (http://www.cancer.gov/drugdictionary/?CdrID=596550). Cripto, a member of the EGF-CFC family of growth factor-like molecules, is overexpressed in carcinomas such as those of the breast, ovary, stomach, lung, and pancreas, while not expressed in normal tissues.

Two drugs, trastuzumab (Herceptin) and lapatinib (Tykerb), are currently approved for the treatment of HER2-positive breast cancer, but they are not effective for all patients suffering from this disease. In efforts to address this problem, trastuzumab linked to DM1 through a non-reducible thioether linkage (trastuzumab-MCC-DM1) was shown to be active in trastuzumab-sensitive and trastuzumab-refractory models of HER2-overexpressing cancer and has shown some clinical responses in a phase I clinical trial [21]. Six studies of trastuzumab-MCC-DM1 in patients with HER2-positive forms of breast cancer have been completed (http://www.cancer.gov/clinicaltrials/search/results?protocolsearchid=8826956), and as of April 2011, four trials (http://www.cancer.gov/drugdictionary/?CdrID=564399) are ongoing. These include a phase III trial in patients with HER2-positive, locally advanced or metastatic breast cancer, a phase III trial in combination with pertuzumab in patients with HER2-positive locally advanced or metastatic breast cancer, a phase I study in combination with docetaxel in patients with locally advanced or metastatic HER2-positive breast cancer, and a phase I study in combination with paclitaxel and pertuzumab in patients with HER2-positive locally advanced or metastatic breast cancer.

The current most complete listing of the preclinical and clinical status of these agents is reported in the 2010 review by Lambert [19]. In summary, through the end of 2009, there have been 14 DM1 or DM4 linked agents that have been reported in preclinical through clinical trials. Of these, four currently fall into the preclinical area, while five are listed in phase I trials. These are SAR-3419 from Sanofi-Aventis, IMGN-388 from Immunogen, IMGN-633 (AVE-9633) from Immunogen/Sanofi-Aventis, BT-062 from Biotest-AG, and BIIB-015 from Biotest-Idec. Two more from Immunogen are in phase I/II trials, namely, IMGN-901 (lorvotuzumab mertansine) and IMGN-242. A more recent review in the "News and Analysis" section of the journal Nature Reviews: Drug Discovery indicates that SAR-3419 and IMGN-901 are now in phase II trials [22].

The fourteenth and most advanced of the DM-linked conjugates, which is currently in phase III trials (April 2011), is from Genentech/Roche and is known as T-DM1 or trastuzumab emtansine. As a result of the data from the phase II trial on patients who had failed at least two HER2 + treatments (trastuzumab and the tyrosine kinase inhibitor lapatinib), a biological license application (BLA, equivalent to the NDA application for small molecules) was submitted to the FDA by Genentech for marketing approval as a treatment for HER2+ breast cancer. Details of the agent and results to the end of 2009 were reported by Lambert [19] and in more detail by Niculescu-Duvaz [23]. Further detailed discussions are given in Ref. 17, and a less technical commentary is given in the article by Hughes [22].

2.2.3 Englerins

The englerins (Fig. 2.2) are new guaiane sesquiterpenes isolated from the Tanzanian plant *Phyllanthus engleri* Pax collected by Z. Mbwambo and MBG staff in 1989. The extract was prepared at NCI in 1992, and testing of the extract in the NCI 60-cancer cell line screen in 1997 showed selective inhibition of growth of the renal cancer cell lines; however, the extract was not pursued at that point. Bioassay-guided fractionation in 2007 yielded englerins A and B, and englerin A showed 1,000-fold selectivity against six of eight renal cancer cell lines with GI(50) values ranging from 1 to 87 nM [24]. These compounds are being developed in collaboration with scientists from the Institute of Traditional Medicine of Muhimbili University of Health and Allied Services in Dar es Salaam, Tanzania (Table 2.1). While the original LOC with ITM had expired at the time of the discovery, an updated LOC was negotiated in 2009 which permitted a multikilogram recollection of plant material to be made. This single collection yielded gram quantities of englerin A for preclinical development at NCI, which is ongoing.

The outstanding activity of englerin A in the NCI 60 cell screen has led many synthetic chemistry groups to attempt the total synthesis of englerin A. The first total synthesis was reported by the Christmann group from Dortmund, Germany [25]. This work also established the absolute configuration of the natural product. Other groups have since reported different synthetic strategies, including Nicolaou [26], Echavarren [27], and Ma [28]. Very recently, a more efficient process has been developed by Chain, in which englerin A was prepared in eight steps and 20% overall yield from readily available starting materials [29].

Structure–activity relationships have been explored for the englerins by both the Chen [30] and Christmann groups [31]. Notably, replacement of the cinnamate ester with a naphthoate moiety led to improved selectivity [31], while a reverse glycol ester and a lactate ester were worthy of note [30]. 9-Deoxy-englerin A showed significantly reduced activity [32].

The mechanism of englerin A's effect on renal cancer has been proposed to involve protein kinase C agonism, with inhibitory effects on the insulin pathway [33]. Given the dependence of renal cancers on glycolysis, this effect of englerin A may lead to its observed experimental therapeutic effects. The availability of labeled compounds (e.g., fluorescent or biotinylated derivatives) derived from synthetic approaches to englerins will likely shed more light on the mechanistic questions.

2.2.4 Schweinfurthins

The schweinfurthins (e.g., schweinfurthins A, B, and F; Fig. 2.2) were isolated in 1996 from the African plant *Macaranga schweinfurthii* Pax, collected in Cameroon in 1987 by MBG. They displayed significant selective activity against central nervous system, renal, and breast cancer cell lines in the NCI 60 cell line anticancer assay [34]. The spectrum of anticancer activity does not match that of any currently used agent, indicating that these compounds may be acting at a previously unrecognized target or through a novel mechanism. Thus far, ten schweinfurthins (A to J) have been isolated from nature by the NCI and Kingston/Virginia Polytechnic groups [35–37]. While successful scale-up recollections and isolation were undertaken in collaboration with colleagues in Cameroon in 1998–1999 and thereafter [38], the isolation of sufficient quantities of the schweinfurthins from the natural source remains challenging, and synthetic strategies have been developed to provide a reliable source of natural schweinfurthins and synthetic analogues for further biological testing [39, 40]. In the case of schweinfurthin F, total synthesis of the (*R*,*R*,*R*) and (*S*,*S*,*S*) enantiomers and comparisons of spectral data, optical rotations, and bioassay data with those reported for the natural product have resulted in assignment of the natural material as the (*R*,*R*,*R*) isomer [41]. These synthetic efforts are continuing, and most of the naturally occurring schweinfurthins have now been obtained by total synthesis [42–47].

Investigations into the mechanism of action of schweinfurthins have yet to identify a proximate molecular target; however, in glioblastoma cell lines, it appears that a defective neurofibromatosis type 1 (NF1) pathway confers sensitivity [48]. Other recent work has implicated an effect on oxysterol binding proteins and other isoprenoid pathways [48a, 48b].

2.2.5 The Pervilleines: Potential Multidrug Resistance Inhibitors

The resistance to treatment with standard anticancer agents developed by many cancer patients is a serious problem encountered in cancer chemotherapy [49]. Resistance to a drug may develop in a cell population through repeated exposure to treatment with that particular drug, and this cell population may subsequently show broad cross-resistance to other anticancer agents even though it has never been exposed to those agents. This phenomenon is called multidrug resistance (MDR) and has been linked to the presence of the *MDR1* gene which encodes P-glycoprotein (Pgp) which effectively pumps the drugs out of the cell, thereby precluding their antitumor actions. A large number of compounds which reverse this effect in vitro in cell line studies (so-called MDR inhibitors) have been discovered, but their effectiveness in the clinic has been mostly disappointing. Thus, there is a continuing search for more effective MDR inhibitors.

The pervilleines, isolated from *Erythroxylum pervillei* Baillon, collected in Madagascar in 2003 have shown promising MDR activity both in vitro and in vivo [50–52]. Pervilleine A (Fig. 2.2) was selected for preclinical development through a collaboration between an NCI-supported National Cooperative Drug Discovery Group (NCDDG) and the Institut Malgache de Recherches Appliquées. The fact that racemic pervilleine A hydrochloride exhibited only weak cholinergic and adrenergic receptor-mediated activities was considered advantageous for the further development of pervilleine A as a new adjuvant in cancer chemotherapy [53].

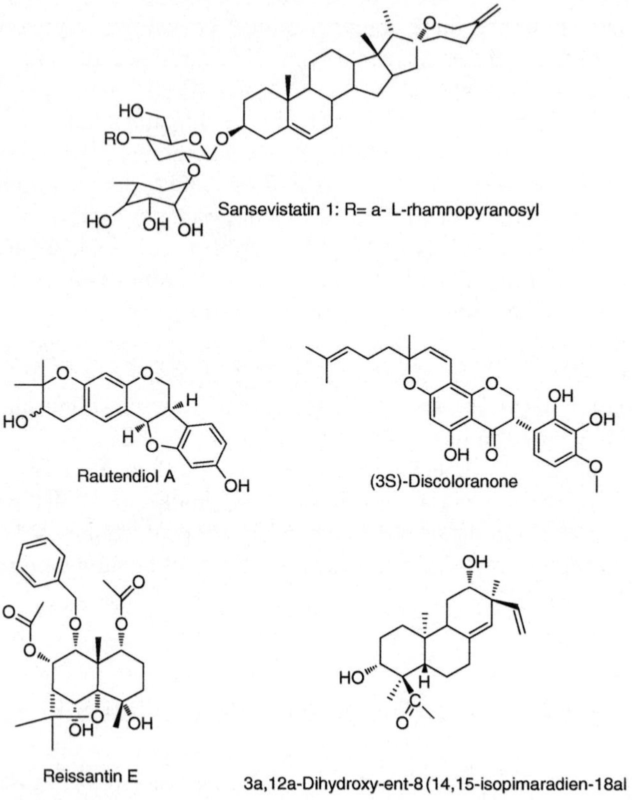

Fig. 2.3 Other novel agents

2.2.6 Other Novel Agents Discovered Through Direct NCI Support

A recent review [54] recorded several other novel antitumor agents discovered through projects directly supported by NCI, either by the early program (1955–1982; Sect. 2.1.1) or by the more recent NCI LOC program (Sects. 2.1.2 and 2.1.3). Sansevistatins 1 (Fig. 2.3) and 2, new cytotoxic spirostanol saponins, were isolated from *Sansevieria ehrenbergii*, initially collected in Kenya in 1966 by the USDA through an interagency agreement with NCI (Sect. 2.1.1) [55]. Several Tanzanian plants collected by the NCI contractor, Missouri Botanical Garden (Sect. 2.1.2), have yielded novel cytotoxic agents. In addition to the englerins (Fig. 2.2) discussed in Sect. 2.2.3, these include new agarofuran sesquiterpenes, reissantins A–E (E; Fig. 2.3), isolated from *Reissantia buchananii* collected in 1989 [56], and rautandiols A (Fig. 2.3) and B, pterocarpans, isolated from *Neorautanenia mitis* collected in 1993 [57]; while the rautandiols did not exhibit significant cytotoxicity, several known rotenone derivatives were isolated as the cytotoxic constituents. During visits of Tanzanian scientists sponsored by the NCI (Table 2.3) to the University of Illinois at Chicago, two new diterpenes, (3α,12α)-dihydroxy-ent-8(14),15-isopimaradien-18-al (Fig. 2.3), and *trans*-9-acetyl-4,9'-di-*O*-methyl-3'-de-*O*-methyldehydrodiconiferyl alcohol were isolated from *Euphorbia quinquecostata* collected by Z. Mbwambo in 1999 [58], while *Berchemia discolor* collected in 1999 yielded new prenylated flavonoids ((3 *S*)-discoloranone; Fig. 2.3) [59].

2.3 Michellamine B: A Potential Anti-HIV Agent from the Cameroonian Liana, *Ancistrocladus korupensis*

From 1987 to 1996, the NCI tested over 60,000 extracts of natural origin in an in vitro cell-based anti-HIV screen which determined the degree of HIV-1 replication in treated infected lymphoblastic cells versus that in untreated infected control cells. Several plant-derived natural products have shown in vitro activity, and of these, michellamine B is the only anti-HIV agent isolated from an African source to advance into preclinical development.

In 1987, a sample of the leaves of a liana identified as an *Ancistrocladus* species was collected in the Korup region of southwestern Cameroon as part of the NCI contract with Missouri Botanical Garden for collections in Africa and Madagascar. Extracts of the leaves exhibited significant in vitro anti-HIV activity, and the dimeric naphthylisoquinoline alkaloid, michellamine B (Fig. 2.2), was isolated as the active agent [60, 61]. Michellamine B, named after the wife of the chemist who performed the initial isolation, showed in vitro activity against both HIV-1 and HIV-2 and was shown to inhibit human immunodeficiency virus-induced cell killing by at least two distinct mechanisms, acting at an early stage of the HIV life cycle by inhibiting reverse transcriptase as well as at later stages by inhibiting cellular fusion and syncytium formation [62]. Thus, it was selected for preclinical development.

The source plant was later identified as a new species and named *A. korupensis* D. Thomas & Gereau [63]. This new species was found only in and around the Korup National Park, and vine densities were very low, on the order of one large vine per hectare. While fallen leaves contained michellamine B, and their collection provided sufficient biomass for the isolation of enough drug to complete preclinical development [64], it was clear that extensive collections of fresh leaves could pose a possible threat to the limited and sparse wild population.

Thus far, no other *Ancistrocladus* species has been found to contain michellamine B, and investigation of the feasibility of cultivation of the plant as a reliable biomass source was initiated in 1993 through a contract with the Center for New Crops and Plant Products of Purdue University working in close collaboration with the University of Yaounde 1, the World Wide Fund for Nature Korup Project, MBG, Oregon State University, and the NCI-Frederick contractor, Science Applications International Corporation (SAIC). An extensive botanical survey was undertaken, and the range and distribution of the species were mapped, and dried leaves were analyzed for michellamine B content. Promising plants were resampled for confirmatory analysis, and those showing reproducible high concentrations were targeted for vegetative propagation. A medicinal plant nursery was established for the *A. korupensis* collection near Korup Park Headquarters in Mundemba, and through selection of promising plants from the wild and their subsequent propagation and growth in the nursery, it was demonstrated that michellamine content well above the wild collected average could be produced routinely. In keeping with the NCI policies of collaboration with source countries, all the cultivation studies were performed in Cameroon and involved the local population in the Korup region where the plant was originally discovered.

Based on the observed activity and the efficient formulation of the diacetate salt, the NCI committed michellamine B to advanced preclinical development, but continuous infusion studies in dogs indicated that in vivo effective anti-HIV concentrations could only be achieved at doses close to neurotoxic levels. Thus, despite in vitro activity against an impressive range of HIV-1 and HIV-2 strains, the difference between the toxic dose level and the anticipated level required for effective antiviral activity (the therapeutic index) was small, and NCI discontinued further studies aimed at clinical development. However, the discovery of novel antimalarial agents, the korupensamines, from the same species [65, 66] added further promise for this species. Michellamines have also been reported to show antioxidant activity [67] and inhibition of protein kinase C [68], while michellamine B has been shown to be a potent but nonselective inhibitor of platelet-type 12-human lipoxygenase [69]. Human lipoxygenases have been implicated in a variety of diseases involving inflammation, immune disorders, and various types of cancers [70, 71]. The publication of several synthetic routes to the michellamines [72, 73] should permit the study of structure–activity relationships, thereby increasing the potential for structural optimization to improve anti-HIV and other bioactivities relative to toxicity.

As mentioned in Sect. 2.1.2, despite negotiations with the appropriate authorities, no formal LOC agreement for plant collections was finalized with the Cameroon government; however, collections by MBG were performed with all of the necessary permits issued by the appropriate government departments. Upon the discovery of the anti-HIV activity of extracts of *A. korupensis* and the isolation of michellamine B as the active agent, the Cameroon government established an Intraministerial Committee for Research on *A. korupensis*, under the chairmanship of Mr. J. Edou of the Prime Minister's Office, to monitor progress in both the development of michellamine B and the cultivation project. The cultivation project was carried out with the full permission and cooperation of the Cameroon government, and close contact was maintained with the Intraministerial Committee regarding progress in all aspects of michellamine development. As can be seen from Table 2.2, several Cameroon government officials and scientists, including Mr. Edou, visited the NCI under the sponsorship of NCI. This collaborative mechanism with the Cameroon government was essential to maintaining progress since, at the time, few legal regulations were in place addressing the study of Cameroonian plants for non-timber uses [74].

2.4 Marine Sources

Marine organisms are proving to be a valuable source of potential anticancer agents, and interested readers are referred to several reviews for details [75, 76]. In the context of drug discovery from African marine resources, it is interesting to note that the first collections of *Dollabella auricularia*, source of the dolastatins [77], were made by Professor G. R. Pettit of Arizona State University off the coast of Mauritius in the early 1960s. Likewise, the first member of the hemiasterlin family of compounds was isolated by Professor Yoel Kashman of Tel Aviv University from the sponge, *Hemiasterella minor* (Kirkpatrick), collected in Sodwana Bay, South Africa [78]. Both the dolastatins and the hemiasterlins have given rise to anticancer agents which have progressed to clinical trials [77, 78]. Collections of the marine worm, *Cephalodiscus gilchristi*, by the Pettit group off the southeastern shores of South Africa in 1988 yielded cephalostatin 1 [79], and further large-scale collections led to the isolation of a further 18 members of this family. These and the closely related ritterazines have been reviewed [80].

In the past decade, the NCI has performed marine invertebrate collections along the coasts of Eastern and Southern Africa through its contract with Coral Reef Research Foundation (see Sect. 2.1.2). Collections off the Southern African coast have been performed in close collaboration with Professor Michael Davies-Coleman through a Memorandum of Understanding with Rhodes University (Table 2.4). Since aspects of marine bioprospecting in Southern Africa are presented in a later chapter in this volume, no further discussion will be included here.

2.5 Conclusions

From the foregoing discussion, it is clear that natural products have made, and continue to make, an indispensable contribution to the discovery and development of effective drugs for the treatment of cancer. This observation applies equally well to many other diseases afflicting humankind. A recent analysis of the new drugs marketed over the past 25 years during the period between 1981 and 2006 shows that some 50% owe their origin in one way or another to natural sources, and in some disease areas well over 60% are derived from natural products [81]. In addition, natural products are an invaluable source of molecular probes in the study of pathways influencing cell cycle progression [82].

While natural products are a proven source of novel bioactive molecules, the actual compound isolated from the natural source is often not suitable for development into an effective drug, but it may be regarded as a lead molecule which can form the basis for further chemical or biochemical modification. The discovery of promising bioactive molecules always involves close collaboration with biologists, firstly in the collection of the source organisms and then in the provision of suitable disease-oriented screens, while the optimization of the lead molecule requires significant input from medicinal and synthetic chemists. The preclinical development of an agent always requires close collaboration with pharmacologists and toxicologists in the determination of the optimal pharmacodynamic and toxicological parameters required for advancement of the agent into clinical trials with human patients. The recent establishment of the Drug Discovery and Development Center under the leadership of Professor Kelly Chibale at the University of Cape Town, South Africa, exemplifies this multidisciplinary collaborative approach and will play a major role in promoting the discovery and development of drug candidates by African scientists for diseases afflicting the continent.

From the NCI experience, African biodiversity has been shown to be the source of several promising anticancer drugs, with some in advanced clinical trials and others in preclinical development. In addition, the discovery of the anti-HIV-active compound michellamine B has emphasized the need for source countries to establish policies for promotion of the exploration of their valuable biological diversity for the discovery and development of non-timber products, including novel bioactive molecules. One of the cornerstones of the NCI drug-discovery program is collaboration with research groups worldwide in the testing of their products, and the development of promising anticancer leads. In the natural products area, the signing of Memoranda of Understanding with qualified research groups is the mechanism most suited for such collaboration, and it is hoped that more such collaborations will be established with African research groups in the years ahead.

Acknowledgments The authors gratefully acknowledge the collaboration of the permitting authorities in source countries where collections for the NCI were performed, both in the early (1960–1982) and more recent (1986–present) programs. These include Cameroon, Central African Republic, Ethiopia, Gabon, Ghana, Kenya, Madagascar, South Africa, Tanzania, and Zimbabwe.

References

1. Cragg GM, Newman DJ (2005) International collaboration in drug discovery and development from natural sources. Pure Appl Chem 77:1923–1942
2. Cragg GM, Newman DJ (2005) Plants as a source of anti-cancer agents. J Ethnopharmacol 100 (1–2):72–79
3. Cragg GM, Newman DJ (2009) Nature: a vital source of leads for anticancer drug development. Phytochem Rev 8:313–331
4. Kaufman D (1993) Botany 2000-ASIA. Newsletter 2:6
5. Hartwell JL (1982) Plants used against cancer. Quarterman, Lawrence, MA
6. Cragg GM, Boyd MR, Cardellina JH II et al (1994) Ethnobotany and drug discovery: the experience of the US National Cancer Institute. In: Chadwick DJ, Marsh J (eds) Ethnobotany

and the search for new drugs, vol 185, Ciba Foundation Symposium. Wiley, Chichester, pp 178–196
7. Cirla A, Mann J (2003) Combretastatins: from natural products to drug discovery. Nat Prod Rep 20:558–564
8. Pinney KG, Pettit GR, Trawick ML et al (2011) The discovery and development of the combretastatins. In: Cragg GM, Kingston DGI, Newman DJ (eds) Anticancer agents from natural products, 2nd edn. CRC/Taylor and Francis, Boca Raton, FL, Chapter 3
9. Siemann DW, Chaplin DJ, Walicke PA (2009) A review and update of the current status of the vasculature-disabling agent combretastatin-A4 phosphate (CA4P). Expert Opin Investig Drugs 18:189–197
10. Eichler W, Yafai Y, Wiedemann P et al (2006) Antineovascular agents in the treatment of eye diseases. Curr Pharm Design 12:2645–60
11. Li Q, Sham HL (2002) Discovery and development of antimitotic agents that inhibit tubulin polymerisation for the treatment of cancer. Expert Opin Ther Pat 12:1663–1702
12. Teicher BA (2009) Antibody-drug conjugate targets. Curr Cancer Drug Targ 9:982–1004
13. Senter PD (2009) Potent antibody drug conjugates for cancer therapy. Curr Opin Chem Biol 13:235–244
14. Zhiqiang A (ed) (2009) Therapeutic monoclonal antibodies: from the bench to the clinic. Wiley, Hoboken, NJ
15. Alley SC, Okeley NM, Senter PD (2010) Antibody-drug conjugates: targeted drug delivery for cancer. Curr Opin Chem Biol 14:529–537
16. Caravella J, Lugovskoy A (2010) Design of next-generation protein therapeutics. Curr Opin Chem Biol 14:520–528
17. Yu T-W, Floss HG, Cragg GM et al (2011) Ansamitocins (maytansinoids). In: Cragg GM, Kingston DGI, Newman DJ (eds) Anticancer agents from natural products, 2nd edn. CRC/Taylor and Francis, Boca Raton, FL, Chapter 17
18. Cassady JM, Chan KK, Floss HG et al (2004) Recent developments in the Maytansanoid antitumor agents. Chem Pharm Bull 52:1–26
19. Lambert JM (2010) Antibody-maytansinoid conjugates: a new strategy for the treatment of cancer. Drugs Future 35:471–80
20. Mita MM, Ricart AD, Mita AC et al (2007) A phase I study of a CanAg-targeted immunoconjugate, huC242-DM4, in patients with Can Ag-expressing solid tumors. J Clin Oncol 25(18S):3062
21. Beeram M, Krop I, Modi S et al (2007) A phase I study of trastuzumab-MCC-DM1 (T-DM1), a first-in-class HER2 antibody-drug conjugate (ADC), in patients (pts) with HER2+ metastatic breast cancer (BC). J Clin Oncol 25(18S):1042
22. Hughes B (2010) Antibody-drug conjugates for cancer: poised to deliver? Nat Rev Drug Discov 9:665–667
23. Niculescu-Duvaz I (2010) Trastuzumab emtansine, an antibody-drug conjugate for the treatment of HER2+ metastatic breast cancer. Curr Opin Mol Ther 12:350–360
24. Ratnayake R, Covell D, Ransom TT et al (2009) Englerin A, a selective inhibitor of renal cancer cell growth, from *Phyllanthus engleri*. Org Lett 11:57–60
25. Willot M, Radtke L, Könning D et al (2009) Total synthesis and absolute configuration of the guaiane sesquiterpene englerin A. Angew Chem Int Ed Engl 48:9105–9108
26. Nicolaou KC, Kang Q, Ng SY et al (2010) Total synthesis of englerin A. J Am Chem Soc 132:8219–8222
27. Molawai K, Delpont N, Echavarren AM (2010) Enantioselective synthesis of (-)-englerins A and B. Angew Chem Int Ed Engl 49:3517–3519
28. Zhou Q, Chen X, Ma D (2010) Asymmetric, protecting-group-free total synthesis of (-)-englerin A. Angew Chem Int Ed Engl 49:3513–3515
29. Li Z, Nakashige M, Chain WJ (2011) A brief synthesis of (-)-englerin A. J Am Chem Soc 133:6553–6556

30. Chan KP, Chen DY (2011) Chemical synthesis and biological evaluation of the englerin analogues. ChemMedChem 6:420–423
31. Radtke L, Willot M, Sun H et al (2011) Total synthesis and biological evaluation of (-)-englerin A and B: synthesis of analogues with improved activity profile. Angew Chem Int Ed Engl 50:1–6
32. Ushakov DB, Navickas V, Ströbele M et al (2011) Total synthesis and biological evaluation of (-)-9-deoxy-englerin A. Org Lett 13:2090–2093
33. Sourbier C, Ratnayake R, Scroggins B et al (2011) Targeting renal carcinoma with englerin A. AACR Annual Meeting, abstract 959
34. Beutler JA, Jato JG, Cragg GM et al (2006) The schweinfurthins. Issues in development of a plant-derived anticancer lead. In: Bogers J, Craker LE, Langa D (eds) Medicinal and aromatic plants. Springer, Amsterdam, Chapter 22
35. Thoison O, Hnawia E, Guéritte-Voegelein F (1992) Vedelianin, a hexahydroxanthene derivative isolated from *Macaranga vedeliana*. Phytochemistry 31:1439–1442
36. Beutler JA, Shoemaker RH, Johnson T et al (1998) Cytotoxic geranyl stilbenes from *Macaranga schweinfurthii*. J Nat Prod 61:1509–1512
37. Yoder B, Norris A, Miller JS et al (2006) Cytotoxic prenylated stilbenes and Flavonoids from *Macaranga alnifolia* from the Madagascar rain forest. J Nat Prod 70:342–346
38. Klausmeyer P, Van QN, Jato JG et al (2010) Schweinfurthins I and J from *Macaranga schweinfurthii*. J Nat Prod 73:479–481
39. Neighbors JD, Beutler JA, Wiemer DF (2005) Synthesis of nonracemic 3-deoxyschweinfurthin. J Org Chem 70:925–931
40. Neighbors JD, Salnikova MS, Beutler JA et al (2006) Synthesis and structure-activity studies of schweinfurthin B analogs: evidence for the importance of a D-ring hydrogen bond donor in expression of differential cytotoxicity. Bioorg Med Chem 14:1771–1784
41. Mente NR, Wiemer AJ, Neighbors JD et al (2007) Total synthesis of (*R*, *R*, *R*)- and (*S*, *S*, *S*)-schweinfurthin F: differences of bioactivity in the enantiomeric series. Bioorg Med Chem Lett 17:911–15
42. Topczewski JJ, Neighbors JD, Wiemer DF (2009) Total synthesis of (+)- schweinfurthins B and E. J Org Chem 74:6965–6972
43. Mente NR, Neighbors JD, Wiemer DF (2008) BF3 · Et_2O-mediated cascade cyclizations: synthesis of schweinfurthins F and G. J Org Chem 73:7963–7970
44. Kuder CH, Neighbors JD, Hohl RJ et al (2009) Synthesis and biological activity of a fluorescent schweinfurthin analogue. Bioorg Med Chem 17:4718–4723
45. Ulrich NC, Kodet JG, Mente NR et al (2010) Structural analogues of schweinfurthin F: probing the steric, electronic and hydrophobic properties of the D-ring substructure. Bioorg Med Chem 18:1676–1683
46. Topczewski JJ, Wiemer DF (2011) First total synthesis of (+)-vedelianin, a potent antiproliferative agent. Tetrahedron Lett 52:1628–1630
47. Topczewski JJ, Kodet JG, Wiemer DF (2011) Exploration of cascade cyclizations terminated by tandem aromatic substitution: total synthesis of (+)-schweinfurthin A. J Org Chem 76:909–919
48. Turbyville TJ, Gürsel DB, Tuskan RG et al (2010) Schweinfurthin A selectively inhibits proliferation and Rho signaling in glioma and neurofibromatosis type 1 tumor cells in an NF1-GRD dependent manner. Mol Cancer Ther 9:1234–124345
48a. Burgett AW, Poulsen TB, Wangkanont K, Anderson DR, Kikuchi C, Shimada K, Okubo S, Fortner KC, Mimaki Y, Kuroda M, Murphy JP, Schwalb DJ, Petrella EC, Cornella-Taracido I, Schirle M, Tallarico JA, Shair MD (2011) Natural roducts reveal cancer cell dependence on oxysterol-binding proteins. Nat. Chem. Biol 7 (9):639–647
48b. Holstein SA, Kuder CH, Tong H, Hohl RJ (2011) Pleiotropic effects of a schweinfurthin on isoprenoid homeostasis. Lipids 46(10):907–921
49. Fojo T, Bates S (2003) Strategies for reversing drug resistance. Oncogene 22:7512–7523

50. Silva GL, Cui B, Chávez D et al (2001) Modulation of the multidrug-resistance phenotype by new tropane alkaloid aromatic esters from *Erythroxylum pervillei*. J Nat Prod 64:1514–1520
51. Mi Q, Cui B, Silva GL (2003) Characterization of tropane alkaloid aromatic esters that reverse the multidrug-resistance phenotype. Anticancer Res 23:3607–3616
52. Chin YW, Jones WP, Waybright TJ et al (2006) Tropane aromatic ester alkaloids from a large-scale re-collection of *Erythroxylum pervillei* stem bark obtained in Madagascar. J Nat Prod 69:414–417
53. Chin YW, Kinghorn AD, Patil PN (2007) Evaluation of the cholinergic and adrenergic effects of two tropane alkaloids from *Erythroxylum pervillei*. Phytochem Res 21:1002–1005
54. Magadula JJ, Erasto P (2009) Bioactive natural products derived from the East African Flora. Nat Prod Rep 26:1535–1554
55. Pettit GR, Zhang Q, Pinilla V et al (2005) Antineoplastic agents. 534. Isolation and structure of sansevistatins 1 and 2 from the African *Sansevieria ehrenbergii*. J Nat Prod 68:729–733
56. Chang FR, Hayashi K, Chen IH et al (2003) Antitumor agents 228. Five new agarofurans, reissantins A–E, and cytotoxic principles from *Reissantia buchananii*. J Nat Prod 66:1416–1420
57. Sakurai Y, Sakurai N, Taniguchi M et al (2006) Rautandiols A and B, pterocarpans and cytotoxic constituents from *Neorautanenia mitis*. J Nat Prod 69:397–399
58. Su BN, Park EJ, Mbwambo ZH et al (2002) New chemical constituents of *Euphorbia quinquecostata* and absolute configuration assignment by a convenient Mosher ester procedure carried out in NMR tubes. J Nat Prod 65:1278–1282
59. Chin YW, Mdee LK, Mbwambo ZH et al (2006) Prenylated flavonoids from the root bark of *Berchemia discolor*, a Tanzanian medicinal plant. J Nat Prod 69:1649–1652
60. Manfredi KP, Blunt JW, Cardellina JH II et al (1991) Novel alkaloids from the tropical plant *Ancistrocladus abbreviatus* inhibit cell-killing by HIV-1 and HIV-2. J Med Chem 34:3402–3405
61. Boyd MR, Hallock YF, Cardellina JH II et al (1994) Anti-HIV michellamines from *Ancistrocladus korupensis*. J Med Chem 37:1740–1745
62. McMahon JB, Currens MJ, Gulakowski RJ et al (1995) Michellamine B, a novel plant alkaloid, inhibits human immunodeficiency virus-induced cell killing by at least two distinct mechanisms. Antimicrob Agents Chemother 39:484–8
63. Thomas DW, Gereau RE (1993) *Ancistrocladus korupensis* (*Ancistrocladaceae*): a new species of liana from Cameroon. Novon 3:494–498
64. Thomas DW, Boyd MR, Cardellina JH II et al (1994) Sustainable harvest of *Ancistrocladus korupensis* (Ancistrocladaceae). Leaf litter for research on HIV. Econ Bot 48:313–414
65. Hallock YF, Manfredi KP, Blunt JW et al (1994) Korupensamines A-D, novel antimalarial alkaloids from *Ancistrocladus korupensis*. J Org Chem 59:6349–6355
66. Hallock YF, Manfredi KP, Dai JR et al (1997) Michellamines D-F, new HIV-inhibitory dimeric naphthylisoquinoline alkaloids, and korupensamine E, a new antimalarial monomer, from *Ancistrocladus korupensis*. J Nat Prod 60:677–683
67. White EL, Ross LJ, Hobbs PD et al (1999) Antioxidant activity of michellamine alkaloids. Anticancer Res 19:1033–5
68. White EL, Chao WR, Ross LJ et al (1999) Michellamine alkaloids inhibit protein kinase C. Arch Biochem Biophys 365:25–30
69. Deschamps JD, Gautschi JT, Whitman S et al (2007) Discovery of platelet- type 12-human lipoxygenase selective inhibitors by high-throughput screening of structurally diverse libraries. Bioorg Med Chem 15:6900–8
70. Samuelsson B, Dahlen SE, Lindgren JA et al (1987) Leukotrienes and lipoxins: structures, biosynthesis, and biological effects. Science 237:1171–1176
71. Ford-Hutchinson AW, Gresser M, Young RN (1994) 5-Lipoxygenase. Annu Rev Biochem 63:383–417
72. Bringmann G, Götz R, Keller PA et al (1998) A convergent total synthesis of the michellamines. J Org Chem 63:1090–1097

73. Bringmann G, Götz R, Harmsen S et al (1996) Acetogenic isoquinoline alkaloids, LXXXII. Biomimetic total synthesis of michellamines A–C. Liebigs Annalen 1996:2045–2058
74. Jato J, Simon JE, Symonds P et al (1996) Rules and regulations on the collection in Cameroon of biological materials for biological testing and drug discovery. J Ethnopharmacol 51:121–125
75. Newman DJ, Cragg GM (2004) Marine natural products and related compounds in clinical and advanced preclinical trials. J Nat Prod 67:1216–1238
76. Newman DJ, Cragg GM (2004) Advanced preclinical and clinical trials of natural products and related compounds from marine Sources. Curr Med Chem 11:1693–1714
77. Flahive E, Srirangam J (2011) The dolastatins: novel antitumor agents from *Dolabella auricularia*. In: Cragg GM, Kingston DGI, Newman DJ (eds) Anticancer agents from natural products, 2nd edn. CRC/Taylor and Francis, Boca Raton, FL, Chapter 11
78. Andersen RJ, Roberge M (2011) A synthetic analog of the antimitotic natural product hemiasterlin. In: Cragg GM, Kingston DGI, Newman DJ (eds) Anticancer agents from natural products, 2nd edn. CRC/Taylor and Francis, Boca Raton, FL, Chapter 14
79. Pettit GR, Inoue M, Kamano Y et al (1988) Antineoplastic agents. 147. Isolation and structure of the powerful cell growth inhibitor cephalostatin 1. J Am Chem Soc 110:2006–2007
80. Moser BR (2008) Review of cytotoxic cephalostatins and ritterazines: isolation and synthesis. J Nat Prod 71:487–491
81. Newman DJ, Cragg GM (2007) Natural products as sources of new drugs over the last 25 years. J Nat Prod 70:461–477
82. Newman DJ, Cragg GM, Holbeck D et al (2002) Natural products and derivatives as leads to cell cycle pathway targets in cancer chemotherapy. Curr Cancer Drug Targets 2:279–308

Chapter 3
Tuberculosis Drug Discovery: Target Identification and Validation

Digby F. Warner and Valerie Mizrahi

3.1 Introduction

Tuberculosis (TB) is a global problem that disproportionately affects a handful of high-burden countries, many of which are located in sub-Saharan Africa. Understandably, efforts to combat TB in endemic countries tend to focus on public health and intervention strategies. Although there are some notable exceptions, research in these regions is limited, and often confined to epidemiological surveillance of strain types and resistance prevalence. Instead, high-burden regions are generally valued primarily as field sites for drug trials and clinical studies that are usually led by international organizations, with local input focused on the collection, processing and analysis of samples, and on patient management. Basic science, which includes the activities related to drug target identification and validation discussed in this chapter, is rare. Consistent with this assessment, our review of the contribution of South African researchers to TB drug discovery, particularly in the area of target identification and validation, reveals the existence of very few research programmes dedicated to this effort. However, the last decade or so has witnessed a shift in international funding mechanisms for TB drug discovery towards multidisciplinary consortia which partner academic laboratories with major pharmaceutical companies that are able to provide access to large libraries of high-quality chemical compounds, as well as technical expertise on the development and application of high-throughput screening systems. These hybrid organizational models remain heavily reliant on specialist (academic) laboratories to perform target identification and validation work, and so hold promise for the expansion of local basic science research programmes that could play an increasing role in future TB drug discovery.

D.F. Warner • V. Mizrahi (✉)
MRC/NHLS/UCT Molecular Mycobacteriology Research Unit, DST/NRF Centre of Excellence for Biomedical TB Research, Institute of Infectious Disease and Molecular Medicine, University of Cape Town, Private Bag X3 Rondebosch 7701, Cape Town, South Africa
e-mail: digby.warner@uct.ac.za; valerie.mizrahi@uct.ac.za

K. Chibale et al. (eds.), *Drug Discovery in Africa*,
DOI 10.1007/978-3-642-28175-4_3,

Our mandate in compiling this chapter is to highlight the contribution of South African researchers to efforts aimed at the identification and validation of drug targets for TB. In order to fulfil this brief, and to provide some suggestions regarding a potential future role for local research, it is necessary to situate current programmes in this area within the context of the global TB problem and to provide a brief summary of trends in TB drug discovery over the past two decades. Several comprehensive reports and reviews have covered the topics of TB research funding [1, 2] and the state of drug discovery for TB [3, 4], and the interested reader is directed to these for a more thorough discussion.

3.2 Tuberculosis in Africa

TB is a global problem: an estimated nine million new cases were reported in 2009, with the disease claiming approximately 1.8 million lives [5]. Moreover, fully one third of the world's population (approximately two billion people) is infected with the causative agent, *Mycobacterium tuberculosis*, and faces a 10% lifelong risk of developing reactivation disease. Even though the global per capita TB incidence rate is declining at around 1% per annum, population growth worldwide is increasing at an estimated 2% per year; the net effect is an increase in the total number of new TB cases, with the figure estimated to approach 9.8 million in 2010 [6]. Although startling, the magnitude of these numbers disguises an important underlying disparity in disease burden: almost 80% of all cases occurred in 22 countries, of which 9 are on the African continent. On its own, Africa accounted for 2.8 million cases in 2009 with an estimated 430,000 deaths, the majority (275,000) of which were reported in the nine high-burden territories where TB continues to devastate public health systems. Perhaps most strikingly, South Africa ranks third in TB burden behind only the world's most populous countries, India and China. By far, the largest proportion (approximately 80%) of HIV-positive incident TB cases are in Africa, with South Africa, which has only 0.7% of the world's population, contributing a huge portion of these at 25% of all HIV–TB co-infections [7].

In addition to the HIV epidemic, multiple other factors continue to undermine TB control measures in these areas, including the increasing emergence of multi- (MDR) and extensively drug-resistant (XDR) *M. tuberculosis* strains that are resistant to the major anti-tubercular agents; the variable availability and inconsistent quality of the frontline anti-TB drugs; declining health infrastructures; the prevalence of other chronic diseases that can increase TB morbidity; and numerous sociological confounders including migration, poverty, political conflict and marginalization. Although a vaccine exists—the widely administered BCG, an attenuated form of the closely related *M. bovis*—it is effective only against the most severe forms of paediatric TB disease; critically, it offers no protection against adult pulmonary TB and, further complicating matters, is characterized by poorly understood regional (geographic) variations in efficacy [8]. Instead, the major thrust of global control efforts is devoted to chemotherapeutic intervention in active disease, utilizing a combination therapy that extends over a minimum 6-month treatment period comprising a 2-month intensive phase with four drugs and a 4-month continuation phase with two drugs [3]. That is for fully susceptible TB. Where MDR-TB is diagnosed, the treatment duration is extended to a minimum of 9–12 months and utilizes drug combinations that are less easily administered, less well tolerated, more expensive and carry a much higher risk of failure. XDR-TB and, more recently, totally drug-resistant (TDR)-TB [9] represent the extreme and are considered largely untreatable, although widely publicized efforts of some dedicated facilities—including groups in South Africa [10]—have had positive results with intensive treatment monitoring by multidisciplinary clinical teams and using experimental combinations of drugs and drug doses that are not commonly applied to TB [11].

3.3 New Drug Discovery for TB

For a disease that ranks number 1 in mortality owing to a single infectious agent, it is sobering that the majority of the drugs used to treat TB were developed 50–60 years ago and, in many cases, were not purpose designed [4]. Moreover, some of these drugs were introduced on the back of limited (or non-existent) clinical trial data with therapeutic regimens (dosages, frequency, combinations) determined subsequently on the basis of empirical (clinical) outcomes (reviewed in [11]). In some cases, regimens remain non-optimal even today, especially for paediatric TB [12], further reinforcing the impression that TB drug development and application has received inadequate attention.

Multiple factors contributed to this neglect, paramount of which were the effective restriction of the TB burden to developing (Third World) economies and the false confidence in the power of antibiotics to eliminate infectious disease. The advent of HIV in the 1980s, however, has fuelled the resurgence of TB as a global threat that is compounded by the emergence of multiply resistant organisms. As Kaufmann and Parida [1] observe in their review of the recent history of global TB research funding, these developments occurred at a time when funds for TB research were limited to small grants for individual laboratories whose drug development imperative might be non-existent or inchoate; for the most part, academic research was not specifically directed to defined drug discovery outcomes, which require that research is performed dispassionately and projects are evaluated at each stage in terms of a well-defined and demanding set of criteria. Antibiotics—and their development—were considered the domain of pharmaceutical companies at this time, and access to compound libraries, and the ability to conduct high-throughput experiments/analyses on cutting-edge equipment, was restricted to these organizations.

Subsequent years have witnessed a significant shift. Therefore, before assessing local (South African) research efforts in this regard, it is important to provide some insight into the dominant trends in international TB drug discovery, particularly in an era that has been defined by the entry into the TB arena of major philanthropic partnerships and initiatives devoted to fighting the disease [1]—for example, *Stop Tuberculosis* (http://www.stoptb.org), *The Global Fund to Fight AIDS, Tuberculosis, and Malaria* (http://www.theglobalfund.org), the *Bill and Melinda Gates Foundation* (http://www.gatesfoundation.org) and the *Global Alliance for TB Drug Development* (http://www.tballiance.org)—together with the established public sector agencies and philanthropic foundations including the US National Institutes of Health (NIH) and National Institutes of Allergy and Infectious Disease (NIAID), the Medical Research Council (UK), the European Commission 6th and 7th Frameworks, and the Wellcome Trust. Under the auspices of these organizations, an estimated 400 million US dollars was dedicated to TB research in 2005 and increased steadily to 619 million US dollars in 2009, with drug development receiving the largest proportion at around 30% of the total R&D spend [1, 2]. Although dramatic, the process of ramping funding to these levels has taken more than two decades and represents a massive lost opportunity: WHO estimates indicate that in South Africa alone, for example, mortality figures for TB (excluding HIV) were in the range of 12,000–29,000 deaths per annum in the period 1990–2009 [5]. Nevertheless, the scale of the grants offered by these initiatives can support large, multidisciplinary research entities, a development which has had a significant impact on the manner in which TB drug discovery is now conducted and, importantly, how it is evaluated. Specifically, the funding levels available have demanded very ambitious projects that, in turn, often require the participation of multiple collaborative partners spanning academia, small biotech and industry. The successful administration of these consortia has, in turn, required that academic labs adopt project management and decision tools that are more commonly associated with industry; in particular, it has resulted in the rigorous adherence to predetermined milestones which, in turn, has necessitated a much more dispassionate assessment of project (target) viability and attractiveness. Here, the goal is delivery of a validated target with corresponding compounds that fulfil stringent pharmacological endpoints; there is no room for sentimental attachment to a project (gene) which, albeit interesting intellectually, fails to satisfy key requirements for progression through the drug delivery pipeline.

Today, that pipeline contains around 10 compounds at different stages of clinical development [3, 4]. This is by far the highest number in the history of TB chemotherapy and provides convincing evidence of the benefit of recent concentrated efforts to deliver an alternative drug candidate(s) to augment the existing therapeutic regimen. Even so, 10 candidates are not nearly sufficient to satisfy the increasing demands of a new TB drug or to overcome the mounting obstacles to effective intervention strategies discussed below. Moreover, it is significant that four of the most promising compounds were approved for other clinical uses and are being repurposed for TB [4].

3.3.1 *What Should a New TB Drug Offer?*

Given that a new drug has not been introduced for 40 years, the criteria that must be satisfied by the new generation TB drugs are unexpectedly onerous. Moreover, in a field that is characterized by strong personalities and dogmatic adherence to preferred theories, there is surprising consensus on the core set of requirements of the new drug [3, 13, 14]: namely, that it satisfy the need to shorten the duration of chemotherapy, either as part of a new combination or as an adjunct to existing drugs; that it be associated with fewer side effects as part of drug regimen that requires fewer tablets and less frequent dosing; that the new drug possess activity against MDR and XDR strains; that it be compatible with other drugs for chronic conditions, especially HIV and diabetes, both of which can significantly impact disease outcomes; and, finally, the requirement that the new drug retain activity against the organism in both intra- and extracellular host environments. Each of these reflects an attempt to address a critical vulnerability that has been directly implicated in the failure of the existing drug regimens, and the list appears to span biological as well as programmatic concerns. It is important to realize, however, that every difficulty associated with tackling TB (perhaps with the exception of drug–drug interactions) is a direct consequence of the inherent characteristics of *M. tuberculosis* as an aetiological agent. Therefore, efforts to improve therapeutic efficacy are wholly dependent on an improved understanding of the physiology of the infecting organism.

3.3.2 *The Challenges Facing a New TB Drug*

The core attributes of a new TB drug that were outlined in the previous section (see Sect. 3.3.1) are daunting, not least because they require that a combination of microbiological and pharmacological hurdles is overcome. Since this chapter is focused on target identification and validation, we will concentrate on those factors—almost exclusively (micro)biological—and will ignore the pharmacological challenges which are significant and include issues related to solubility, stability, specificity, toxicity, bioavailability, permeability and whole-cell activity. Additional concerns include the need to minimize drug–drug interactions, as well as the need to assess the effect of new drugs on host metabolic pathways that might impact therapeutic efficacy [13]. Again, the reader is directed to recent reviews of the special obstacles that TB drug development faces [3, 13] as well as the revealing exposition by David Pampliano and colleagues [15] of the general difficulties associated with antimicrobial development from the perspective of the pharmaceutical industry. In the context of target identification and validation, there is little that can be done to influence pharmacological outcomes. This is predominantly the domain of the hit-to-lead component of the development pipeline and requires the application of specialized chemistry expertise. If there is a role for target identification and validation here, it is to reapply selected assays to hit compounds that have undergone chemical modification and derivatization to ensure that these retain inhibitory activity and, importantly, remain "on target".

3.3.3 Mycobacterial Physiology and Drug Discovery

M. tuberculosis is an obligate pathogen whose viability within the human population is dependent on its inherent capacity to initiate successive (and repetitive) cycles of infection, disease, transmission and clinical latency [14]. We argued previously [16] (and see Sect. 3.4) that all research into the fundamental physiology of *M. tuberculosis* necessarily, and inevitably, carries a drug discovery mandate since it holds the promise of revealing critical vulnerabilities in the bacillary make-up that can be exploited. While this might suggest that the identification of potential drug targets should be relatively straightforward, fundamental gaps in our knowledge of the basic biology of *M. tuberculosis* and the complexities of the host–pathogen interaction render these efforts extraordinarily difficult and ensure that the translation of experimental observation into practical interventions remains elusive. A detailed review of *M. tuberculosis* physiology and its impact on TB pathogenesis is beyond the scope of this chapter; instead, the reader is directed to several recent articles which provide valuable insight into the status of the field [14, 17–20]. However, it is important to introduce the salient concepts here since these underpin most of the major barriers to new drug discovery for TB; moreover, many of the characteristics of the organism that continue to confound drug development inform the discussion of current research in South African laboratories (see Sect. 3.6).

M. tuberculosis is a formidable adversary. For an obligate pathogen, the organism is unusual in possessing an expanded metabolic repertoire: analysis of the genome suggests a prototroph that is able to synthesize all essential amino acids, vitamins and enzyme cofactors [21]. It has the ability to oxidize multiple carbon substrates and can utilize a variety of alternate electron transport chains during adaptation to host environments characterized by different oxygen and nutrient availabilities. A defining feature of *M. tuberculosis* is its complex, impermeable cell wall that has been linked to the intrinsic resistance of the organism to host immune effectors and numerous antibiotic agents. It is not surprising, therefore, that the bacillus devotes considerable genetic resources to lipid metabolism, including both synthetic and degradative pathways, many of which have undergone large-scale expansion. The *M. tuberculosis* genome also contains an array of sigma factors and transcriptional regulators which indicate the ability to respond rapidly to environmental cues. Moreover, the bacillus is armed with multiple defence and detoxification pathways, as well as DNA repair and maintenance functions [17].

This limited catalogue is intended to provide some sense of *M. tuberculosis* as a pathogen that is well equipped to survive variable and hostile intracellular environments; what the list fails to convey is the increasing appreciation of

M. tuberculosis as a mycobacterium that has adapted exquisitely to host colonization and persistence within the human population [22]. This concept is crucial to understanding the difficulties associated with new drug development for TB and critically undermines target selection. It is perhaps best illustrated by the profound disconnect between the prevailing knowledge of mycobacterial physiology gained from various in vitro and in vivo models and observed pathogenesis: it remains extremely difficult to link the clinical phenotypes defining a specific disease stage with an underlying bacterial or physiological state. Although a correlation has been inferred between active disease (a clinical phenomenon) and aggressive bacterial replication (similarly, latent disease is intuitively associated with limited—or no—mycobacterial replication and limited metabolic activity), active infections are characterized by heterogeneity in bacterial populations at discrete physiological loci within lesions [14, 18]. Moreover, the facile association of metabolic activity with disease pathology obscures a crucial requirement of a new TB drug, namely, that it target both active and non-replicating bacilli (either alone or as part of a new drug combination) in order to eradicate tolerant organisms in a patient who, by definition, will be presenting with active disease.

As noted above, a key challenge facing a new TB drug is that it must be developed in the context of limited knowledge about bacterial metabolism during host infection [19]. A recent report describing the development of novel pyrimidine–imidazoles for antimycobacterial use provides a cogent (even sobering) example of the dangers inherent in extrapolating in vitro observations to the in vivo environment [23]. In turn, the limited insight into infection-relevant physiology impacts the extent to which the various disease models, in vitro and in vivo, can be considered optimal [24]. Compounding this problem is the fact that approximately 50% of the *M. tuberculosis* genome encodes hypothetical proteins, conserved hypothetical proteins, PE/PPE-family members [25] and other proteins of unknown function [26, 27]. The function of these genes and their contribution to pathogenesis therefore constitutes a major research priority and, as discussed below, is readily incorporated into target identification efforts.

3.3.4 The Properties of a Good Target

As discussed in previous sections, recent developments in funding mechanisms available for TB drug discovery have prioritized the formation of multicentre (often multinational) consortia whose members are drawn from academic laboratories, small to medium biotechnology firms and pharmaceutical companies, each of which has a different perspective on the preferred attributes of a good target. A consequence of the amalgamation of expertise under a single drug discovery banner is that the criteria for target selection have been refined to reflect biological as well as pharmacological (chemical) considerations. So, it is now generally accepted that the target should be essential for growth or survival under a chosen condition, that it should be amenable to chemical inhibition—a property commonly referred to as "druggability" and that it should be accessible to the inhibiting agent [28]. Critically, this list recognizes that the first of these attributes (essentiality) is not a guarantee of the other two. This shift in thinking perhaps best reflects the increasing sophistication of TB drug discovery efforts and, in turn, demands the application of more well-defined target identification and validation criteria.

3.4 Target Identification and Validation

In general terms, target identification and validation describes the set of activities designed to identify reactions or pathways that are critical for some aspect of pathogenesis (bacterial viability, virulence, transmission, latency, drug tolerance, etc.) and the subsequent demonstration (validation) by genetic or chemical means that abrogation or inhibition of target function profoundly impairs the identified pathogenic property under disease-relevant conditions.

3.4.1 Target Identification

Although the term "target identification and validation" implies that these activities are conducted sequentially, and by the same research entity, in practice the process often deviates from this simplistic model. For an obvious example, one needs only to consider the impact on recent TB drug discovery efforts of a series of landmark studies performed in the laboratory of Eric Rubin in the early 2000s. Utilizing whole-genome random mutagenesis, the authors were able to identify those genes whose disruption severely attenuates the proliferative ability or virulence of *M. tuberculosis* under specific conditions [29, 30]. In some ways, this approach effectively combines the identification and validation steps (but see Sect. 3.4.2), and the essentiality data inferred from these experiments have informed most target selection algorithms—they are even applied as the default "target identification" criterion in some cases despite the acknowledged limitations regarding the inference of essentiality from mutant pools [31]. Where "Sassetti and Rubin essentiality" governs target selection, validation efforts have been focused primarily on verifying the inferred essentiality—and so attractiveness—of a specific gene by subjecting the equivalent targeted gene deletion (knockout) mutant to a battery of phenotypic screens. Although additional target identification criteria have also been proposed [32], the Sassetti and Rubin lists have been very successful in elucidating multiple potential drug targets. However, validation remains critical, as we and others have reported on genes whose essentiality is limited to specific in vitro conditions [33]. Moreover, as noted below, confirmation of essentiality in vitro does not on its own constitute validation. In a previous section (see Sect. 3.3.3), we reiterated the concept that *M. tuberculosis* physiology is inextricably linked to pathogenesis. It is not surprising, therefore, that many studies elucidating fundamental aspects of mycobacterial physiology have, in retrospect, yielded important target prediction data and validation data even though they might have been initiated in the absence of any drug targeting motive. This is easily appreciated where a study demonstrates essential gene function in a specific disease model [34]. Perhaps less obvious, however, is the very important—but often undervalued—contribution of "basic science" studies of mycobacterial physiology to the *elimination* of an otherwise attractive target owing to the unpredicted activity of complementary or alternate pathway function. This applies in particular to central metabolism [19], where incomplete knowledge of the metabolic states adopted by *M. tuberculosis* during host infection continues to plague predictions of essential pathways. Even where a specific metabolite (or cofactor) is known to be crucial to the infection cycle, key questions remain: for example, how many pathways are available to the organism to synthesize the particular metabolite? Can the organism scavenge the metabolite (or precursors) from the host? Is the metabolite required throughout the course of infection, or only in certain disease states? The final sections of this chapter provide some examples of these ideas in the discussion of work conducted in our laboratory. Therefore, although our evaluation of the critical role of basic science approaches to target identification and validation might in some respects betray our own bias, recent developments—in particular, the major advances in techniques for in vitro [35] and in vivo imaging [14, 35–37] as well as bacterial cell content (metabolite) profiling [38, 39]—have already challenged prevailing beliefs about the physiology of *M. tuberculosis* during host infection. It seems reasonable to conclude, therefore, that future target identification and validation efforts will demand the application of these and other techniques—for example, systems biology [40, 41] and metabolic modelling [42]—in order to gain a level of insight into the "essentiality" of a specific target that genetic and chemical genetic studies have been unable to provide.

An additional, unintended consequence of the prioritization of essentiality data in the selection of candidate targets is that that any work on "essential" genes can claim a drug discovery imperative, whereas studies of "non-essential" genes can be perceived as less valuable. This is a criticism that is often levelled against "basic science", whose principal mandate is discovery [1]. Several authoritative reviews have cautioned against this simplistic evaluation in pointing out that a good target need not be essential [18]. For example, recent work in the field of antifungal drug discovery has demonstrated the enormous value of synthetic lethal screens to identify synergistic drug interactions [43]. In addition, there is an increasing recognition that, for infectious diseases such as TB, the best strategy might not be to target the pathogen itself but rather to target host factors required for pathogenesis [44].

Another common deviation from the classic target identification and validation paradigm relates to those studies designed to identify the target(s) of compounds with demonstrated antibacterial activity, but whose mode of action is unknown. This is becoming an increasingly important component of modern drug discovery programmes which are designed to identify lead molecules/compound classes with whole-cell activity from screens of compound and natural product libraries—in fact, four of the most promising compounds in the drug discovery pipeline were identified in whole-cell screens [13]. Here, the goal of target identification and validation is effectively one of reverse engineering: the target is identified utilizing a combination of techniques that include the isolation and sequencing of resistant mutants, whole-genome transcriptional profiling and assays of macromolecular synthesis (RNA, protein, DNA, peptidoglycan and fatty acids) [45, 46] and then validated through targeted gene modification or chemical genetic approaches. While the failure to identify the target does not necessarily disqualify a compound as a candidate chemical scaffold, it does complicate subsequent efforts towards enhancement of the pharmacological properties of the active compound. For example, knowledge of the target is critical to the generation of crystallographic information that can aid chemical derivatization, as well as the design and subsequent refinement of small molecule inhibitors. Also, in initiatives to re-engineer known compounds, it is critical that the target is known in order to assess the effect of the re-engineering process on activity. It is interesting to note, therefore, that several mainstay TB drugs were approved for public use in the absence of a defined mechanism of action; moreover, for some of these—for example, pyrazinamide, isoniazid and ethambutol—a considerable portion of the post-genomic era has been devoted to the unequivocal elucidation of their respective targets [47]. According to current criteria, each of these antibiotics would have been downgraded as a candidate drug, a sobering thought which again highlights the importance of this form of *post hoc* target identification in drug discovery.

Section 3.3.4 highlighted the key features of a new TB drug. It is noteworthy that several of these are critically dependent on the success of target identification and validation efforts. For example, in order to be active against MDR and XDR strains, a new compound must have a novel mechanism of action. Similarly, reducing treatment duration requires the identification of compounds that are potently bactericidal and active against heterogeneous populations in different growth (physiological) states. This in turn demands improved knowledge of the bacterial processes that must be overcome to effect sterilization, most important of which is the so-called persistent sub-population that survives antimicrobial treatment even under optimal conditions [48]. Perhaps the greatest obstacle facing efforts to identify sterilizing compounds is the absence of a validated in vitro screen for chronic, recalcitrant infection. Although several models are in use that are intuitively sensible [49–51], and notwithstanding the ability of these screens to elucidate compounds with special activities, none has been validated. Moreover, as discussed below, validation poses an intractable problem in its own right since there is currently no conclusive evidence of the contribution of persister organisms to treatment failure.

3.4.2 *When Is a Target Validated?*

Target validation requires the demonstration that abrogation of function of a specific gene product results in a loss of bacterial viability (or phenotypic alteration) that is sufficiently dramatic to suggest that gene product as an attractive target for antibacterial compounds. Validation is therefore critically dependent on genetic information. For TB researchers, the post-genomic era began in 1998 following the publication of the complete genome sequence of the virulent laboratory strain, H37Rv [21]. The period since then has been characterized by a rapid advance in mycobacterial genetics which has enabled the development and application of various tools to disrupt protein or pathway function and to regulate gene expression. Again, the interested reader is directed to several thorough reviews of these technologies and their application to target identification and validation [31, 52, 53]. In this section, we will provide a brief introduction to the three major approaches and evaluate the impact of recent innovations in regulated gene expression and protein degradation on target evaluation.

The requirement for demonstrated essentiality under specific conditions in vitro or, better still, at some stage of the infection process in an in vivo model, has meant that target validation has been dominated by the generation of mutants which are then subjected to phenotypic screening. Reverse genetics by targeted gene deletion involves preselection of the target; because it is hypothesis-driven, this approach is subject to the limitations of prevailing knowledge about mycobacterial physiology. It is perhaps not surprising, therefore, that the yield of antibacterial compounds from target-based approaches has been poor [13, 15] (and see Sect. 3.6.1.6).

Techniques for targeted mutagenesis have improved, and a variety of tools have been described [53]; however, as with experimental manipulations of *M. tuberculosis*, the generation of mutants is subject to the slow growth rate of the organism, and so is time-consuming. For this reason, targeted mutagenesis remains inapplicable to large gene sets. Nevertheless, the precise nature of the mutation introduced ensures that the target gene alone is disrupted; moreover, with recent advances in sequencing technologies, it is likely that whole-genome sequencing of deletion mutants will be performed routinely to confirm that the targeted disruption alone differentiates parental from derivative knockout strains [54]. It is important to note as well that the same techniques employed to generate the mutant can be applied in complementation analysis, which is a critical component of the target validation process and can be used to establish gene essentiality in cases in which knockout mutants cannot be recovered.

In contrast to targeted approaches, random mutagenesis enables the generation and analysis of a large pool of mutants in a single experiment. No prior knowledge of individual gene function is required, so the potential for bias inherent in target selection is avoided. As described above (see Sect. 3.4.1), the most compelling example of the power of this approach was provided by the generation of whole-genome essentiality data [29, 30]. There are limitations, such as the fact that coverage of mutant libraries may be limited by peculiarities in DNA sequence (e.g. G + C content), as well as the requirement that the transposon insertion site must be accurately determined in order to reduce the risk of false conclusions as a result of disruptions that are non-inactivating or confer polar effects on neighbouring genes. Also, in certain randomized approaches, the specific mutant cannot be isolated from the pool for further analysis in pure culture and so requires that the corresponding knockout strain is constructed by a targeted approach. Nevertheless, these techniques remain extremely useful: for example, modified protocols, in which mutant libraries are generated in a specific strain (or mutant) background or are applied in screens designed to mimic environmental stimuli or induce metabolic states thought to be relevant to specific aspects of the infection process, have the potential to elucidate novel persistence targets as well as other classes of conditionally essential genes, genes of complementary function and genes in related pathways [55–57].

Both targeted and random mutagenesis approaches are limited by the fact that abrogation of function is permanent. Although this might not always impact on an observed phenotype, there are instances—especially where in vivo models are used—in which a greater level of control is required in order to identify the disease stage-specific contribution of a gene product to bacterial pathogenesis. The development of systems for the conditional regulation of gene expression in *M. tuberculosis* under both in vitro and in vivo conditions provides an elegant solution to this problem and has provided unprecedented insights into otherwise intractable metabolic pathways [58–60]. Moreover, recent refinements of these systems have enabled the simultaneous assessment of multiple conditional mutants in a single experiment [50]—a modification that holds some promise for the development of high-throughput in vivo screens.

By definition, conditional expression systems are dependent on the ability to titrate the expression of the target gene in response to a particular stimulus—usually a chemical such as tetracycline [61]. It can be expected, therefore, that for an essential gene (or antibiotic target), a critical threshold must exist below which a phenotype (growth retardation, loss of cell wall integrity, etc.) will be manifest. Moreover, it is likely that this threshold will differ for different targets [62]. In turn, this suggests the potential application of regulated expression systems to the question of target vulnerability, a concept that might loosely be defined as the degree of inhibition that is required to impact cellular function detrimentally [63]. This is not merely an intellectual exercise: in contrast to gene disruption, chemical inhibition rarely achieves complete elimination of target function [14]. The possibility of prioritizing according to vulnerability could, therefore, ensure the allocation of maximum resources to the most tractable targets. Recently, a system for regulated protein degradation has been developed which is also aimed at assessing vulnerability but, in this case, by depleting the target protein [63]. This method, which overcomes the criticisms associated with the potential disconnect between RNA transcript levels and actual (translated) protein [64], has provided convincing evidence that the level of inhibition required to impact cellular function is target-specific and so reiterates the idea that vulnerability might offer a more sophisticated measure in the validation process.

Finally, in selecting the approach to target validation, it is necessary to consider the desired mechanism of action of a new compound in the context of the clinical application of the drug: the current therapeutic regimen entails the administration of anti-TB agents to patients presenting with active disease, that is, after the manifestation of symptoms associated with an established (and often increasing) bacterial population. For compounds that target active disease, the test is relatively simple—does the drug reduce bacillary load?—and often assessed by sputum colony count. A more difficult question arises where compounds are designed specifically to target non-replicating and/or persister organisms. What models are appropriate to validate these compounds? Similarly, how is the efficacy of those compounds that are designed for use as adjunct therapy to limit the emergence of resistant organisms (see Sect. 3.6) to be assessed?

3.4.3 How Many Targets Are There?

The well-documented failure of large pharmaceutical enterprises to deliver new anti-infectives [15], including for TB, has prompted questions regarding the total number (already identified plus those not yet identified) of (myco)bacterial pathways that satisfy the criteria of a good target [65] (and see Sect. 3.3.4). For *M. tuberculosis*, the number of fully validated antibiotic targets is small and includes RNA polymerase (the target of rifamycins), DNA gyrase (fluoroquinolones), NADH-dependent enoyl-acyl-carrier protein reductase (isoniazid) and ATP synthetase (diarylquinolines) [3, 13, 52]. The last of these, ATP synthetase, is especially exciting since it validates recent drug discovery efforts aimed at identifying novel targets, and suggests that additional, unexpected targets might exist that simply require the application of the "right" screen (e.g. compound library) in order to be elucidated. In support of this conclusion, it is important to note that a number of additional targets have also been identified recently [52] and are at various stages of validation. One of these is an enzyme required for arabinan biosynthesis, decaprenylphosphoryl-β-D-ribose 2′-epimerase, which is vulnerable to inhibition by at least two different drug classes, the benzothiazinones [66] and dinitrobenzamides [67], and so might reveal a special vulnerability in *M. tuberculosis*.

3.5 A Research Imperative in a TB Endemic Country?

In arguing for increased levels of funding for TB research, Kaufmann and Parida [1] argued that "…the lure of TB research lies in targeting highly sophisticated basic research to meet specific societal needs." Is this "lure" more pressing in endemic regions? Does (should) it acquire an urgency that transforms it into an imperative? Together with Swaziland, the TB notification rate in South Africa is the highest in the world at around 1% of the population, a statistic which points to the failure of control efforts [68]. TB ranks third in mortality among men and women in South Africa, despite the existence of a national control programme and the availability of clinics and antibiotics. In addition, recent analyses of the incidence of primary drug resistance suggest that MDR isolates constitute an increasing proportion of circulating (transmitted) strains in sub-Saharan Africa [69] with corresponding studies in South Africa predicting an emerging epidemic [70].

Given the current tools for diagnosis and treatment, it might be argued that the control of TB depends solely on reforming TB control programmes and ensuring the political commitment required for their sustainability. An alternative view is that the future of global TB control rests instead on the ability of basic and applied science to deliver new and improved tools for diagnosis, prevention and treatment, irrespective of how daunting the challenges may seem. For a country that is defined as a middle-income emerging market and which possesses an established tertiary education sector comprising 24 state-funded universities as well as considerable scientific, technological and medical resources at its disposal, South Africa seems to have a special responsibility to participate in such research. We would argue further that there is a requirement for the country to make its commitment to disease eradication manifest by prioritizing all research activities that are directed against an endemic problem.

In addition to the profound health imperative, there are practical benefits in support of an increased investment in basic TB research. In their article, Kaufmann and Parida [1] point out that TB offers an excellent platform for the training of scientific and clinical staff since it involves the application of a range of research and technical skills that spans a number of disciplines, including microbiology, cell biology, immunology, chemistry, mathematics and informatics. The practical and financial challenges are, however, significant. *M. tuberculosis* is classified as a group 3 pathogen, which requires that all laboratory manipulations of this organism are performed in purpose-built biosafety level 3 containment facilities by specialist personnel. The investment needed to construct, equip and maintain such laboratories is considerable, as are the costs of the specialist consumables and personal protective equipment necessary to ensure continued compliance, all of which are incurred in addition to standard running costs. Unlike many other bacterial pathogens, *M. tuberculosis* is characterized by a very slow growth rate: experiments are arduous, results can be very slow in coming, and the translation of observations made from basic research into practical outcomes might require years. Moreover, the stringent safety requirements and potential risks attendant on working with a human pathogen constitute a major impediment to recruiting scientific staff and students. Although this barrier applies globally, the problem is acute in areas such as southern Africa which lacks critical mass in the requisite scientific skills.

3.6 Target Identification and Validation in South Africa

The remainder of this chapter details research performed in South Africa which is directed at target identification and validation for TB. Where possible, we have attempted to highlight the insights into mycobacterial pathogenesis derived from these activities which might inform future drug discovery efforts. For a country that is so badly affected by TB, the number of groups whose research programmes engage any aspect of fundamental biomedical TB research is small. The analysis presented here is, therefore, heavily biased to recent work conducted in our own laboratory. However, we have also attempted to capture research efforts in other local laboratories that fall within the domain of TB research but might not necessarily satisfy a strict definition of target identification and validation.

3.6.1 Fundamental Mycobacterial Metabolism: The MMRU

The Molecular Mycobacteriology Research Unit (MMRU) was created in 2000 out of the Molecular Biology Unit of the South African Institute for Medical Research where a programme in molecular mycobacteriological research had been established in 1992. The MMRU's existence overlaps with the most dramatic era in the modern history of TB drug discovery research. It is not surprising, therefore, that changes in priorities in the Unit parallel the major developments described above (Sect. 3.3). So, from an early focus on genetics and the development of genetic tools to elucidate fundamental mycobacterial metabolism, the agenda of the MMRU has shifted to the study of metabolic processes within a drug discovery framework. The central motivating concern underlying all research has, however, remained unaltered: to understand (and counteract) the disease caused by *M. tuberculosis*, it is necessary to describe key pathways defining mycobacterial physiology and metabolism [16, 17]. This is a considerable task, and in many respects, the variety of projects tackled since 1992 reinforces the diversity of mechanisms that contribute to mycobacterial survival and the complexity of the underlying metabolic pathways.

For many of the projects summarized below, target identification and validation was not the primary outcome; nevertheless, the information gained can (and has) impacted TB drug discovery efforts. Here, it is also important to reiterate that studies of mycobacterial metabolism, whether in vitro or in any of the (imperfect) in vivo models, are intended to establish the *capacity* of the organism to perform a specific metabolic function. Observed metabolic activity in any model should not be conflated with the functional contribution of a specific target to pathogenesis in the natural (human) host: validation of relevance is obtained solely from the demonstration under clinical conditions that abrogation of specific protein function alters natural disease aetiology.

A feature of the pathways described below is that many comprise multiple, potentially redundant, homologs. Genetic complexity characterizes many physiological processes in *M. tuberculosis*, a pathogen whose genome reveals the expansion of many gene families. This complexity has, in turn, hindered genetic approaches to resolving questions of redundancy and, in the context of drug discovery, reinforces the need to confirm essentiality of a specific target (gene) under a given condition. Here, it is worth recalling that the functions of almost half the genes in the *M. tuberculosis* genome cannot, as yet, be assigned. Moreover, a similar proportion of the 194 genes identified as important for in vivo growth of

M. tuberculosis are of unknown function, and 25% of these have no obvious homologues outside of mycobacteria and closely related species [29].

3.6.1.1 DNA Metabolism

A decade of mycobacterial genomics has revealed that chromosomal rearrangements and point mutations drive the microevolution of *M. tuberculosis*—there is little evidence of horizontal gene transfer and the organism carries no epigenetic information in the form of plasmids [71]. However, while genome plasticity in *M. tuberculosis* is well documented [72], the mechanisms underlying genome diversification remain largely unexplored. This is of special relevance to drug resistance which, in *M. tuberculosis*, is exclusively associated with mutations in target or related genes [73]. Like all pathogens, *M. tuberculosis* is exposed to multiple genotoxic stresses during host infection [17]. In turn, this observation implies that DNA repair pathways might be critical to both bacillary survival and the adaptive evolution of *M. tuberculosis* within its human host. It also suggests that DNA repair might represent a key area in efforts to combat disease and, by extension, raises the possibility that compounds might be designed to inhibit repair pathways in *M. tuberculosis*, especially those associated with induced mutagenesis [74]. Moreover, the fact that the pathways involved in the maintenance, repair and replication of mycobacterial DNA are central to every stage of disease [75, 76] reinforces the idea that the essential components of these processes are a potentially untapped source of novel targets. This last possibility has gained increased credence owing to two recent observations. Until recently, DNA replication was considered especially relevant to the early phases of infection in which bacillary propagation occurs at an exponential rate prior to the onset of the host's adaptive immune response, at which time repair functions become increasingly important to bacillary survival in the face of antibacterial immune effectors. However, evidence obtained with a "molecular clock" indicates that replication continues throughout the disease process—at least in mice—and implies a continual role for DNA replication and repair mechanisms in the maintenance of genome integrity throughout the course of prolonged infection, including periods of clinical latency [77]. Secondly, the significant conservation of DNA metabolic functions across eukaryotes and prokaryotes has ensured that these are considered difficult and unattractive targets. The notable exception, of course, are the bacterial DNA gyrases—the targets of the fluoroquinolone class of antibiotics—which are sufficiently different from the corresponding host topoisomerases to render them amenable to selective targeting. However, the identification of a mycobacterium-specific inhibitor of ATP synthetase [46] and, separately, the continued development of increasingly sophisticated informatic and chemical techniques that enable refinements to the drug design process [78] have established the possibility of targeting prokaryote enzymes with close eukaryotic (human) homologues, for example, DNA polymerases which, until recently, were considered non-viable targets.

A seminal bioinformatic analysis of the DNA repair gene complement in *M. tuberculosis* identified components of most major DNA repair pathways, the most notable exception being the predicted absence of mismatch repair [79]. This analysis also revealed the characteristic complexity in mycobacterial DNA metabolism, identifying multiple homologues of specific DNA repair genes. Consistent with its predicted role in pathogenesis, multiple gene expression studies have identified DNA metabolic pathway components in various *M. tuberculosis* infection models [80, 81], as well as in bacilli isolated from patients with pulmonary tuberculosis [82], all of which support a role for active DNA repair throughout the course of infection [75]. As for many other bacterial pathogens, however, direct examples of the dominant contribution of a single DNA repair pathway to *M. tuberculosis* pathogenesis are rare. Among numerous similar examples, separate studies in the MMRU of alkylation repair and predicted lesion bypass pathways failed to correlate in vitro phenotypes with loss of viability in vivo [83, 84].

The major contribution of the MMRU to the DNA repair field was the demonstration that *M. tuberculosis* possesses a DnaE-type C family polymerase that is required for damage-induced base substitution mutagenesis arising from error-prone bypass of DNA lesions [85]. Although specialist polymerases have been implicated in the emergence of drug resistance in other model systems, the observation that *dnaE2* deletion attenuates virulence and reduces the frequency of drug resistance in vivo [85] constitutes the only evidence to date of a direct role for an inducible mutagenic mechanism in both pathogenesis and the adaptive evolution of drug resistance. As such, these observations identify DnaE2 as a potential drug target. Moreover, follow-up work in the MMRU has recently elucidated key molecular details underlying DnaE2 function, specifically that the polymerase operates in a novel mutagenic pathway comprising two additional accessory factors whose combined action is essential for DNA damage tolerance and induced mutagenesis [86].

Allied work in the area of DNA metabolism has focused on the role of the mycobacterial ribonucleotide reductases (RNRs) in pathogenesis. RNRs catalyse the first committed (and essential) step in DNA synthesis, the conversion of ribonucleotides to deoxyribonucleotides, and so are considered attractive targets for antibiotics. As is the case for so many other metabolic functions, *M. tuberculosis* possesses an expanded complement of class I and class II RNR-encoding genes whose contribution to pathogenesis remains unresolved. The differential oxygen requirements of these enzymes suggested the tantalizing possibility that specialist RNRs were required for the maintenance of dNTP pools for repair and replication during periods of active disease and extended clinical latency. However, in sequential genetic studies [87, 88], researchers in the MMRU have provided convincing evidence of the essentiality of the *nrdEF*-encoded class Ib RNR in bacterial replication, validating this enzyme as a candidate target for new anti-TB drugs.

3.6.1.2 Cofactors

Our interest in the mycobacterial RNRs (see Sect. 3.6.1.1) prompted a detour into cofactor metabolism that has, in turn, informed both new and existing projects.

Class II RNRs are dependent on a vitamin B_{12}-derived cofactor (adenosylcobalamin) for activity. In addition to this RNR, the *M. tuberculosis* genome encodes two other B_{12}-dependent enzymes, a methionine synthase [33] and a methylmalonyl CoA mutase [89]. Moreover, *M. tuberculosis* is unusual in being included among a limited number of prokaryotes which possess a complete pathway for de novo vitamin B_{12} synthesis (only prokaryotes synthesize vitamin B_{12}, all eukaryotes including humans acquire the essential cofactor exogenously). In a series of studies, we have established that none of the B_{12}-dependent pathways is essential for growth of *M. tuberculosis* in vitro [33, 87, 89]; however, the demonstration of a role for the MutAB-dependent methylmalonyl pathway in detoxifying propionate as by-product of fatty acid catabolism [89] might provide some insight into the apparent dispensability of the methylcitrate cycle in mice [90], a result which cast some doubt on the biological function of the bi-functional isocitrate lyase, originally identified as one of the most promising new drug targets following its characterization as a persistence factor in 2000 [34]. These observations raise important questions regarding bacillary metabolism in vivo—in this case, the possibility that *M. tuberculosis* switches to B_{12}-driven pathways as a result of an ability to synthesize vitamin B_{12} and/or scavenge the cofactor from its host—and, as with many other studies of core metabolic functions, suggest the possible application of mutant strains as bioprobes to establish the nature of the in vivo environment. As we have argued before, the extent to which metabolic pathway and substrate utilization are defined by stage of infection, the tissue-specific distribution of nutrients and the ability of the bacillus to access those nutrients remain fundamental questions in mycobacterial pathogenesis.

A related study on the biosynthesis of the molybdenum cofactor in *M. tuberculosis* confirmed the activity of multiple functional molybdopterin synthase homologs [91]. Moreover, there are nine predicted molybdenum cofactor-dependent enzymes in this organism, two of which—nitrate reductase and type I NADH dehydrogenase—have previously been implicated in pathogenesis and virulence. As for vitamin B_{12}, however, further work is required to elucidate the role of molybdenum cofactor biosynthesis and the corresponding molybdenum cofactor-dependent enzymes in pathogenesis.

3.6.1.3 Resuscitation Promoting Factors

A major portion of MMRU research efforts has been devoted to the investigation of the role of the mycobacterial homologues of resuscitation promoting factor (Rpf), a muralytic enzyme which, in other organisms, has been directly implicated in the increased culturability of bacteria from a state of dormancy [92]. Despite the considerable progress made to date, the physiological role of Rpfs in bacterial growth and culturability remains unresolved; as a result, the tantalizing link between bacterial dormancy and resuscitation, on the one hand, and long-term clinical latency, on the other, remains unproven. Reiterating a common theme in mycobacterial genetics, *M. tuberculosis* possesses five distinct Rpf homologues.

Although specific combinations of *rpf* gene deletions are associated with loss of viability in vivo, these genes are neither individually or collectively essential in vitro [93]. However, depletion of Rpf's results in profound growth attenuation in the mouse model and renders the organism susceptible to other drugs, observations which suggest the potential utility of targeting Rpfs in an adjunct therapy [94], for example, as growth stimulants to increase drug susceptibility [95], a concept recently established for other organisms [96]. Moreover, in an unexpected development, the potential addition of recombinant Rpf(s) to sputum samples is attracting increasing attention as a potential enhancement to clinical diagnostics as well as an alternative to cell counts (colony-forming units; CFUs) for assessing early bactericidal activity (EBA) in clinical trials [97].

3.6.1.4 Carbon Metabolism

Of all the projects undertaken in the MMRU, the investigation of the role of two genes of unknown function, Rv2557 and Rv2558, provides perhaps the best illustration of the difficulties inherent in target identification [98]. In a cautionary example of the risks associated with the use of transcription data to identify putative targets, these genes were prioritized owing to their significant upregulation under carbon-starved conditions in vitro and in human granulomas. Single and double deletion mutants of Rv2557 and/or Rv2558 were generated, neither of which affected the long-term survival of *M. tuberculosis* under carbon-starved conditions in vitro nor its virulence in a mouse model. In some ways, this experience is reminiscent of the observed disconnect between transcriptional responsiveness and inferred gene essentiality [99] and reiterates the difficulties associated with the development of in vitro and in vivo models to recapitulate specific aspects of the host environment.

3.6.1.5 Auxotrophs in Target Validation

As noted above, an initial research focus in the MMRU was on the development of genetic tools for mycobacterial manipulation. Two of these studies resulted in the generation of auxotrophic mutants of *M. tuberculosis* that were defective in amino acid biosynthesis [100, 101]. Although not the original objective of this work, the observations made in these early studies suggests a possible additional application for these mutants. A major hurdle confronting microbiologists pertains to the problem of measuring bacterial cell death; more specifically, the ability to determine rapidly and unequivocally whether a single (myco)bacterial cell has lost its potential for regeneration and replication. This problem is of special relevance to TB, a disease known for long periods of clinical latency and caused by an organism that is notorious for its recalcitrance to antibiotic treatment as well as its capacity to exist for extended periods in the absence of replication. Without the ability to identify "death" in this pathogen, efforts to develop novel anti-TB drugs—especially those active against non-replicating bacilli—will continue to suffer doubts regarding real efficacy. Therefore, the relative (and differing) rates at which viability declines in auxotrophic mutants (the *metE* mutant is another of these, [33]) in unsupplemented growth medium suggests the potential application of these strains in the development of assays to classify bacterial death.

3.6.1.6 Target-Based Whole-Cell Screening

A recent analysis of all new molecular entities (NMEs) approved for use by the US Food and Drug Administration between 1999 and 2008 reiterates the conclusion that, despite the strong focus on target-based approaches, the contribution of these methods to new drug discovery is exceeded by phenotypic screening strategies [102]. The poor success rate of target-based discovery has prompted the development of hybrid approaches that combine the benefits of target-based methods (predetermined target selection and predicted mode of action) with those of whole-cell screens (demonstrated activity against viable cells) to identify antimicrobial compounds that possess activity against essential gene products. In ongoing work, the feasibility of performing target-based whole-cell screens in *M. tuberculosis* is being assessed through the application of a promoter-replacement strategy to generate tetracycline-responsive mutant strains that conditionally express selected target genes (G. Abrahams and V. Mizrahi, in press). This methodology has been used to confirm the essentiality of selected genes for the growth of *M. tuberculosis* in vitro. Moreover, by demonstrating that a decrease in the intracellular concentration of the target gene product renders the conditional mutant strain hypersensitive to target-specific inhibitors, these assays have confirmed the potential applicability of this method in validation studies designed to confirm the mode of action of antimicrobial agents directly in whole *M. tuberculosis* cells. In turn, this suggests the potential of this approach to identify molecules that are target-specific and able to exert their inhibitory effect compounds on whole cells. This is an important result: in their review of the difficulties associated with drug discovery, Payne et al. [15] acknowledged that it was "easier to find the cellular target of an antibacterial compound than it [was] to engineer permeability into an enzyme inhibitor", and concluded that "whole-cell assays are favoured for finding a lead compound that has a modicum of antibacterial activity" [15]. In combining these two outcomes, the target-based whole-cell screen thus offers a potential means to fast-track the identification of new drugs for priority targets.

3.6.2 Research in Other South African Laboratories

In this final section, we have attempted to identify reports emanating from other local laboratories that fall within the domain of TB research but might not necessarily satisfy a strict definition of target identification and validation.

3.6.2.1 Inhibiting Mycothiol Biosynthesis

Actinobacteria including *M. tuberculosis* are unusual in producing mycothiol as functional equivalent of glutathione. Although the full range of metabolic processes utilizing mycothiol has not yet been established, the role of this metabolite in detoxification [17] suggests the biosynthetic pathway as potential drug target. Recently, researchers at the University of Cape Town reported that conjugates of plumbagin and phenyl-2-amino-1-thioglucoside are active against a deacetylase that functions in mycothiol biosynthesis in vitro [103].

3.6.2.2 Targeting Glutamine Synthetase

The Innovation Fund of the Department of Science and Technology (DST) supported a research programme on nitrogen metabolism in mycobacteria [104, 105] that involved groups at the Council for Scientific and Industrial Research (CSIR) and Stellenbosch University. This project led to the design of benzimidazole-derived ATP analogues as potential inhibitors of the glutamine synthetase enzyme from *M. tuberculosis* [106, 107].

3.6.2.3 Efflux Mechanisms

Recently, researchers from Stellenbosch University have shown that activation of efflux pumps is critical to the decreased susceptibility of rifampicin-resistant mutants to treatment [108]. Notably, activation of efflux in specific rifampicin-resistant backgrounds also impacts the efficacy of other drug classes, an effect that can be reversed by inhibiting efflux mechanisms. In turn, these data support the idea that an efflux pump inhibitor(s) might be added to treatment regimens in order to preserve the efficacy of existing and future drugs—a possibility that is under active investigation at Stellenbosch University.

3.6.2.4 Improved Diagnostics

A major obstacle confronting TB control efforts is the current lack of rapid and reliable point-of-care diagnostics for the accurate identification of active TB disease in high-incidence settings. Although these activities do not fall within the scope of target identification and validation, the critical dependence of therapeutic efficacy on the ability to implement appropriate treatment regimens at the time of presentation ensures that drug and diagnostic development efforts are inextricably linked. Several South African researchers are active in this domain. For example, a group at the University of Pretoria has focused on the development of methods to diagnose active tuberculosis based on the detection of mycolic acids and their antibodies in TB patients [109, 110], while a team at the CSIR is pursuing the application of aptamer technology to the development of a rapid and reliable point-of-care diagnostic for TB [111]. This last example is especially interesting given the recent precedent for the use of aptamers for both targeted inhibition of *M. tuberculosis* [112–114] as well as for diagnosis [115], which suggests the potential therapeutic use of aptamers as a potential future line of research.

3.6.2.5 Drug Delivery

Another research thrust that does not fall within the ambit of target identification and validation, but which is intimately related to drug efficacy, involves the development of improved modes of delivery for TB drugs. To this end, research groups at the CSIR have focused on the development of nanoparticle-based drug delivery systems for respiratory diseases [116], as well as on the design of systems that will ensure slow-release of TB drugs. Both approaches are aimed at improved compliance by reducing the heavy pill burden on patients who are currently required to take multiple drugs several times a week for a minimum of 6 months.

3.6.2.6 Drug Uptake

Riminophenazines are a class of compounds with known antimycobacterial activity, but unknown mode of action. In collaboration with the MMRU, researchers at the MRC/UP Inflammation and Immunity Research Unit at the University of Pretoria investigated the possibility that the representative compound, clofazimine, disrupts potassium transport mechanisms [117]. Although this and subsequent work failed to identify a specific target, it is worth noting that a recent study has implicated the corresponding potassium transporter in drug tolerance in *M. smegmatis* [118], possibly reiterating the idea (Sect. 3.6.2.2) that transport and efflux mechanisms should be added to the list of high-priority targets for inhibition by adjunct therapeutics.

3.7 Future Prospects for Local TB Drug Discovery Programmes

In this chapter, we have given a brief overview of the technical, scientific and biological challenges facing new TB drug development and have described the critical role of target identification and validation efforts in approaches to overcome some of these obstacles. We have also provided a short history of the field of TB drug discovery, highlighting in particular the problems of limited funding and lack of coordination that hampered research in this domain until the very recent entry of large philanthropic foundations and institutions whose financial muscle has prompted a radical change in the structure and composition of drug discovery consortia and their modes of operation and has ensured the critical participation of major pharmaceutical enterprises.

From our summary of South African research efforts, it is evident that, apart from one or two notable exceptions, the involvement of local research groups in the TB drug discovery arena has been limited. Importantly, the lack of a coherent and sizeable basic research programme seems inconsistent with our status among those countries with the highest burdens of TB and HIV. This lack is even more glaring in light of South Africa's position of international leadership in clinical and TB research established through the work of world-class groups based at the Universities of Cape Town (http://www.cidri.uct.ac.za; http://www.iidmm.uct.ac.za; http://www.satvi.uct.ac.za; http://www.desmondtutuhivcentre.org.za; http://web.uct.ac.za/depts/pha/tbpk.php), KwaZulu-Natal (http://www.caprisa.org), Stellenbosch (http://sun025.sun.ac.za/portal/page/portal/Health_Sciences/English/Centres/dttc), Witwatersrand (http://www.witshealth.co.za/Pages/default.aspx) and the Aurum Institute (http://www.auruminstitute.org/index.php), most of which have a focus on TB chemotherapy. So, what is being done to address this gap, and how might it be tackled in future?

In many respects, the problem as it stands might be considered representative of the international TB drug discovery field in the 1990s. That is, the fragmentary and highly circumscribed approach adopted by South African research groups is largely a function of the limited availability of local funding for research of this nature as well as a shortage of trained personnel. The remedies appear simple: greater access to internal and external funding sources, on the one hand, and increased efforts to develop and retain young scientists with the requisite skills, on the other. However, realizing these aims is complex and requires creative endeavours such as the formation of partnerships with international universities and even the pharmaceutical industry for the training of scientists from developing countries. Of critical importance is the ability of initiatives such as those described above to provide a realistic prospect of a sustainable career in an expanding industry supported nationally and thereby act as a significant incentive to attract young scientists to this task.

It is noteworthy that a number of local and international programmes have already made significant forays towards this end with the South African government playing an increasing role in supporting such endeavours through various funding instruments of the DST, the National Research Foundation (NRF) and the South African Medical Research Council (MRC). For example, the creation in 2004 of the DST/NRF Centre of Excellence for Biomedical TB Research (CBTBR) provided significant, long-term support for research and training in TB research at the laboratory–clinic–community interface. The CBTBR now comprises nodes based at Stellenbosch University, the University of the Witwatersrand and the University of Cape Town and is in its second 5-year cycle of funding. In addition to supporting postgraduate student training, this funding mechanism has also facilitated capacity retention by making available career opportunities for postdoctoral fellows and early-career scientists. Another exciting development is the recent establishment of the UCT Drug Discovery and Development Centre (H3-D) (http://www.h3-d.uct.ac.za/) as a partnership between various pharmaceutical companies, the Medicines for Malaria Venture (MMV) and the DST's Technology Innovation Agency (TIA). This centre is led by Professor Kelly Chibale, who also holds a Research Chair in Drug Discovery under the South African Research Chairs Initiative (SARChI) of the NRF. The inclusion of an earmarked Chair in Drug Discovery Sciences among the 62 new Chairs on offer in the NRF's 2011/12 SARChI call further underscores the strategic intent of the DST and NRF to "strengthen research and innovation capabilities in the country" (http://hicd.nrf.ac.za/sarchi/sarchi_overview.htm). Moreover, the capacity for TB drug discovery research in the CBTBR and H3-D are in turn significantly strengthened by support from the MRC, which funds the MRC/SU Centre for Molecular and Cell Biology—a component of which forms the Stellenbosch University node of the CBTBR—as well as the MMRU and the MRC/UCT Drug Discovery and Development Research Unit. TIA has also made another major investment in TB drug discovery in the form of the newly inaugurated South African Tuberculosis Research and Innovation Initiative (SATRII), a TB drug discovery partnership with the NIAID that involves a number of chemistry and biology groups in South Africa. In terms of international support, the major investment by the Howard Hughes Medical Institute (HHMI) in the newly established KwaZulu-Natal Research Institute for TB and HIV (K-RITH) at the University of KwaZulu-Natal promises to strengthen significantly TB drug discovery research in South Africa (http://www.k-rith.org).

Buttressing these initiatives are capacity-development schemes that provide training opportunities for South African researchers in laboratories abroad. The NRF Drug Discovery Training Programme (http://www.nrf.ac.za/projects.php?pid=46) was established by the NRF in partnership with Emory and other universities in the USA to "support and accelerate pharmaceutical research in South Africa through capacity building and knowledge transfer". This scheme supports pre- and postdoctoral training, advancement of research expertise, exchange visits and training workshops and seminars in South Africa. In addition, for more than a decade, the Columbia University-Southern African Fogarty AIDS Training and Research Program (CU-SA Fogarty AITRP) has provided opportunities for pre- and postdoctoral fellows from southern Africa to undergo training in the TB basic sciences in eminent TB laboratories in the USA.

In combination, these and similar developments should ensure that South Africa maintains momentum towards the establishment of a continuum of research capabilities ranging from basic through preclinical and clinical that will enable this country to assume an increasingly important role in global TB drug discovery and development efforts.

3.8 Conclusions

Target identification and validation seek to exploit knowledge about the fundamental physiology of the organism in order to locate potential weaknesses. As is evident from the discussion throughout this chapter, it is tempting to include any research aimed at elucidating physiological (metabolic) properties of a pathogen such as *M. tuberculosis* under the banner, "target identification and validation". However, an all-inclusive definition such as this is not helpful since it fails to recognize the fact that genuine (good) target identification studies are specifically designed to test the effects on viability (or pathogenesis) of loss of (disruptions to) functions which prior analyses (bioinformatic or biological, e.g. transposon mutagenesis, or a combination of both) have identified as "core". More importantly, there is the very real risk that grouping all studies into a single category can undervalue important research into "(myco)bacterial physiology" that is not conducted with a predetermined drug targeting goal in mind—particularly those studies investigating "non-essential" pathways. Conversely, by assessing these studies in terms of a "targeting" metric, there is a risk of consigning studies which reveal that a specific pathway/function is not essential for viability or virulence—but which are nonetheless valuable for the insights gained into an otherwise intractable and enigmatic adversary—to the category of disqualified targets. The potential consequences of this error for basic research funding cannot be overstated. Similarly, the increased impetus to identify targets, and specific inhibitors of those targets, is likely to yield a pool of compounds that do not progress through the development pipeline owing to their failure to satisfy defined pharmacological parameters. However, such compounds can serve as extremely valuable tools for probing the biology of the system under investigation [119].

The drug discovery pipeline for infectious diseases such as TB is critically dependent on the identification and validation of new targets [120]. As described above, increased funding for these activities has already yielded a number of technical and technological advances that promise a greater level of sophistication to these processes. The advent of conditional knockdown technology, in particular, suggests that a more rigorous definition of a validated target might therefore be one whose depletion by targeted knockdown or chemical inhibition is associated with a demonstrated effect on bacillary function in a suitable disease model. Selection of the most appropriate model itself poses a difficult problem: existing in vitro and in vivo platforms are thought to reproduce certain facets of the host environment, but the tools to assess their accuracy are limited. In this regard, it seems pertinent to consider the recent observation [24] that a model will, by definition, recapitulate only some aspect of the greater disease; instead, improved knowledge of TB in humans is required in order to ensure that an appropriate selection is made. That is, to appreciate the model, we need to characterize the disease. It is expected that target identification and validation and the activities surrounding these efforts, will continue to fulfil a critical role to this end.

Acknowledgements Research in the MMRU is supported by the DST, the NRF, TIA, the MRC, the National Health Laboratory Service and the University of Cape Town.

References

1. Kaufmann SH, Parida SK (2007) Changing funding patterns in tuberculosis. Nat Med 13:299–303
2. Salazar EJ (2011) Tuberculosis Research and Development: 2010 Report on Tuberculosis Research Funding Trends, 2005–2009
3. Koul A, Arnoult E, Lounis N, Guillemont J, Andries K (2011) The challenge of new drug discovery for tuberculosis. Nature 469:483–490
4. Ma Z, Lienhardt C, McIlleron H, Nunn AJ, Wang X (2010) Global tuberculosis drug development pipeline: the need and the reality. Lancet 375:2100–2109
5. WHO (2010) Global Tuberculosis Control: WHO Report 2010
6. Dye C, Williams BG (2010) The population dynamics and control of tuberculosis. Science 328:856–861
7. Lawn SD, Zumla AI (2011) Tuberculosis. Lancet 57:57–72
8. Andersen P, Doherty TM (2005) The success and failure of BCG – implications for a novel tuberculosis vaccine. Nat Rev Microbiol 3:656–662
9. Velayati AA, Masjedi MR, Farnia P, Tabarsi P, Ghanavi J, Ziazarifi AH, Hoffner SE (2009) Emergence of new forms of totally drug-resistant tuberculosis bacilli: super extensively drug-resistant tuberculosis or totally drug-resistant strains in Iran. Chest 136:420–425
10. Dheda K, Shean K, Zumla A, Badri M, Streicher EM, Page-Shipp L, Willcox P, John MA, Reubenson G, Govindasamy D, Wong M, Padanilam X, Dziwiecki A, van Helden PD, Siwendu S, Jarand J, Menezes CN, Burns A, Victor T, Warren R, Grobusch MP, van der Walt M, Kvasnovsky C (2010) Early treatment outcomes and HIV status of patients with extensively drug-resistant tuberculosis in South Africa: a retrospective cohort study. Lancet 375:1798–1807
11. Gandhi NR, Nunn P, Dheda K, Schaaf HS, Zignol M, van Soolingen D, Jensen P, Bayona J (2010) Multidrug-resistant and extensively drug-resistant tuberculosis: a threat to global control of tuberculosis. Lancet 375:1830–1843
12. Donald PR, Maritz JS, Diacon AH (2011) The pharmacokinetics and pharmacodynamics of rifampicin in adults and children in relation to the dosage recommended for children. Tuberculosis (Edinb) 91:196–207
13. Balganesh TS, Alzari PM, Cole ST (2008) Rising standards for tuberculosis drug development. Trends Pharmacol Sci 29:576–581
14. Barry CE 3rd, Boshoff HI, Dartois V, Dick T, Ehrt S, Flynn J, Schnappinger D, Wilkinson RJ, Young D (2009) The spectrum of latent tuberculosis: rethinking the biology and intervention strategies. Nat Rev Microbiol 7:845–855
15. Payne DJ, Gwynn MN, Holmes DJ, Pompliano DL (2007) Drugs for bad bugs: confronting the challenges of antibacterial discovery. Nat Rev Drug Discov 6:29–40
16. Warner DF, Mizrahi V (2008) Physiology of *Mycobacterium tuberculosis*. In: Kaufmann SH, Rubin H (eds) Handbook of tuberculosis: Molecular biology and biochemistry. Wiley-VCH GmbH & Co., Weinheim, pp 53–70
17. Warner DF, Mizrahi V (2006) Tuberculosis chemotherapy: the influence of bacillary stress and damage response pathways on drug efficacy. Clin Microbiol Rev 19:558–570
18. Russell DG, Barry CE 3rd, Flynn JL (2010) Tuberculosis: what we don't know can, and does, hurt us. Science 328:852–856
19. Rhee KY, Carvalho LP, Bryk R, Ehrt S, Marrero J, Park SW, Schnappinger D, Venugopal A, Nathan C (2011) Central carbon metabolism in Mycobacterium tuberculosis: an unexpected frontier. Trends Microbiol 19:307–314

20. Boshoff HI, Barry CE 3rd (2005) Tuberculosis – metabolism and respiration in the absence of growth. Nat Rev Microbiol 3:70–80
21. Cole ST, Brosch R, Parkhill J, Garnier T, Churcher C, Harris D, Gordon SV, Eiglmeier K, Gas S, Barry CE 3rd, Tekaia F, Badcock K, Basham D, Brown D, Chillingworth T, Connor R, Davies R, Devlin K, Feltwell T, Gentles S, Hamlin N, Holroyd S, Hornsby T, Jagels K, Krogh A, McLean J, Moule S, Murphy L, Oliver K, Osborne J, Quail MA, Rajandream MA, Rogers J, Rutter S, Seeger K, Skelton J, Squares R, Squares S, Sulston JE, Taylor K, Whitehead S, Barrell BG (1998) Deciphering the biology of Mycobacterium tuberculosis from the complete genome sequence. Nature 393:537–544
22. Djelouadji Z, Raoult D, Drancourt M (2011) Palaeogenomics of Mycobacterium tuberculosis: epidemic bursts with a degrading genome. Lancet Infect Dis 11:641–650
23. Pethe K, Sequeira PC, Agarwalla S, Rhee K, Kuhen K, Phong WY, Patel V, Beer D, Walker JR, Duraiswamy J, Jiricek J, Keller TH, Chatterjee A, Tan MP, Ujjini M, Rao SP, Camacho L, Bifani P, Mak PA, Ma I, Barnes SW, Chen Z, Plouffe D, Thayalan P, Ng SH, Au M, Lee BH, Tan BH, Ravindran S, Nanjundappa M, Lin X, Goh A, Lakshminarayana SB, Shoen C, Cynamon M, Kreiswirth B, Dartois V, Peters EC, Glynne R, Brenner S, Dick T (2010) A chemical genetic screen in Mycobacterium tuberculosis identifies carbon-source-dependent growth inhibitors devoid of in vivo efficacy. Nat Commun 1:57
24. Young D (2009) Animal models of tuberculosis. Eur J Immunol 39:2011–2014
25. Gey van Pittius NC, Sampson SL, Lee H, Kim Y, van Helden PD, Warren RM (2006) Evolution and expansion of the Mycobacterium tuberculosis PE and PPE multigene families and their association with the duplication of the ESAT-6 (esx) gene cluster regions. BMC Evol Biol 6:95
26. Camus JC, Pryor MJ, Medigue C, Cole ST (2002) Re-annotation of the genome sequence of Mycobacterium tuberculosis H37Rv. Microbiology 148:2967–2973
27. Lew JM, Kapopoulou A, Jones LM, Cole ST (2011) TubercuList – 10 years after. Tuberculosis (Edinb) 91:1–7
28. Sacchettini JC, Rubin EJ, Freundlich JS (2008) Drugs versus bugs: in pursuit of the persistent predator Mycobacterium tuberculosis. Nat Rev Microbiol 6:41–52
29. Sassetti CM, Rubin EJ (2003) Genetic requirements for mycobacterial survival during infection. Proc Natl Acad Sci USA 100:12989–12994
30. Sassetti CM, Boyd DH, Rubin EJ (2003) Genes required for mycobacterial growth defined by high density mutagenesis. Mol Microbiol 48:77–84
31. Warner DF, Mizrahi V (2004) Mycobacterial genetics in target validation. Drug Discov Today Technol 1:93–98
32. Hasan S, Daugelat S, Rao PS, Schreiber M (2006) Prioritizing genomic drug targets in pathogens: application to Mycobacterium tuberculosis. PLoS Comput Biol 2:e61
33. Warner DF, Savvi S, Mizrahi V, Dawes SS (2007) A riboswitch regulates expression of the coenzyme B12-independent methionine synthase in Mycobacterium tuberculosis: implications for differential methionine synthase function in strains H37Rv and CDC1551. J Bacteriol 189:3655–3659
34. McKinney JD, Honer zu Bentrup K, Munoz-Elias EJ, Miczak A, Chen B, Chan WT, Swenson D, Sacchettini JC, Jacobs WR Jr, Russell DG (2000) Persistence of Mycobacterium tuberculosis in macrophages and mice requires the glyoxylate shunt enzyme isocitrate lyase. Nature 406:735–738
35. Russell DG (2011) Mycobacterium tuberculosis and the intimate discourse of a chronic infection. Immunol Rev 240:252–268
36. Davis SL, Be NA, Lamichhane G, Nimmagadda S, Pomper MG, Bishai WR, Jain SK (2009) Bacterial thymidine kinase as a non-invasive imaging reporter for Mycobacterium tuberculosis in live animals. PLoS One 4:e6297
37. Young DB, Perkins MD, Duncan K, Barry CE 3rd (2008) Confronting the scientific obstacles to global control of tuberculosis. J Clin Invest 118:1255–1265

38. Venugopal A, Bryk R, Shi S, Rhee K, Rath P, Schnappinger D, Ehrt S, Nathan C (2011) Virulence of Mycobacterium tuberculosis depends on lipoamide dehydrogenase, a member of three multienzyme complexes. Cell Host Microbe 9:21–31
39. de Carvalho LP, Fischer SM, Marrero J, Nathan C, Ehrt S, Rhee KY (2010) Metabolomics of Mycobacterium tuberculosis reveals compartmentalized co-catabolism of carbon substrates. Chem Biol 17:1122–1131
40. Stumpf MP, Robertson BD, Duncan K, Young DB (2007) Systems biology and its impact on anti-infective drug development. Prog Drug Res 64(1):3–20
41. Young D, Stark J, Kirschner D (2008) Systems biology of persistent infection: tuberculosis as a case study. Nat Rev Microbiol 6:520–528
42. Beste DJ, McFadden J (2010) System-level strategies for studying the metabolism of Mycobacterium tuberculosis. Mol Biosyst 6:2363–2372
43. Spitzer M, Griffiths E, Blakely KM, Wildenhain J, Ejim L, Rossi L, De Pascale G, Curak J, Brown E, Tyers M, Wright GD (2011) Cross-species discovery of syncretic drug combinations that potentiate the antifungal fluconazole. Mol Syst Biol 7:499
44. Schwegmann A, Brombacher F (2008) Host-directed drug targeting of factors hijacked by pathogens. Sci Signal 1:re8
45. Manjunatha U, Boshoff HI, Barry CE (2009) The mechanism of action of PA-824: novel insights from transcriptional profiling. Commun Integr Biol 2:215–218
46. Andries K, Verhasselt P, Guillemont J, Gohlmann HW, Neefs JM, Winkler H, Van Gestel J, Timmerman P, Zhu M, Lee E, Williams P, de Chaffoy D, Huitric E, Hoffner S, Cambau E, Truffot-Pernot C, Lounis N, Jarlier V (2005) A diarylquinoline drug active on the ATP synthase of Mycobacterium tuberculosis. Science 307:223–227
47. Riccardi G, Pasca MR, Buroni S (2009) Mycobacterium tuberculosis: drug resistance and future perspectives. Future Microbiol 4:597–614
48. Keren I, Minami S, Rubin E, Lewis K (2011) Characterization and transcriptome analysis of Mycobacterium tuberculosis persisters. MBio 2:e00100-11
49. Cho SH, Warit S, Wan B, Hwang CH, Pauli GF, Franzblau SG (2007) Low-oxygen-recovery assay for high-throughput screening of compounds against nonreplicating Mycobacterium tuberculosis. Antimicrob Agents Chemother 51:1380–1385
50. Blumenthal A, Trujillo C, Ehrt S, Schnappinger D (2010) Simultaneous analysis of multiple Mycobacterium tuberculosis knockdown mutants in vitro and in vivo. PLoS One 5:e15667
51. Bryk R, Gold B, Venugopal A, Singh J, Samy R, Pupek K, Cao H, Popescu C, Gurney M, Hotha S, Cherian J, Rhee K, Ly L, Converse PJ, Ehrt S, Vandal O, Jiang X, Schneider J, Lin G, Nathan C (2008) Selective killing of nonreplicating mycobacteria. Cell Host Microbe 3:137–145
52. Lamichhane G (2011) Novel targets in M. tuberculosis: search for new drugs. Trends Mol Med 17:25–33
53. Wei JR, Rubin EJ (2008) The many roads to essential genes. Tuberculosis (Edinb) 88(Suppl 1):19–24
54. Ioerger TR, Feng Y, Ganesula K, Chen X, Dobos KM, Fortune S, Jacobs WR Jr, Mizrahi V, Parish T, Rubin E, Sassetti C, Sacchettini JC (2010) Variation among genome sequences of H37Rv strains of Mycobacterium tuberculosis from multiple laboratories. J Bacteriol 192:3645–3653
55. Joshi SM, Pandey AK, Capite N, Fortune SM, Rubin EJ, Sassetti CM (2006) Characterization of mycobacterial virulence genes through genetic interaction mapping. Proc Natl Acad Sci USA 103:11760–11765
56. Baek SH, Li AH, Sassetti CM (2011) Metabolic regulation of mycobacterial growth and antibiotic sensitivity. PLoS Biol 9:e1001065
57. Beste DJ, Espasa M, Bonde B, Kierzek AM, Stewart GR, McFadden J (2009) The genetic requirements for fast and slow growth in mycobacteria. PLoS One 4:e5349
58. Stallings CL, Stephanou NC, Chu L, Hochschild A, Nickels BE, Glickman MS (2009) CarD is an essential regulator of rRNA transcription required for Mycobacterium tuberculosis persistence. Cell 138:146–159

59. Gandotra S, Schnappinger D, Monteleone M, Hillen W, Ehrt S (2007) In vivo gene silencing identifies the Mycobacterium tuberculosis proteasome as essential for the bacteria to persist in mice. Nat Med 13:1515–1520
60. Marrero J, Rhee KY, Schnappinger D, Pethe K, Ehrt S (2010) Gluconeogenic carbon flow of tricarboxylic acid cycle intermediates is critical for Mycobacterium tuberculosis to establish and maintain infection. Proc Natl Acad Sci USA 107:9819–9824
61. Blokpoel MC, Murphy HN, O'Toole R, Wiles S, Runn ES, Stewart GR, Young DB, Robertson BD (2005) Tetracycline-inducible gene regulation in mycobacteria. Nucleic Acids Res 33:e22
62. Korycka-Machala M, Rychta E, Brzostek A, Sayer HR, Rumijowska-Galewicz A, Bowater RP, Dziadek JMM (2007) Evaluation of NAD(+) -dependent DNA ligase of mycobacteria as a potential target for antibiotics. Antimicrob Agents Chemother 51:2888–2897
63. Wei JR, Krishnamoorthy V, Murphy K, Kim JH, Schnappinger D, Alber T, Sassetti CM, Rhee KY, Rubin EJ (2011) Depletion of antibiotic targets has widely varying effects on growth. Proc Natl Acad Sci USA 108:4176–4181
64. Taniguchi Y, Choi PJ, Li GW, Chen H, Babu M, Hearn J, Emili A, Xie XS (2010) Quantifying E. coli proteome and transcriptome with single-molecule sensitivity in single cells. Science 329:533–538
65. Overington JP, Al-Lazikani B, Hopkins AL (2006) How many drug targets are there? Nat Rev Drug Discov 5:993–996
66. Makarov V, Manina G, Mikusova K, Mollmann U, Ryabova O, Saint-Joanis B, Dhar N, Pasca MR, Buroni S, Lucarelli AP, Milano A, De Rossi E, Belanova M, Bobovska A, Dianiskova P, Kordulakova J, Sala C, Fullam E, Schneider P, McKinney JD, Brodin P, Christophe T, Waddell S, Butcher P, Albrethsen J, Rosenkrands I, Brosch R, Nandi V, Bharath S, Gaonkar S, Shandil RK, Balasubramanian V, Balganesh T, Tyagi S, Grosset J, Riccardi G, Cole ST (2009) Benzothiazinones kill Mycobacterium tuberculosis by blocking arabinan synthesis. Science 324:801–804
67. Christophe T, Jackson M, Jeon HK, Fenistein D, Contreras-Dominguez M, Kim J, Genovesio A, Carralot JP, Ewann F, Kim EH, Lee SY, Kang S, Seo MJ, Park EJ, Skovierova H, Pham H, Riccardi G, Nam JY, Marsollier L, Kempf M, Joly-Guillou ML, Oh T, Shin WK, No Z, Nehrbass U, Brosch R, Cole ST, Brodin P (2009) High content screening identifies decaprenyl-phosphoribose 2' epimerase as a target for intracellular antimycobacterial inhibitors. PLoS Pathog 5:e1000645
68. Wood R, Lawn SD, Johnstone-Robertson S, Bekker LG (2011) Tuberculosis control has failed in South Africa – time to reappraise strategy. S Afr Med J 101:111–114
69. Zager EM, McNerney R (2008) Multidrug-resistant tuberculosis. BMC Infect Dis 8:10
70. Cox HS, McDermid C, Azevedo V, Muller O, Coetzee D, Simpson J, Barnard M, Coetzee G, van Cutsem G, Goemaere E (2010) Epidemic levels of drug resistant tuberculosis (MDR and XDR-TB) in a high HIV prevalence setting in Khayelitsha, South Africa. PLoS One 5:e13901
71. Hershberg R, Lipatov M, Small PM, Sheffer H, Niemann S, Homolka S, Roach JC, Kremer K, Petrov DA, Feldman MW, Gagneux S (2008) High functional diversity in *Mycobacterium tuberculosis* driven by genetic drift and human demography. PLoS Biol 6:e311
72. Borrell S, Gagneux S (2011) Strain diversity, epistasis and the evolution of drug resistance in Mycobacterium tuberculosis. Clin Microbiol Infect 17:815–820
73. Sandgren A, Strong M, Muthukrishnan P, Weiner BK, Church GM, Murray MB (2009) Tuberculosis drug resistance mutation database. PLoS Med 6:e2
74. Smith PA, Romesberg FE (2007) Combating bacteria and drug resistance by inhibiting mechanisms of persistence and adaptation. Nat Chem Biol 3:549–556
75. Gorna AE, Bowater RP, Dziadek J (2010) DNA repair systems and the pathogenesis of Mycobacterium tuberculosis: varying activities at different stages of infection. Clin Sci (Lond) 119:187–202
76. Warner DF (2010) The role of DNA repair in M. tuberculosis pathogenesis. Drug Discov Today Dis Mech 7:e5

77. Gill WP, Harik NS, Whiddon MR, Liao RP, Mittler JE, Sherman DR (2009) A replication clock for Mycobacterium tuberculosis. Nat Med 15:211–214
78. Ciulli A, Abell C (2007) Fragment-based approaches to enzyme inhibition. Curr Opin Biotechnol 18:489–496
79. Mizrahi V, Andersen SJ (1998) DNA repair in *Mycobacterium tuberculosis*. What have we learnt from the genome sequence? Mol Microbiol 29:1331–1339
80. Schnappinger D, Ehrt S, Voskuil MI, Liu Y, Mangan JA, Monahan IM, Dolganov G, Efron B, Butcher PD, Nathan C, Schoolnik GK (2003) Transcriptional Adaptation of *Mycobacterium tuberculosis* within Macrophages: insights into the Phagosomal Environment. J Exp Med 198:693–704
81. Talaat AM, Lyons R, Howard ST, Johnston SA (2004) The temporal expression profile of *Mycobacterium tuberculosis* infection in mice. Proc Natl Acad Sci USA 101:4602–4607
82. Rachman H, Strong M, Ulrichs T, Grode L, Schuchhardt J, Mollenkopf H, Kosmiadi GA, Eisenberg D, Kaufmann SH (2006) Unique transcriptome signature of *Mycobacterium tuberculosis* in pulmonary tuberculosis. Infect Immun 74:1233–1242
83. Durbach SI, Springer B, Machowski EE, North RJ, Papavinasasundaram KG, Colston MJ, Bottger EC, Mizrahi V (2003) DNA alkylation damage as a sensor of nitrosative stress in Mycobacterium tuberculosis. Infect Immun 71:997–1000
84. Kana BD, Abrahams GL, Sung N, Warner DF, Gordhan BG, Machowski EE, Tsenova L, Sacchettini JC, Stoker NG, Kaplan G, Mizrahi V (2010) Role of the DinB homologs Rv1537 and Rv3056 in Mycobacterium tuberculosis. J Bacteriol 192:2220–2227
85. Boshoff HI, Reed MB, Barry CE 3rd, Mizrahi V (2003) DnaE2 polymerase contributes to in vivo survival and the emergence of drug resistance in *Mycobacterium tuberculosis*. Cell 113:183–193
86. Warner DF, Ndwandwe DE, Abrahams GL, Kana BD, Machowski EE, Venclovas C, Mizrahi V (2010) Essential roles for imuA'- and imuB-encoded accessory factors in DnaE2-dependent mutagenesis in Mycobacterium tuberculosis. Proc Natl Acad Sci USA 107:13093–13098
87. Dawes SS, Warner DF, Tsenova L, Timm J, McKinney JD, Kaplan G, Rubin H, Mizrahi V (2003) Ribonucleotide reduction in Mycobacterium tuberculosis: function and expression of genes encoding class Ib and class II ribonucleotide reductases. Infect Immun 71:6124–6131
88. Mowa MB, Warner DF, Kaplan G, Kana BD, Mizrahi V (2009) Function and regulation of class I ribonucleotide reductase-encoding genes in mycobacteria. J Bacteriol 191:985–995
89. Savvi S, Warner DF, Kana BD, McKinney JD, Mizrahi V, Dawes SS (2008) Functional characterization of a vitamin B12-dependent methylmalonyl pathway in Mycobacterium tuberculosis: implications for propionate metabolism during growth on fatty acids. J Bacteriol 190:3886–3895
90. Munoz-Elias EJ, Upton AM, Cherian J, McKinney JD (2006) Role of the methylcitrate cycle in *Mycobacterium tuberculosis* metabolism, intracellular growth, and virulence. Mol Microbiol 60:1109–1022
91. Williams MJ, Kana BD, Mizrahi V (2011) Functional analysis of molybdopterin biosynthesis in mycobacteria identifies a fused molybdopterin synthase in Mycobacterium tuberculosis. J Bacteriol 193:98–106
92. Kana BD, Mizrahi V (2010) Resuscitation-promoting factors as lytic enzymes for bacterial growth and signaling. FEMS Immunol Med Microbiol 58:39–50
93. Kana BD, Gordhan BG, Downing KJ, Sung N, Vostroktunova G, Machowski EE, Tsenova L, Young M, Kaprelyants A, Kaplan G, Mizrahi V (2008) The resuscitation-promoting factors of Mycobacterium tuberculosis are required for virulence and resuscitation from dormancy but are collectively dispensable for growth in vitro. Mol Microbiol 67:672–684
94. Kana BD, Mizrahi V, Gordhan BG (2010) Depletion of resuscitation-promoting factors has limited impact on the drug susceptibility of Mycobacterium tuberculosis. J Antimicrob Chemother 65:1583–1585
95. Garton NJ, Waddell SJ, Sherratt AL, Lee SM, Smith RJ, Senner C, Hinds J, Rajakumar K, Adegbola RA, Besra GS, Butcher PD, Barer MR (2008) Cytological and transcript analyses reveal fat and lazy persister-like bacilli in tuberculous sputum. PLoS Med 5:e75

96. Allison KR, Brynildsen MP, Collins JJ (2011) Metabolite-enabled eradication of bacterial persisters by aminoglycosides. Nature 473:216–220
97. Mukamolova GV, Turapov O, Malkin J, Woltmann G, Barer MR (2010) Resuscitation-promoting factors reveal an occult population of tubercle Bacilli in Sputum. Am J Respir Crit Care Med 181:174–180
98. Gordhan BG, Smith DA, Kana BD, Bancroft G, Mizrahi V (2006) The carbon starvation-inducible genes Rv2557 and Rv2558 of Mycobacterium tuberculosis are not required for long-term survival under carbon starvation and for virulence in SCID mice. Tuberculosis (Edinb) 86:430–437
99. Rengarajan J, Bloom BR, Rubin EJ (2005) Genome-wide requirements for Mycobacterium tuberculosis adaptation and survival in macrophages. Proc Natl Acad Sci USA 102:8327–8332
100. Gordhan BG, Smith DA, Alderton H, McAdam RA, Bancroft GJ, Mizrahi V (2002) Construction and phenotypic characterization of an auxotrophic mutant of Mycobacterium tuberculosis defective in L-arginine biosynthesis. Infect Immun 70:3080–3084
101. Davies BW, Kohanski MA, Simmons LA, Winkler JA, Collins JJ, Walker GC (2009) Hydroxyurea induces hydroxyl radical-mediated cell death in Escherichia coli. Mol Cell 36:845–860
102. Swinney DC, Anthony J (2011) How were new medicines discovered? Nat Rev Drug Discov 10:507–519
103. Gammon DW, Steenkamp DJ, Mavumengwana V, Marakalala MJ, Mudzunga TT, Hunter R, Munyololo M (2010) Conjugates of plumbagin and phenyl-2-amino-1-thioglucoside inhibit MshB, a deacetylase involved in the biosynthesis of mycothiol. Bioorg Med Chem 18:2501–2514
104. Harper CJ, Hayward D, Kidd M, Wiid I, van Helden P (2010) Glutamate dehydrogenase and glutamine synthetase are regulated in response to nitrogen availability in Myocbacterium smegmatis. BMC Microbiol 10:138
105. Hayward D, van Helden PD, Wiid IJ (2009) Glutamine synthetase sequence evolution in the mycobacteria and their use as molecular markers for Actinobacteria speciation. BMC Evol Biol 9:48
106. Salisu S, Kenyon C, Kaye PT (2011) Studies towards the synthesis of ATP analogs as potential glutamine synthetase inhibitors. Synthetic Commun 41:2216–2225
107. Gxoyiya BSB, Kaye PT, Kenyon C (2010) Benzimidazole-derived ATP analogues as potential glutamine synthetase inhibitors. Synthetic Commun 40:2578–2587
108. Louw GE, Warren RM, Gey van Pittius NC, Leon R, Jimenez A, Pando RH, McEvoy CR, Grobbelaar M, Murray M, van Helden PD, Victor TC (2011) Rifampicin reduces susceptibility to ofloxacin in rifampicin resistant mycobacterium tuberculosis through efflux. Am J Respir Crit Care Med 184:269–276
109. Beukes M, Lemmer Y, Deysel M, Al Dulayymi JR, Baird MS, Koza G, Iglesias MM, Rowles RR, Theunissen C, Grooten J, Toschi G, Roberts VV, Pilcher L, Van Wyngaardt S, Mathebula N, Balogun M, Stoltz AC, Verschoor JA (2010) Structure-function relationships of the antigenicity of mycolic acids in tuberculosis patients. Chem Phys Lipids 163:800–808
110. Lemmer Y, Thanyani ST, Vrey PJ, Driver CH, Venter L, van Wyngaardt S, ten Bokum AM, Ozoemena KI, Pilcher LA, Fernig DG, Stoltz AC, Swai HS, Verschoor JA (2009) Chapter 5 – Detection of antimycolic acid antibodies by liposomal biosensors. Methods Enzymol 464:79–104
111. Khati M (2010) The future of aptamers in medicine. J Clin Pathol 63:480–487
112. Chen F, Zhang X, Zhou J, Liu S, Liu J (2011) Aptamer inhibits Mycobacterium tuberculosis (H37Rv) invasion of macrophage. Mol Biol Rep. doi: doi: 10.1007/s11033-011-0963-3
113. Shum KT, Lui EL, Wong SC, Yeung P, Sam L, Wang Y, Watt RM, Tanner JA (2011) Aptamer-mediated inhibition of Mycobacterium tuberculosis polyphosphate kinase 2. Biochemistry 50:3261–3271
114. Chen F, Zhou J, Luo F, Mohammed AB, Zhang XL (2007) Aptamer from whole-bacterium SELEX as new therapeutic reagent against virulent Mycobacterium tuberculosis. Biochem Biophys Res Commun 357:743–748

115. Qin L, Zheng R, Ma Z, Feng Y, Liu Z, Yang H, Wang J, Jin R, Lu J, Ding Y, Hu Z (2009) The selection and application of ssDNA aptamers against MPT64 protein in Mycobacterium tuberculosis. Clin Chem Lab Med 47:405–411
116. Swai H, Semete B, Kalombo L, Chelule P, Kisich K, Sievers B (2009) Nanomedicine for respiratory diseases. Wiley Interdiscip Rev Nanomed Nanobiotechnol 1:255–263
117. Cholo MC, Boshoff HI, Steel HC, Cockeran R, Matlola NM, Downing KJ, Mizrahi V, Anderson R (2006) Effects of clofazimine on potassium uptake by a Trk-deletion mutant of Mycobacterium tuberculosis. J Antimicrob Chemother 57:79–84
118. Castaneda-Garcia A, Do TT, Blazquez J (2011) The K + uptake regulator TrkA controls membrane potential, pH homeostasis and multidrug susceptibility in Mycobacterium smegmatis. J Antimicrob Chemother 66:1489–1498
119. Falconer SB, Czarny TL, Brown ED (2011) Antibiotics as probes of biological complexity. Nat Chem Biol 7:415–423
120. Aguero F, Al-Lazikani B, Aslett M, Berriman M, Buckner FS, Campbell RK, Carmona S, Carruthers IM, Chan AW, Chen F, Crowther GJ, Doyle MA, Hertz-Fowler C, Hopkins AL, McAllister G, Nwaka S, Overington JP, Pain A, Paolini GV, Pieper U, Ralph SA, Riechers A, Roos DS, Sali A, Shanmugam D, Suzuki T, Van Voorhis WC, Verlinde CL (2008) Genomic-scale prioritization of drug targets: the TDR Targets database. Nat Rev Drug Discov 7:900–907

Chapter 4
Targeting Conserved Pathways as a Strategy for Novel Drug Development: Disabling the Cellular Stress Response

Adrienne L. Edkins and Gregory L. Blatch

4.1 Introduction to the Cellular Stress Response

The biological activity of proteins is dependent on the ability to assume and maintain the appropriate three-dimensional biophysical structure. Despite the fact that the primary amino acid sequence is sufficient to define the three-dimensional structure of a protein [1], the crowded cellular environment and molecular stress often result in protein misfolding and aggregation. Stress at the biological level can be defined as any stimulus or condition, such as extreme temperature, oxidative radicals, enhanced growth rate and xenobiotics, that perturbs the correct function of the cell, the result of which is the disruption of protein homeostasis. To cope with these conditions, the organism responds by inducing the expression of a series of highly conserved proteins, known as heat shock proteins (HSP), which function as molecular chaperones to overcome the effects of protein misfolding and aggregation. A molecular chaperone is defined as a protein that is capable of interacting with and stabilising non-native protein structures or nascent polypeptides to prevent aggregation and promote the formation of correct, functional conformations [2]. Since the discovery of HSP in heat-treated *Drosophila* cells in the early 1960s, a significant amount of research has been devoted to the characterisation of the role these proteins play in physiological and stressful processes with the cell [3–5]. The response to stress is conserved across organisms; the ability to respond to stress involves molecular chaperones that regulate the integrity of cellular proteins. As protein homeostasis is central to all organisms, cellular stress and molecular chaperone function has been linked to a range of human disorders. HSP have been implicated in a range of pathologies that are linked by the core element of cellular stress [6].

A.L. Edkins (✉) • G.L. Blatch (✉)
Biomedical Biotechnology Research Unit (BioBRU), Department of Biochemistry, Microbiology and Biotechnology, Rhodes University, Grahamstown 6140, South Africa
e-mail: a.edkins@ru.ac.za; g.blatch@ru.ac.za

K. Chibale et al. (eds.), *Drug Discovery in Africa*,
DOI 10.1007/978-3-642-28175-4_4,

4.2 Heat Shock Proteins as Molecular Chaperones

The ability of HSP to act as molecular chaperones is integral to their protective roles in the cell [7]. Molecular chaperones catalyse the refolding of stress-denatured or nascent polypeptides in addition to regulating the assembly of oligomeric proteins, protein transport and degradation and protein–protein interactions. Molecular chaperones are catalysts in the folding process, in the sense that they do not actively fold proteins, but rather assist folding by preventing non-productive or aggregation reactions and do not constitute part of the final protein structure [8]. HSP may be expressed constitutively or induced by the binding of heat shock transcription factors (HSF) to specific heat shock elements (HSE). Stress induces the phosphorylation of the HSF, leading to the formation of trimeric HSF species that bind the HSE and stimulate expression of the HSP [9]. HSP have been identified in almost all organisms, from bacteria to lower and higher eukaryotes, and show a high level of conservation across species. They are classified and named on the basis of their molecular size, for example, HSP70 will have a molecular weight of approximately 70 kDa. To date, a number of HSP families have been described. These are the HSP100, HSP90, HSP70, HSP60, HSP40 and the small HSP (sHSPs; ~18–43 kDa) families [8, 10]. The link between cell stress, molecular chaperones and human disease has made HSP attractive drug targets. In particular, the focus has been on the development of inhibitors of the HSP90 and HSP70 families of molecular chaperones as drug targets for the treatment of cancer and parasitic diseases.

4.2.1 Heat Shock Protein 90 (HSP90)

In eukaryotic cells, HSP90 species are some of the most abundant molecular chaperones [11]. The human genome encodes five distinct HSP90 genes. Cytosolic HSP90 has two isoforms, the α and β isoforms, which display 85% sequence identity and are encoded for by separate genes that arose from duplication of the original HSP90A gene [12]. HSP90α expression is defined as more inducible than that of HSP90β [13]. HSP90 isoforms are also expressed in the mitochondria (HSP75/TRAP-1), the endoplasmic reticulum (GRP94/GRP96) and under certain conditions, HSP90α, HSP90β and GP96 isoforms have been identified in the plasma membrane or extracellular matrix. The genomes of the parasites *Plasmodium falciparum* and *Trypanosoma cruzi* encode 11 and 4 different HSP90 genes, respectively [14]. While both parasites encode cytosolic and ER isoforms of HSP90, they do not encode for the α and β cytosolic isoforms that are found in humans [15]. While eukaryotic HSP90 is essential for cell survival, bacterial HSP90 (HTPG) is dispensable for cell growth. HSP90 is also subject to post-translational modification, including *s*-nitrosylation, phosphorylation, methylation and acetylation, which may influence its function and cellular localisation [11, 16–20].

HSP90 functions as a dimer, with each monomer consisting of three highly conserved functional domains, namely an N-terminal domain (25 kDa), a middle (M) domain (35 kDa) and a C-terminal domain (12 kDa) [21, 22]. The N-terminal and M-domain are connected by a charged linker region, which varies in length and amino acid composition according to the species or isoform (and is entirely absent from the prokaryotic HSP90) [23, 24]. The N-terminal domain contains the primary binding site for ATP/ADP [25], as determined by crystallisation studies on both yeast and human HSP90. This ATP/ADP binding site is the same as that bound by the natural HSP90 inhibitor, geldanamycin [26], which can also bind to HSP90 from other species, including *P. falciparum* [27] and *T. cruzi* [28]. The C-terminal domain is the site of dimerisation of HSP90 [29] and contains the MEEVD motif which is the primary binding site for the TPR-containing co-chaperones, such as HOP (HSP70/HSP90 organising protein) [30, 31]. The structure of full-length HSP90 has been determined at low resolution by cryo-EM [32], while each of these functional domains has been crystallised independently.

4.2.2 Heat Shock Protein 70 (HSP70)

The HSP70 molecular chaperone family is composed of members that are essential for correct protein folding, transport of proteins to different sub-cellular compartments and degradation of unstable proteins in the cell [33]. HSP70 isoforms are found in most species. We have demonstrated that even the ancient coelacanth [34] encodes a functional HSP70, as do protozoan parasites such as *P. falciparum* and *T. cruzi*. There are 13 different isoforms of HSP70 in humans, with 6 isoforms in *P. falciparum* and 12 in *T. brucei* [35, 36]. HSP70 have three functional domains, the N-terminal ATPase domain (~44 kDa), substrate binding region (~18 kDa) and a variable C-terminal region (~10 kDa) that is the site of specific interactions with other proteins, such as HOP [31, 37]. The ATPase domain of HSP70 forms two distinct, equally sized structural lobes, separated from each other by a deep cleft, which is the binding site of ATP [38, 39]. ATP hydrolysis by the ATPase domain of HSP70s is central to the role of the protein as a chaperone during assisted protein folding [40]. Despite the basic HSP70 structure being conserved in HSP70 from different organisms [36, 39, 41], there is evidence that there are differences between HSP70 from different organisms that may be exploited to selectively inhibit the HSP70 from certain organisms [42].

4.3 Molecular Chaperones as Drug Targets in Africa

Biological molecules are required to conform to a set of criteria if they are to be considered potential drug targets [43]. Molecular chaperones, and in particular HSP90 and HSP70, are currently considered as bona fide biological targets for drug development as they conform to many of these criteria [43]. Ideally, a drug target should be essential for the development of the disorder, but absent under normal conditions, to allow selective targeting of diseased and non-healthy cells. HSP are important in the development of human disorders including cancer and are conserved across different species of human parasites, probably because they are often vital or essential to survival of the parasite. HSP70 and HSP90 are among the most well-characterised chaperones, with functional and structural data available for different protein isoforms from a range of species. In addition, the naturally occurring inhibitor of HSP90, the ansamycin antibiotic geldanamycin (GA), has facilitated the rational development of HSP90 inhibitors, many of which are analogues of GA (e.g. 17-AAG, 17-DMAG), as well as the search for numerous other natural products that have the potential to be HSP90 inhibitors. The targeting of such highly conserved proteins of human parasites may not resonate with current dogmas of drug discovery, especially in terms of selectivity for the diseased state over the normal condition (such as parasite over human host or cancer cell versus normal cell). However, highly conserved proteins that are essential for cell survival are likely to evolve considerably slower than other less conserved protein families, making them less susceptible to variation under selection pressure, a desirable attribute for any potential drug target. There is also evidence that despite the high levels of conservation, these HSP are not identical and that there are functional and structural differences that may be exploited. In fact, there is even evidence that disease states can generate HSP with different biophysical and biochemical characteristics to their counterparts in normal cells, meaning that the *same* protein may be biochemically distinct from the normal cell. For example, 17-AAG was found to have a higher affinity for HSP90 in transformed cells compared to normal cells and is currently in clinical trials as an anti-cancer agent [44]. Our research into the identification of inhibitors of HSP90 and HSP70 in protozoan parasites and in cancer is based on our fundamental understanding of these chaperones as a result of comparative studies of the chaperones in humans and parasites. We have identified HSP70 and HSP90 as potential drug targets in the treatment of cancer, malaria and trypanosomiasis and have used the biochemical differences between these proteins to screen potential inhibitors.

4.3.1 Molecular Chaperones of Protozoan Parasites

We and others have begun to scrutinise the chaperone machineries of the causative agents of certain infectious diseases of humans, such as malaria (*P. falciparum;* reviewed in [36, 45, 46]) and the various trypanosomiases (*Trypanosoma brucei* and *T. cruzi;* reviewed in [35]). These parasites have evolved chaperone systems to cope with the extreme demands of cyclical development through physiologically diverse host and vector systems [14]. HSP70 and HSP90, in particular, are expressed at critical developmental stages and persist at high levels in the infectious stages of these parasites. Therefore, it is not surprising that inhibitors of HSP90 (e.g. GA) have

Table 4.1 HSP70 small molecule modulators with anti-plasmodial activity[a]

Compound	Binding site	Biological activity	Reference
15-Deoxyspergualin (DSG)	C-terminus	Anti-plasmodial	[52, 61–63]
Pyrimidinone–peptoid (e.g. MAL3-101)	ND[b]	Anti-cancer and anti-plasmodial	[64–66]
Pyrimidinone–peptoid (e.g. MAL3-39, DMT3024, DMT2264)	ND	Anti-plasmodial	[65, 67, 68]
Dihydropyrimidines (e.g. 116-9e)	HSP40-binding site	ND	[69]
Lapachol derivatives	ND	Anti-cancer and anti-plasmodial	[42, 70, 71]
Malonganenones	ND	Anti-plasmodial	[42]

[a]Compounds with known anti-plasmodial activity, or structurally related compounds yet to be shown to have anti-plasmodial activity
[b]ND = not determined

been shown to arrest the growth of these parasites [27, 28, 47]. Therefore, plasmodial HSP90 (PfHSP90) is an obvious anti-plasmodial drug target, and promising pre-clinical tests have been conducted using GA and its derivatives [48]. We are discovering that while there is a high degree of conservation at the sequence level between parasitic and human HSPs, there are major structural and functional differences that suggest that the parasitic HSP can be selectively targeted [36, 49].

We and others have conducted extensive biochemical and cell biological studies on the major cytosolic and inducible plasmodial HSP70, PfHSP70-1. As expected, it is a bona fide chaperone, exhibiting ATPase and protein aggregation suppression activities [50–52], and capable of functionally substituting for both a bacterial [50] and a yeast HSP70 [53]. It is also a highly abundant protein expressed at all erythrocytic stages of the *P. falciparum* life cycle, with increased expression after heat shock [54, 55]. Importantly, there is strong evidence that PfHSP70-1 is relatively thermostable compared to human HSP70 (hHSP70) and capable of functioning optimally at temperatures around 50°C [56]. In addition, there are biochemical differences between PfHSP70-1 and hHSP70 at the level of ATP affinity and ATPase activity, and structural modelling has identified differences at their HSP40-binding sites [51, 57].

While HSP70 is considered an emerging drug target [58], including PfHSP70-1 [59, 60], there have been limited studies on the development of small molecule modulators of HSP70 with anti-plasmodial activity (Table 4.1). The Brodsky Laboratory has pioneered the development of HSP70 small molecule modulators [64] and recently published promising findings on potential modulators of PfHSP70-1 derived from pyrimidinone-based compounds [65, 67]. We tested some of these compounds in our recent side-by-side study of PfHSP70-1 and hHSP70, confirming that certain pyrimidinones differentially inhibited PfHSP70-1 and hHSP70 (MAL3-39 and DMT002264; [68]). We have started to screen for modulators of PfHSP70-1, using libraries of compounds known to have anti-cancer or anti-plasmodial activity. We discovered that lapachol and certain of its derivatives were able to inhibit the chaperone activity of PfHSP70-1, with lapachol exhibiting specificity towards PfHSP70-1 [42]. We have also screened compounds of marine origin and identified a novel class of inhibitors of PfHSP70-1 (malonganenones) that also showed anti-plasmodial activity [42]. These studies represent a platform for the benchmarking of further hit compounds from recent large-scale screens [72]. All of these PfHSP70-1 inhibitors represent potential hits for the development of leads for anti-plasmodial drug discovery. However, they also represent useful molecular probes for elucidating the mechanism of action of plasmodial chaperones compared to their human homologues.

4.3.2 Human Molecular Chaperones in Cancer

HSP90 is known anecdotally as the 'cancer chaperone' due to its ability to exclusively mediate the folding and stability of numerous transcription factors and signalling intermediates *in vivo*. Many of these so-called client proteins of HSP90 are oncogenes that are either mutated or upregulated in a range of cancers. HSP90 maintains its client proteins in an inactive, but easily inducible, state [73]. As these states are often inherently unstable and labile in the absence of substrate, the role of the HSP90 multi-chaperone complex is to enhance client protein stability [74]. Therefore, using HSP90 as an anti-cancer target allows the simultaneous inhibition of multiple signalling pathways [75]. The molecular chaperone activity of HSP90 is regulated by conformational changes which are dependent on two factors. The first is the intrinsic ATPase activity of HSP90, the importance of which is demonstrated by the fact that mutations that result in either a loss of ATP binding or ATP hydrolysis inhibit both the in vitro and in vivo functions of HSP90. This fact is exploited in drug discovery as many of the naturally occurring or synthetic HSP90 inhibitors bind to the ATP binding site within the N terminus of HSP90. The second factor is the association of HSP90 with a range of co-chaperones (HSP70, HSP40, HOP, p23, immunophilins) into a multi-chaperone complex [11, 76]. HSP90, in association with client protein and co-chaperones, is considered 'complexed' or activated, conditions under which HSP90 displays an enhanced sensitivity to and binding of anti-HSP90 drugs compared to free/uncomplexed HSP90 [77, 78].

The inhibition of a number of these pathways through targeting of HSP90 is already being harnessed to develop anti-cancer agents. GA is a specific inhibitor of HSP90 ATPase activity, and its synthetic derivatives (17-AAG, 17-DMAG etc.) are leading the current focus in HSP90 inhibitors as anti-cancer agents [79–82]. The HSP90-directed drug development process has been greatly facilitated by the fact that many natural products, including GA, novobiocin, gambogic acid, radicicol, epilgallocatechin-3-gallate (EGCG) and taxol, have been found to be HSP90 inhibitors [83]. Many of these compounds show structural similarity in that they possess multiple quinone and/or coumarin ring structures (Fig. 4.1), structural similarity that has facilitated the rational design of analogues of these natural inhibitors. The screening of indigenous natural products from Africa has the potential to identify a number of putative HSP90 inhibitors with anti-cancer or anti-parasite activity and is a major research focus for many groups [84, 85].

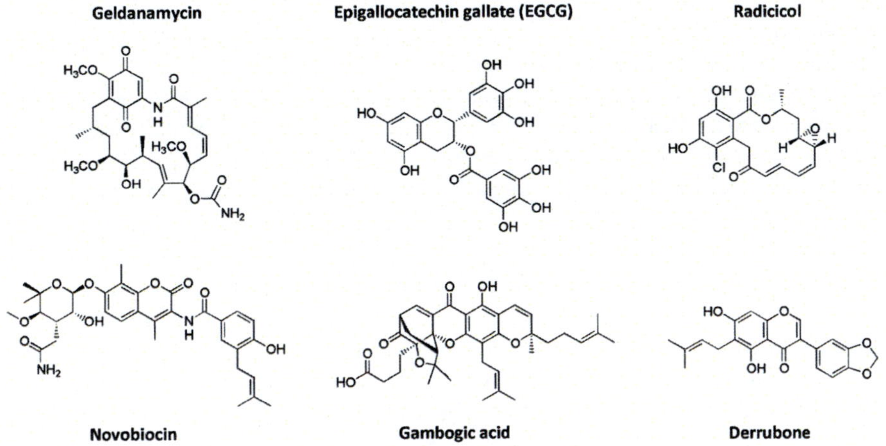

Fig. 4.1 Structural similarity between known natural product inhibitors of HSP90. Many of the current HSP90 inhibitors are natural products, which show structural similarity. A number of indigenous compounds isolated from South Africa sources show similarity to these compounds and therefore are potential inhibitors of HSP90 function

In addition to its potential as a general anti-cancer drug target, HSP90 has recently been considered as a treatment for certain cell subtypes within cancers. Extracellular HSP90 protein is expressed on the surface of a range of cell types, including melanoma, fibrosarcoma, breast adenocarcinoma and neuronal cells [86–89] and linked to cell migration, metastasis and invasion of cancer cells. Selective inhibition of only surface HSP90 reduced migration of a range of cancer cell types [81]; this may offer a unique therapeutic opportunity for the treatment of metastasis, particularly since over 90% of deaths from cancer are as a result of cancer spread (metastasis). Similarly, HSP90 is emerging as a potential drug target for the removal of a specific subpopulation of cancer cells, known as cancer stem cells (CSC). The CSC hypothesis describes cancer development and maintenance as being driven by cancer cells that display stem-like properties (reviewed by us in [90]). Cancer stem cells are cancer cells that, like normal tissue stem cells, are capable of both self-renewal and differentiation, which accounts for tumour heterogeneity. These cells are thought to control tumour development, just as normal tissue stem cells are required for organ development [91–93]. CSC populations are relevant to cancer drug discovery due to their potential role in metastasis and cancer recurrence, the latter as a result of CSC's apparent resistance to chemotherapy and radiotherapy [94, 95]. We have linked HSP90 function to stem-cell-associated pathways that may be important in CSC function [96]. We and others have shown that STAT3 is an HSP90 client protein and the LIF/STAT3 pathway is important in self-renewal of embryonic stem cells [97, 98]. The HSP90 inhibitor 17-AAG was effective at inhibiting growth of both glioma cells and glioma cancer stem cells [99]. In addition, cancer stem cells from medulloblastoma were more sensitive to inhibition of the kinase Akt, which requires HSP90 for activity, than the cancer cells from the same tumour mass [100].

However, inhibition of HSP90 as a drug strategy is often associated with the compensatory upregulation of other molecular chaperones, especially HSP70, in cells treated with HSP90 inhibitors. Inhibition of HSP90 causes activation of HSF-1 that induces the expression of other members of the HSP family [101, 102]. While upregulation of other HSP has been associated with drug resistance to HSP90 inhibitors [103, 104], some researchers have used this fortuitous induction of other chaperones as a strategy to treat protein folding diseases such as Alzheimer's disease [105] and as a mechanism to promote cytoprotection [106]. The most effective strategy, with respect to anti-cancer therapy, therefore may be the simultaneous inhibition of HSP90 and HSP70 [107] or even the inhibition of the heat shock response by direct inhibition of HSF-1 [108]. We and others have identified a number of small molecule inhibitors of HSP70 function in parasites, which may also have future potential as co-inhibitors of HSP70 in human cancers.

4.4 Future Prospects for HSP-Directed Drug Discovery in Africa

Despite extensive research into the function of HSP in cancer and parasitic diseases, there are still relatively few anti-HSP compounds that have made it to the market as drugs [109–111]. The most successful compound at present is 17-AAG, currently in clinical trials as an anti-cancer agent and also with potential as a treatment for parasite diseases demonstrated by in vitro studies. However, there is still great potential for drug discovery in this area, particularly as our fundamental knowledge of the structure and function of different HSP systems grows.

4.4.1 Identification of New Indigenous Compounds That Target HSP

A number of small molecules have been found by us and others to modulate the activity of the conserved HSP70 and HSP90 proteins from different organisms. What is striking is the number of these compounds that are either natural products or analogues of natural products. Natural products are a good source of compounds that are difficult to synthesise by traditional chemical methods and may also serve as the basis for rational drug design. Many natural compounds may exhibit activity as the parent compound and may be of value as scaffolds for the development of novel compounds. The diversity of South African flora and fauna has yielded many natural products that show anti-cancer [84, 112, 113] and anti-plasmodial activity [114] that have structural similarity to known HSP inhibitors [115, 116]. Therefore, the analysis of the effect of these indigenous compounds has the potential to identify novel modulators of chaperone function.

4.4.2 Evaluation of Co-chaperones as Drug Targets

Molecular chaperones function as part of multi-chaperone complexes, complexes which often include other proteins known as co-chaperones. Co-chaperones are a diverse group of proteins that modulate the function of and promote the interaction between the major molecular chaperones. The most well-known co-chaperones of both the HSP90 and HSP70 families are the HSP40 family and the tetratricopeptide repeat containing co-chaperone HOP. The central role played by co-chaperones in the modulation of the function of other chaperones means that these proteins may also be good drug targets.

HSP40s are the major co-chaperones in the HSP70-assisted protein folding process and act via regulation of the ATPase activity of HSP70. HSP40 are the largest and most diverse group of co-chaperones. This diversity is conserved across organisms; there are 49 HSP40 genes in humans, 43 in *P. falciparum* and 67 genes in *T. brucei* [46, 117, 118]. HSP40 proteins are classified as type I, II or III, depending on the degree of conservation of the functional motifs they contain, with respect to *Escherichia coli* (*E. coli*) DnaJ [119]. We have recently proposed an additional group, type IV HSP40, which, besides a few isoforms in humans and yeast, appears specific to malaria parasites [46]. The type and number of motifs contained in an HSP40 will govern its function, with most of the HSP70 chaperone interactions being modulated by type I HSP40s. More specific functions are likely to be mediated by type III or type IV HSP40s [120, 121]. HSP40 may prove to be the better chaperone drug target, as unlike HSP90 and HSP70 where the same chaperone is responsible for chaperoning numerous clients, many HSP40 are selective and will only bind a restricted number of proteins. This may allow for the preferential targeting of specific signalling pathways through the targeting of a single HSP40 isoform [46].

HOP is a co-chaperone that is involved in mediating interactions between HSP70 and HSP90. HOP acts as an adapter between the two chaperones and controls the transfer of protein substrates between the two systems. HOP is essential for the chaperone function of both Hsp70 and Hsp90 as part of the Hsp90 complex. Therefore, inhibition of HOP has the theoretical potential not only to inhibit complexed Hsp90 (the state of Hsp90 observed in cancerous cells) but also to facilitate the simultaneous dual inhibition of both Hsp90 and Hsp70, thereby preventing any redundancy between the two chaperone systems that may occur when these proteins are targeted individually. In addition, it is now apparent that HOP is also expressed by parasites, such as *P. falciparum* and *T. cruzi*. We and others are therefore in the process of characterising human and parasitic HOP as future potential drug targets [122–125].

4.4.3 HSP as Biotechnological Tools in Drug Discovery

In addition to the well-described role as drug targets, there is also the potential to use molecular chaperones as tools in biotechnology. The protein folding and quality control functions of HSP can be harnessed to synthesise other drug targets for drug characterisation studies. The limiting factor for many drug development programmes is the availability of sufficient quantities of active, folded drug targets for use in drug characterisation studies. We have recently published on the use of co-expression of molecular chaperones for the synthesis of putative drug targets [126]. We were able to demonstrate the production of active, folded cyclohydrolase enzyme from *P. falciparum* for use in *in vitro* enzymatic assays for inhibitor characterisation. Similar techniques were used to successfully produce large quantities of other malarial drug targets, such as PfDXR [127, 128]. To the best of our knowledge, these publications are the first to demonstrate the use of a homologous chaperone–client system, where the drug target and the chaperone used to refold it are from the same organism (in this case, *P. falciparum*), for biotechnological production of a drug target. This research is ongoing and has great potential to develop a drug discovery platform technology in Africa.

References

1. Anfinsen CB (1973) Principles that govern the folding of protein chains. Science 181 (96):223–230
2. Hendrick JP, Hartl FU (1993) Molecular chaperone functions of heat-shock proteins. Annu Rev Biochem 62:349–384
3. Ritossa FM, Vonborstel RC (1964) Chromosome puffs in *Drosophila* induced by ribonuclease. Science 145:513–514
4. Li Z, Srivastava P (2004) Heat-shock proteins. Curr Protoc Immunol Appendix 1:Appendix 1T
5. Morimoto RI, Kline MP et al (1997) The heat-shock response: regulation and function of heat-shock proteins and molecular chaperones. Essays Biochem 32:17–29
6. Dudek J, Benedix J et al (2009) Functions and pathologies of BiP and its interaction partners. Cell Mol Life Sci 66(9):1556–1569
7. Palotai R, Szalay MS, Csermely P (2008) Chaperones as integrators of cellular networks: changes of cellular integrity in stress and diseases. IUBMB Life 60(1):10–18
8. Fink AL (1999) Chaperone-mediated protein folding. Physiol Rev 79(2):425–449
9. Park HO, Craig EA (1991) Transcriptional regulation of a yeast HSP70 gene by heat shock factor and an upstream repression site-binding factor. Genes Dev 5(7):1299–1308
10. Frydman J (2001) Folding of newly translated proteins *in vivo*: the role of molecular chaperones. Annu Rev Biochem 70:603–647
11. Csermely P, Schnaider T et al (1998) The 90-kDa molecular chaperone family: structure, function, and clinical applications. A comprehensive review. Pharmacol Ther 79(2):129–168
12. Krone PH, Sass JB (1994) HSP 90 alpha and HSP 90 beta genes are present in the zebrafish and are differentially regulated in developing embryos. Biochem Biophys Res Commun 204 (2):746–752

13. Terasawa K, Minami M, Minami Y (2005) Constantly updated knowledge of Hsp90. J Biochem 137(4):443–447
14. Shonhai A, Maier AG et al (2011) Intracellular protozoan parasites of humans: the role of molecular chaperones in development and pathogenesis. Protein Pept Lett 18(2):143–157
15. Chen B, Zhong D, Monteiro A (2006) Comparative genomics and evolution of the HSP90 family of genes across all kingdoms of organisms. BMC Genomics 7:156
16. Rao R, Fiskus W et al (2008) HDAC6 inhibition enhances 17-AAG–mediated abrogation of hsp90 chaperone function in human leukemia cells. Blood 112(5):1886–1893
17. Aoyagi S, Archer TK (2005) Modulating molecular chaperone Hsp90 functions through reversible acetylation. Trends Cell Biol 15(11):565–567
18. Yang Y, Rao R et al (2008) Role of acetylation and extracellular location of heat shock protein 90α in tumor cell invasion. Cancer Res 68(12):4833–4842
19. Martinez-Ruiz A, Villanueva L et al (2005) S-nitrosylation of Hsp90 promotes the inhibition of its ATPase and endothelial nitric oxide synthase regulatory activities. Proc Natl Acad Sci USA 102(24):8525–8530
20. Duval M, Le Boeuf F et al (2007) Src-mediated phosphorylation of Hsp90 in response to vascular endothelial growth factor (VEGF) is required for VEGF receptor-2 signaling to endothelial NO synthase. Mol Biol Cell 18(11):4659–4668
21. Nemoto T, Sato N (1998) Oligomeric forms of the 90-kDa heat shock protein. Biochem J 330 (2):989–995
22. Nemoto T, Sato N et al (1997) Domain structures and immunogenic regions of the 90-kDa heat-shock protein (HSP90). Probing with a library of anti-HSP90 monoclonal antibodies and limited proteolysis. J Biol Chem 272(42):26179–26187
23. Hainzl O, Lapina MC et al (2009) The charged linker region is an important regulator of Hsp90 function. J Biol Chem 284(34):22559–22567
24. Scheibel T, Siegmund HI et al (1999) The charged region of Hsp90 modulates the function of the N-terminal domain. Proc Natl Acad Sci USA 96(4):1297–1302
25. Prodromou C, Roe SM et al (1997) Identification and structural characterization of the ATP/ADP-binding site in the Hsp90 molecular chaperone. Cell 90(1):65–75
26. Grenert JP, Sullivan WP et al (1997) The amino-terminal domain of heat shock protein 90 (hsp90) that binds geldanamycin is an ATP/ADP switch domain that regulates hsp90 conformation. J Biol Chem 272(38):23843–23850
27. Kumar R, Musiyenko A, Barik S (2003) The heat shock protein 90 of *Plasmodium falciparum* and antimalarial activity of its inhibitor, geldanamycin. Malar J 2:30
28. Graefe SE, Wiesgigl M et al (2002) Inhibition of HSP90 in *Trypanosoma cruzi* induces a stress response but no stage differentiation. Eukaryot Cell 1(6):936–943
29. Wayne N, Bolon DN (2007) Dimerization of Hsp90 is required for *in vivo* function. Design and analysis of monomers and dimers. J Biol Chem 282(48):35386–35395
30. Brinker A, Scheufler C et al (2002) Ligand discrimination by TPR domains. Relevance and selectivity of EEVD-recognition in Hsp70 x Hop x Hsp90 complexes. J Biol Chem 277 (22):19265–19275
31. Odunuga OO, Hornby JA et al (2003) Tetratricopeptide repeat motif-mediated Hsc70-mSTI1 interaction. Molecular characterization of the critical contacts for successful binding and specificity. J Biol Chem 278(9):6896–6904
32. Southworth DR, Agard DA (2011) Client-loading conformation of the Hsp90 molecular chaperone revealed in the Cryo-EM structure of the human Hsp90:Hop complex. Mol Cell 42(6):771–781
33. Agashe VR, Hartl FU (2000) Roles of molecular chaperones in cytoplasmic protein folding. Semin Cell Dev Biol 11(1):15–25
34. Modisakeng KW, Jiwaji M et al (2009) Isolation of a *Latimeria menadoensis* heat shock protein 70 (Lmhsp70) that has all the features of an inducible gene and encodes a functional molecular chaperone. Mol Genet Genomics 282(2):185–196
35. Louw CA, Ludewig MH et al (2010) The Hsp70 chaperones of the Tritryps are characterized by unusual features and novel members. Parasitol Int 59(4):497–505

36. Shonhai A, Boshoff A, Blatch GL (2007) The structural and functional diversity of Hsp70 proteins from *Plasmodium falciparum*. Protein Sci 16(9):1803–1818
37. James P, Pfund C, Craig EA (1997) Functional specificity among Hsp70 molecular chaperones. Science 275(5298):387–389
38. Flaherty KM, DeLuca-Flaherty C, McKay DB (1990) Three-dimensional structure of the ATPase fragment of a 70K heat-shock cognate protein. Nature 346(6285):623–628
39. Wisniewska M, Karlberg T et al (2010) Crystal structures of the ATPase domains of four human Hsp70 isoforms: HSPA1L/Hsp70-hom, HSPA2/Hsp70-2, HSPA6/Hsp70B′, and HSPA5/BiP/GRP78. PLoS One 5(1):e8625
40. Wawrzynow A, Banecki B et al (1995) ATP hydrolysis is required for the DnaJ-dependent activation of DnaK chaperone for binding to both native and denatured protein substrates. J Biol Chem 270(33):19307–19311
41. Karlin S, Brocchieri L (1998) Heat shock protein 70 family: multiple sequence comparisons, function, and evolution. J Mol Evol 47(5):565–577
42. Cockburn IL, Pesce ER et al (2011) Screening for small molecule modulators of Hsp70 chaperone activity using protein aggregation suppression assays: inhibition of the plasmodial chaperone PfHsp70-1. Biol Chem 392(5):431–438
43. Bakheet TM, Doig AJ (2009) Properties and identification of human protein drug targets. Bioinformatics 25(4):451–457
44. Chiosis G, Neckers L (2006) Tumor selectivity of Hsp90 inhibitors: the explanation remains elusive. ACS Chem Biol 1(5):279–284
45. Acharya P, Kumar R, Tatu U (2007) Chaperoning a cellular upheaval in malaria: heat shock proteins in *Plasmodium falciparum*. Mol Biochem Parasitol 153(2):85–94
46. Botha M, Pesce ER, Blatch GL (2007) The Hsp40 proteins of *Plasmodium falciparum* and other apicomplexa: regulating chaperone power in the parasite and the host. Int J Biochem Cell Biol 39(10):1781–1803
47. Banumathy G, Singh V et al (2003) Heat shock protein 90 function is essential for *Plasmodium falciparum* growth in human erythrocytes. J Biol Chem 278(20):18336–18345
48. Pallavi R, Roy N et al (2010) Heat shock protein 90 as a drug target against protozoan infections: biochemical characterization of HSP90 from *Plasmodium falciparum* and *Trypanosoma evansi* and evaluation of its inhibitor as a candidate drug. J Biol Chem 285 (49):37964–37975
49. Edkins AL, Ludewig MH, Blatch GL (2004) A *Trypanosoma cruzi* heat shock protein 40 is able to stimulate the adenosine triphosphate hydrolysis activity of heat shock protein 70 and can substitute for a yeast heat shock protein 40. Int J Biochem Cell Biol 36(8):1585–1598
50. Shonhai A, Boshoff A, Blatch GL (2005) *Plasmodium falciparum* heat shock protein 70 is able to suppress the thermosensitivity of an *Escherichia coli* DnaK mutant strain. Mol Genet Genomics 274(1):70–78
51. Matambo TS, Odunuga OO et al (2004) Overproduction, purification, and characterization of the *Plasmodium falciparum* heat shock protein 70. Protein Expr Purif 33(2):214–222
52. Ramya TN, Surolia N, Surolia A (2006) 15-Deoxyspergualin modulates *Plasmodium falciparum* heat shock protein function. Biochem Biophys Res Commun 348(2):585–592
53. Bell SL, Chiang AN, Brodsky JL (2011) Expression of a malarial Hsp70 improves defects in chaperone-dependent activities in ssa1 mutant yeast. PLoS One 6(5):e20047
54. Kumar N, Koski G et al (1991) Induction and localization of *Plasmodium falciparum* stress proteins related to the heat shock protein 70 family. Mol Biochem Parasitol 48(1):47–58
55. Pesce ER, Acharya P et al (2008) The *Plasmodium falciparum* heat shock protein 40, Pfj4, associates with heat shock protein 70 and shows similar heat induction and localisation patterns. Int J Biochem Cell Biol 40(12):2914–2926
56. Misra G, Ramachandran R (2009) Hsp70-1 from *Plasmodium falciparum*: protein stability, domain analysis and chaperone activity. Biophys Chem 142(1–3):55–64
57. Shonhai A, Botha M et al (2008) Structure-function study of a *Plasmodium falciparum* Hsp70 using three dimensional modelling and in vitro analyses. Protein Pept Lett 15(10):1117–1125

58. Evans CG, Chang L, Gestwicki JE (2010) Heat shock protein 70 (hsp70) as an emerging drug target. J Med Chem 53(12):4585–4602
59. Pesce ER, Cockburn IL et al (2010) Malaria heat shock proteins: drug targets that chaperone other drug targets. Infect Disord Drug Targets 10(3):147–157
60. Shonhai A (2010) Plasmodial heat shock proteins: targets for chemotherapy. FEMS Immunol Med Microbiol 58(1):61–74
61. Nadeau K, Nadler SG et al (1994) Quantitation of the interaction of the immunosuppressant deoxyspergualin and analogs with Hsc70 and Hsp90. Biochemistry 33(9):2561–2567
62. Brodsky JL (1999) Selectivity of the molecular chaperone-specific immunosuppressive agent 15-deoxyspergualin: modulation of Hsc70 ATPase activity without compromising DnaJ chaperone interactions. Biochem Pharmacol 57(8):877–880
63. Nadler SG, Dischino DD et al (1998) Identification of a binding site on Hsc70 for the immunosuppressant 15-deoxyspergualin. Biochem Biophys Res Commun 253(1):176–180
64. Fewell SW, Smith CM et al (2004) Small molecule modulators of endogenous and co-chaperone-stimulated Hsp70 ATPase activity. J Biol Chem 279(49):51131–51140
65. Huryn DM, Brodsky JL et al (2011) Chemical methodology as a source of small-molecule checkpoint inhibitors and heat shock protein 70 (Hsp70) modulators. Proc Natl Acad Sci USA 108(17):6757–6762
66. Wright CM, Chovatiya RJ et al (2008) Pyrimidinone-peptoid hybrid molecules with distinct effects on molecular chaperone function and cell proliferation. Bioorg Med Chem 16 (6):3291–3301
67. Chiang AN, Valderramos JC et al (2009) Select pyrimidinones inhibit the propagation of the malarial parasite, *Plasmodium falciparum*. Bioorg Med Chem 17(4):1527–1533
68. Botha M, Chiang AN et al (2011) *Plasmodium falciparum* encodes a single cytosolic type I Hsp40 that functionally interacts with Hsp70 and is upregulated by heat shock. Cell Stress Chaperones 16(4):389–401
69. Wisen S, Bertelsen EB et al (2010) Binding of a small molecule at a protein-protein interface regulates the chaperone activity of hsp70-hsp40. ACS Chem Biol 5(6):611–622
70. Bonifazi EL, Rios-Luci C et al (2010) Antiproliferative activity of synthetic naphthoquinones related to lapachol. First synthesis of 5-hydroxylapachol. Bioorg Med Chem 18 (7):2621–2630
71. Pérez-Sacau E, Estévez-Braun A et al (2005) Antiplasmodial activity of naphthoquinones related to lapachol and beta-lapachone. Chem Biodivers 2(2):264–274
72. Guiguemde WA, Shelat AA et al (2010) Chemical genetics of *Plasmodium falciparum*. Nature 465(7296):311–315
73. Whitesell L, Lindquist SL (2005) HSP90 and the chaperoning of cancer. Nat Rev Cancer 5 (10):761–772
74. Buchner J (1999) Hsp90 & Co. – a holding for folding. Trends Biochem Sci 24(4):136–141
75. Zhang H, Burrows F (2004) Targeting multiple signal transduction pathways through inhibition of Hsp90. J Mol Med 82(8):488–499
76. Odunuga OO, Longshaw VM, Blatch GL (2004) Hop: more than an Hsp70/Hsp90 adaptor protein. Bioessays 26(10):1058–1068
77. Kamal A, Thao L et al (2003) A high-affinity conformation of Hsp90 confers tumour selectivity on Hsp90 inhibitors. Nature 425(6956):407–410
78. Onuoha SC, Mukund SR et al (2007) Mechanistic studies on Hsp90 inhibition by ansamycin derivatives. J Mol Biol 372(2):287–297
79. Taldone T, Gozman A et al (2008) Targeting Hsp90: small-molecule inhibitors and their clinical development. Curr Opin Pharmacol 8(4):370–374
80. Jensen MR, Schoepfer J et al (2008) NVP-AUY922: a small molecule HSP90 inhibitor with potent antitumor activity in preclinical breast cancer models. Breast Cancer Res 10(2):R33
81. Tsutsumi S, Scroggins B et al (2008) A small molecule cell-impermeant Hsp90 antagonist inhibits tumor cell motility and invasion. Oncogene 27(17):2478–2487
82. Sydor JR, Normant E et al (2006) Development of 17-allylamino-17-demethoxygeldanamycin hydroquinone hydrochloride (IPI-504), an anti-cancer agent directed against Hsp90. Proc Natl Acad Sci USA 103(46):17408–17413

83. Donnelly A, Blagg BS (2008) Novobiocin and additional inhibitors of the Hsp90 C-terminal nucleotide-binding pocket. Curr Med Chem 15(26):2702–2717
84. van der Merwe E, Huang D et al (2008) The synthesis and anticancer activity of selected diketopiperazines. Peptides 29(8):1305–1311
85. Bisi-Johnson MA, Obi CL et al (2011) Evaluation of the antibacterial and anticancer activities of some South African medicinal plants. BMC Complement Altern Med 11:14
86. Eustace BK, Sakurai T et al (2004) Functional proteomic screens reveal an essential extracellular role for hsp90 alpha in cancer cell invasiveness. Nat Cell Biol 6(6):507–514
87. Becker B, Multhoff G et al (2004) Induction of Hsp90 protein expression in malignant melanomas and melanoma metastases. Exp Dermatol 13(1):27–32
88. Sidera K, Samiotaki M et al (2004) Involvement of cell surface HSP90 in cell migration reveals a novel role in the developing nervous system. J Biol Chem 279(44):45379–45388
89. Sims JD, McCready J, Jay DG (2011) Extracellular heat shock protein (Hsp)70 and Hsp90α assist in matrix metalloproteinase-2 activation and breast cancer cell migration and invasion. PLoS One 6(4):e18848
90. Lawson JC, Blatch GL, Edkins AL (2009) Cancer stem cells in breast cancer and metastasis. Breast Cancer Res Treat 118(2):241–254
91. Clarke MF, Fuller M (2006) Stem cells and cancer: two faces of eve. Cell 124(6):1111–1115
92. Dalerba P, Cho RW, Clarke MF (2007) Cancer stem cells: models and concepts. Annu Rev Med 58:267–284
93. Burger PE, Gupta R et al (2009) High aldehyde dehydrogenase activity: a novel functional marker of murine prostate stem/progenitor cells. Stem Cells 27(9):2220–2228
94. Ma S, Lee TK et al (2008) CD133+ HCC cancer stem cells confer chemoresistance by preferential expression of the Akt/PKB survival pathway. Oncogene 27(12):1749–1758
95. Glinsky GV (2007) Stem cell origin of death-from-cancer phenotypes of human prostate and breast cancers. Stem Cell Rev 3(1):79–93
96. Prinsloo E, Setati MM et al (2009) Chaperoning stem cells: a role for heat shock proteins in the modulation of stem cell self-renewal and differentiation? Bioessays 31(4):370–377
97. Kim HL, Cassone M et al (2008) HIF-1alpha and STAT3 client proteins interacting with the cancer chaperone Hsp90: therapeutic considerations. Cancer Biol Ther 7(1):10–14
98. Setati MM, Prinsloo E et al (2010) Leukemia inhibitory factor promotes Hsp90 association with STAT3 in mouse embryonic stem cells. IUBMB Life 62(1):61–66
99. Sauvageot CM, Weatherbee JL et al (2008) Efficacy of the HSP90 inhibitor 17-AAG in human glioma cell lines and tumorigenic glioma stem cells. Neuro Oncol 11(2):109–121
100. Hambardzumyan D, Becher OJ et al (2008) PI3K pathway regulates survival of cancer stem cells residing in the perivascular niche following radiation in medulloblastoma in vivo. Genes Dev 22(4):436–448
101. Ali A, Bharadwaj S et al (1998) HSP90 interacts with and regulates the activity of heat shock factor 1 in *Xenopus* oocytes. Mol Cell Biol 18(9):4949–4960
102. Conde R, Belak ZR et al (2009) Modulation of Hsf1 activity by novobiocin and geldanamycin. Biochem Cell Biol 87(6):845–851
103. McCollum AK, Teneyck CJ et al (2006) Up-regulation of heat shock protein 27 induces resistance to 17-allylamino-demethoxygeldanamycin through a glutathione-mediated mechanism. Cancer Res 66(22):10967–10975
104. McCollum AK, Lukasiewicz KB et al (2008) Cisplatin abrogates the geldanamycin-induced heat shock response. Mol Cancer Ther 7(10):3256–3264
105. Nagai Y, Fujikake N et al (2010) Induction of molecular chaperones as a therapeutic strategy for the polyglutamine diseases. Curr Pharm Biotechnol 11(2):188–197
106. Westerheide SD, Bosman JD et al (2004) Celastrols as inducers of the heat shock response and cytoprotection. J Biol Chem 279(53):56053–56060
107. Massey AJ, Williamson DS et al (2010) A novel, small molecule inhibitor of Hsc70/Hsp70 potentiates Hsp90 inhibitor induced apoptosis in HCT116 colon carcinoma cells. Cancer Chemother Pharmacol 66(3):535–545

108. Whitesell L, Lindquist S (2009) Inhibiting the transcription factor HSF1 as an anticancer strategy. Expert Opin Ther Targets 13(4):469–478
109. Trepel J, Mollapour M et al (2010) Targeting the dynamic HSP90 complex in cancer. Nat Rev Cancer 10(8):537–549
110. Wang Y, Trepel JB et al (2010) STA-9090, a small-molecule Hsp90 inhibitor for the potential treatment of cancer. Curr Opin Investig Drugs 11(12):1466–1476
111. Kim YS, Alarcon SV et al (2009) Update on Hsp90 inhibitors in clinical trial. Curr Top Med Chem 9(15):1479–1492
112. Brauns SC, Dealtry G et al (2005) Caspase-3 activation and induction of PARP cleavage by cyclic dipeptide cyclo(Phe-Pro) in HT-29 cells. Anticancer Res 25(6B):4197–4202
113. Brauns SC, Milne P et al (2004) Selected cyclic dipeptides inhibit cancer cell growth and induce apoptosis in HT-29 colon cancer cells. Anticancer Res 24(3a):1713–1719
114. Afolayan AF, Mann MG et al (2009) Antiplasmodial halogenated monoterpenes from the marine red alga *Plocamium cornutum*. Phytochemistry 70(5):597–600
115. Afolayan AF, Bolton JJ et al (2008) Fucoxanthin, tetraprenylated toluquinone and toluhydroquinone metabolites from *Sargassum heterophyllum* inhibit the in vitro growth of the malaria parasite *Plasmodium falciparum*. Z Naturforsch C 63(11–12):848–852
116. Antunes EM, Beukes DR et al (2004) Cytotoxic pyrroloiminoquinones from four new species of South African latrunculid sponges. J Nat Prod 67(8):1268–1276
117. Kampinga HH, Hageman J et al (2009) Guidelines for the nomenclature of the human heat shock proteins. Cell Stress Chaperones 14(1):105–111
118. Louw CA, Ludewig MH, Blatch GL (2010) Overproduction, purification and characterisation of Tbj1, a novel Type III Hsp40 from *Trypanosoma brucei*, the African sleeping sickness parasite. Protein Expr Purif 69(2):168–177
119. Cheetham ME, Caplan AJ (1998) Structure, function and evolution of DnaJ: conservation and adaptation of chaperone function. Cell Stress Chaperones 3(1):28–36
120. Hennessy F, Boshoff A, Blatch GL (2005) Rational mutagenesis of a 40 kDa heat shock protein from *Agrobacterium tumefaciens* identifies amino acid residues critical to its in vivo function. Int J Biochem Cell Biol 37(1):177–191
121. Hennessy F, Cheetham ME et al (2000) Analysis of the levels of conservation of the J domain among the various types of DnaJ-like proteins. Cell Stress Chaperones 5(4):347–358
122. Walsh N, Larkin A et al (2011) RNAi knockdown of Hop (Hsp70/Hsp90 organising protein) decreases invasion via MMP-2 down regulation. Cancer Lett 306(2):180–189
123. Longshaw VM, Chapple JP et al (2004) Nuclear translocation of the Hsp70/Hsp90 organizing protein mSTI1 is regulated by cell cycle kinases. J Cell Sci 117(Pt 5):701–710
124. Longshaw VM, Baxter M et al (2009) Knockdown of the co-chaperone Hop promotes extranuclear accumulation of Stat3 in mouse embryonic stem cells. Eur J Cell Biol 88 (3):153–166
125. Daniel S, Bradley G et al (2008) Nuclear translocation of the phosphoprotein Hop (Hsp70/ Hsp90 organizing protein) occurs under heat shock, and its proposed nuclear localization signal is involved in Hsp90 binding. Biochim Biophys Acta 1783(6):1003–1014
126. Stephens LL, Shonhai A, Blatch GL (2011) Co-expression of the *Plasmodium falciparum* molecular chaperone, PfHsp70, improves the heterologous production of the antimalarial drug target GTP cyclohydrolase I, PfGCHI. Protein Expr Purif 77(2):159–165
127. Bodill T, Conibear AC et al (2011) Synthesis and evaluation of phosphonated N-heteroarylcarboxamides as DOXP-reductoisomerase (DXR) inhibitors. Bioorg Med Chem 19(3):1321–1327
128. Goble JL, Adendorff MR et al (2010) The malarial drug target *Plasmodium falciparum* 1-deoxy-D-xylulose-5-phosphate reductoisomerase (PfDXR): development of a 3-D model for identification of novel, structural and functional features and for inhibitor screening. Protein Pept Lett 17(1):109–120

Chapter 5
Natural Product-Based Drug Discovery in Africa: The Need for Integration into Modern Drug Discovery Paradigms

Eric M. Guantai and Kelly Chibale

Abbreviations

ABPP	Activity-based protein profiling
ACE	Angiotensin-converting enzyme
ADME	Absorption, distribution, metabolism, and excretion
ALNAP	African Laboratory for Natural Products
CIS	Chemical Information System
CYP	Cytochrome
DNDi	Drugs for Neglected Diseases Initiative
HAART	Highly active antiretroviral therapy
HTS	High-throughput screening
MMV	Medicines for Malaria Venture
NABSA	Network for Analytical and Bioassay Services in Africa
NAPRECA	Natural Products Research Network for East and Central Africa
RITAM	Research Initiative on Traditional Antimalarial Methods
SANBI	South African National Biodiversity Institute
SAR	Structure-activity relationship
TB	Tuberculosis
UCT	University of Cape Town

E.M. Guantai
Division of Pharmacology, School of Pharmacy, University of Nairobi, Box 19676-00202, Nairobi, Kenya
e-mail: eguantai@uonbi.ac.ke

K. Chibale (✉)
Department of Chemistry and Institute of Infectious Disease and Molecular Medicine, University of Cape Town, Rondebosch, South Africa
e-mail: Kelly.Chibale@uct.ac.za

K. Chibale et al. (eds.), *Drug Discovery in Africa*,
DOI 10.1007/978-3-642-28175-4_5, © Springer-Verlag Berlin Heidelberg 2012

VAO Vanillyl-alcohol oxidase
WRAIR Walter Reed Army Institute of Research
wwPDB Worldwide Protein Data Bank

5.1 Introduction

The economically challenged populations of sub-Saharan Africa continue to bear a substantial burden of infectious diseases such as malaria, tuberculosis (TB), and HIV/AIDS, which continue to bring about extraordinary morbidity and mortality among these populations. In their attempts to seek relief from this debilitating disease burden, these populations resort to either conventional (modern) medicine or traditional medicine, or quite commonly a combination of both. It has been reported that in excess of 80% of the African population resort to traditional medicine to help meet their health care needs [1]. This may be the result of their inability to afford conventional medicines or a genuine belief in the efficacy of traditional medicine.

Traditional medicine refers to those practices and knowledge, defined by culture, beliefs, and environment, which were historically applied to satisfy the health needs of communities before the arrival of modern conventional medicine. Herbal treatments form the largest and single most important component of traditional medicine. The extensive use of traditional medicine in Africa underscores the potential role of African natural products in modern health care, a fact that has not gone unnoticed within the scientific community. Considerable efforts have been made to scientifically corroborate claims of efficacy made by traditional African healers. A systematic review by the Research Initiative on Traditional Antimalarial Methods (RITAM) reported on several randomized clinical trials and other comparative studies across Africa that were aimed at demonstrating the clinical efficacy of selected antimalarial herbal preparations (in the form used by traditional healers) [2].

The apparent importance of traditional medicine/natural products in health care should not come as a surprise seeing that substantial components of conventional medicine are actually based on therapeutic agents derived from plants and other natural sources. The antimalarial drugs quinine and artemisinin, the anticancer agent paclitaxel, and the antibiotics penicillin and erythromycin are just a few of the many outstanding examples of therapeutic products of natural origin (Fig. 5.1) [3–6]. Indeed, several newly discovered antibiotics and other bioactive natural products with unique scaffolds and/or novel mechanisms of action are currently in advanced clinical trials or have since been approved for clinical use [7, 8]. In addition, natural products have also provided leads for development of clinically important semisynthetic drugs. The first-line anti-TB drug rifampicin, for example, is a semisynthetic derivative of rifamycin (Fig. 5.1), an antimycobacterial polyketide natural product isolated from the soil bacterium *Amycolatopsis mediterranei* [9].

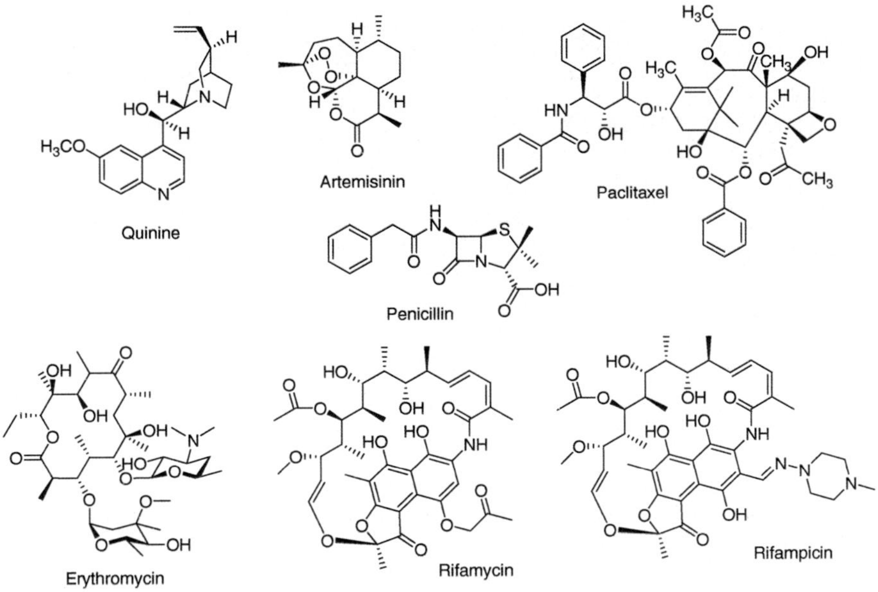

Fig. 5.1 Natural products that have found extensive clinical application: the antimalarials quinine and artemisinin, the anticancer agent paclitaxel, and the antibiotics penicillin and erythromycin. Also, the naturally occurring anti-TB agent rifamycin and its semisynthetic derivative rifampicin

Based on this long-standing precedent of natural products serving as a source of novel drugs and drug-like compounds, it is reasonable to conclude that the exceptionally diverse and as yet relatively unexplored biological resources/natural products in Africa are uniquely poised to provide new, pharmacologically active chemical entities that could themselves be developed as drugs or otherwise serve as the basis for mainstream drug discovery efforts. The above (and numerous other) examples clearly demonstrate how science can be used to effectively bridge the "gap" between herbal medicine/natural products and conventional/modern medicine, thereby allowing the full therapeutic potential of natural products to be tapped.

5.2 Generic Approaches to Natural Product Drug Discovery

The identification of a biologically active natural material may be serendipitous (such as the discovery of penicillin, an antibiotic principle produced by *Penicillium notatum*, by Sir Alexander Fleming [10, 11]), but has more often been the outcome of the scientific follow-up of historical/traditional accounts and involves bioassays of extracts of plant and other material used in traditional medicine.

Bioassay-guided fractionation of extracts derived from such material allows for the isolation and subsequent characterization of pure, active compounds. Medicinal

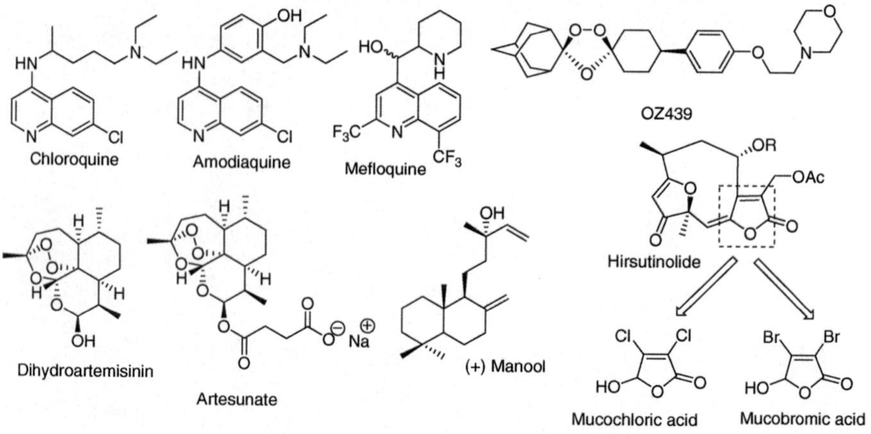

Fig. 5.2 Structures of the quinoline antimalarials chloroquine, amodiaquine, and mefloquine and the artemisinin derivatives dihydroartemisinin and sodium artesunate; OZ439 is a promising synthetic peroxide antimalarial drug candidate; (+) Manool is a naturally occurring compound useful as a substrate for semisynthesis; general structure of the hirsutinolides showing the furanone substructure; and the structures of mucochloric and mucobromic acids

chemistry efforts geared toward the total synthesis and/or the generation of analogues usually follow the identification of promising compounds; these efforts are primarily aimed at identifying superior/more potent analogues, but may also be geared toward providing a supplementary source of the primary compound for further study, revealing structure-activity relationships (SAR), and/or overcoming challenging physicochemical and biological properties.

The history of antimalarial drug discovery provides several illustrative examples. Quinine is an extremely potent antimalarial compound isolated from the barks of trees of the *Cinchona* genus; this compound was first isolated and purified in the early nineteenth century and is still in clinical use today [6]. Attempts to synthesize quinine led to the development of the well-known 4-aminoquinolines (e.g., chloroquine and amodiaquine) and quinoline amino alcohols (e.g., mefloquine) which were the mainstay of malarial chemotherapy for much of the latter half of the twentieth century (Fig. 5.2) [12]. On the other hand, artemisinin is a sesquiterpene lactone first isolated from the Chinese plant *Artemisia annua* in 1971 [6, 13]; this highly active antimalarial drug has since been derivatized to yield a very diverse range of semisynthetic analogues such as the clinically established dihydroartemisinin and artesunate [14] and has also inspired the design and synthesis of very promising, longer-acting analogues that are now in clinical development such as the synthetic peroxide antimalarial compound OZ439 (Fig. 5.2) [15, 16].

These developments can be seen to follow a distinct pattern—the initial identification of a bioactive natural product followed by its direct modification (or modification of templates derived from it) through medicinal chemistry—and show that more potent and long-acting compounds can be discovered if the bioactive entity of natural origin is identified and used as the basis of derivatization/modification by medicinal chemistry [12].

Some laboratories in Africa have also adopted this approach to natural product drug discovery. Researchers in South Africa have used natural products as substrates for the semisynthesis of novel, bioactive compounds or to provide pharmacophores for their design. For example, the naturally occurring labdane diterpenoid (+)-Manool (Fig. 5.2) was derivatized to yield antiplasmodial 17-tri- and tetracyclic norditerpenoids; the resultant semisynthetic derivatives all exhibited moderate in vitro antiplasmodial activity against the malarial parasite *Plasmodium falciparum* [17]. In another study, the 2(5H)-furanone unit was among the privileged substructures proposed as pharmacophores responsible for the observed antiplasmodial activity of two structurally related hirsutinolides isolated through bioassay-guided fractionation of the dichloromethane extracts of the leaves of *Vernonia staehelinoides* (Fig. 5.2). To investigate this potential pharmacophore, mucochloric and mucobromic acids (Fig. 5.2) were identified as appropriate 2(5H)-furanone substructures, and the antiplasmodial data obtained in respect of these two acids suggested that the 2(5H)-furanone substructure was indeed a key pharmacophore in the observed antiplasmodial activity. These acids exhibited comparable or superior activity against various *P. falciparum* strains relative to the hirsutinolide natural products and also showed selective cytotoxicity to the malaria parasites compared to mammalian (Chinese hamster ovarian) cells in vitro [18].

5.3 Challenges to Natural Product Drug Discovery in Africa

Natural product drug discovery efforts in most parts of Africa still remain centered around the traditional approach of bioassay-guided fractionation of extracts, culminating in attempts at structural elucidation of pure, bioactive entities that may be isolated in this way. Literature suggests that this is by and large the terminal stage of such discovery efforts.

The challenges faced are many and can be loosely considered as being of two general types: those challenges inherent to all natural product drug discovery and those challenges emanating from the African setting.

Challenges inherent to natural product drug discovery include basic problems such as possible seasonal or environmental variations in the content of the bioactive principle, problems of guaranteed access and supply of the source material, and loss of source through extinction or legislation. More practical challenges include the complexity of the extracts for fractionation (resulting in difficulties in purification and isolation of the bioactive entity), the frequent loss of biological activity following fractionation or purification, the isolation of very small quantities of bioactive substance, and challenging physicochemical properties such as solubility and stability [19].

In addition, African natural product drug discovery has struggled to make an impact due to a combination of various additional factors including inadequate financial resources, poor infrastructure, and a lack of skills and competency in several key areas. Furthermore, drug development efforts are fragmented across the continent, and these efforts are further compounded by limited access to technological platforms necessary to conduct robust drug discovery and development research [20–22].

It is rather encouraging that bioassay-guided fractionation and comparable approaches continue to yield an impressive range of structurally diverse bioactive compounds from a variety of chemical classes. Unfortunately, in recent years relatively few drug candidates that have made it to preclinical development or further have originated from a natural product start point. This is despite the many promising natural products that have shown pharmacological effects in biological assays. For a variety of reasons, including moderate biological activity, chemical tractability issues, and challenging physicochemical and pharmacokinetic properties, these compounds have not been pushed forward into modern hit-to-lead drug discovery projects and beyond.

An illustrative example of this is the disappointingly low number of new antimalarial natural products that have been identified as being active enough for clinical development in recent years. Approximately 480 new antimalarial compounds were reportedly isolated over the period 2005–2008 from plants used in traditional medicine, and of these only 30–45 new structures exhibited promising in vitro antiplasmodial activity [23]. This is in stark contrast to the thousands of promising compounds, many exhibiting submicromolar in vitro inhibitory concentrations, which have been identified through high-throughput screening (HTS) of databases of synthetic compounds from the pharmaceutical industry [24].

In light of these challenges, and in a bid to improve on the overall output from natural product drug discovery, it has been suggested that there is an urgent need to reevaluate the current conventional strategies of natural product drug discovery and to embrace a more elaborate and modern approach [12]. This call is particularly timely for Africa for a variety of reasons, including the current overreliance on traditional approaches to natural product research, the existing strong interest from African researchers to participate in and support such a transition, the presence of a rich biodiversity capable of supporting such initiatives, and the establishment of para-national institutions that encourage and support innovative natural products research.

African scientists have a powerful resource available to them in the form of the abundant natural products present on the continent (many of which possess known medicinal properties) and which they could leverage on to improve the scope and quality of their research, thereby enhancing the potential for meaningful findings and eventual success.

In this chapter, we introduce some of the components of the proposed paradigm shift in natural product drug discovery in Africa and discuss some of the strategies that would ensure that such a transformation is successful and sustainable.

5.4 Integration of Natural Product Drug Discovery into Modern Drug Discovery Paradigms

Though the possibility still exists that a compound of natural origin that could serve as a clinically useful drug may be identified by extraction and fractionation techniques, most compounds isolated from natural sources in this way are usually only moderately active, or possess challenging physicochemical and biological properties, and as such represent "hits" rather than actual drug candidates. This implies the need for further development of these compounds into "leads" and ultimately into drug candidates.

To boost the chances of the successful identification of quality lead compounds from natural products, it is necessary to complement conventional natural product drug discovery techniques with what may be considered as more modern drug discovery strategies; these would arguably increase the chances of successful identification and/or design of new potential leads as well as considerably enhance the range and depth of knowledge and information that is derived from these discovery efforts [25, 26].

Modern drug discovery involves a range of sciences including pharmacology, biochemistry, medicinal chemistry, and molecular biology, as well as drug metabolism and pharmacokinetic (DMPK) studies and the application of chemo- and bioinformatics tools. These various disciplines take advantage of modern techniques such as HTS, ligand- and fragment-based virtual screening, ligand docking tools, in silico (computational) and in vitro ADME (absorption, distribution, metabolism and excretion) profiling, among others.

As already alluded to, it is necessary to reevaluate the traditional approach that has hitherto been applied in natural product research in Africa in order for this research to translate into tangible modern pharmaceuticals or at the very least contribute more positively to the drug discovery value chain. There is a need for a total paradigm shift to a truly multidisciplinary approach that involves the incorporation of modern drug discovery strategies into natural product research in Africa, and the establishment of collaborations to achieve this. As elaborated by Guantai et al. 2011 [26], such a reevaluation would involve the incorporation of a number of key components:

- Leveraging of the abundant natural products present in Africa (many of which possess known medicinal properties) to improve the scope and quality of drug discovery research on the continent. This would involve the creation of physical and virtual repositories of known, bioactive, naturally derived chemical entities of African origin. These libraries may be expanded by biotransformation technologies and medicinal chemistry.
- The urgent adoption, across Africa, of a scientifically integrated approach to drug discovery, incorporating the entire complement of drug discovery sciences; this may be readily based on African natural products libraries. These libraries would facilitate the identification of natural product hits by the use of HTS

techniques and/or the application of appropriate chemo- and bioinformatics tools, to be followed by the optimization of these hits through ADME-guided medicinal chemistry. These modern strategies and techniques can be applied in the development of the primary natural products into leads or be used to prompt the identification and/or design of new potential leads based on these primary bioactive entities, with the overall aim being to improve the flow of quality leads for onward development into drug candidates.

- The laying out of strategies on how the proposed multinational, natural product-based drug discovery efforts can be financed, implemented, and managed in a sustainable fashion [26].

5.5 Harnessing African Natural Products for Drug Discovery

5.5.1 Development of Natural Product Databases and Repositories

A good number of bioactive molecules have already been isolated and characterized from natural products of African origin, with potentially many more to be discovered. It would therefore make sense to develop comprehensive databases of these compounds that include all the known information for each entry. Such information would ideally include the source (species of origin, geographical location), characterization data, chemical structure (if known), and all available biological data on the various compounds. Repositories of actual samples of these compounds could also be established.

The setting up of such databases would probably require a collective drive involving academic institutions and other research organizations currently involved in natural product research across Africa, which in turn stand to benefit from access to the combined wealth of knowledge and information that such resources would provide.

Presently, limited collections, records, and databases of natural products are scattered across Africa and include collections of natural products hosted by established research networks such as the Natural Products Research Network for East and Central Africa (NAPRECA) and the Network for Analytical and Bioassay Services in Africa (NABSA). The South African National Biodiversity Institute (SANBI) is also home to extensive records of South African flora. However, these collections and databases represent only a fraction of the total biodiversity present on the continent. They are limited in scope within the context of modern drug discovery paradigms and poorly adapted for utilization by modern drug discovery tools.

What is required is the development of more comprehensive databases of African natural products, which should contain associated chemical and biological data entered in appropriate formats. Existing databases such as the African

Laboratory for Natural Products (ALNAP) Bibliographic Database, which contains nearly 26,000 entries derived from a range of scientific publications on African plants, may be useful as starting points or templates for such efforts. To ensure that these databases serve as a collective resource available to feed into modern paradigms of drug discovery, they should be hosted on a platform of online servers across the continent, interlinked *via* networking applications that allow for the mining and manipulation of data. The precedent for natural product databases available online for research purposes exists on other continents in examples such as the The University of Arizona's Natural Products Database [27].

Concurrently, an elaborate campaign to pool and store African natural product extracts and pure compounds in central repositories should be implemented, beginning with those natural products samples that have been isolated and characterized over the years of natural products research in Africa. Avid bioprospecting initiatives of plants, marine organisms, bacteria, and fungi should be undertaken in an effort to expand these repositories, deliberately increasing both genetic (within-species) and biological (interspecies) diversity. Such collections should be stored under optimum conditions and should be readily retrievable for research purposes. It is acknowledged that such an undertaking to build libraries directly from purified natural products may be quite laborious and time-consuming, but great value will in turn be derived from the various applications of such resources.

The creation of such physical and virtual natural product-based platforms would help to keep track of these compounds and simultaneously create a rich resource that can be tapped for drug discovery efforts alongside classical combinatorial libraries. Chemical characterization information contained in such databases could also make it easier to elucidate the structures of newly isolated natural products, which in many cases are structurally related to known ones.

In order to develop a natural products repository in Africa, it may be necessary to borrow from existing models such as the Queensland Compound Library (Eskitis Institute, Griffith University, Australia) [28]. The Queensland Compound Library is a dedicated, fully automated compound management facility. This compound library serves as a central repository that hosts over 375,000 samples of small molecules and natural product extracts placed in a format that facilitates biomedical research. The samples are stored in both pure (dry) and solution (liquid) forms conveniently formatted in microtubes for long-term storage and screen-friendly microplates for HTS.

5.5.2 *Expansion of Natural Product Libraries by Biotransformation Technologies and Medicinal Chemistry*

The generation of compound libraries has, by and large, ignored natural products as a compound source. This is despite the recognition that natural biodiversity is a highly attractive and only partly explored resource with the potential to generate chemical diversity that exceeds the degree of diversity that can be accessed through synthesis. This situation is partly due to the prevalent (and probably well-founded) opinion that natural products are often highly complex and not amenable to medicinal chemistry efforts [29]. In this regard, biotransformation would therefore offer itself as a convenient method by which collections of bioactive African natural products can be enriched.

Panels of recombinant bacterial and/or human cytochrome P450 enzymes (in bioreactor setups) can be applied for the biotransformation of natural products, including but not limited to chemically complex and intractable compounds, and in this way ably complement synthetic and semisynthetic efforts at their derivatization [30, 31].

The cytochrome P450 enzymes represent the most versatile biological catalyst with respect to the diverse reactions they can catalyze. These include aliphatic and aromatic hydroxylations, *O*-dealkylations, *N*-dealkylations, *N*-oxidations, *S*-oxidations, reductions, and numerous other reactions [30]. In addition, some members of this superfamily of enzymes, such as CYP1, 2, and 3, are postulated to have evolved to detoxify plant metabolites [32] and as such appear to be uniquely suited for the biotransformation of natural products. This can be exploited to broaden the chemical diversity around naturally derived molecules, generating sufficient amounts of the analogues/metabolites to allow for their characterization and structural elucidation as well as for biological evaluation and derivation of SAR associated with pharmacological activity.

Interestingly, active metabolites in many cases have been known to have superior pharmacological, pharmacokinetic and safety profiles compared to their respective parent molecules [31]; metabolites of some established drugs such as chloroquine contribute significantly to overall observed activity [33, 34]. In addition, the metabolism of natural products may be required for activity in certain instances. For example, extracts of *Nauclea pobeguinii* exhibited antimalarial activity in both rodent models and patients despite strictosamide and other constituent alkaloids not showing any in vitro activity, suggesting possible metabolic activation of the active principles of the extract [35]. Such findings add to the allure of this approach to chemical diversification and further underline the potential of metabolism/biotransformation studies in identifying possible promising novel compounds, even for otherwise chemically accessible hit compounds.

Metabolism/biotransformation studies on promising naturally derived compounds would:

- Allow for the generation of analogues that can be collated into secondary databases that would be complementary to the primary repositories of natural products and be available for the same applications.
- Enable the elucidation of SAR information through preliminary bioassays of the generated structural analogues against relevant disease models and molecular targets. Such information would be useful in designing third-generation synthetic analogues that still possess the key descriptors and pharmacophores associated with the biological activity of the parent natural product.

For those promising and chemically tractable natural products, medicinal chemistry would offer a convenient diversification approach either by their direct chemical derivatization or by the synthesis of structural analogues and near neighbors (see below). Chemoinformatics tools (see below) can also be applied to bridge the gap between nontractable natural products and synthetic analogues. By carrying out molecular "descriptor space" similarity calculations, these tools can search for and identify synthetic compounds (from synthetic compound libraries) that are similar to the nontractable natural products of interest (i.e., "synthetic mimics") which can be synthesized, derivatized, and investigated in lieu of the nontractable natural product [29]. In this way, these tools would allow the medicinal chemists to systematically capture, and extend, structural features of natural products.

5.5.3 *Current Capacity for Biotransformation in Africa*

Well-established African research institutions, mainly in South Africa, already possess the necessary expertise in biotechnology to undertake biotransformation activities and are currently involved in very promising research aimed at developing these technologies. The expertise available in the research group of Professor Martie Smit in the Department of Microbial, Biochemical and Food Technology at the University of the Free State is noteworthy. Smit and coworkers have been investigating the expression of different prokaryotic and eukaryotic P450s (as well as other non-P450 oxidative enzymes such the *Penicillium simplicissimum* vanillyl-alcohol oxidase, VAO) in different *E. coli* strains. This research is aimed at identifying suitable microbial strains and plasmids that would facilitate the quick identification of suitable systems for high-level expression and excellent whole-cell activity of any P450 or VAO of interest [36]. Upon the identification of such suitable and efficient enzyme expression systems, it would then be necessary to scale up these biotransformation reactions and optimize for substrate and enzyme conditions. In this regard, the University of Cape Town Centre for Bioprocess Engineering Research [37] is a welcome development. This center contributes fundamental insights and knowledge in bioprocess engineering and related processes and allows for the transfer and application of this knowledge across fields in which bioprocesses play a role—drug discovery is one such field.

In recognition of the important role that biotransformation could play in drug discovery, the recently launched University of Cape Town Drug Discovery and Development Centre, also known as Holos 3ple-D or H3-D (http://www.h3-d.uct.ac.za/), has established a drug metabolite generation platform specifically to generate, isolate, characterize, scale up (in bioreactor setups), and pharmacologically evaluate drug metabolites. This is with a view to studying the contribution of metabolites to pharmacological activity and toxicity of parent drugs, including natural products.

5.6 Integration of Modern Drug Discovery Techniques

5.6.1 *Virtual Screening, Molecular Docking Studies, and High-Throughput Screening*

Chemo- and bioinformatics tools are increasingly playing a central role in modern drug discovery [38]. For this reason, any databases of African natural products, including secondary (metabolite and analogue) databases, must be developed in appropriate formats that allow for their mining and the application of these tools.

Basic pharmacophoric unit(s) and privileged scaffolds can be identified from the structures of either synthetic compounds or compounds of natural origin and can be validated and/or refined by preliminary SAR data where such data is available. Chemoinformatics tools can then be applied for the systematic fragment-based virtual screening of the databases of African natural product based on chemical descriptors obtained from these privileged scaffolds and pharmacophores, thereby facilitating the identification of potentially active structural analogues that can be investigated further [39]. Bioinformatics tools can be useful for ligand-based virtual screening of these databases, including for ligand docking studies (intended for mechanism of action and SAR studies) against panels of crystal structures or homology models of novel and validated druggable molecular targets of infectious diseases [40]. In this way, hitherto unknown biological activities of the natural products present in such databases could be identified and followed up.

The resources needed to achieve this are available. A variety of informatics tools exist that can be useful for this type of application, including widely used software packages such as AutoDock, DOCK FlexX, and GOLD [41]. Some of these applications, such as AutoDock, are freely available, while others may be freely licensed for academic research. The genomes of a variety of human parasites are now available as databases for vaccine development and drug discovery applications, such as the Sanger Institute's Pathogen Genomics project database [42]. Projects such as the TDR Targets database [43, 44]—an open-access database for tropical disease pathogens—go a step further and utilize the emerging genetic, biochemical, and pharmacological data to facilitate the identification and prioritization of candidate drug targets for pathogens. Furthermore, crystal structures of protein targets from a variety of infectious microorganisms are also freely available from such sources as the Worldwide Protein Data Bank (wwPDB) [45], making wide-ranging ligand-based virtual screening attractively feasible.

The activities of in silico (virtual or computational) hits derived in this way would then be verified and/or validated through experimental in vitro studies; these compounds can be acquired/synthesized and assayed alongside the primary compound to confirm the presence of biological activity. The physical repositories of crude and purified natural products (and their analogues) would be quite useful in this regard as they would conveniently provide the samples required for this kind of confirmatory testing. The physical repositories of natural products would also be

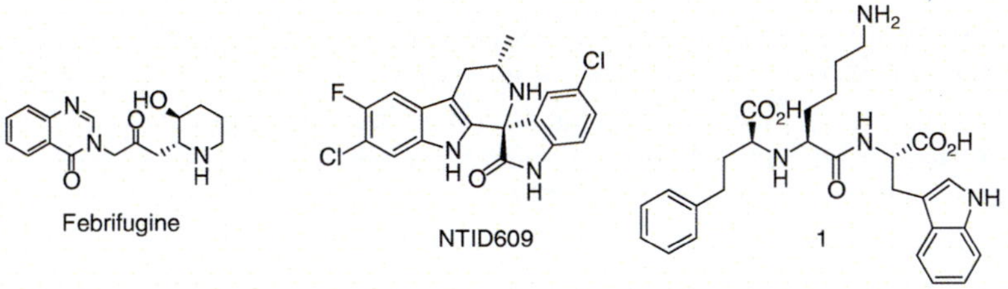

Fig. 5.3 Febrifugine, a potent antimalarial natural product; NTID609, an extremely active and highly promising antimalarial spiroindolone; structure of the C-domain-selective compound **1**

useful for providing samples for HTS and for more elaborate bioassays as appropriate.

Selected examples from other continents illustrate just how useful such resources can be. An example of the application of fragment-based screening using a template derived from a natural product is the use of the piperidinyl-(acetonyl)quinazoline moiety, a structural unit derived from the antimalarial natural product febrifugine (Fig. 5.3), as a template in the search for analogues in the Walter Reed Army Institute of Research (WRAIR) Chemical Information System (CIS) database. Potent analogues of febrifugine were identified that also had significantly reduced in vitro cytotoxicity against a variety of mammalian cell lines [46].

The application of natural product repositories for screening and identification of lead compounds is aptly illustrated by the identification of the antimalarial lead compound NITD609 (Fig. 5.3), which followed the screening of 12,000 pure natural products and synthetic compounds with structural features found in natural products using whole-cell antimalarial proliferation assays. This primary screening allowed for the identification of 250 hits with submicromolar activity against *P. falciparum*. These were further filtered based on in vitro antiplasmodial activity, cytotoxicity against mammalian cells, and physicochemical and pharmacokinetic properties, and this exercise led to the identification of a compound related to the spiroazepineindole class as a starting point for medicinal chemistry efforts. Further development led to the identification of NITD609, a compound found to be as effective as artesunate, showed no evidence of diminished potency against drug-resistant strains, exhibited favorable in vitro solubility and permeability, and did not show cytotoxicity across several human cell lines [47, 48].

Molecular docking expertise is already present on the African continent. For example, scientists at the University of Cape Town (UCT), South Africa, reported the use of molecular docking techniques to complement their efforts to design more C-domain-selective angiotensin-converting enzyme (ACE) inhibitors, which are envisaged to be better tolerated than the nonselective ACE inhibitors currently available clinically. In this study, an inhibitor **1** (Fig. 5.3) that demonstrated a significantly increased specificity for the C-domain as compared to the commercially available ACE inhibitor lisinopril was identified. Application of molecular docking tools revealed the basis of this selectivity to be increased hydrophobic interactions of the inhibitor with two valine residues within the S_2' subunit of the C-domain of ACE as well as strong H-bonding interactions between the inhibitor and an aspartate residue of the C-domain [49]. This information has been used to inform the design and synthesis of further analogues in a bid to identify more potent, C-domain-selective ACE inhibitors.

5.6.2 Lead Optimization Through ADME-Guided Medicinal Chemistry

The databases of African natural products and their exploitation using the modern technology platforms described above would ultimately provide a stream of actual hits or lead compounds, or SAR information vital for their design. This would be useful for the design of analogues that could be further optimized to yield quality leads for development into drug candidates.

The main considerations during the early drug discovery stages are potency, physicochemical properties, and ADME characteristics, all of which are structure dependent. It is widely acknowledged that challenging ADME properties significantly hamper the development of many promising compounds into leads and drug candidates [50, 51]. As structural changes are the most fundamental changes on a molecule, affecting all its properties and characteristics, these three considerations need to be addressed as early as possible in drug development.

Furthermore, for drugs used to treat tropical diseases, oral administration is greatly preferred as huge numbers of patients across the developing world must be treated in facilities with limited resources or infrastructure. The proportion of an orally administered drug that is delivered into the systemic circulation (i.e., its oral bioavailability) is primarily dependent on three physicochemical and ADME factors: solubility, permeability, and metabolic stability [52, 53]. Therefore, derivatization and synthesis of analogues of promising and chemically tractable natural products by medicinal chemistry should be guided by predicted and experimental physicochemical and ADME data that can be availed by the complementary application of both in silico and in vitro techniques [51, 54, 55].

In silico tools refer to software applications such as VolSurf and MetaSite which provide in silico platforms for the prediction of a wide range of physicochemical and ADME properties including aqueous solubility, log *P*/log *D*, pKa, passive and active membrane permeability, metabolic stability, sites and routes of metabolism, protein binding, and volume of distribution [56, 57]. The in silico ADME prediction tools provide information that enables the drug discovery scientists to design analogues and derivatives of the natural product hits that would most likely possess acceptable bioavailability profiles, thereby saving time and resources by focusing synthesis and experimental evaluation efforts on a smaller number of more promising compounds.

Other general recommendations such as the Astex "Rule of Three" [58] and the Oprea lead-like criteria [59] for lead compounds and the Lipinski's "Rule of Five" [60, 61] for oral absorption can also be used as guides in the design of the natural product-based leads. However, caution must be exercised in following these guidelines so as not to unnecessarily exclude potentially promising candidate natural products. It is noteworthy that even sections of the pharmaceutical industry recognize the importance of exercising caution when applying the Lipinski and related guidelines as valuable discoveries can indeed be made at the margins or even outside areas considered mainstream [62]. An illustration of this is provided by the recent analyses of 24 unique natural products discovered during the period 1970–2006 that led to an approved drug. These clinically approved natural products could be divided into two equal subsets: one that fell in the so-called Lipinski universe and complied with the Lipinski "Rule of Five" and another subset that fell in the so-called parallel universe that violated the Lipinski guidelines [63]. This underscored the fact that promising natural products can very well be exceptions to the Lipinski and related guidelines. However, notwithstanding the noncompliance of the latter subset, the conclusion reached through the aforementioned analysis was that compounds in this subset remained largely Lipinski "Rule of Five" compliant vis-à-vis log *P* and H-bond donors, highlighting the importance of these two parameters in predicting bioavailability.

In vitro ADME determinations would then provide actual experimental data to verify the in silico predictions and which can be used to prioritize compounds for in vivo evaluation. The experimental data would also provide insights that the medicinal chemistry teams can use as a guide for the design and synthesis of further analogues predicted to have favorable ADME properties. In vitro platforms enable the determination of physicochemical properties such as log *P*, solubility, and pKa; they also provide ADME models for the prediction of compound permeability using Caco2 cells and vesicles for the role of transporters, metabolic stability using liver microsomes and hepatocytes, and biotransformation routes by metabolite identification [52]. Recombinant enzymes and liver microsomes can also be used to evaluate the new chemical entities for potential for drug-drug interactions.

In general, in silico and in vitro ADME studies are primarily aimed at assessing the pharmacokinetic properties of promising hit and lead compounds, thereby guiding the design of superior analogues while also providing useful information to guide and/or help in the interpretation of the results from subsequent pharmacokinetic studies and in vivo assays [52]. These approaches are discussed in greater detail in Chapter 7.

Furthermore, the molecular modeling and docking tools alluded to earlier may also be applied at this stage to enrich the process of derivatization of natural products by guiding or complementing the ADME-guided rational design and selection of more promising derivatives for subsequent synthesis and biological evaluation.

There already exists a decent capacity for medicinal chemistry/organic synthesis in Africa, implying that it is quite realistic for drug discovery scientists in Africa to generate late leads and possibly drug candidates by adopting an ADME-guided

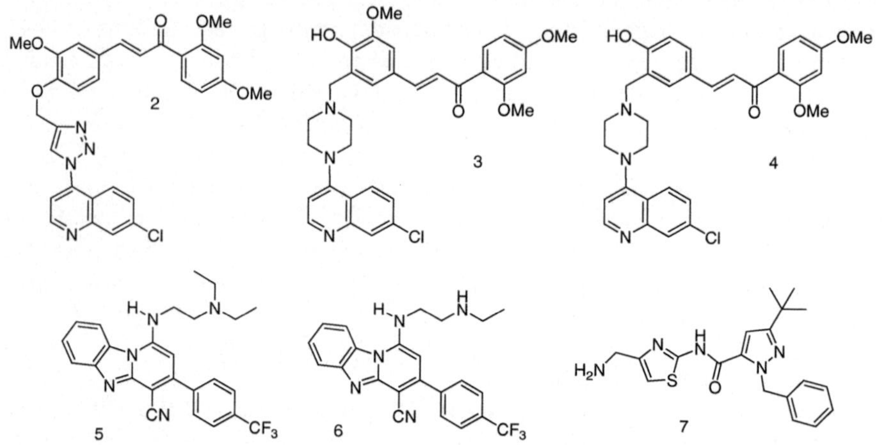

Fig. 5.4 Structures of the promising antimalarial hybrid compound **2** and its structural analogues **3** and **4**; promising antimalarial pyridobenzimidazoles **5** and **6** and aminomethylthiazole pyrazole carboxamide **7**

approach to the medicinal chemistry of natural product-derived hits. Maintaining an emphasis on low-cost synthetic strategies would potentially afford cost-effective drug candidates that are relatively simple to synthesize.

Excellent examples of ADME-guided medicinal chemistry programs in Africa are described below. Having previously identified the highly active antimalarial hybrid compound **2** (Fig. 5.4), scientists at the University of Cape Town set about designing, synthesizing, and biologically evaluating analogues of this hit with the aim of identifying compounds that showed improved ADME profiles and which retained antimalarial potency. The design process was guided by in silico prediction of physicochemical and ADME parameters of the proposed analogues and led to the identification of analogues with predicted improved solubilities, particularly at low pH; these analogues retained acceptable predicted permeability properties but were predicted to be susceptible to hepatic metabolism. Based on this predicted data, these promising analogues were synthesized and found to retain notable in vitro antimalarial activity, with analogues **3** and **4** (Fig. 5.4) being the most active. To corroborate the computational predictions, experimental solubility and log *D* determinations were carried out on the compounds, and the results were found to be consistent with the predicted data. In vitro metabolism studies were carried out in rat and human microsomes, and the results indicated susceptibility of the analogues to hepatic metabolism—oxygenation, O-demethylation, and N- dealkylation appeared to be the predominant metabolic pathways—which was also consistent with the in silico site of metabolism predictions. There was also evidence of primary glucuronidation for some of the analogues [55].

As mentioned previously, in silico and in vitro ADME studies aid in assessing the pharmacokinetic properties of promising hit and lead compounds as well as provide useful information to guide and/or help in the interpretation of the results from subsequent pharmacokinetic studies and in vivo assays. As illustrated by the examples below, this component of drug discovery is now being embraced by drug discovery scientists in Africa and who are involved in collaborative research projects that apply these techniques. In a recent study of 3-aryl-pyridobenzimidazoles as potential antimalarial agents, in vitro metabolism studies of a number of analogues were conducted using human, mouse, and rat liver microsomes. This was in a bid to provide data to guide the selection of the most appropriate analogues for in vivo investigations and to guide in the interpretation of results from pharmacokinetic studies. Routes of metabolism and putative metabolites predicted by the in vitro investigations were found to be remarkably consistent with the findings from subsequent pharmacokinetic studies. For example, **5** was predicted to be quite susceptible to N-dealkylation by hepatic cytochromes, with the more metabolically stable analogue **6** predicted as the primary metabolite (Fig. 5.4 for structures). A similar pattern was observed in vivo, with **6** identified as a primary metabolite of **5** in plasma samples, conversion of **5** to **6** estimated to be approximately 70%. As suggested by the in vitro data, **6** also exhibited a long apparent half-life and low plasma clearance in vivo [64]. In a similar study on the promising aminomethylthiazole pyrazole carboxamide antimalarial lead compound **7** (Fig. 5.4), the blood clearance was consistent with the results of the in vitro metabolic stability studies in rat liver microsomes which predicted that **7** would be subjected to moderate in vivo hepatic clearance [65].

All in all, these studies provide examples of how in silico and in vitro physicochemical and/or ADME predictive techniques have been applied in African drug discovery initiatives, undoubtedly aided by the establishment of strong collaborations within and beyond the continent.

5.6.3 Dual Drugs and Drug Combinations

The design of derivatives of promising natural products could borrow from established strategies such as the design of dual drugs, also known as hybrid compounds. These types of compounds consist of two drugs/active compounds/ pharmacophoric units linked together covalently by a linker [66], designed to take advantage of the observed (or anticipated) synergistic or additive pharmacological activities of the hybrid components and enable the identification of highly active novel chemical entities (possibly exhibiting dual modes of action).

The incorporation of components derived from natural products into hybrid compounds may be considered to be at the border between bio-inspired design and rational drug design. This is an attractive derivatization strategy as, theoretically, millions of hybrids bearing components of natural products can be prepared.

Interestingly, naturally occurring, bioactive hybrid compounds have been isolated. For example, the antimicrobial antibiotic thiomarinol isolated from the marine bacterium *Alteromonas* was shown to be a hybrid of the pseudomonic acid C and holothin (Fig. 5.5). This hybrid shows characteristics of both parent

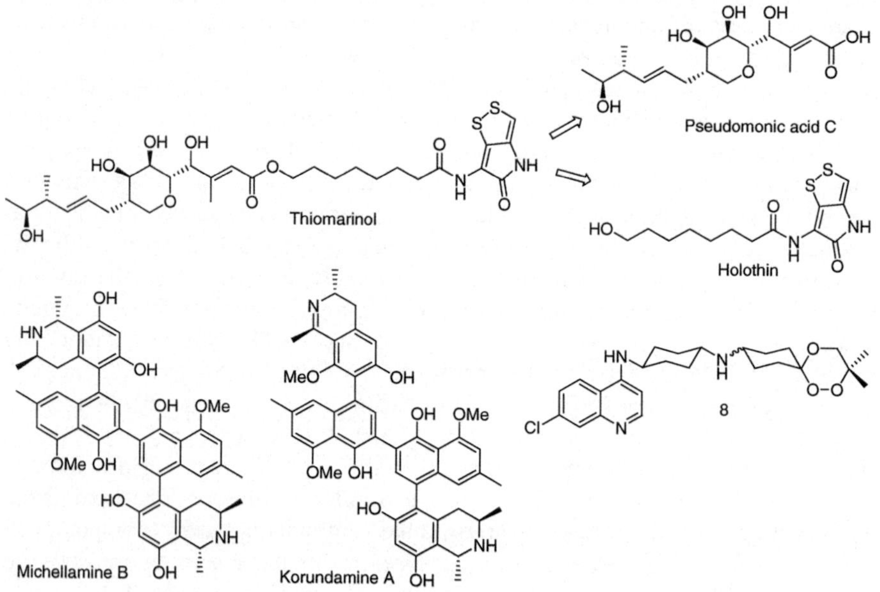

Fig. 5.5 Structures of the naturally occurring hybrid compounds thiomarinol, michelline B and korundamine A, and the synthetic hybrid trioxane **8**

compounds, and its effects are more pronounced than those of either parent compound. Michellamine B and korundamine A (Fig. 5.5) are dimeric naphthylisoquinoline alkaloids isolated from the plant *Ancistrocladus korupensis*. These natural hybrids show strong anti-HIV activity; in addition, korundamine A (16) is a potent antiplasmodial compound [67].

Hybridization has led to the identification of very promising compounds. In the previously described University of Cape Town study that applied ADME-guided medicinal chemistry, it should be noted that the promising compound **2** and the subsequent analogues were all actually hybrid compounds that incorporated components inspired by natural products [55]. Hybridization has also led to the identification of the highly active antimalarial trioxaquines. These are potent antimalarial compounds that have been developed by covalently linking a 1,2,4-trioxane motif (a peroxide entity derived from the highly active natural product artemisinin) to a 4-aminoquinoline moiety (borrowed from chloroquine and other chloroquinoline-based antimalarials) *via* an appropriate spacer. Several of these trioxaquine derivatives were found to be extremely potent antiplasmodial agents, and the trioxaquine **8** (Fig. 5.5) has been selected as a candidate for preclinical development [68]. Similar hybridization strategies involving established natural product-based drugs have also been applied in the search for drugs against diseases such as cancer where, for example, in an effort to improve drug targeting and/or tissue distribution, taxoids have been conjugated with various other compounds, including 3,17b-estradiol [69] and various fatty acids [70].

Another strategy that takes advantage of possible synergistic or additive pharmacological activities is the development of suitable combination therapies, i.e., two (or more) active compounds that can be rationally and advantageously coadministered to yield overall biological effects superior to the effects of the individual agents when administered singly. Drugs selected for combination in this way usually have a demonstrated synergistic effect when applied together, thereby allowing for the use of lower doses and achievement of more rapid therapeutic outcomes while taking advantage of the lower risk of the concurrent development of resistance by the microorganisms to the coadministered agents.

The emphasis is currently on the use of combination therapy in the management of various infectious diseases. For example, due to the rapid development of resistance by the respective causative microorganisms, the current drug therapy of malaria, TB, and HIV/AIDS is principally by combination therapy. Treatment regimens for TB consist of various combinations of drugs like rifampicin, pyrazinamide, streptomycin, and ethambutol [71]; highly active antiretroviral therapy (HAART) regimens currently recommended for HIV chemotherapy consist of various combinations of nucleoside and nonnucleoside reverse transcriptase inhibitors as well as protease inhibitors [72]. Artemisinin-based combination therapies (ACTs) are now first-line treatments in most malaria-endemic countries [73–75] and are prime examples of the application of a natural product in combination therapy.

It is in this regard that the identification of new and efficacious drug combinations for the treatment of infectious diseases remains a priority, and the identification of new, active, naturally derived compounds could provide additional possibilities. Numerous reports have been made of synergy between natural products and established anti-infective agents [76–79], making this a very promising approach for the identification of new, highly effective treatment regimens incorporating components derived from African natural products.

5.6.4 *Application of Natural Products in Other Aspects of Drug Discovery: Target Identification*

In addition to providing hits or leads for onward drug development, natural products can also contribute to identifying protein targets, unraveling mechanisms of action, and studying biological systems through the application of such techniques as metabolic profiling and activity-based protein profiling (ABPP) [9]. Newly discovered targets that are validated (i.e., are essential for organism survival, biochemically and structurally characterized, susceptible to inhibition, druggable, and amenable to scaled up production to facilitate screening) can then be used as a basis for drug discovery efforts.

Metabolic profiling involves the use of chromatographic and spectroscopic tools such as NMR [80, 81] and mass spectroscopy [82, 83] in the assessment of the metabolic profile of an organism/cell/parasite and is useful in evaluating gene function as well as assessing the metabolic responses of an organism following exposure to a bioactive molecule. Any changes in the metabolic profile of the organism may indicate the metabolic pathway interrupted by the bioactive molecule and may even point to the actual protein target of the compound [84, 85]. On the other hand, protein-reactive natural products (or privileged structures derived from them) can be useful in the development of active site-directed chemical probes; these small-molecule probes are able to covalently label the active site of a given enzyme or enzymes, thereby aiding the identification of specific druggable enzyme/receptor/protein targets from complex proteomes in a process termed as activity-based protein profiling [86, 87].

These techniques offer exciting approaches for the identification of (oftentimes new) molecular targets. Bioactive natural products, by virtue of their complexity and chemical diversity, are particularly suited to act as probes for the identification of new molecular targets in this way [84, 85], providing another possible application for diversity-oriented libraries of African natural product-like compounds described previously.

Proteins/enzyme targets identified in this way can then be investigated further as potential therapeutic targets by both classical mechanism of action studies or reverse pharmacology [88]. Reverse pharmacology, as opposed to classical pharmacology, begins with the initial identification of potential protein targets (such as enzymes and receptors) either by the methods described above or by the application of bioinformatics tools that exploit the vast DNA-sequence databases provided by intensive genomic research. The potential targets identified in this way are then cloned and used to screen candidate ligands, which may include natural products [89], in a process known as "ligand fishing." In this way, reverse pharmacology offers an alternative means by which the molecular targets of biologically active compounds may be identified.

Physical repositories of crude and purified African natural products can provide compound libraries for metabolic profiling, reverse pharmacology, and related endeavors. Furthermore, the reported successful metabolic profiling of the human malaria parasite *P. falciparum* [81, 90] as well as the complete sequencing of the genomes of the malaria parasite [91, 92], *Mycobacterium tuberculosis* [93], and the HIV virus [94] all serve to emphasize the applicability of these approaches in drug discovery endeavors for these major infectious diseases [95].

5.7 Financing, Managing, and Sustaining Such Natural Product-Based Drug Discovery Efforts

Financing, managing, and sustaining such an integrated natural product-based drug discovery effort on the African continent pose arguably the biggest challenge to such an initiative.

A model similar to that of the Queensland Compound Library (see Sect. 5.5.1) could be adapted for the African natural products database, to be hosted by leading academic or research institutions in Africa that already possess reliable first-world research infrastructure. Such institutions can play an integral role in hosting and expanding the primary repositories and in the application of the various modern technologies to these compound libraries.

The resource as proposed above would be quite extensive and convenient to use, allowing access to a broad diversity of natural products from Africa on a scale not seen previously. This would make it attractive to the pharmaceutical industry, research institutions, and academia. Storage and access to compounds by clients may be on a fee basis to generate income; members or partners could make annual contributions and/or pay a per-access fee that may vary for the different categories of clients.

Income generated in this way would be used to maintain the operation of the resource and sustain transcontinental access and transfer of knowledge. Such income could be supplemented by grant applications to global research-funding organizations and funding from African governments. Partnerships with such entities as Medicines for Malaria Venture (MMV), Drugs for Neglected Diseases Initiative (DNDi), and the Global Alliance for Tuberculosis Drug Development (TB Alliance) could further strengthen the initiative and help position it within the global research sphere.

Expertise in the various scientific fields and their required synergy is not yet established in Africa. Expertise and research capacity of more African institutions needs to be developed to address current deficiencies in, for example, in vitro ADME and biotransformation studies as well as the expression and crystallization of recombinant target proteins. Eminent African natural product scientists and scholars could manage the establishment of the proposed natural product databases and repositories. They could be partnered by medicinal chemists, clinicians, pharmacologists, and other scientists to form productive collaborations, thereby stimulating the desired synergy between the various fields.

5.8 Conclusion

The proposals made above (and illustrated in Fig. 5.6) of how natural product research in Africa can embrace modern drug discovery strategies are primarily aimed at boosting the chances of the successful identification of quality lead compounds from natural products.

Similar approaches have been applied in the developed world and possess the potential to impact greatly and positively on drug discovery initiatives in Africa if appropriately adapted to this setting, thereby providing the potential for exciting new discoveries and hopefully bringing to fruition the enduring prospect of new cures for the infectious diseases affecting Africa's populations.

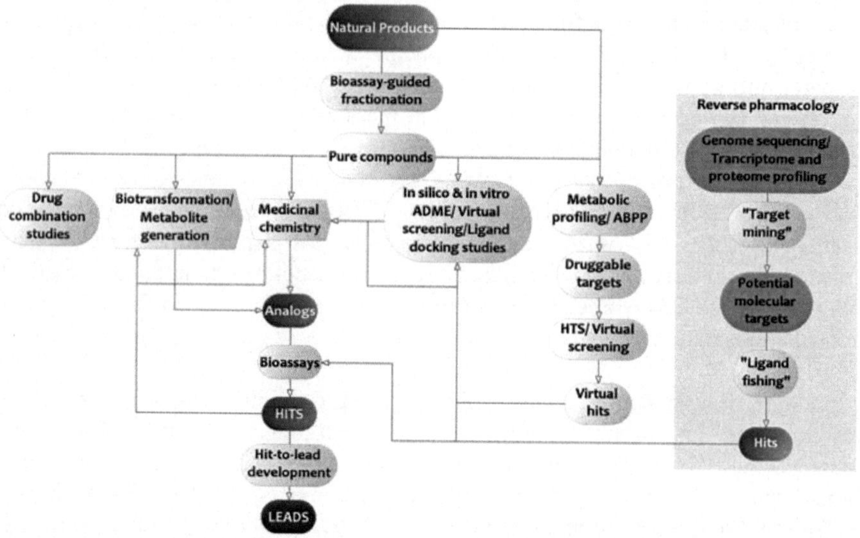

Fig. 5.6 Development of leads from natural products (reproduced from Guantai and Chibale 2011 [25], with author consent)

The approaches described may not be novel in their own right, but most of the aspects need to be embraced and/or developed in Africa in a manner that is both practical and self-sustaining. As already noted, limited expertise in a number of key areas already exists in Africa, such as in ADME-guided medicinal chemistry and the application of computational tools. However, there is still much room for development, and thus the proposed paradigm shift is likely to stimulate the transfer of knowledge and contribute immensely to the development of a research infrastructure on the continent.

References

1. World Health Organization. Traditional Medicine Strategy. 2002–2005.
2. Willcox ML, Bodeker G (2004) Traditional herbal medicines for malaria. Br Med J 329:1156–1159
3. Newman DJ, Cragg GM (2007) Natural products as sources of new drugs over the last 25 years. J Nat Prod 70:461–477
4. Itokawa H, Morris-Natschke SL, Akiyama T et al (2008) Plant-derived natural product research aimed at new drug discovery. J Nat Med 62:263–280
5. Wang M-W, Hao X, Chen K (2007) Biological screening of natural products and drug innovation in China. Phil Trans R Soc Lond B Biol Sci 362:1093–1105
6. Saxena S, Pant N, Jain DC et al (2003) Antimalarial agents from plant sources. Curr Sci 85:1314–1329
7. Clardy J, Fischbach MA, Walsh CT (2006) New antibiotics from bacterial natural products. Nat Biotechnol 24:1541–1550

8. Butler MS (2005) Natural products to drugs: natural product derived compounds in clinical trials. Nat Prod Rep 22:162–195
9. Gutierrez-Lugo M-T, Bewley CA (2008) Natural Products, Small Molecules, and Genetics in Tuberculosis Drug Development. J Med Chem 51:2606–2612
10. Fleming A (1922) On a remarkable bacteriolytic element found in tissues and secretions. Phil Trans R Soc Lond B Biol Sci 93:306–317
11. Hare R (1982) New light on the history of Penicillin. Med Hist 26:1–24
12. Wells TNC (2011) Natural products as starting points for future antimalarial therapies: going back to our roots? Malar J 10(Suppl 1):S3
13. Klayman DL (1985) Quinghaosu (artemisinin): an antimalarial drug from China. Science 228:1049–1055
14. Haynes RK (2006) From artemisinin to new artemisinin antimalarials: biosynthesis, extraction, old and new derivatives, stereochemistry and medicinal chemistry requirements. Curr Top Med Chem 6:509–537
15. Charman SA, Arbe-Barnes S, Bathurst I et al (2011) Synthetic ozonide drug candidate OZ439 offers new hope for a single-dose cure of uncomplicated malaria. Proc Natl Acad Sci USA 108:4400–4405
16. Vennerstrom JL, Arbe-Barnes S, Brun R et al (2002) Identification of an antimalarial synthetic trioxolane drug development candidate. Nature 430:900–904
17. van Wyk AWW, Lobb KA, Caira MR et al (2007) Transformations of manool. Tri- and tetracyclic norditerpenoids with in vitro activity against plasmodium falciparum. J Nat Prod 70:1253–1258
18. Pillay P, Vleggaar R, Maharaj VJ et al (2007) Antiplasmodial hirsutinolides from Vernonia staehelinoides and their utilization towards a simplified pharmacophore. Phytochemistry 68:1200–1205
19. Li JW-H, Vederas JC (2009) Drug discovery and natural products: end of an era or an endless frontier? Science 325:161–165
20. Trouiller P, Olliaro P, Torreele E et al (2002) Drug development for neglected diseases: a deficient market and a public-health policy failure. Lancet 359:2188–2194
21. Nyigo VA, Malebo HM (2005) Drug discovery and developments in developing countries: bottlenecks and way forward. Tanzan Health Res Bull 7:154–158
22. Chibale K (2010) Discovering Africa's drug potential. http://www.scidev.net/en/opinions/discovering-africa-s-drug-potential.html. Accessed June 2011
23. Bero J, Frederich M, Quetin-Leclercq J (2009) Antimalarial compounds isolated from plants used in traditional medicine. J Pharm Pharmacol 61:1401–1433
24. Gamo F-J, Sanz LM, Vidal J et al (2010) Thousands of chemical starting points for antimalarial lead identification. Nature 465:305–310
25. Guantai EM, Chibale K (2011) How can natural products serve as a viable source of lead compounds for the development of new/novel anti-malarials? Malar J 10(Suppl 1):S2
26. Guantai EM, Masimirembwa C, Chibale K (2011) Extracting molecular information from African natural products to facilitate unique African-led drug-discovery efforts. Future Med Chem 3:257–261
27. The University of Arizona's Natural Products Database. http://npd.chem.arizona.edu/. Accessed June 2011.
28. Queensland Compound Library. http://www.griffith.edu.au/science/queensland-compound-library. Accessed June 2011.
29. Bajorath J (2002) Chemoinformatics methods for systematic comparison of molecules from natural and synthetic sources and design of hybrid libraries. J Comput Aided Mol Des 16:431–439
30. Coon MJ (2005) Cytochrome P450: nature's most versatile biological catalyst. Annu Rev Pharmacol Toxicol 45:1–25
31. Fura A, Shu Y-Z, Zhu M et al (2004) Discovering drugs through biological transformation: role of pharmacologically active metabolites in drug discovery. J Med Chem 47:4339–4350

32. Gonzalez FJ, Nebert DW (1990) Evolution of the P450 gene superfamily: animal-plant 'warfare', molecular drive and human genetic differences in drug oxidation. Trends Genet 6:182–186
33. Projean O, Baune B, Farinotti R et al (2003) *In vitro* metabolism of chloroquine: identification of CYP2C8, CYP3A4 and CYP2D6 as the main isoforms catalyzing *N*-desethylchloroquine formation. Drug Metab Dispos 31:748–754
34. Fu S, Björkman A, Wåhlin B et al (1986) *In vitro* activity of chloroquine, the two enantiomers of chloroquine, desethylchloroquine and pyronaridine against *Plasmodium falciparum*. Br J Clin Pharmacol 22:93–96
35. Mesia K, Cimanga RK, Dhooge L et al (2010) Antimalarial activity and toxicity evaluation of a quantified Nauclea pobeguinii extract. J Ethnopharmacol 131:10–16
36. Smit MS (2011) Comparison of recombinant *E. coli*- and yeast-based whole cell hydroxylating biocatalysts (Abstract). http://lamp3.tugraz.at/~acib/index.php/wbNews/detail/44. Accessed June 2011
37. University of Cape Town Centre for Bioprocess Engineering Research. http://www.chemeng.uct.ac.za/research/bioprocess/. Accessed June 2011
38. Claus BL, Underwood DJ (2002) Discovery informatics: its evolving role in drug discovery. Drug Discov Today 7:957–966
39. Terstappen GC, Reggiani A (2001) *In silico* research in drug discovery. Trends Pharmacol Sci 22:23–26
40. Bajorath J (2001) Rational drug discovery revisited: interfacing experimental programs with bio- and chemo-informatics. Drug Discov Today 6:989–995
41. Moitessier N, Englebienne P, Lee D et al (2008) Towards the development of universal, fast and highly accurate docking/scoring methods: a long way to go. Br J Pharmacol 153:S7–S26
42. Sanger Institute GeneDB Project. http://www.genedb.org/Homepage. Accessed June 2011
43. The TDR Targets Database v5 - A chemogenomics resource for neglected tropical diseases. http://tdrtargets.org/. Accessed September 2011
44. Agüero F, Al-Lazikani B, Aslett M et al (2008) Genomic-scale prioritization of drug targets: the TDR Targets database. Nat Rev Drug Discov 7:900–907
45. Worldwide Protein Data Bank (wwPDB). http://www.wwpdb.org/. Accessed June 2011
46. Jiang S, Zeng Q, Gettayacamin M et al (2005) Antimalarial Activities and Therapeutic Properties of Febrifugine Analogs. Antimicrob Agents Chemother 49:1169–1176
47. Yeung BKS, Zou B, Rottmann M (2010) Spirotetrahydro b-carbolines (spiroindolones): a new class of potent and orally efficacious compounds for the treatment of malaria. J Med Chem 53:5155–5164
48. Rottmann M, McNamara C, Yeung BKS (2010) Spiroindolones, a potent compound class for the treatment of malaria. Science 329:1175–1180
49. Nchinda AT, Chibale K, Redelinghuys P et al (2006) Synthesis and molecular modeling of a lisinopril-tryptophan analogue inhibitor of angiotensin I-converting enzyme. Bioorg Med Chem Lett 16:4616–4619
50. Butina D, Segall MD, Frankcombe K (2002) Predicting ADME properties *in silico*: methods and models. Drug Discov Today 7:S83–S88
51. Yu H, Adedoyin A (2003) ADME–Tox in drug discovery: integration of experimental and computational technologies. Drug Discov Today 8:852–861
52. Kerns EK, Di L (2008) Drug-like properties: concepts, structure, design and methods: from ADME to toxicity optimization. Academic Press/Elselvier, Amsterdam
53. Masimirembwa CM, Bredberg U, Andersson TB (2003) Metabolic stability for drug discovery and development: pharmacokinetic and biochemical challenges. Clin Pharmacokinet 42:515–528
54. Eddershaw PJ, Beresford AP, Bayliss MK (2000) ADME/PK as part of a rational approach to drug discovery. Drug Discov Today 5:409–414
55. Guantai EM, Ncokazi K, Egan TJ et al (2011) Enone- and chalcone-chloroquinoline hybrid analogues: *in silico* guided design, synthesis, antiplasmodial activity, *in vitro* metabolism, and mechanistic studies. J Med Chem 54:3637–3649

56. Cruciani G, Pastor M, Guba W (2000) VolSurf: a new tool for the pharmacokinetic optimization of lead compounds. Eur J Pharm Sci 11:S29–S39
57. Cruciani G, Carosati E, de Boeck B et al (2005) Understanding metabolism in human cytochromes from the perspective of the chemist. J Med Chem 48:6970–6979
58. Congrieve M, Carr R, Murray C (2003) A "Rule of Three" for fragment-based lead discovery? Drug Discov Today 8:876–877
59. Oprea TI, Davis AM, Teague SJ et al (2001) Is there a difference between leads and drugs? a historical perspective. J Chem Inf Comput Sci 41:1308–1315
60. Lipinski CA, Lombardo F, Dominy BW et al (2001) Experimental and computational approaches to estimate solubility and permeability in drug discovery and development settings. Adv Drug Deliv Rev 46:3–26
61. Lipinski CA (2004) Lead- and drug-like compounds: the rule-of-five revolution. Drug Discov Today Technol 1:337–341
62. Fotouhi N, Gillespie P, Goodnow R Jr (2008) Lead generation: reality check on commonly held views. Expert Opin Drug Discov 3:733–744
63. Ganesan A (2008) The impact of natural products upon modern drug discovery. Curr Opin Chem Biol 12:306–317
64. Ndakala A, Gessner RK, Gitari PW et al (2011) Antimalarial pyrido[1,2-a]benzimidazoles. J Med Chem 54:4581–4589
65. Cabrera DG, Douelle F, Feng T-S et al (2011) Novel orally active antimalarial thiazoles. J Med Chem 54:7713–7719
66. Walsh JJ, Bell A (2009) Hybrid drugs for malaria. Curr Pharm Des 15:2970–2985
67. Tietze LF, Bell HP, Chandrasekhar S (2003) Natural product hybrids as new leads for drug discovery. Angew Chem Int Ed Engl 42:3996–4028
68. Coslédan F, Fraisse L, Pellet A et al (2008) Selection of a trioxaquine as an antimalarial drug candidate. Proc Natl Acad Sci USA 105:17579–17584
69. Liu C, Strolb JS, Schilling JK et al (2004) Design, synthesis, and bioactivities of steroid-linked taxol analogues as potential targeted drugs for prostate and breast cancer. J Nat Prod 67:152–159
70. Kuznetsova L, Chen J, Sun L et al (2006) Synthesis and evaluation of novel fatty acid-second genaration taxoid conjugates as promising anticancer agents. Bioorg Med Chem Lett 16:974–977
71. World Health Organization (2003) Treatment of tuberculosis: guidelines for national programs, 3rd edn. World Health Organization, Geneva, Switzerland
72. World Health Organization (2009) Rapid advice: antiretroviral therapy for HIV infection in adults and adolescents. World Health Organization, Geneva, Switzerland
73. Edwards G, Biagini GA (2006) Resisting resistance: dealing with the irrepressible problem of malaria. Br J Clin Pharmacol 61:690–693
74. Mutabingwa TK (2005) Artemisinin-based combination therapies (ACTs): best hope for malaria but inaccessible to the needy. Acta Trop 95:305–315
75. White N (1999) Antimalarial drug resistance and combination chemotherapy. Phil Trans R Soc Lond B Biol Sc 354:739–749
76. Heamiswarya S, Kruthiventi AK, Doble M (2008) Synergism between natural products and antibiotics against infectious diseases. Phytomedicine 15:639–652
77. Lin R-D, Chin Y-P, Lee M-H (2005) Antimicrobial activity of antibiotics in combination with natural flavonoids against clinical extended-spectrum b -lactamase (ESBL)-producing *klebsiella pneumoniae*. Phytother Res 19:612–617
78. Hu Z-Q, Zhao W-H, Hara Y et al (2001) Epigallocatechin gallate synergy with ampicillin/sulbactam against 28 clinical isolates of methicillin-resistant *Staphylococcus aureus*. J Antimicrob Chemother 48:361–364
79. Zhao W-H, Hu Z-Q, Okubo S et al (2001) Mechanism of synergy between epigallocatechin gallate and b-lactams against methicillin-resistant *staphylococcus aureus*. Antimicrob Agents Chemother 45:1737–1742

80. Lian L-Y, Al-Helal M, Roslaini AM et al (2009) Glycerol: an unexpected major metabolite of energy metabolism by the human malaria parasite. Malar J 8:38
81. Teng R, Junankar PR, Bubb WA et al (2009) Metabolite profiling of the intraerythrocytic malaria parasite *Plasmodium falciparum* by 1H NMR spectroscopy. NMR Biomed 22:292–302
82. Holmes E (2010) The evolution of metabolic profiling in parasitology. Parasitology 137:1437–1449
83. Cheng K-W, Wong C-C, Wang M et al (2010) Identification and characterization of molecular targets of natural products by mass spectrometry. Mass Spectrom Rev 29:126–155
84. Pucheault M (2008) Natural products: chemical instruments to apprehend biological symphony. Org Biomol Chem 6:424–432
85. Harrigan GG, Brackett DJ, Boros LG (2005) Medicinal chemistry, metabolic profiling and drug target discovery: a role for metabolic profiling in reverse pharmacology and chemical genetics. Mini Rev Med Chem 5:13–20
86. Cravatt BF, Wright AT, Kozarich JW (2008) Activity-based protein profiling: from enzyme chemistry to proteomic chemistry. Annu Rev Biochem 77:383–414
87. Harvey AL (2008) Natural products in drug discovery. Drug Discov Today 13:894–901
88. Böttcher T, Pitscheider M, Sieber SA (2010) Natural products and their biological targets: proteomic and metabolomic labeling strategies. Angew Chem Int Ed Engl 49:2680–2698
89. Takenaka T (2001) Classical vs reverse pharmacology in drug discovery. BJU Int 88:7–10
90. Cassera MB, Merino EF, Peres VJ et al (2007) Effect of fosmidomycin on metabolic and transcript profiles of the methylerythritol phosphate pathway in *Plasmodium falciparum*. Mem Inst Oswaldo Cruz 102:377
91. Gardner MJ, Hall N, Fung E et al (2002) Genome sequence of the human malaria parasite *Plasmodium falciparum*. Nature 419:498–511
92. The Plasmodium Genome Database Collaborative (2001) PlasmoDB: an integrative database of the *Plasmodium falciparum* genome.Tools for accessing and analyzing finished and unfinished sequence data. Nucleic Acids Res 29:66–69
93. Fleischmann RD, Alland D, Eisen JA et al (2002) Whole-genome comparison of *Mycobacterium tuberculosis* clinical and laboratory strains. J Bacteriol 184:5479–5490
94. Watts JM, Dang KK, Gorelick RJ et al (2009) Architecture and secondary structure of an entire HIV-1 RNA genome. Nature 460:711–716
95. Pink R, Hudson A, Mouriés M-A et al (2005) Opportunities and challenges in antiparasitic drug discovery. Nat Rev Drug Discov 4:727–740

Chapter 6
Searching for Drugs That Target Multiple Receptors for Anthelmintics from African Natural Products

Timothy G. Geary and Eliane Ubalijoro

Abbreviations

ADME	Absorption, distribution, metabolism, and excretion
DMSO	Dimethyl sulfoxide
FLPs	FMRFamide-like peptides
FLP-GPCRs	FMRFamide-like peptide G-protein-coupled receptors
GPCR	G-protein-coupled receptor
HTS	High-throughput screening
IC_{50}	Half maximal inhibitory concentration
IP	Intellectual property
MTA	Material transfer agreements
NTDs	Neglected tropical diseases
p450	Cytochrome p450
pANPL	pan-African Natural Product Library
R&D	Research and development
SAR	Structure–activity relationships
SOP	Standard operating procedure
TK	Traditional knowledge
UB	University of Botswana
UCT	University of Cape Town

T.G. Geary (✉)
Institute of Parasitology, McGill University, 21111 Lakeshore Road, Ste-Anne-de-Bellevue, QC, Canada H9X 3V9
e-mail: timothy.g.geary@mcgill.ca

E. Ubalijoro
Institute of Parasitology, McGill University, 21111 Lakeshore Road, Ste-Anne-de-Bellevue, QC, Canada H9X 3V9

Institute for the Study of International Development, Peterson Hall, McGill University, 3460 McTavish Street, Montreal, QC, Canada H3A1X9
e-mail: eliane.ubalijoro@mcgill.ca

K. Chibale et al. (eds.), *Drug Discovery in Africa*,
DOI 10.1007/978-3-642-28175-4_6,

Table 6.1 Estimates of the prevalence of global nematode infections[a]

Nematode group	Estimates of infection
Ascaris lumbricoides	800,000,000
Trichuris trichiura	600,000,000
Hookworms	570,000,000
Lymphatic filariasis	120,000,000
Onchocerca volvulus	37,000,000
Dracunculus medinensis	10,000

[a]Estimates derived from [3]. It should also be noted that human infections with the pinworm *Enterobius vermicularis* and *Strongyloides* spp. are considered to be very common [4], but recent surveys of global incidence are lacking

6.1 Introduction and Background

Africa is home to most of the one billion people affected by neglected tropical diseases (NTDs). These infectious diseases "*are neglected because they persist exclusively in the poorest and most marginalized communities, and have been largely eliminated elsewhere and thus are often forgotten*" [1]. They perpetuate cycles of poverty and limit economic growth in the very countries that most face issues of food security, lack of sanitation infrastructure, effective local health care and dependence on foreign aid. Of the 17 NTDs identified by the World Health Organization [2], infections with helminths figure prominently, including the nematode infections (*Ascaris lumbricoides*, *Trichuris trichiura*, hookworm, lymphatic filariasis, onchocerciasis and dracunculiasis; Table 6.1). The driving motivation behind our approach has been the recognition that scientists working in areas most affected by such diseases must be empowered to lead the search for solutions to these infections.

An extremely important historical source of compounds used in chemotherapy has been natural products, including those obtained from botanical extracts and microbial fermentations. Importantly, Africa hosts ~30% of the world's flora, represented in ~72,000 species [5], yet relatively little investment has been made in exploiting this resource for the discovery of new drugs compared to that expended on global pharmaceutical research. Nonetheless, for centuries, highly active natural products have been widely used in medicine, especially for treatment of infectious diseases and cancer. Natural products are still being investigated by many research groups worldwide for their potential to affect protozoan and helminth parasites and other infectious agents responsible for neglected diseases, but such materials have been de-emphasized in most pharmaceutical companies over the past 10–20 years [6, 7]. The principal reason has been that pressures to replenish dwindling pipelines have pushed for quick wins through company mergers and acquisitions, as well as in-house synthesis of synthetic chemical libraries. Natural products are seen as less amenable to high-throughput screening compared to synthetic product libraries. They present contentious intellectual property landscapes and can be difficult to source and restock while having more complex structures is not always favourable to manufacturing [6]. Against this background, it is clear that integrated, networked efforts are needed to

Mass Drug Administration (MDA) Campaigns

Table 2
Major global helminthic disease control initiatives

Disease(s)	Public-private partnership	Major drug or control tool	Target date control by WHA resolution
LF	Global Alliance to Eliminate LF (GAELF)	Diethylcarbamazine and albendazole; ivermectin and albendazole	2020
Onchocerciasis	African Programme for Onchocerciasis Control (APOC): Onchocerciasis Elimination Program of the Americas (OEPA)	Ivermectin	2010
Schistosomiasis	Schistosomiasis Control Initiative (SCI): Partnership for Parasite Control (PPC)	Praziquantel	2010
Soil-transmitted helminth infections	Partnership for Parasite Control (PPC): Schistosomiasis Control Initiative (SCI): Human Hookworm Vaccine Initiative (HHVI)	Albendazole or mebendazole	2010
Ascariasis, trichuriasis, hookworm, schistosomiasis, LF, onchocerciasis, and trachoma	Global Network for Neglected Tropical Diseases (GNNTDC)	Ivermectin or diethylcarbamazine; praziquantel; albendazole or mebendazole; and azithromycin	2015

Fig. 6.1 MDA campaigns for human helminth infections. Figure obtained from [3]

accelerate drug discovery from natural products. Our primary goal in this regard is to use new assay technology to push the characterization of African natural products to become highly valued intellectual property, with nematode infections as the initial therapeutic target. The concept was supported with a Phase I award from the initial Gates Grand Challenge Explorations programme in 2008, followed by Phase II funding from this programme, supported by the Grand Challenges Explorations programme in concert with Grand Challenges Canada and the Canadian Institutes for Health Research in 2011. We used this funding to create a research network involving experts in African natural product chemistry and drug discovery and development technology to address human nematodiases common in developing regions but essentially absent from wealthy populations. Nematode infections have attracted discovery investment from the animal health industry as major problems persist in livestock and companion animals, but they are rarely on the radar of global pharmaceutical companies or North/West governments.

Our focus is to discover drug candidates that can be developed and produced in Africa to meet local human health and veterinary needs. Many nematode infections are currently controlled through mass drug administration (MDA) campaigns that rely on donated drugs (Fig. 6.1) [8]. While of extraordinary value, sustainability of these programmes cannot be guaranteed, and strategies to address the possible selection and spread of drug-resistant parasite strains remain undeveloped. In addition, many gaps remain in the anthelmintic arsenal for the treatment of human infections [9], and the near-term pipeline contains few examples of molecules that could easily be adapted for use in humans in resource-limited areas [10, 11]. Against this background, we propose the development of a drug discovery programme harnessing African biodiversity to stimulate local markets for efficacious and safe locally produced anthelmintics as a novel model for innovation and economic growth for Africa. Artemisinin production in Africa has proved to be successful from the agricultural cultivation of *Artemisia* to the crystallization of the active ingredient against malaria. More models of locally manufactured drugs from natural products that enhance local solutions to local problems are needed [12]. Plant genetic resources form the basis of a dynamic, diverse and adaptable agriculture as well as a source of traditional medicine and valuable plant products; they have fundamental roles in health, food security and economic development. Primarily using African natural products, our testing stream has focused on neuropeptide receptor-based screening in recombinant yeast as part of an integrated anthelmintic discovery effort.

This unique system can provide a pull-through focus on neglected tropical diseases to build new local markets for human and veterinary drugs, while the lead molecules should attract investment in African pharmaceutical and biotechnology industries based on the new intellectual property (IP) generated.

This research approach could be duplicated for other disease indications in Africa, which provides added value to it. However, our primary aims are (1) discovery of new anthelmintic leads, (2) development of a transnational chemistry resource of purified African natural products, (3) development of bioassay-guided fractionation protocols to identify receptor ligands in extracts of African plants and microbial fermentations, (4) support the economic and ecological viability of African biodiversity preservation efforts to alleviate diseases of poverty and improve income security, and (5) promote African scientists as leaders in antiparasitic drug discovery and development. The long-term outcomes of this project will enhance the resilience of local environmental and innovation systems by directly addressing local health issues that affect the livelihood of populations.

6.2 Mechanism-Based Screening with Recombinant Microbes: A New Way Forward

The transition from whole organism to target-based strategies transformed the process of drug discovery from a chemically and labour-demanding, low-throughput endeavour to one that requires extremely small amounts of chemicals and can operate at an astonishingly high throughput. This transition was made possible by the development and implementation of expensive and technically sophisticated, miniaturized robotic assay platforms that sharply reduced the chemical and labour requirements for primary screening. It is arguable whether this transition has enhanced the success of the enterprise, but it is nonetheless clear that the future of drug discovery will continue to rely on sophisticated technology platforms, whether for mechanism-based or higher level organism-based assays (such as recent high-throughput antimalarial screens against parasites in culture; [13, 14]). For many reasons, it is not feasible to screen with parasitic nematodes in high-throughput systems, so mechanism-based approaches will continue to have value for these indications [15, 16].

Recombinant microbes as screening tools

Advantages	Disadvantages
Minimize equipment costs	Microbial permeability barrier
Minimize training costs	Not always useful (no useful phenotype; inexpressible)
Simple, robust endpoint	
Basic selectivity included	
Platforms available (*yeast, E. coli*)	
Multiple kinds of targets accessible	

Fig. 6.2 Characteristics of recombinant microbe screens. Benefits and disadvantages of recombinant microbes as a screening platform. Although microbial permeability barriers may affect the penetration of xenobiotics into these cells, similar permeability barriers are encountered in parasitic nematodes. It must be recognized that not all drug targets have a convenient phenotype for screening in microbes, and some proteins may be refractory to functional expression in these organisms

The need for expensive and sophisticated equipment, enabling multiple assay formats and read-outs, precludes the adoption of cutting-edge drug discovery operations in resource-limited areas without reliable power resources or ready access to (expensive) technical support personnel and parts. Facing similar obstacles in screening for new antiparasitic agents for veterinary indications, a novel platform employing recombinant microbes as a screening platform was developed in the early 1990s [17, 18].

Based on remarkable advances in genomics, proteomics and cell physiology, model organisms such as *Escherichia coli* and *Saccharomyces cerevisiae* were engineered to express heterologous proteins in a functional context. Culture and reporter gene systems could be modified to create conditions in which the function of the heterologous protein is required for microbial growth; in this situation, growth can be used as an indicator of drug effects on the parasite protein [16–18]. Considerations of the value of this general approach are outlined in Fig. 6.2. The system has advantages for operation in resource-limited settings, especially the fact that a single format (microbial growth) can be used to screen multiple kinds of targets (enzymes, receptors, ion channels, etc.) with a single, easy-to-measure read-out. It can operate with minimal and very simple laboratory equipment. Because viability of the recombinant microbe requires the normal function of many proteins, non-selective "biocides" are not detected. Selectivity can be easily assessed by growing the microbe in conditions that do not require the function of the parasite target protein, or in antiparasitic screening, by comparing the effects of the drug on microbes expressing the host homolog. Selectivity can then be proven in direct assays of target protein function.

The primary advantage of mechanism-based screens lies in their ability to focus chemistry and biological follow-up resources on the most promising hits, those with a defined mechanism of action. The ability of mechanism-based screening to detect actives that might otherwise be "masked" by more frequent and higher-titer biocides makes them very attractive for natural product screening. The space represented by natural products has not been sampled by such assays though they have been successfully screened using synthetic compounds [19] Anthelmintic screens using whole organisms typically identify very large numbers of hits, the vast majority of which are inactive in follow-up assays and are difficult to prioritize for additional investment. In contrast, hits with a known mechanism of action can be investigated in hypothesis-based experiments, more quickly leading to go/no-go decisions for development [16].

6.2.1 Choice of the Target

The genesis of this project lies in the recognition that mechanism-based approaches to anthelmintic discovery are a valuable tool for new antiparasitic discovery [15–20] and that the nematode nervous system contains ideal targets for such an approach, in particular, receptors for a family of neuropeptides related to the invertebrate peptide FMRFamide (FMRFamide-like peptides, FLPs) [21, 22] (Fig. 6.3). FLPs are exceptionally abundant and widespread in invertebrate phyla but are very rarely found in mammals and other vertebrates. A variety of studies demonstrated potent and profound bioactivity of FLPs in nematodes and the essential nature of the –RFamide moiety for bioactivity [21, 23]. It was also apparent that nematodes express a large number of GPCRs that recognize FLPs [22, 24–26]. Based on the apparently conserved receptor structures associated with recognition of the RFamide component, we hypothesize that non-peptide compounds that mimic the chemical space of RFamide (shape, charge, lipophilicity) should interact with many or most FLP-GPCRs with similar affinity. The driving force for this project is to discover non-peptide ligands which act on more than one receptor in a parasitic nematode FLP-GPCR. Drugs based on such a ligand are expected to exhibit lethal antiparasitic activity and to less readily select for receptor-mediated resistance in treated populations than drugs that act on single receptors.

GPCRs have been targeted in innumerable drug discovery projects in the pharmaceutical industry, including receptors that recognize neuropeptides [27]. Non-peptide ligands are of crucial importance for therapeutic exploitation of these targets, since peptides per se exhibit poor pharmacokinetic properties; the best example of non-peptide ligands for peptide receptors is found in the opiate alkaloids and their derivatives, which act on GPCRs that respond to endogenous enkephalin/endorphin neuropeptides [27]. The pharmacology of mammalian peptidergic GPCRs is very complex, with many peptides acting on multiple subtypes with distinct physiological roles. Attaining exquisite sub-type selectivity

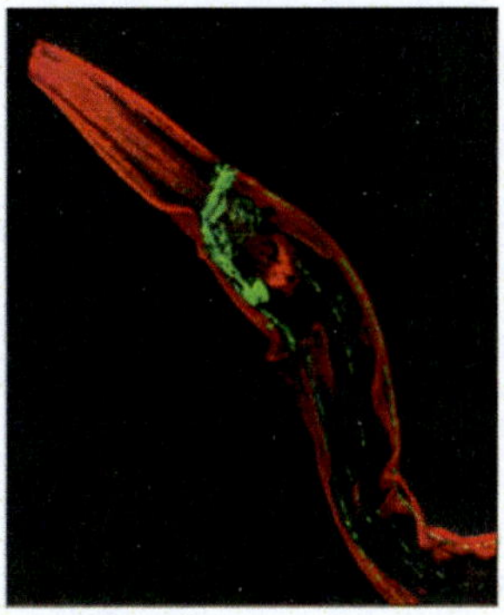

FMRFamide-Related peptides (FaRPs)

F M R F – NH_2
Y L
I
V

***FaRPs mediate key functions in helminths & insects**
(eating, moving, egg laying)
***Nematode FaRPs are active in insects & flatworms**
(conserved ligand-receptor recognition features)
***FaRPs present in all nematodes**
-same peptides found in all worms
***FaRPs present in all arthropods**
***FaRPs present in flukes, tapeworms**
***Very rare in vertebrates**

Fig. 6.3 Pharmacological characteristics of FMRFamide-like peptides. The *inset* shows a specimen of the free-living nematode *Panagrellus redivivus* stained with phalloidin-rhodamine (actin in muscle—*dark grey*) and anti-FMRFamide-fluorescein (FLP-containing nerves—*light grey*) and observed under confocal laser microscopy; courtesy of A. Maule, Queen's University Belfast, Northern Ireland, UK. Peptides in this class are characterized by a C-terminal pentapeptide comprised of F (or another aromatic residue)–M (or another aliphatic residue)–R–F–amide. Many kinds of experimental evidence, summarized here, suggest that receptors for FLPs are ideal anthelmintic targets

may be a prerequisite for therapeutic utility, but this is not an issue for chemotherapy of helminth infections; in contrast, we wish to discover ligands that affect many nematode FLP-GPCRs (while of course lacking affinity for human GPCRs).

6.2.2 Screening Strategy

In the current case, we selected a set of yeast (*S. cerevisiae*) strains that functionally express invertebrate peptidergic GPCRs. Work on developing a yeast-based assay

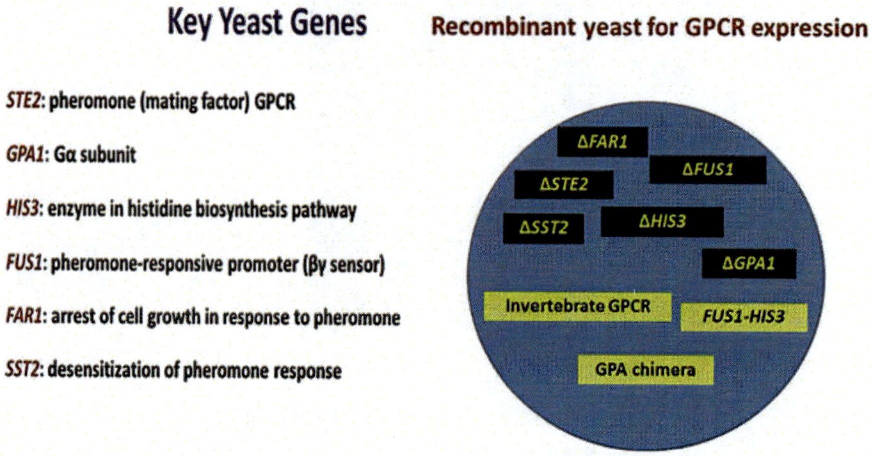

Fig. 6.4 Yeast strains. Constructs used in these experiments are characterized by changes in genes important for GPCR signalling as shown

for screening parasite peptidergic GPCRs originated at The Upjohn Company in the early 1990s [TG Geary, unpublished observations; 24, 25, 28], based on previous work and licensing of a set of yeast strains for this purpose from Cadus, Inc. [29, 30]. Yeast strains were modified for use with GPCRs from the free-living nematode *Caenorhabditis elegans* and the model arthropod *Drosophila melanogaster*. We see little risk at this point in using the nematode model organism *C. elegans* in our anthelmintic research despite it not being a parasitic nematode. Nematode GPCRs are highly conserved, and no nematode genome has been characterized as well as *C. elegans* [31]. Thus, the choice of target favours identification of wide spectrum drug candidates that would span parasitic nematodes from different classes. Leads can then be further characterized in model parasitic in vitro and animal systems. Because of the availability of so many tools and knowledge for genetic engineering using *C. elegans* compared to parasitic nematodes, we have been able to initiate screening quickly to maximize time to results within the time frame of the current funding. Nematode GPCRs matched to known FLP ligands were of primary interest; the system was also modified to express chimeric yeast-nematode G_α subunits to enhance receptor–G-protein interactions. A description of the yeast system is shown in Fig. 6.4.

The assay is based on the existence of a peptide-GPCR system in yeast, normally functioning for detection of the mating pheromone, a 39-amino-acid peptide. The gene encoding this receptor is knocked out, replaced with the target GPCR of interest. In addition, the yeast gene encoding the G_α subunit is replaced with a chimeric gene comprising the yeast sequence except for the terminal pentapeptide, which is derived from mammalian or invertebrate G_α sequences. Genes that down-regulate the mating response to receptor activation are also deleted. In wild-type yeast, the signal pathway following pheromone receptor activation is transduced by interaction of the βγ complex with the FUS1 transcription factor. In the assay strain, the wild-type FUS1 locus is deleted and is replaced by a construct in which FUS1 is coupled to a reporter gene. In our constructs, we chose this reporter to be HIS3, an essential enzyme for histidine biosynthesis. The strain is deleted for the wild-type HIS3 locus so that it can grow in culture in the absence of exogenous histidine only if the heterologous GPCR is activated by an agonist. A very sensitive way to monitor growth is provided by the vital dye Alamar blue, which undergoes colorimetric and fluorometric changes in the presence of living cells. Changes in Alamar blue colour in yeast growing in 96-well plates can easily be detected by eye at the lab bench and can be quantified in a fluorimeter (520 nm excitation/560 nm emission) if so desired.

Running the screening assay in culture medium lacking histidine permits the sensitive and facile detection of agonists, as even slight increases in growth against a null background can be detected. This property also allows multiplexing of the assay strains, with as many as ten individual strains added to each well. Detection of growth in a well is followed by dereplication, with each strain present in the mixture assayed separately.

Based on physiological observations of the profound and potent activity of FLPs in parasitic nematodes [21–23] and the lack of profound phenotypes found in some RNA interference assays for peptide precursor genes in *C. elegans* [TG Geary, unpublished observations], we prioritized the discovery of non-peptide FLP agonists over antagonists. This prioritization is consistent with peptide pharmacology in other systems [26]. For instance, the best-described peptide system in pharmacology, the opiates, is characterized by agonists that are profoundly active, but antagonists that are almost without detectable effects except in addicts or overdose cases. In many peptidergic systems, antagonists seem to do little to disrupt normal physiological function at "steady state". There are exceptions: angiotensin II receptor antagonists in hypertension, oxytocin antagonists in premature labour and vasopressin antagonists for enuresis [27]. Non-peptide agonists, like morphine, are therapeutically active because they persist at the receptor far longer than the native peptides. Prolonged stimulation of nematode GPCRs by a non-peptide FLP agonist is expected to be highly detrimental to the parasite, leading to elimination from the host.

Historically, it has been proven that it is easier to discover non-peptide antagonists than agonists; this may reflect the fact that there are many more sites at which binding of a ligand can disrupt receptor function than those that can activate it [27]. From the discovery perspective, it should be relatively common to find a receptor antagonist that binds far from the ligand binding site, but it would not necessarily be easy to find one that would bind similarly to multiple FLP receptors. However, if a "plug" for the RFamide binding site were found, it could exhibit activity against multiple receptors, an attractive option. Modelling approaches may be able to identify compounds that act in this way and will be investigated. To fully exploit the potential of the system, the assay will be run in antagonist mode after the completion of the agonist screens.

Operation in the antagonist mode involves growing the yeast strain in the presence of the endogenous FLP matched to the expressed GPCR and measuring the ability of chemicals from the library to reduce ligand-induced growth. This assay cannot be run in multiplex. Selectivity is assessed by testing the ability of a hit compound to reduce growth of the same strain in the presence of histidine; a substance that inhibits growth in histidine as much as in the ligand cannot be a selective receptor antagonist. As for the agonist mode, standard operating procedures (SOPs) were established previously for the antagonist mode and are available for adaptation. Alternative reporter genes, such as green fluorescent protein or beta-galactosidase, can also be used to detect antagonist-induced inhibition of ligand activation of the GPCR, but growth is more useful in resource-limited settings [29, 30].

Although many FLP-GPCRs have been matched in reverse pharmacology experiments [22, 24, 26], we selected a subset of the matched receptors for inclusion in the screen. Although we seek promiscuous FLP receptor ligands, it is prudent to start with a relatively broad net. We therefore chose 3 *C. elegans* FLP-GPCRs for inclusion in the multiplex agonist assay: Y58G8A.4, the receptor for FLPs encoded on the *flp*18 precursor gene [32]; C16D6.2, also a receptor for flp-18 peptides [24]; and the recently matched receptor T19F4.1 (recognizes AF2, KHEYLRFamide; TG Geary and ER Lancheros, unpublished data). These two FLPs cause potent and profound changes in neuromuscular physiology in *Ascaris suum* [21–23] and in cut preparations of *C. elegans* (TG Geary and ER Lancheros, unpublished observations), making them excellent targets for the discovery of non-peptide ligands. The inclusion of two distinct GPCRs that recognize the same ligands, as well as a GPCR that recognizes a distinct FLP, allows us to assess promiscuity at two different levels simultaneously.

As a control in the agonist format, a strain of *S. cerevisiae* functionally expressing a *D. melanogaster* allatostatin receptor [33] is used to prove selectivity; allatostatin is structurally distinct from FLPs, and we consider it very unlikely that a non-peptide FLP agonist would also activate this insect GPCR. The screen is initiated by the addition of a mixture of each strain (300–3,000 log phase cells per well of a 96-well plate) in minimal medium; an advantage of the system is that results are independent of inoculum in this range (data not shown). The total volume of 200 μl contains various amounts of test chemicals in 1% DMSO (final concentration); initial test concentration is 10 μM. Plates are kept in a 20°C incubator for 48 h, at which time an aliquot of Alamar blue is added and fluorescence measured 3 h later. Another advantage of the format is the ability of the assay to be conducted on a laboratory bench as long as the ambient temperature remains within the range tolerated by yeast (~15–28°C in this case). Using only standard pipettors, it is generally feasible for 10–100 plates to be run per day (data not shown) with two researchers engaged. It is important to note that a similar assay was developed and run at Pharmacia/Pfizer Animal Health (Kalamazoo, Michigan, USA) from 2002–2008, and SOPs have been thoroughly vetted. It is also important to note that use of the Cadus technology was negotiated to permit operation for the discovery of anthelmintic lead compounds for possible application to human chemotherapy, and the collection of nematode GPCR yeast strains was donated without restriction on use for this purpose by Pfizer Animal Health.

6.3 Chemistry

Although many important drugs have roots in natural products, screening of collections of pure compounds derived from botanical or microbial sources, or of extracts from biological sources, has fallen into disfavour in the pharmaceutical industry as noted above [6, 7]. This shift occurred for many reasons, including difficulty in synthesizing complex natural products to provide enough material for therapy, poorly understood variations in production by plant or microbial materials and the lack of compliance of many natural products with the physical–chemical "rules" associated with oral bioavailability [34]. However, the proposer of those rules specifically noted that natural products were outliers, presumably because evolution has selected secondary metabolites specifically for their bioactivity. In addition, the development of combinatorial chemistry methods to generate chemical diversity from relatively simple building blocks in a high-throughput strategy suggested that chemists could populate the areas of chemical space previously restricted to natural products. Whether we have been able to sample the bioactive chemical space exemplified by natural products with strictly synthetic collections remains to be proven.

Natural products and their synthetic derivatives are richly represented among antiparasitic drugs, including anthelmintics [35]. It is therefore logical to assume that the discovery of new anthelmintics has been hampered by the diminished availability of natural products for HTS operations. The combination of high-value mechanism-based HTS using nematode FLP-GPCRs with collections of novel natural products thus offers a promising and too rarely explored route to the discovery of new lead anthelmintics. Drugs derived from them should succeed in clinical and field use for many years before encountering resistance. We strongly believe that the chances for finding solutions for the treatment of neglected diseases will be greatly improved if the effort involves scientists from affected regions at the most basic levels. In that regard, African natural product chemists have an impressive history of generating novel compounds, particularly from botanical sources [36–38], but characterization of these compounds in systematic HTS campaigns for antiparasitic drugs has been rare. This opportunity was a driving motivator for building the project team.

The lack of a centralized resource for access to pure natural products derived from African biodiversity is a challenge for systematic drug discovery efforts. This lack makes the inclusion of natural product extracts in our screening campaign an advantage, expanding the chemistry space explored, despite the additional steps that arise to further characterize potential leads from extracts. The potentially high value of such a collection for licensing by pharmaceutical companies (in a process controlled by the African partners) is an economic incentive for its creation; it is also better for HTS operations to run in connection with a centralized chemistry resource, especially if HTS efforts are to expand around the core already in place. The combination of a mechanism-based HTS platform suitable for operation in

Table 6.2 Chemical matter resources

Partner	Product	Number	Availability	Compounds to follow
McGill	Synthetic compounds	~120,000	Available	Yes
	Natural products	~2,000	Available	Yes
UCT/RSA[a]	Botanical, marine organisms and microbial fermentations	~21,000	Available	Yes
UB[b]	Natural products	~900	Available	Yes
Univ. Mich.[c]	Microbial fermentations	~22,000 extracts	Available	Yes
GIBEX[d]	Plant extracts	~2,000 initially	Available	Yes
pANPL[e]	Natural products	~200 per partner lab	Potential	Yes

[a]UCT/RSA refers to collections available at the University of Cape Town and elsewhere in the Republic of South Africa
[b]Pure compounds available from the collection of B. Abegaz at the University of Botswana, Gaborone, Botswana
[c]Extracts of marine microbe fermentations in the laboratory of D. Sherman, University of Michigan, Ann Arbor, MI, USA
[d]See [39]
[e]pan-African Natural Products Library, to be formed when infrastructure support is available

resource-limited settings with a new collection of structurally diverse natural products in Africa offers great benefits only if we can efficiently screen the library; this remains a challenge [38] (see below).

An initial estimate of the number of purified, structurally characterized natural products that could be made available for HTS in Africa was ~10,000 (unpublished observations), but this was overly optimistic. As we pursue the genesis of a pan-African Natural Products Library (pANPL) that houses the collection, and expand the number of compounds available, alternative approaches to provide chemical matter for screening have been advanced (Table 6.2). To fully exploit the potential of the yeast screening system for the discovery of new anthelmintics, we are currently screening non-African sources of natural products and synthetic chemicals held at academic centres, including McGill University [40]. As another alternative, we elected to pursue bioassay-guided fractionation of plant extracts and microbial fermentations provided by collaborating scientists; this policy will add bioactive molecules to the pANPL (see below).

Although an aggressive medicinal chemistry campaign is beyond the scope of operations funded by the Phase II award, we will provide preliminary characterization of the chemical space around a hit by acquiring structurally related compounds by purchase or request and also by subjecting hits to cytochrome P450-catalysed biotransformations [e.g. 41] in an ongoing project at the University of Cape Town; activity of metabolites will be used to provide initial information on structure–activity relationships (SAR) that can be used to request additional funding from other sources for a medicinal chemistry campaign and additional early development work.

6.3.1 Secondary Assays: Moving from Hit to Lead

Hits from the screens accrue added value by evaluation through in vitro and in vivo antiparasitic testing and preclinical pharmacology as well as through medicinal chemistry. Decisions about hit prioritization for additional studies will be made by open consultation among the partners in the project. Following confirmation and titration of activity in the yeast screen, the chemical structure of the hit will be verified and additional supplies sought as necessary to enable a thorough and timely evaluation of its potential as an anthelmintic candidate. Simultaneous investigations will be made into the biological and pharmacological properties of the hit following initial chemical triage. The focus in these experiments will be to measure activity of the hit against representative nematodes in culture and in animal models of infection, coupled with measurements of pharmacological properties related to absorption, distribution, metabolism and excretion (ADME), especially with regard to bioavailability and metabolic stability.

6.3.2 Chemical Prioritization

Compound prioritization will be pursued based on literature and patent searches, evaluating structural similarity to compounds with known pharmacology. This will eliminate the least interesting series and select others to progress with a view to the early identification of liabilities, which will be addressed by further testing through chemical modification or purchase of close-in analogues to establish SAR as resources permit.

Botanical and microbial extracts will be prioritized for bioassay-guided fractionation based on activity in the yeast assay format. Extracts with selective activity related to activation or inhibition of FLP receptors will be considered for fractionation at the site of screening. Preliminary experiments have demonstrated that standard fractionation procedures are compatible with the yeast assay format (data not shown). The yeast system provides a sensitive and relatively rapid assay to monitor separation and purification of active molecules from a mixture. We do not anticipate storing extracts *per se* as part of this project, but will archive pure compounds as they become available; these would be an important source for expansion of the nascent pANPL.

Specific criteria for prioritization include:

- Drug-like chemotypes, structurally unrelated to known anthelmintics, will receive the highest priority.
- Hits with a proven pedigreed structural template as validated compounds with in vitro potency and selectivity in other assays. Hits with evidence of initial SAR from their respective screening libraries. This would ensure that substitution around the core scaffold will present an opportunity to optimize potency and

modify physicochemical properties, impacting key in vivo properties such as stability to drug metabolism and bioavailability.

- Hits that possess generally acceptable physicochemical properties, including solubility and an ADME profile suitable for in vivo testing.
- Hits with well-developed synthetic protocols for further analogue generation, which will facilitate broader SAR studies, or natural products that offer a feasible strategy for synthesis or semi-synthetic modification.

These criteria would equally apply for prioritization for artemisinins, avermectins and depsipeptide classes, thus having a wide applicability to other antiparasitics programmes.

6.3.3 Biological Prioritization

Assuming adequate supplies are available, hit compounds will be tested in additional biological and pharmacological assays to identify the most promising for additional investment. The initial steps will involve testing in whole organism assays, followed by evaluation for pharmacokinetic suitability. For this assay, a "hit" is defined as a compound showing activity against $\geq$1 FLP receptor in yeast and inactivity against the allatostatin receptor at a concentration of 10 μM. This activity must be confirmed. Potency will be an important variable for prioritization in this assay, with compounds active at concentrations $<$1 μM being of greatest interest. Results from secondary tests will contribute to prioritization as follows:

- Activity against whole or cut *C. elegans* preparations at concentrations similar to or lower than those active in the yeast assays; potency is a variable for further prioritization.
- Activity against intact adult parasites in culture (*Brugia malayi* and *Heligmosomoides polygyrus* or *Nippostrongylus brasiliensis*), as judged by reductions in motility; quick onset of action and potency are variables for prioritization.
- Compounds active against adult parasites in culture with IC_{50} values $<$1 μM will be subjected to preliminary ADME assays to support testing in rodent models of nematodiases. Bioavailability and plasma half-life will be measured parameters for prioritization for in vivo analysis in birds infected with *B. malayi* and/or mice infected with *H. polygyrus*; compounds with inadequate bioavailability or metabolic stability will be submitted for chemical optimization work to generate active analogues with suitable pharmaceutical properties prior to in vivo evaluation.
- Compounds with suitable pharmaceutical properties and activity in whole organism assays in culture will be submitted for evaluation in animal models as noted above; these models are in place at McGill, but we intend to establish them in Africa to support this project. A compound exhibiting activity in one or both of these systems will be at a highly advantageous stage for exploring partnerships for development for human and/or veterinary applications.

6.4 Establishing an Effective IP Strategy

Concerns over the perception of "biopiracy"—the practice of Western companies developing patented medications based on the biodiversity and traditional knowledge of developing nations—have led some countries to block access to genetic resources and associated traditional knowledge [e.g. 42]. Such policies may have the unfortunate consequence of hampering drug discovery R&D that draws on and benefits local biological resources. By limiting access to its biological and human resources, Africa would lose not only the opportunity to attract pharmaceutical R&D investment based on its local human and biological wealth, but also the prospect of creating a phytopharmaceutical industry that would enable its traditional knowledge holders and ethno-scientists to find innovative, low-cost solutions to pressing local health problems.

Our project is intended to discover new compounds that have utility for the treatment of human nematode infections. This work may be divided into the biological evaluation of compounds and the chemical preparation and collection of compounds, including the synthesis of derivatives or analogues, for evaluation. The first principle of this collaboration illustrates the philosophy of the entire project: that ownership of IP rights should be retained to the maximum extent possible on the African continent. The partners recognize that the tangible economic value of the project lies primarily in the chemistry derived from it. Accordingly, the natural product chemistry to be screened during the project is primarily derived from African sources. Hits identified from screens in non-African laboratories will be transferred for follow-up evaluation and medicinal chemistry to African chemistry laboratories to the maximum extent possible.

Recognizing that the available chemical supply of African natural products in pure form in a centralized facility like pANPL is insufficient to sustain an HTS campaign, we have also undertaken HTS on collections of natural products and synthetic chemicals available in academic institutions in Africa and elsewhere, as described above. An essential component of this process is ensuring that the chemistry resources to be screened in non-African centres are non-proprietary and free from downstream IP entanglements; added value from these substances, without additional medicinal chemistry efforts, can be realized only through the use of patents (as opposed to composition of matter claims).

6.4.1 IP Related to Biology

IP for the screens used to evaluate compounds relates to the recombinant microbial systems used. Downstream assays of compounds against whole organisms and in animal models are entirely in the public domain. In the case of recombinant microorganism-based screens, IP may be assigned to the microbial systems themselves or to the genes used to create the screens. In the case of this project, the genes currently involved are (1) in the public domain or (2) the subject of abandoned patents and, so, now in the public domain. In the case of the microbes used to express the heterologous genes, these are (1) in the public domain, (2) the subject of IP held by outside parties (viz., Cadus Corp.) or (3) donated without claim or restriction by outside parties (viz., Pfizer, Inc).

In this context, there are no background IP considerations specific to McGill University, to Pfizer, Inc., or to other academic collaborators. With regard to the IP relating to the system for expression of heterologous G-protein-coupled receptors in *S. cerevisiae* owned by Cadus Corp., the parties involved in this project have agreed to operating principles that enable use of the technology in Africa to discover drugs for the indication of human nematode infections. Should compounds be discovered using Cadus technology that can be licensed for other applications in human medicine or for parasite infections important to animal health, or for any non-parasite indications, the licensing/royalty arrangement will include fair compensation as determined by good-faith negotiation for the use of the yeast technology. If targets from non-neglected diseases of concern in Africa are developed for screening, rights for discoveries made using the Cadus technology will be negotiated prior to the introduction of such screens and are the responsibility of the African investigator pursuing those applications.

6.4.2 IP Related to Chemistry

Chemistry derived from random (i.e. not directed by TK) collections of locally generated compounds, botanical material, microbial fermentations or marine organisms will generate IP rights that will be divided among the responsible chemist(s) and the screening and chemistry centres based on relative contributions, including the generation of semi-synthetic or synthetic derivatives. The distribution will reflect the contributions of the respective parties, demands by the chemist's institution for IP rights and the potential assignment of partial IP claims to the screening centres. Income derived from licences or royalties by the project will be distributed according to agreements which will be developed in consideration of ongoing efforts in this area [e.g. 43, 44].

6.5 Organizational and Operational Issues

A common library of purified natural products for screening requires management policies that serve rather than hinder networked research. Policies must include ways to ensure fair IP incentives that encourage compound sharing across a highly diverse continent. Support for acquisition of more compounds and continued chemical validation of deposited compounds will increase the value of the library over time. A plan related to negotiation of costs for use and disbursement of income will be constructed as part of this project in consultation with similar projects elsewhere in the world. Generating sustained operating funds for the chemistry and screening centres may be the sternest challenge to overcome. It will require recruitment of new researchers in Africa, capacity building for infrastructure and training of researchers in grant writing. Spin-off/growth opportunities could use the chemical collection and extracts to engage targets with economic value outside of Africa, such as insecticides, plant parasitic nematocides, antibiotics and antifungals, in association with the nematode projects but supported through separate venture funding. This strategy could provide a more rapidly developing revenue stream for sustaining work on targets of primarily African interest, but would also be of direct local benefit.

Nonetheless, discovery of new antiparasitic drugs remains the primary criterion of success. In the long term, success will be measured by the ability of this system to attract sustainable sources of revenue for continued operation with expansion and local training based on the economic value of the chemistry and the screening capacity and expertise. This prospect is based on the belief that economic value is embodied in providing external parties access and expertise in novel chemical space for drug discovery in mechanism-based screens. The platform must be adapted to local conditions for operation and improvement and be amenable to expansion to other African centres. Local leadership is required to develop this approach to ensure access to innovations from local, regional and international sources, while respecting TK and promoting equitable community profit sharing. For new graduates in areas important for R&D, engineering, regulatory and commercialization aspects of the pharmaceutical industry, this can represent a critical alternative to becoming a brain drain statistic.

The essential initial step was to recruit partners in Africa. The first were located at the University of Botswana in Gaborone (UB; Profs. Berhanu Abegaz and Kerstin Marobela) and at the University of Cape Town (UCT; Prof. Kelly Chibale). Profs. Abegaz and Marobela, along with Dr. Eliya Madikane from UCT, visited McGill in January 2009 for a meeting to set the basic organizing principles of a project to establish pANPL, to collect, manage and supply natural product compounds for screening in novel assays to discover new anthelmintics. Initial training in assay performance was also conducted during this visit. The meeting was followed by a trip to Botswana and South Africa by V. Muise to deliver the recombinant yeast and reagents needed for screening and to continue training in both locations in screen performance. This trip resulted in the establishment of a functional screening laboratory at UB. Additional natural product chemists from Africa were recruited to join as founding members of pANPL, which was formally organized at a meeting at UB in April 2009. This meeting resulted in the establishment of a founding board, including the following: B. Abegaz; K. Marobela; E. Dagne, Addis Ababa University; B.T. Ngadjui, Yaoundé University; J. Magadula, Muhimbili University of Health and Allied Sciences; S. Khalid, University of Khartoum; J.O. Midiwo, University of Nairobi; K. Chibale; B. Mudenda, Bristol-Myers Squibb, USA; E. Ubalijoro; and T.G. Geary. The members drafted a preliminary policy on IP issues to guide further development of the consortium.

It was agreed that, since the screens and the screening concepts are either in the public domain or the freely donated property of non-member laboratories, institutions or companies, IP resulting from this work, whether based on novel use or novel composition of matter, would reside with the chemistry contributors. The project was approved and endorsed by the UB administration, which committed space for work enabled by a Phase II award.

An advisory board composed of African scientists who have experience in the pharmaceutical industry was established to help the consortium keep on milestone targets, decide on compound progression, and, serving as "honest brokers", help negotiate licensing/partnership agreements for compounds of potential commercial interest. Their unique perspective as African scientists with extensive experience in the pharmaceutical sector built trust among the collaborators that their interests will be fairly represented in any commercial negotiations. The board is composed of Dr. P. Atadja, Novartis; Dr. B. Mudenda, Bristol-Myers Squibb; Dr. B. Nare, Scynexis; Dr. E. WoldeMussie, Pfizer; Dr. B. Mekonnen, Hager Biosciences, Bethlehem, PA, USA; and Dr. D. Alemayehu, Columbia University/Pfizer. The experience of the advisory board will be leveraged to ensure that progress in Phase II attracts downstream investment for early development, scale-up and possible licensing or local development of new anthelmintics.

Initial work in Phase I utilized ~100 pure natural products from Prof. Abegaz for proof of concept. This included confirmatory retests and concentration titrations of active compounds. To maximize the appeal of the collection to potential partners from the pharmaceutical sector, it was agreed that the pANPL would be restricted to pure compounds and that the highest priority for screening would be placed on these compounds. However, as noted above, we will also use the screening platform for bioassay-guided fractionation and purification of active compounds from extracts and fermentations. pANPL members also established the conceptual basis for an IP policy to proportionately direct revenues to submitting partners and the screening centres. While it was agreed that the basic structure of the IP and the Materials Transfer Agreement (MTA) policies were to attract and support chemical submission to pANPL, these policies have yet to be finalized.

6.6 An Innovation Network Is Critical to Support This Approach

Our goal is to assemble a collection of African natural products, with the long-term aim to develop a programme that will have sufficient value to convince African scientists, global health granting agencies and venture capitalists to invest funding into a drug discovery and development programme, to bring long-term sustainability for growth, maintenance and further utilization of pANPL.

The pilot partnership established a network of scientists using IP from pharmaceutical and biotechnology companies and funding from Canadian and US agencies

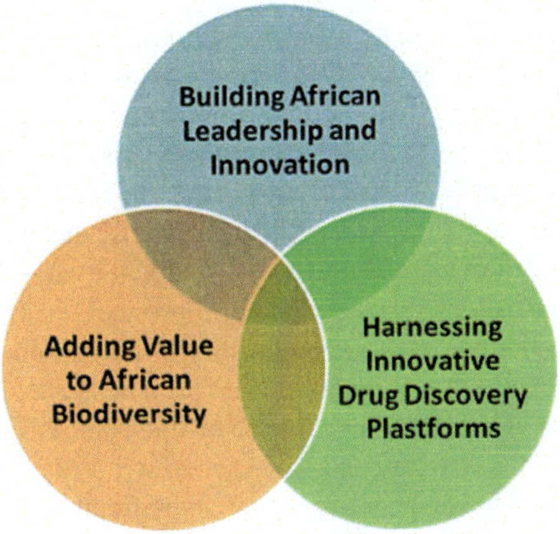

Fig. 6.5 Our approach to Integrated Innovation™. See [45]

to provide an opportunity to discover new drugs via Integrated Innovation™ (Fig. 6.5). We share information within the partnership that spans social, business and scientific/technological innovation. The areas covered are network building, leadership development, private public partnership and technology transfer, all evolving from the implementation of screening platforms that strengthen Africa-wide collaboration in natural product chemistry to address local health issues.

The current exploitation of African biodiversity for drug discovery, especially for indications that are indigenous to the continent and not shared with Western countries (e.g. parasitic infections), has been largely confined to a simple model in which resources from Africa are mined and evaluated in developed countries, with minimal participation or direction from African collaborators. To enable progress in antiparasitic drug discovery in the countries that suffer most from them, African scientists and institutions must be empowered to lead the process. To develop a more accurate global health perspective, science institutions in developed countries must be made aware of the significant contributions African scientists can make to combat major health problems using their unique resources. Aligning South-North partnerships can grow sustainable natural product-based drug discovery programmes in Africa. Our team includes scientists with strong research profiles in drug discovery, synthetic chemistry, natural product chemistry, pharmacology, molecular biology, IP and management. The diverse backgrounds of the researchers provide training and networking opportunities that open new avenues for joint projects.

6.7 The Future

We have focused on peptide GPCRs as targets for screening for novel broad-spectrum anthelmintics with a low propensity for resistance development, but it is clear that the platform of recombinant microbes has unique value in the African setting for screening against other kinds of targets for helminths and other parasites and pathogens. This status allows expansion of the recombinant microbe technology platform to address many other discovery targets at the discretion and initiative of scientists in Africa. On-site discovery and development must be encouraged to ensure better drugs are part of the answer for improved parasite control; this effort needs to be built in places where the parasites are. This ensures that long-term solutions are championed by local ownership of those most affected by a problem. Expanding available natural product chemical space available is crucial to bringing value to local chemical libraries. Characterization of these libraries through evaluation of local screen performance, medicinal chemistry progression and early development is required in the context of local leadership to ensure appropriate focus and intensity of the effort. Very importantly, this kind of drug discovery enterprise has long-term, beneficial implications for biodiversity conservation, with long-term benefits to the sustainability of natural product-driven economic activity [46].

6.8 Conclusions

The yeast-based platform is highly robust and suitable for routine operation in resource-limited environments, comprising a strong and unique advantage. The application of this technology to discover non-peptide ligands that could act on multiple parasites GPCRs won support in the initial Gates Grand Challenges Phase I programme: the simplicity and reliability of the yeast screens, which require minimal infrastructure for maintenance and performance, enables mechanism-based HTS in many African laboratories.

Importantly, this project put control of the initial phase of the process in the hands of investigators from helminth disease-endemic countries. We defined our criteria for success in Phase I as the following: (1) develop and initiate processes leading to the creation of a centralized collection of African natural products in African centres; (2) introduce HTS platforms for FLP-GPCRs based on recombinant strains of yeast to African screening centres, including training; (3) initiate programmes for the creation of novel microbe-based screening systems by recruiting and training local scientists; and (4) advance African scientists into leadership roles in the project. Proof of concept was obtained for all the goals in Phase I, although the number of novel natural products available from African laboratories was lower than anticipated.

The Phase I experiments were an opportunity to "test drive" the concept that the recombinant microbe screening platform could be usefully deployed in Africa. It was not unexpected that challenges to its implementation would be encountered. The major challenges were in the area of chemistry, including access to the number of African natural products needed to justify HTS and establish a viable pANPL. We did not reach "no-go" challenges in the establishment of the microbe-based screening platform at UB and UCT. Despite the challenges, work in Phase

I established mechanism-based screening expertise and performance in a disease-endemic region.

India and China have thriving, sustainable pharmaceutical-based bioeconomies, including strong natural product-based contributions. One must ask, "Why not Africa?" What will be required of Africa to compete? A strong natural product-based drug discovery pipeline could play a critical role in building a new African pharmaceutical industry. Initially, enforceable agreements that provide equitable distribution of costs and rewards will be critical to ensure that bioprospecting and not biopiracy thrives. Our challenges remain the establishment of a pan-African Natural Product Library with significant numbers of compounds. Local control of the economic value of chemical resources through local ownership of IP must predominate to ensure negotiating power and willingness of African chemists and traditional medicine practitioners to contribute to growing the library. Crucially, complete transparency in operation of transnational projects is essential to maintain trust and enthusiastic progress. Key milestones have been establishing screening centres at UB and UCT. Local leadership of the project, combined with a common library of purified natural products for screening, in true partnership with international institutions, can sustain and grow discovery centres while stimulating investment from philanthropic and venture capital sources to expand local innovation-based industries. To fully harness this potential, the project will build capacity for biotechnology-guided characterization of local biodiversity and promote indigenous plant species and microbial biodiversity as sources of new drugs to meet local needs. These resources can also lead to drug candidates for the control of livestock parasites in Africa, not only addressing human health concerns but also becoming a lever for improving local food security. We see the ultimate criteria of success will be attracting funding to move potential leads to marketable drugs while making the leadership more and more locally driven.

Through this project, African partners benefit from new screening technologies, new antiparasitic candidates and growth of local research capacity, which we see as key milestones. These milestones will assist promising African researchers with options beyond going abroad to continue research careers. Cultivating excellence in African scientific leadership and developing and promoting cross African research networks are fundamental to expanding the role of natural products in African drug discovery operations. An ongoing challenge for the project remains integration with existing efforts devoted to the discovery of new antiparasitic drugs from African chemical sources. The integration of the Phase II Gates project with existing or nascent natural product-based drug discovery efforts in Africa is critical to long-term sustainability of pANPL. In particular, synergies with efforts such as ANDI [47] and GIBEX [39, 48] must be cultivated. Given the chronic limitation of funds available to support research in Africa, it is imperative that we cooperate instead of compete with projects which are already underway or soon to be launched. Both ANDI and GIBEX are notable for their efforts to extract value from African natural products and to support drug discovery and development projects in the continent. Established connections with both programmes should add to, rather than compete with, all respective efforts. This is particularly relevant for bioassay-guided fractionation. Additional R&D investment based on the value of hits identified in Phase II will be required to complete the hit-to-lead process and recruit development partners as warranted, to support capacity building around the pANPL to develop sustainable programmes in drug discovery in Africa. This will transcend the scope of this work to empower local scientists to discover agents which could be of use worldwide.

References

1. http://www.who.int/features/factfiles/neglected_tropical_diseases/ntd_facts/en/index.html. Accessed July 2011
2. http://www.who.int/neglected_diseases/diseases/en/. Accessed July 2011.
3. Hotez PJ, Brindley PJ, Bethony JM et al (2008) Helminth infections: the great neglected tropical diseases. J Clin Invest 118:1311–1321
4. Crompton DWT, Savioli L (2007) Helminthiasis for public health. CRC/Taylor and Francis, Boca Raton, FL, 362pp
5. Klopper RR (2006) Inventory of the African flora: a world first for the forgotten continent. S Afr J Sci 102:185–186
6. Li JW-H, Vederas JC (2009) Drug discovery and natural products: end of an era or an endless frontier? Science 325:161–165
7. Molinari G (2009) Natural products in drug discovery: present status and perspective. In: Guzman CA, Feuerstein GZ (eds) Pharmaceutical biotechnology. Landes Bioscience and Springer Science+Business Media, Austin, TX, pp 13–27
8. Hotez PJ, Molyneux DH, Fenwick A et al (2007) Control of neglected tropical diseases. N Engl J Med 357:1018–1027
9. Geary TG, Woo K, McCarthy JS et al (2010) Unresolved issues in anthelmintic pharmacology for helminthiases of humans. Int J Parasitol 40:1–13
10. Geary TG, Gauvry N (2011) Anthelmintic discovery for human infections. In: Palmer MJ, Wells TNC (eds) Neglected diseases and drug discovery. Royal Society of Chemistry Press, Cambridge
11. Geary TG, Mackcbrrenzie CD (2011) Progress and challenges in the discovery of macrofilaricidal drugs. Expert Rev Anti Infect Ther 9(8):681–695
12. http://www.irinnews.org/report.aspx?reportid=82486. Accessed Oct 2011
13. Gamo F-J, Sanz LM, Vidal J et al (2010) Thousands of chemical starting points for antimalarial lead identification. Nature 465:305–310
14. Rottmann M, McNamara C, Yeung BKS et al (2010) Spiroindolones, a potent compound class for the treatment of malaria. Science 329:1175–1180
15. Geary TG, Thompson DP, Klein RD (1999) Mechanism-based screening: discovery of the next generation of anthelmintics depends upon more basic research. Int J Parasitol 29:105–112
16. Geary TG (2011) Mechanism-based screening strategies for anthelmintic discovery. In: Caffrey C (ed) Parasitic helminths: targets, drugs and vaccines. Wiley-VCH, Weinheim
17. Klein RD, Geary TG (1997) Recombinant microorganisms as tools for high-throughput screening for non-antibiotic compounds. J Biomol Screen 2:41–49
18. Geary TG (2001) Screening for parasiticides using recombinant microorganisms. In: Kirst HA, Yeh W-K, Zmijewski M, Bronson DB (eds) Enzyme technology for pharmaceutical and biotechnological applications. Dekker, New York, pp 323–341
19. Woods D, Butler C, Williams T, Greenwood K (2010) Receptor-based discovery strategies for insecticides and parasiticides. In: Geary TG, Maule AG (eds) Neuropeptide systems as targets

for parasite and pest control. Landes Bioscience and Springer Science+Business Media, Austin, TX, pp 1–9
20. Geary TG, Woods DJ, Williams T, Nwaka S (2009) Target identification and mechanism-based screening for anthelmintics: application of veterinary antiparasitic research programmes to search for new antiparasitic drugs for human indications. In: Selzer PM (ed) Drug discovery in infectious diseases. Wiley-VCH, Weinheim, pp 1–16
21. Maule AG, Mousley A, Marks NJ et al (2002) Neuropeptide signaling systems – potential drug targets for parasite and pest control. Curr Top Med Chem 2:733–758
22. McVeigh P, Geary TG, Maule AG (2006) On the FLP-side of nematode neuropeptides. Trends Parasitol 22:385–396
23. Geary TG, Marks NJ, Maule AG et al (1999) Pharmacology of FMRFamide-related peptides (FaRPs) in helminths. Ann N Y Acad Sci 897:212–227
24. Lowery DE, Geary TG, Kubiak TM, Larsen MJ (2007) G protein-coupled receptor-like receptors and modulators thereof. US Patent No. 7,208,591
25. Greenwood K, Williams T, Geary T (2005) Nematode neuropeptide receptors and their development as anthelmintic screens. Parasitology 131:S169–S177
26. Husson SJ, Mertens I, Janssen T et al (2007) Neuropeptidergic signaling in the nematode *Caenorhabditis elegans*. Prog Neurobiol 82:33–55
27. Geary TG (2010) Non-peptide ligands for peptidergic G protein-coupled receptors. In: Geary TG, Maule AG (eds) Neuropeptide systems as targets for parasite and pest control. Landes Biosciences, Austin, TX, pp 10–26
28. Woods DJ, Butler C, Williams T, Greenwood K (2010) Receptor-based discovery strategies for insecticides and parasiticides: a review. In: Geary TG, Maule AG (eds) Neuropeptide systems as targets for parasite and pest control. Landes Bioscience, Austin, TX, pp 1–9
29. Wang ZX, Broach JR, Peiper SC (2006) Functional expression of CXCR4 in *Saccharomyces cerevisiae* in the development of powerful tools for the pharmacological characterization of CXCR4. Methods Mol Biol 332:115–127
30. Minic J, Sautel M, Salesse R, Pajot-Augy E (2005) Yeast system as a screening tool for the pharmacological assessment of G protein coupled receptors. Curr Med Chem 12:961–969
31. Brooks DR, Isaac RE (2002) Functional genomics of parasitic worms: the dawn of a new era. Parasitol Int 51:319–325
32. Kubiak TM, Larsen MJ, Bowman JW et al (2008) FMRFamide-like peptides (FLPs) encoded by the *flp18* precursor gene activate two isoforms of the orphan *Caenorhabditis elegans* G-protein-coupled receptor Y58G8 heterologously expressed in mammalian cells. Biopolymers 90:339–348
33. Birgül N, Weise C, Kreienkamp HJ, Richter D (1999) Reverse physiology in *Drosophila:* identification of a novel allatostatin-like neuropeptide and its cognate receptor structurally related to the mammalian somatostatin/galanin/opioid receptor family. EMBO J 18:5892–5900
34. Lipinski CA, Lombardo F, Dominy BW et al (2001) Experimental and computational approaches to estimate solubility and permeability in drug discovery and development settings. Adv Drug Deliv Rev 46:3–26
35. Tagboto S, Townson S (2001) Antiparasitic properties of medicinal plants and other naturally occurring products. Adv Parasitol 50:199–295
36. Juliani HR, Simon J, Ho C-T (eds) (2010) African natural plant products. New discoveries and challenges in chemistry and quality. Oxford University Press, Kettering, 616pp
37. http://www.napreca.net/. Accessed July 2011
38. Chibale K, Masimirembwa CM, Guantai EM (2011) Extracting molecular information from African natural products to facilitate unique African-led drug discovery efforts. Future Med Chem 3:257–261
39. http://www.gibex.org/index.php?suj=99. Accessed July 2011
40. http://www.mcgill.ca/lifesciencescomplex/core/hts-hcs/chemical-libraries. Accessed July 2011

41. Fura A, Shu Y-Z, Zhu M et al (2004) Discovering drugs through biological transformation: role of pharmacologically active metabolites in drug discovery. J Med Chem 47:4339–4351
42. Hamilton C (2006) Biodiversity, biopiracy and benefits: what allegations of biopiracy tell us about intellectual property. Dev World Bioeth 6:158–173
43. Frew SE, Liu VY, Singer PA (2009) A business plan to help the global South in its fight against neglected diseases. Health Aff (Millwood) 28:1760–1773
44. Daar AS, Berndtson K, Persad DL, Singer PA (2007) How can developing countries harness biotechnology to improve health? BMC Public Health 7:346–355
45. http://www.grandchallenges.ca/wp-content/uploads/integratedinnovation_EN.pdf. Accessed July 2011
46. Kingston DG (2011) Modern natural products drug discovery and its relevance to biodiversity conservation. J Nat Prod 74:496–511
47. Nwaka S, Ilunga TB, Da Silva JS et al (2010) Developing ANDI: a novel approach to health product R&D in Africa. PLoS Med 7:e1000293
48. Dushenkov V, Raskin I (2008) New strategy for the search of natural biologically active substances. Russ J Plant Physiol 55:564–567

Chapter 7
Application of In Silico, In Vitro and In Vivo ADMET/PK Platforms in Drug Discovery

Collen Masimirembwa and Roslyn Thelingwani

Abbreviations

ADMET	Absorption, distribution, metabolism, excretion and toxicity
AiBST	African Institute of Biomedical Science and Technology
AUC	Area under the curve
CL	Clearance
C_{max}	Maximum plasma concentration
DME	Drug-metabolising enzymes
DMPK	Drug metabolism and pharmacokinetics
F_a	Fraction absorbed
f_u	Fraction unbound
HPGL	Hepatocytes per gram of liver
HTS	High-throughput screening
IVIVE	In Vitro to in vivo extrapolation
K_{el}	Elimination rate constant
LD	Lead discovery
LO	Lead optimisation

C. Masimirembwa (✉)
Department of DMPK/PD and Toxicology, African Institute of Biomedical Science and Technology, 211 Herbert Chitepo Street, Harare, Zimbabwe

Department of Clinical Pharmacology, University of Cape Town, Cape Town, South Africa
e-mail: collenmasimirembwa@yahoo.com

R. Thelingwani
Department of DMPK/PD and Toxicology, African Institute of Biomedical Science and Technology, 211 Herbert Chitepo Street, Harare, Zimbabwe

Department of Chemistry, University of Cape Town, Cape Town, South Africa

K. Chibale et al. (eds.), *Drug Discovery in Africa*,
DOI 10.1007/978-3-642-28175-4_7,

MPPG	Microsomal protein per gram
NCEs	New chemical entities
PD	Pharmacodynamics
PK	Pharmacokinetics
PSA	Polar surface area
QSAR	Quantitative structure activity relationship
SAR	Structure activity relationship
$t_{1/2}$	Half-life
TDI	Time-dependent inhibition
T_{max}	Time to reach maximum plasma concentration
V_d	Volume of distribution

7.1 Background

Prior to the 1990s, studies in pharmacokinetics (PK) were completed late in drug discovery and during drug development. The preclinical studies were mainly animal in vivo studies used for safety evaluation and human dose predictions by allometric scaling. A landmark paper by Prentis [1] reviewed the factors responsible for the failure of new chemical entities (NCEs) in British pharmaceutical companies and found that between 1964 and 1985, inadequate PK was the leading cause of failure in 39% of new chemical entities (NCEs) followed by lack of efficacy and by toxicity. PK has an intricate impact on efficacy and toxicity, emphasising the role of PK in the failure of NCEs. Further reviews corroborated Prentis's findings [2]. Drugs for the treatment of infectious diseases failed more due to poor PK than those for non-communicable diseases, strongly suggesting the necessity for the incorporation of PK into drug discovery programmes.

During the 1990s, industry responded by setting up many assays that enabled projects to characterise the process that determined the PK of compounds. These included in vitro methods for compound permeability, metabolism and excretion. Application of these assays was frontloaded to early phases of the drug discovery process from hit identification, lead discovery and lead optimisation through to candidate drug selection [3]. By 2000, the pharmaceutical industry had reduced the failure rate of NCEs due to PK issues from 40% to less than 10% [2]. Figure 7.1 shows the relative contribution of various factors in the attrition rates of new chemical entities [4]. It is important to note the rapid increase in the cost of goods as a cause of drug discovery and development failure followed by poor efficacy and clinical safety. Most drug discovery projects in Africa do not have access to the PK platforms which major pharmaceutical industries are using to reduce attrition rates. It is therefore likely that the failure or slow progress of some of the discovery projects in Africa could be due to lack of PK support.

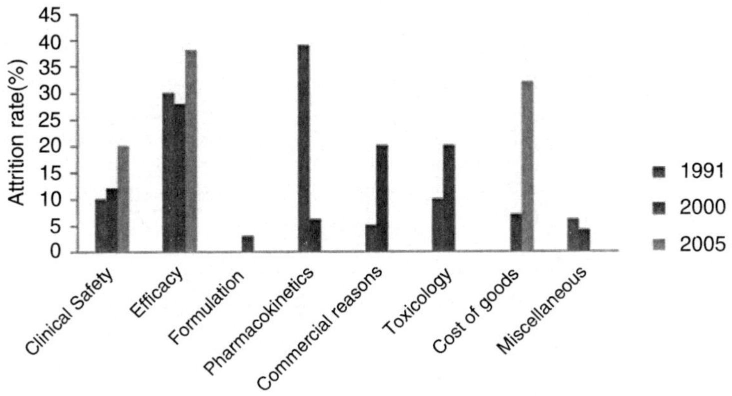

Fig. 7.1 Evolution of the principal reasons for drug development failure between 1991 and 2005. Adapted from [4]

7.2 Pharmacokinetic Challenges in Drug Discovery

The overall objective of drug discovery is to "find a *potent and selective compound*, reaching to the *site of action* at *sufficient concentration* for a *sufficiently long time* to elicit the desired effect with a *reasonable dose* that can be administered orally to humans in an *acceptable form once or twice a day*" (Ulf Bredberg, personal communication). In addition to this, the compound must also be safe for human use. It is in the context of achieving these objectives that pharmacokinetics can play an important role. Poor drug systemic exposure could be due to low absorption from the intestine caused by low compound permeability or low solubility. Low bioavailability can also be due to high clearance and short half-life caused by extensive liver metabolism and excretion. Absorption and metabolism therefore play a key role in determining drug exposure levels. Figure 7.2 depicts how these factors determine drug exposure levels as a measure of bioavailability.

With respect to safety, pharmacokinetic factors can contribute through risk for drug–drug interactions due to enzyme and/or transporter inhibition or induction. Inhibition of the metabolism of one drug (victim drug) by another (perpetrator drug) could result in overexposure of the victim drug, resulting in either exaggerated pharmacologic effects or unselective effects on other receptors. Induction, on the other hand, could result in underexposure of the victim drug associated with sub-therapeutic effects. Toxicity or idiosyncratic reactions could also arise from increased production of reactive and toxic metabolites of the new chemical entity. Toxicity can be mediated through formation of protein adducts (cytotoxicity) or nucleic acid adducts (genotoxicity).

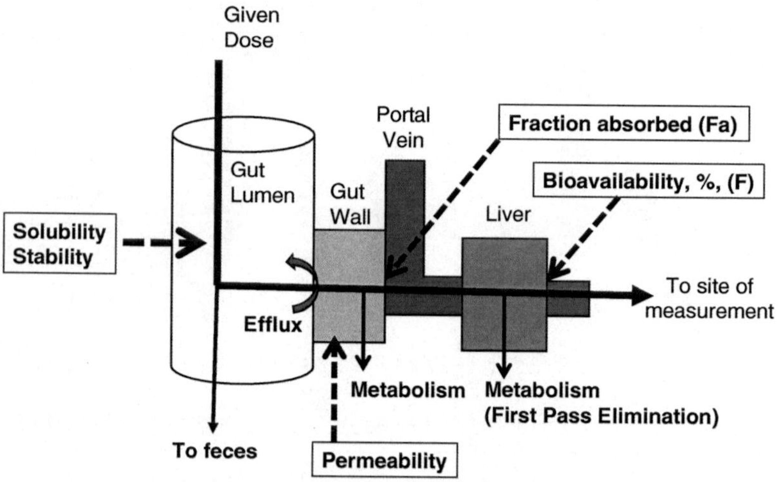

Fig. 7.2 Pharmacokinetic processes that affect drug bioavailability

7.3 Overview of Pharmacokinetics

Pharmacokinetics can be defined as "*the study of the time course of drug and metabolite levels in different fluids, tissues, and excreta of the body, and of the mathematical relationships required to develop models to interpret such data*" [5]. The key PK parameters that enable one to determine how much and how often a drug should be administered to achieve a desired therapeutic effect in a safe, efficacious and convenient manner are absorption (F_a), volume of distribution (V_d), clearance (CL), half-life ($t_{1/2}$) and bioavailability (F). Figure 7.3 illustrates the pharmacokinetic profiles of a drug after an intravenous (i.v) and oral (p.o) dose. The figure also shows the importance of drug concentrations with respect to effective concentration ranges and toxic ranges, parameters relevant in the safe and efficacious use of medicines. Table 7.2 shows the equations used to calculate the important pharmacokinetic parameters used in dose regimen determinations.

These PK parameters (Table 7.1) can all be determined from in vivo studies in preclinical animals and in humans when the drug is administered enterally (oral, buccal or rectal) and/or parenterally (intravenous, intramuscular, peritoneal and subcutaneous). Most studies in discovery use oral and intravenous drug administration to determine the PK parameters indicated in Table 7.1.

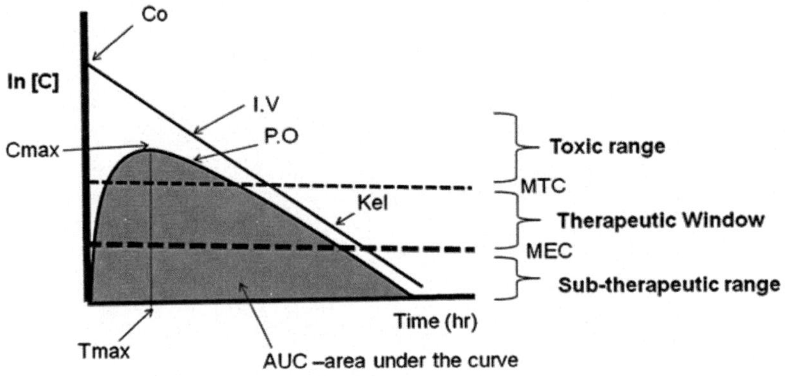

Fig. 7.3 Schematic representation of a plot of the natural logarithm of drug plasma concentration (In[*D*]) with time (hours) after intravenous (i.v) and oral (p.o.) administration of a drug. Co—extrapolated initial dose, C_{max}—maximum drug concentration achieved, T_{max}—time to reach C_{max}, K_{el}—elimination rate constant, AUC—total area under the curve, a measure of drug exposure. *MEC* minimum effective (therapeutic) concentration, *MTC* minimum toxic concentration

7.4 The Drug Absorption, Distribution, Metabolism and Excretion Process

Traditionally, pharmacokinetic studies had been done either as a documentation process or as part of toxicokinetic studies in drug discovery and development. As understanding of the potential role of PK in selecting compounds with favourable PK and pharmacodynamics (PD) increased, the preclinical animal studies where used to predict human pharmacokinetics using allometric scaling. This method is based on the hypothesis that physiological parameters are proportional to body size and weight. Allometric scaling has been used to predict various PK parameters indicated in Table 7.1 with mixed success.

Important biomedical breakthroughs in the 1990s allowed scientists to develop in vitro systems to study the individual processes that constituted the PK of a compound (Table 7.2). For absorption, systems to study drug permeability and transport (e.g. Caco2, intestinal segments and transfected cell lines) enabled the prediction of the fraction of drug absorbed, F_a. For metabolism, liver microsomes, hepatocytes, recombinant enzymes and cell lines enabled the humanisation of in vitro drug metabolism towards the prediction of drug clearance (CL) and drug–drug interactions [6]. These revolutions in biological systems were also accompanied by technological advancements such as LC-MS/MS and automated assays for high-throughput screening (HTS).

The deconvolution of PK to discrete processes of drug absorption, distribution, metabolism and excretion (ADME) allowed for mechanistic insights that were not

Table 7.1 Key pharmacokinetic parameters, definitions, their determination and use

Parameter	What is it?	How is it measured?	How do you apply it?
K_a	Absorption rate constant	Residual method Residual slope $= -K_a = \text{Ln}(R_1)\text{–Ln}(R_2)/(t_1–t_2) = \text{min}^{-1}$	Assess rate of absorption
C_{max}	Maximum drug concentration attained	Read off the Ln[*D*]/Time plot μM or mg/L	Used to relate drug levels attained (C_{ss}) and therapeutic concentrations
T_{max}	Time to reach C_{max}	Read off the Ln[*D*]/Time plot	Assess rate of absorption
T_{lag}	Absorption lag time	Residual method	For correct estimation of K_a and K_{el}
K_{el} (terminal slope)	First-order kinetics indicating the rate of compound elimination	$K_{el} = (\text{Ln}C_1 - \text{Ln}C_2)/(t_1 - t_2) = \text{min}^{-1}$	Calculation of drug clearance
F (bioavailability)	Fraction of dose that reaches systematic circulation	$\frac{\text{AUC p.o} \times \text{Dose i.v}}{\text{AUC i.v} \times \text{Dose p.o}}$	Assess extent of absorption
$t_{1/2}$	Time taken for drug concentration to decline to half of its original value	$0.693/K_{el}$ $0.693\ V_d/\text{CL}$	Gives you the intervals of repeated doses, 4, which give C_{ss}. Inversely, gives you time it takes to eliminate drug from the body, 4 half-lives
V_d	Apparent volume in which the compound is dissolved in the body	$V_{ss} = \text{Dose/Co}$ (for i.v) also $V/F = \text{Dose}/C_{max}$ (for p.o)	Gives idea of location of drug: 3L-plasma, 12L-interstitial fluid, 25L-intracellualr water, >40L-tissue binding. Estimation of loading dose: $\text{DL} = V_{ss} \cdot C_{ss}$
AUC	Area under the curve—measure of drug exposure	Use the trapezoidal $\Sigma(T_2–T_1)^*(C_1 + C_2)/2 + C\text{–last}/\lambda z = \text{ng h/ml}$	Used to calculate clearance (CL) and bioavailability (*F*)
Clearance (CL), i.v	Volume of blood from which the drug is completely removed per unit time	$\text{CL} = \text{Dose/AUC}$ (for i.v) or $\text{CL}/F = \text{Dose/AUC}$	$C_{ss} = \frac{\text{Dose rate}}{\text{CL}}$

*K_{el}, $t_{1/2}$, and CL relate to first-order processes. V_{ss}—volume of distribution at stead state

Table 7.2 ADMET parameters and systems used to study them

Parameter	System
CL_{int} (intrinsic metabolic stability)	Microsomes, hepatocytes, recombinant enzymes
Enzyme inhibition	Recombinant enzymes, microsomes
Enzyme induction	HepaRG cell line, hepatocytes, reporter gene assay
Enzymes involved (reaction phenotyping)	Liver microsomes, recombinant enzymes, hepatocytes
Metabolite identification	Hepatocytes, microsomes, recombinant enzymes
Permeability/absorption	Caco2 cell line, intestinal segments
Transporters	MDCK cells (Pgp), transfected cell lines (different human transporters), Caco2 cell lines

possible from the holistic in vivo PK studies. They also facilitated the evolution of computational methods towards the derivation of qualitative and quantitative structure–activity relationships (SAR and QSAR) for the prediction of ADME properties. Such SAR and QSAR have resulted in very fruitful collaborations between medicinal chemists and pharmacokinetic scientists which are enabling the rapid optimisation of compounds with respect to PK and PD properties. Table 7.3 summarises the physicochemical and in vitro ADME that can now be determined and used to effect chemical modification or predict in vivo PK, thus in-calculate rational design in the drug discovery process.

There are pros and cons of using in vitro or in vivo preclinical models. The advantages of in vivo models are that you get integrated information on permeability, metabolism, secretion and transport. One can sometimes make in vitro to in vivo correlations for the animal model to support human in vitro to in vivo predictions, which can lead to the observation of the effect of unknown PK mechanisms. The major disadvantage of preclinical animal studies is that there can be great inter-species difference between the model animal and humans. This is particularly so with respect to metabolic processes. The advantages of using in vitro systems are that one can use both human and animal tissue, study the individual components of PK, derive insights into the mechanism of drug disposition and test for specific properties such as enzyme inhibition and amenability to high-throughput screening (HTS). The disadvantages of the in vitro systems are that one needs to know all mechanisms involved in a drug's PK in order to study them individually, assays for some mechanism might not be there or are too complex to study. A major disadvantage of in vitro system is the many physiological assumptions one has to make from recombinant, sub-cellular, cellular and organ to whole animal/human systems. In general, as one progresses from simple recombinant enzymes, microsomes, to complex hepatocyte systems, success in the prediction of in vivo metabolic clearance increases, but the costs of doing the studies also increase.

Table 7.3 Drug absorption, distribution, metabolism and excretion (ADME) parameters and physicochemical properties that can be determined in vitro and used to predict and or rationalise in vivo pharmacokinetics

Process	Definition	Primary PK parameters	Determinants	Measured parameters
Absorption	Movement of unchanged drug from the site of administration to the site of measurement	*F*—bioavailability K_a—absorption rate constant	• Solubility • Permeability • First-pass extraction • Transporters	Chemical stability Solubility (So) pKa, HBD and HBA PSA Log *P* or Log *D* P_{app} Efflux/influx ratio Metabolic stability
Distribution	Reversible transfer of drug from and to the site of measurement	V_d, f_{uB}, f_{uT}, *B*/*P*	• Permeability • Transporters • Partition equilibria • Plasma protein binding	Log *P* or Log *D* f_u (fraction unbound) Efflux/influx ratio P_{app}
Metabolism	Irreversible loss of drug from the site of measurement via formation of other species	CL_H	• Permeability • Interaction with enzymes • Plasma protein binding	Log *P* or Log *D* Intrinsic clearance Identification of DME and percent contribution to metabolism (f_m) Metabolite identification
Excretion	Irreversible loss of unchanged drug	CL_R, CL_B	• Permeability • Interaction with transporters • Plasma protein binding	Log *P* or Log *D* pKa Efflux/influx ratio
Disposition	All the processes that occur subsequent to drug absorption, these are mainly distribution and elimination			

f_{uB} fraction unbound in blood, f_{uT} fraction unbound in tissue, CL_R renal clearance, CL_B blood clearance, *B*/*P* blood plasma partition ratio, Log *P* or *D* lipophilicity, pKa ionisation constant, *HBD* hydrogen bond donor, *HBA* hydrogen bond acceptor, *PSA* polar surface area, *DME* drug-metabolising enzymes, P_{app} apparent permeability

7.5 In Silico ADMET

Before the development of the elaborate in vitro ADMET toolkit, efforts to predict physicochemical and pharmacokinetic parameters were solely based on the physicochemical properties of new chemical entities. The methods used properties such as lipophilicity (Log *P* and Log *D*), pKa, Mwt, melting points, binding energy, number of atoms, etc.) to predict physicochemical properties such as solubility and metabolic properties such as Michaelis–Menten constant (K_m), clearance and toxicity. As the in vitro ADME assays on HTS platforms started churning out large amounts of data which were initially meant for in vitro–in vivo correlation studies, the data provided an opportunity for SAR and QSAR studies. Metabolic stability, enzyme and metabolite identification, and cytochrome P450 inhibition data using recombinant enzymes produced mechanistically clean data that were used to develop SAR and QSAR models. In addition to the in vitro data, starting with the publication of the soluble bacterial CYP crystal structures in the 1980s, homology models of CYPs were developed and docking studies used to infer ligand–enzyme active site interaction studies. This work led to the derivation of pharmacophore models for substrates and inhibitors of the major drug-metabolising enzymes, CYP1A2, 2C9 and CYP2D6. Advances in molecular biology assisted this effort by providing site-directed mutagenesis functional studies towards mapping the geometry and chemistry of proposed active sites of CYPs. After years of failed efforts to crystallise membrane-bound CYPs, success was achieved in 2000 with the crystallisation of the rabbit CYP2C5 enzyme. This work was followed by successful crystallisation of several human CYPs including CYP21A2, 2C8, 2C9, 2D6 and 3A4. Some of the enzymes were co-crystallised with substrates and/or inhibitors to give the first direct information on enzyme–ligand interactions. Advances in computational tools in quantum mechanical calculations, molecular dynamic simulations, GRID-based calculation of physicochemical properties of substrates/inhibitors and enzyme active site, and docking and scoring functions to understand orientation and affinity of enzyme–ligand interactions were successfully applied to ADMET computational modelling. This strengthened the link between Medicinal Chemistry and DMPK scientists in addressing ADMET issues at molecular level. Software such as Flex X, GOLD, GLIDE and AutoDock are being used for docking substrates/inhibitors into active sites of drug-metabolising enzymes towards studies to identify involved enzymes and predict sites of metabolism. The Volsurf software (http://www.moldiscovery.com) is being used to predict permeability, solubility and metabolic stability. Metasite (http://www.moldiscovery.com) is being used for the identification of site metabolism and when linked to mass spectrometry in metabolite identification studies. There are now many software programmes on the market, some free and others commercial. The choice of software to use is very important; they are based on different modelling techniques and different data quality and might require secondary processing for meaningful interpretation of the results. Interpretation of most of the results requires a good understanding of the biology of the ADME properties being predicted.

The initial purpose of the in vitro data of making in vitro–in vivo correlations also experienced significant advances. Notable products on the market being Simcyp (http://www.simcyp.com), Gastroplus (http://www.simulations-plus.com), WinNONlin (http://www.pharsight.com/products/prod_winnonlin_home.php) and Aureus (http://www.aureus-pharma.com), which, through various ways, try to address some of the challenges of in vitro to in vivo extrapolation (IVIVE) of ADME properties such as (a) inter-individual variability in metabolic processes, (b) interplay of drug-metabolising enzymes and drug transporters, (c) interethnic differences in PK properties and (d) estimation of concentration of test drug reaching the enzyme active site. These developments have given birth to the field of pharmacometrics where PK simulations and physiologically based pharmacokinetic modelling (PBPK) are being used to predict the likely pharmacokinetics of a candidate drug when given to humans. This is contributing to the efficient design of clinical studies and the projection of data from few patients to whole populations, something which could result in reduced costs of clinical trials which are currently the most expensive phase of the whole drug discovery and development process.

7.6 ADMET and PK Preclinical Models in Drug Discovery

Various in silico, in vitro and in vivo models are being applied in the design, characterisation and selection of new chemical entities with ADMET and PK properties predicted to result in a safe and efficacious product for clinical use in humans. Over the past 20 years, the pharmaceutical industry has developed an elaborate generic "ADMET/PK toolkit" that is used across most disease areas and is amenable to modification to address some project-specific questions. The models provide different pieces of information of varying complexity throughout the hit evaluation, lead discovery, lead optimisation and candidate drug nomination value chain (Fig. 7.4).

Given the background of drug discovery activities by African research groups being driven by medicinal chemistry (mainly herbal extracts, purified natural products from herbs with reported medicinal properties and to a lesser extent conventional organic synthesis) and pharmacology (mainly screens against in vitro parasite cultures, animal disease model and to a lesser extent recombinantly expressed molecular targets), integration of DMPK was virtually absent. Before the establishment of drug discovery DMPK expertise at the African Institute of Biomedical Science and Technology (AiBST), the only PK studies done by some institutions were in vivo animal and human PK on drugs already on the market with either the aim of evaluating them for general toxicity, PK variability in African populations or PK/PD relationships in patients. The DMPK/PD-Tox Department at AiBST was therefore established with the aim of setting up an industrial DMPK platform benchmarked against the in silico, in vitro and in vivo models being used by leading pharmaceutical industry. The in silico, in vitro and in vivo models for

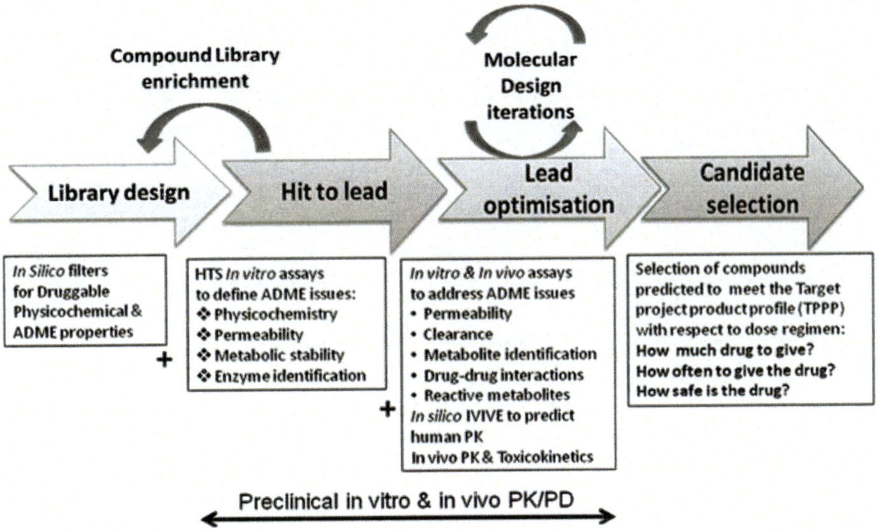

Fig. 7.4 Schematic representation of how in silico, in vitro and in vivo preclinical platforms for ADMET characterisation are being frontloaded in the drug discovery value chain

drug discovery will therefore be described in the context of those set up at AiBST and being applied in the characterisation of anti-parasitic drugs in use and new chemical entities from various drug discovery research groups in Africa. To make the concepts relevant for the African case, drugs indicated in Fig. 7.5 will be used to demonstrate the potential benefits of integrating such tools in the discovery and development of safe and efficacious drugs in the fight against infectious diseases endemic to Africa.

7.7 Compound Library Characterisation and Hit Identification and Evaluation

During the early phases of target identification and validation, the approach could involve the use of tool compounds (e.g. inhibitory analogues of the target substrate) and screening large compound libraries against a high-throughput in vitro assay for the molecular target. The compound libraries to be screened can be in-house resources or can be purchased from external sources. Before purchase of compounds and during the in vitro pharmacological screens, the compound libraries should be screened for general druggable properties. A number of such general properties have been proposed and the most widely used being the Lipinski rule of 5 for compound solubility and permeability, the major determinants of

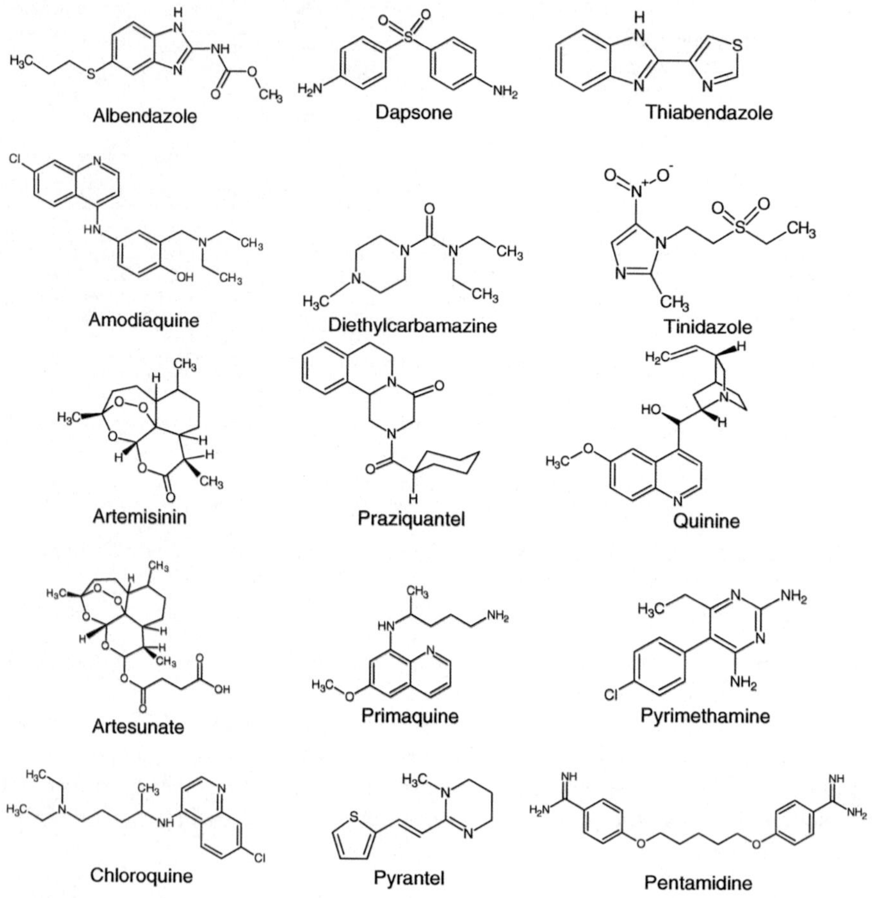

Fig. 7.5 Chemical structures of some anti-parasitic drugs whose ADME properties have been characterised

compound absorption [17]. It is important that the application of such in silico filters be done with caution since they are only guidelines which if used dogmatically, could result in the loss of novel compounds that could have unique PK and PD properties, with the possibility of intellectual property (IP) and market competitive advantages for a company. During this phase, the tractability of the biological target is evaluated with respect to PK and PD issues. PD issues involve the identification of compounds with reproducible biologic effect for the desired type of activity (e.g. agonist, antagonist or enzyme inhibition). PK issues could relate to the accessibility of the target (e.g. central nervous system), knowledge of competing endogenous ligands and intended dose formulation (e.g. i.v, p.o. or inhalation). If oral administration is the intended formulation, the following PK parameters should be predicted at this stage and be used to judge the quality of the compound library to be purchased or the hits obtained from in vitro and in silico screens:

Table 7.4 Predicted physicochemical properties of some anti-parasitic drugs (properties calculated using the Marvin sketch calculator plug-ins, http://www.chemaxon.com, predicted solubility using ALOGpS and predicted protein binding (PB) in VolSurf)

Compound	Mwt	Log *P*	Log $D_{7.4}$	HBD	HBA	PSA (A^2)	pKa	Solubility (predicted) (g/L)	Protein binding (%)
Albendazole	265.33	3.20	3.20	2	5	92.31	4.21, 9.68	0.02284	79.57
Amodiaquine	355.86	4.53	3.19	2	5	48.39	6.90, 8.57, 10.19	0.00880	100
Artemisinin	282.33	3.11	3.11	0	4	53.99	–	1.27	50.72
Artesunate	384.42	3.10	−0.17	1	7	100.52	3.77, 6.97	0.68	40.15
Chloroquine	319.87	3.93	0.65	1	4	28.16	7.73, 10.33	0.01746	96.44
Dapsone	248.30	1.27	1.27	2	4	94.56	1.57, 2.39	0.28	79.22
Diethylcarbamazepine	199.29	0.09	−0.03	0	2	26.79	6.90	240	49.69
Praziquantel	312.41	2.30	2.30	0	2	40.62	–	0.38	69.39
Primaquine	259.35	1.64	−0.96	2	4	60.17	0.62, 4.31, 10.20	0.05639	73.42
Pyrantel	206.31	1.96	−0.40	0	2	43.84	10.71	0.12	68.22
Thiabendazole	201.25	2.25	2.24	1	2	69.81	4.08, 10.28	0.14	66.67
Tinidazole	247.27	−0.58	−0.58	0	5	106.16	3.10	3.03	41.08
Quinine	324.42	2.51	0.86	1	4	45.59	4.02, 9.05	0.33	74.34
Pyrimethamine	248.71	2.75	2.23	2	5	77.82	0.10, 7.77	0.18	51.71
Pentamidine	340.42	2.32	−2.50	4	6	118.20	11.53, 12.13	0.02364	75.90

Fig. 7.6 Predictive projection of Caco2 permeability of novel 4-aminoquinolines in a permeability model in Volsurf

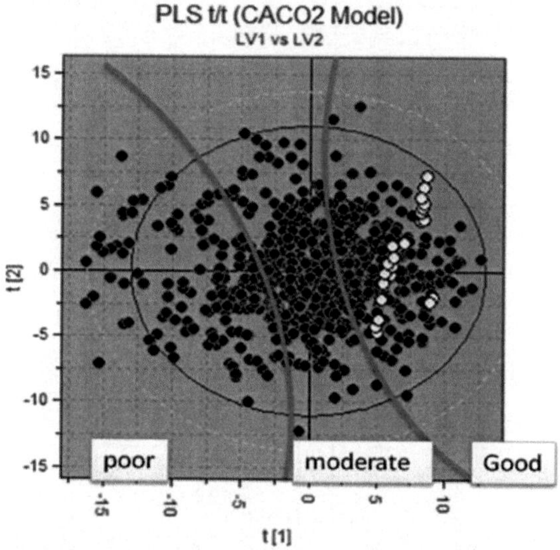

(a) Lipinski rule of 5 (Mwt, Log *P*, HBD, HBA), (b) PSA, (c) permeability, pKa and (d) solubility. Table 7.4 shows these predictions for some anti-parasitic drugs done using various softwares (Volsurf, MOKA, Marvin Sketch, etc.).

Given that a Lipinski score is predictive of oral absorption and that this PK parameter is a function of solubility and permeability, the predictions of permeability using Volsurf (Fig. 7.6) are supportive of the role of the physicochemical properties predicted in Table 7.4. The prediction tool in Volsurf was developed using Caco2 permeability data. Representative compounds from the various clusters for the predicted parameters are then experimentally determined for (a) water solubility, (b) lipophilicity, (c) pKa, (d) protein binding, (e) metabolic stability in human liver microsomes and microsomes from PK or PD animal model, (f) glutathione trapping of reactive metabolites in human liver microsomes, (g) inhibition of major CYPs (reversible and irreversible), (h) stability in human and PK or PD animal model plasma for drugs with moieties that could undergo hydrolysis and (i) permeability in Caco2 cells. This is done to check the predictive success of the computational tools and to generate real experimental values. The latter can be used to derive structure–activity relationship (SAR) models for the various parameters for the specific chemical series under study. At this stage, HTS approaches are used, and a ranking system is employed to make decisions on which series or cluster of compounds to proceed with and what potential ADMET and PK liabilities one needs to explore further at lead discovery stage.

Table 7.4 shows that none of the anti-parasitic drugs violates Lipinski's rule of 5; $Mwt \leq 500$, $\text{Log } P \leq 5$, $HBD \leq 5$, $HBA \leq 10$ and $PSA \leq 100$ Å^2. The compounds are therefore predicted to be permeable. The predicted solubility is, however, low, a reason why most of these drugs were made salts to enhance solubility. Due to the importance of solubility and permeability for absorption, a biopharmaceutical classification system, BCS, based on permeability and solubility, has been developed which classifies compounds into four major categories: Class 1—high permeability and high solubility, Class 2—high permeability and low solubility, Class 3—low permeability and high solubility and Class 4—low permeability and low solubility.

The compounds are predicted not to be highly protein bound except for chloroquine and amodiaquine. The prediction success can vary, e.g. artemisinin is predicted to be 50% bound and the measured value is 40%, compared to less successful, e.g. thiabendazole, predicted to be 67% bound and the measured value is 89%.

Predictions and measurements of some physicochemical and ADME properties were also researched for a novel series of 4-aminoquinolines synthesised in Prof. Kelly Chibale's laboratory (University of Cape Town, South Africa) (Table 7.5). None of the compounds violate more than 2 of Lipinski's rule and are therefore predicted to have good permeability. This is also supported by predictions using the software Volsurf (Fig. 7.6). In Table 7.4, it is evident that measured and predicted properties are not always in agreement which serves as a cautionary note not to be over-dependent on in silico methods in deciding the fate of NCEs. It is to be always kept in mind that predictions are only as good as the assumptions, and data quality and quantity that are used to make the predictive tools. Poor correlations on the other hand might simply mean that some unique aspects of the series one is working on are not well captured in the computational method one is using. Whether one gets good or bad predictions, the exercise demands that the drug discovery teams seek to understand the chemical basis of the ADME behaviour of new chemical entities.

7.7.1 *Determination of Compound Lipophilicity*

At this stage of drug discovery, Log $D_{7.4}$ is determined using a reverse phase chromatography HPLC method with the mobile phase at pH 7.4. In this method, a number of standard compounds of previously determined lipophilicity (using the shake flask, water–octanol partitioning method) are run on the reverse phase column, and their retention times are noted (k'). These are then plotted against their known Log $D_{7.4}$ values to make a standard curve. The standard compounds covering a wide span of Log $D_{7.4}$ values are used: metoprolol (0.02), propranolol (1.15), testosterone (3.19) and felodipine (5.20). The unknown compounds are then run on the same column, and their retention times are noted. These are used to read off the Log $D_{7.4}$ values of the test compounds from the standard curve. Table 7.5 shows the predicted and measured Log $D_{7.4}$ values of a series of 4-aminoquinolines with demonstrated antimalarial activity (Thelingwani et al. unpublished). Figure 7.7 shows the retention times of standard compounds and the standard curve (plot of

Table 7.5 Predicted and measured physicochemical and ADME properties of some novel 4-aminoquinolines with anti-malaria activity

Compound	W2 (μM)	D10 (μM)	3D7 (μg/ml)	K1 (μg/ml)	In silico Log $D_{7.4}$	In vitro Log $D_{7.4}$	In silico solubility (μM)	In vitro solubility (μM)	$CL_{int, app}$ (ml/min/kg)	CYP3A4 inhibition, (K_i μM)	Type of inhibition
CHQ			0.003	0.14							
11A	23.7	>2	ND	ND	4.02	1.53	2.41	>100	7.24	9.66	Non-competitive
11A4	0.70	>2	ND	ND	1.94	0.26	55.97	>100	19.84	39.60	Non-competitive
11B	5.97	>2	ND	ND	4.54	1.47	1.87	>100	13.43	0.97	Mixed
11B2	2.9	>2	ND	ND	ND	ND	44.05	>100	ND	70.99	Non-competitive
12A	4.95	>2	ND	ND	4.07	1.63	1.54	>100	2.59	0.36	Non-competitive
12A1	ND	ND	ND	ND	ND	ND	41.23	>100	ND	5.42	competitive
12AB	5.2	>2	ND	ND	4.59	2.62	1.13	>100	7.45	1.14	Non-competitive
14A	2.5	0.16	0.023	0.97	4.59	2.27	0.74	>100	5.04	1.19	Non-competitive
14AB	0.22	>2	0.002	0.06	5.11	3.66	0.60	50	7.96	27.59	Non-competitive
14AB1	0.37	>2	0.024	0.12	3.11	1.27	20.70	1	15.62	6.70	Non-competitive
13A	0.83	>2	ND	ND	5.48	4.77	0.33	25	3.46	3.73	Non-competitive
13A1	0.15	0.15	ND	ND	3.52	1.30	7.59	>100	26.93	5.81	Non-competitive
13AB	0.03	0.14	ND	ND	6.00	5.82	0.25	25	14.04	0.79	Non-competitive
13AB.2HCl	ND	ND	0.004	0.05	6.00	5.82	0.25	50	11.63	3.39	Mixed
13AB1	0.19	0.11	0.037	0.289	4.05	2.40	5.79	>100	6.77	0.34	Non-competitive
17A	ND	ND	0.052	0.11	2.86	0.22	14.92	10	11.99	3.69	mixed
17A1	ND	ND	ND	ND	ND	ND	11.34	>100	ND	3.95	Non-competitive
18A	ND	ND	ND	ND	3.75	1.63	4.63	>100	13.82	4.38	Non-competitive
18A1	ND	ND	0.026	0.08	ND	ND	3.48	>100	ND	3.58	Non-competitive

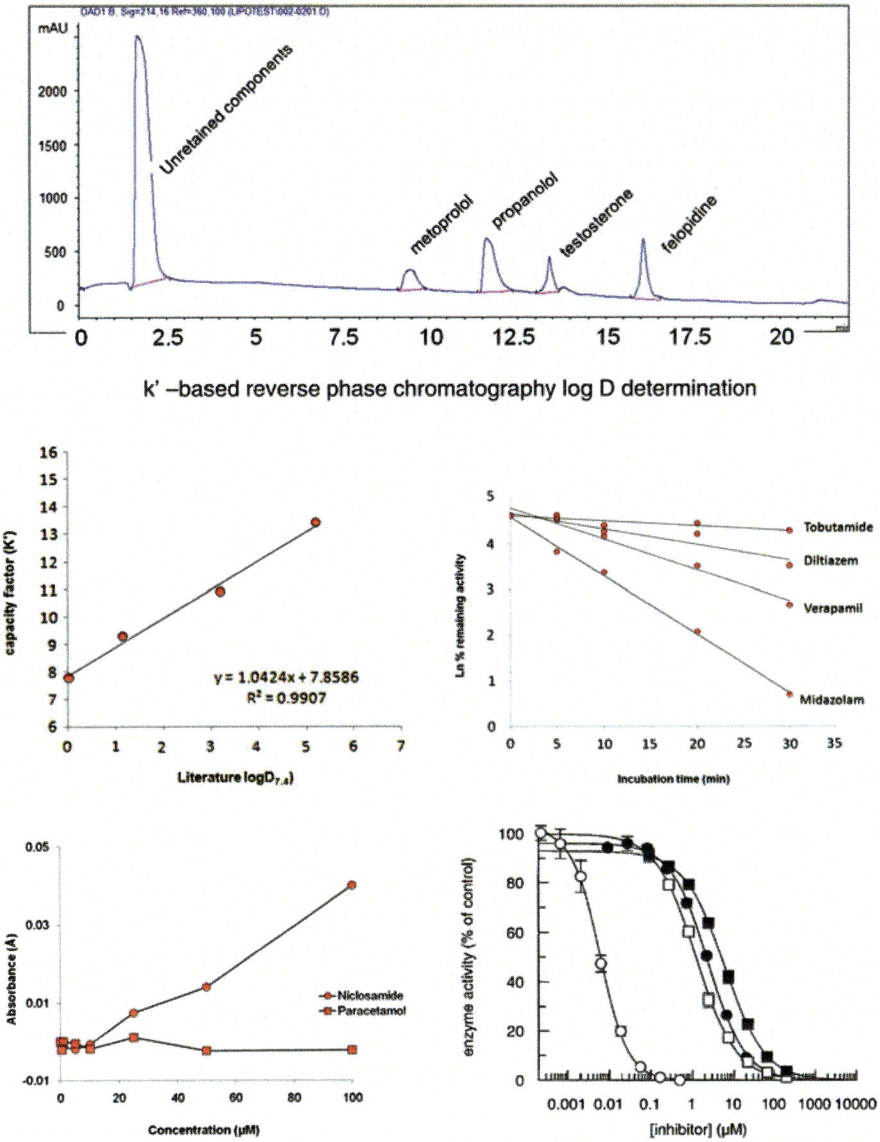

Fig. 7.7 Typical results outputs/presentation of various physicochemical and ADME determinations available at AiBST

literature Log $D_{7.4}$ against measured capacity factor of standard compounds) from which the capacity factors of unknown compounds are used to read off their predicted Log $D_{7.4}$.

7.7.2 Determination of Compound Solubility

Solubility in water is determined using a turbidimetric assay. In this assay, test compounds dissolved in DMSO are diluted in water, and the different solutions are evaluated for precipitation spectrophotometrically at a wavelength of 595 nm. In this method, a highly water-soluble compound, paracetamol, and a poorly soluble compound, niclosamide, are used as controls. In the absence of a precipitate, the absorbance remains below 0.05 units and will increase as precipitate is formed. Compounds are then classified as soluble (>100 μM), intermediate (<50 μM), partial (<20 μM) and insoluble (<10 μM). Solubility is very important in the design of in vitro experiments as it informs one on the maximum concentrations one can reliably work with in aqueous solutions. Figure 7.7 shows typical solubility/precipitation results of paracetamol, niclosamide and a test 4-aminoquinoline.

7.7.3 Determination of Compound Permeability

The Caco2 cell line derived from the human colon carcinoma has proved a very good predictive model for the fraction of drug absorbed, F_a. This is because it has many of the morphological and physiological attributes of the small intestine, such as microvilli, and various mechanisms of permeability (transcellular, paracellular and active transport based). The cells are plated in a cell culture inset device in which the cells grow to confluence on a porous filter. Compounds are loaded on the apical side of the cell culture insert device, and samples are collected over time on the basolateral side. These data are then used to calculate the apparent permeability, P_{app}, as cm/s. Compounds with $P_{app} < 1 \times 10^{-6}$ cm/s being classified as low permeability, $<20 \times 10^{-6}$ cm/s—moderate permeability and $>20 \times 10^{-6}$ cm/s—high permeability. To investigate the potential role of efflux transport, the test compound can also be put in the basolateral side, and samples can be collected and measured on the apical side. If the ratio of A to B/B to A is greater than 2, this points to the possible existence of active transport of the compound.

7.7.4 Determination of Compound Metabolic Stability

Metabolic stability, a measure of a compound's extent of biotransformation, is determined in liver microsomes (for oxidative metabolism) and plasma (for compounds likely to undergo hydrolysis). Microsomes are the most commonly used in vitro system due to the fact that most drugs are mainly metabolised by cytochrome P450s which have the highest concentration in liver microsomes. The study involves incubating 1.0 μM of test compound in 0.5 mg/ml liver microsomes for 0, 5, 10, 15, 20 and 30 min. The elimination rate constant, K_{el}, is estimated from the plot of the natural logarithm of remaining drug concentration versus incubation time. The K_{el} is then used to calculate the elimination half-life, which in turn, is used to calculate the intrinsic metabolic clearance (ml/min), $CL_{int} = (0.693 \times \text{volume})/t_{1/2}$. Figure 7.7 shows a typical metabolic stability result for low, high and intermediate clearance compounds. Such studies are done with liver microsomes and cytosols from preclinical animal models and humans. To control the assays, a set of standard compounds is run with each batch of test compounds. Table 7.5 shows the metabolic stability of a novel series of 4-aminoquinolines with demonstrated antimalarial activity.

7.7.5 Determination of Cytochrome P450 Inhibition

The modern clinical setup is characterised by the practice of poly-pharmacy where patients are taking many drugs at the same time, either to treat multiple co-ailments or due to the need of combination therapy for increased efficacy. The latter is very common in the treatment of infectious disease in order to reduce the risk for drug resistance. At hit identification and characterisation, the test compounds are evaluated for inhibitory effects on recombinant CYPs, 1A2, 2C9, 2C19, 2D6 and 3A4. These are among the most important CYPs responsible for the metabolism of most drugs on the market and are associated with a risk for drug–drug interactions if inhibited. High-throughput assays using substrates which are metabolised to fluorescent metabolites can be used. In the assay, each test compound is tested for inhibitory effects on each rCYP at different concentrations. The inhibitor concentration resulting in 50% reduction in enzyme activity (IC_{50}) is determined by plotting the remaining enzyme activity as a measure of fluorescence against the varying concentrations of the test compound. For each rCYP, a positive control inhibitor is used, napthaflavone for CYP1A2, sulfaphenazole for CYP2C9, ticlopidine for CYP2C19, quinidine for CYP2D6 and ketoconazole for CYP3A4. Figure 7.7 shows typical inhibitory effects of some 4-aminoquinolines on CYP1A2. Table 7.5 shows the inhibitory effects of 4-aminoquinolines on the major drug-metabolising enzyme, CYP3A4.

While most inhibitory effects of compounds are of a reversible mode (competitive or non-competitive), some compounds demonstrate time-dependent inhibition (TDI). TDI is usually associated with the bioactivation of a test compound to a reactive metabolite which in turn forms a covalent bond with the CYP, resulting in irreversible (referred to as mechanism-based inhibition—MBI) or quasi-irreversible (referred to as metabolite intermediate complex inhibition—MIC) inhibition. Mechanism-based inhibition can, in addition, result in idiosyncratic reactions due to immunogenic reactions caused by proteins covalently modified by the reactive metabolites. The assay for the determination of TDI involves pre-incubation of the test compound with the enzyme (that would result in the inactivation of enzyme molecules) followed by a second incubation in which the enzyme substrate is added (to measure the remaining enzyme activity). Control reactions with pre-incubation without the test compounds or without NADPH, the cofactor required for CYP activity, are also run in parallel. The final result is given as a ratio, which if below 0.7 is indicative of TDI effects and if >0.9 is indicative of no TDI effects. Control TDI compounds for each of the enzymes are used, furafylline for CYP1A2, tielinic acid for CYP2C9, ticlopidine for CYP2C19, paroxetine for CYP2D6 and troleandomycin for CYP3A4.

The in silico and in vitro studies at this compound library and hit identification and characterisation stage help projects in identifying potentially important trends that can affect the feasibility of pursuing certain compound classes to the lead discovery phase. For example, in the presence of alternatives, strong correlations between Log *P*, potency, metabolic stability and CYP inhibition could be a "show stopper" as they point to a very difficult chemical space to separate and optimise for these important PK/PD/safety variables. A series exhibiting very strong TDI can also be a serious liability as the possibility of idiosyncratic reactions that could be caused by such compounds is difficult to predict clinically. High Log *P* values at hit identification stage are also not favourable since medicinal chemists generally increase lipophilicity to increase potency during lead optimisation. It is therefore important to invest in highly competent teams to exploit the large amounts of data generated in a manner that extracts information and translates it to innovative project strategies for successful hit identification.

7.8 Lead Discovery

During lead discovery (LD), clusters identified in hit identification stage are worked on to select two to three series which meet the lead compound criteria set for the project. The main purpose of this phase is to identify key liabilities associated with the selected series and to suggest ways they could be addressed during the subsequent lead optimisation (LO) stage. Series carrying severe DMPK problems associated with the pharmacophore, e.g. an overlap of the pharmacophore for pharmacologic effects and that of a DMPK liability such as metabolic instability or CYP inhibition, are usually discontinued. The LD phase is a learning phase where the selected scaffolds from LI are expanded to identify any structure activity relationship in physicochemical and ADMET properties. Assays performed during LI continue to be performed to keep track of any changes in physicochemical and ADME properties as the exploration of the chemical space of the selected series is conducted.

Additional assays are also conducted at this stage. These include metabolite identification (to understand the possible basis of metabolic instability and/or mechanism of reactivity), enzyme identification and in vivo PK in preclinical PK and PD animal models. These data are used to understand the relationship between structural and DMPK properties (QSPR—quantitative structure property relationship). At this stage, one also investigates in vitro–in vivo correlation for DMPK properties such as absorption and clearance. In vitro and in vivo DMPK and PD data are also modelled at this stage to derive PK/PD models. Such models at this early stage are important as they can rescue a chemical series with seemingly poor DMPK or PD properties. There are many successful drugs on the market which would not have made it based on their DMPK or their PD properties alone. Addressing each of these properties separately is therefore risky in drug discovery and development. The ideal situation in the selection of lead series is where one can optimise the DMPK issues without compromising pharmacological properties. The difficult situation and usually a show stopper is when these key properties covariate and cannot be optimised independently.

7.8.1 Metabolite Identification

The major determinant of drug clearance is metabolism. To understand and possibly address metabolic instability or chemical reactivity, identification of metabolites formed is important. There are many software programmes which have been developed for the prediction of metabolites. In our laboratory, we use the software called Metasite (Molecular Discovery Ltd) and various docking software (Auto Dock, Flex X, or GOLD). Metasite works by using GRID (Molecular Discovery Ltd) derived chemical interaction energy maps of the test compound and that of the CYP's active site. The maps of the compounds are overlayed on the maps of the CYP's active sites, and a similarity index is used to identify compound-CYP matches. To suggest the site of metabolism, the distance of the nearest hydrogen to the CYP reactive centre is measured and used to rank all hydrogen in the compound. Distances (3–5 Å) associated with the possibility of hydrogen abstraction (the major mechanism of compound oxidation by CYPs) are used to identify and rank metabolic hot spots. The prediction engine can also take into account reactivity factors (i.e. in addition to the optimal distance, it will also identify functional groups associated with increased easy of oxidation). The result is therefore a ranking of sites for possible metabolism and hence the likely metabolites formed. Figure 7.8 shows the successful use of Metasite in the prediction of sites of metabolism of thiabendazole. We also use various docking algorithms to deduce possible metabolites. Figure 7.8 also shows the application of the docking programme, GOLD, in the prediction of the enzyme and site of metabolism in the biotransformation of amodiaquine.

In vitro, metabolite identification is done from incubations of test compound with human and preclinical PK and PD animal model liver microsomes or hepatocytes or human recombinant enzymes. Metabolite identification is then done by LC-MSMS. The MSMS product ion is interpreted by biotransformation scientists. It can also be done through assistance by software such as MetaboLynx (http://www.waters.com/waters), LighSight (http://info.appliedbiosystems.com/metaboliteID) and Mass-Metasite (http://www.moldiscovery.com). Most CYP-mediated reactions are aromatic hydroxylations, aliphatic hydroxylations, N-dealkylation and O-dealkylation and epoxidations. Metabolite structure can be

P450s	Cavity Radius	Hits	Distance (Å)[α]
CYP2C8	15 Å	5	2.47–7.94
	12 Å	7	2.65–4.10
	10 Å	2	2.29–3.44
CYP2C9	15 Å	1	
	12 Å	0	
	10 Å	1	
CYP2C18	15 Å	0	
	12 Å	0	
	10 Å	2	
CYP2C19 (many different solutions of binding modes)	15 Å	0	
	12 Å	2	
	10 Å	0	

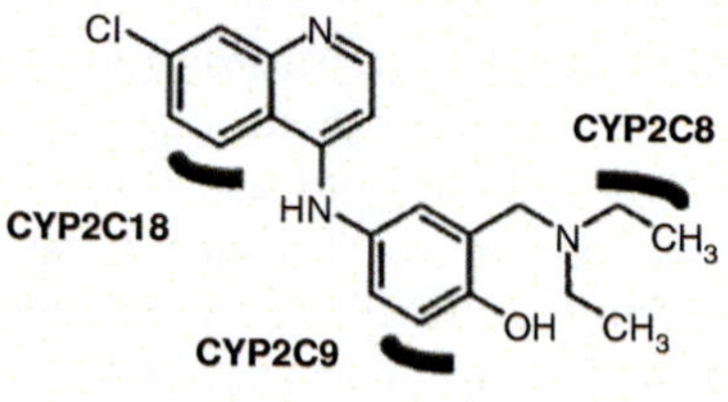

Amodiaquine

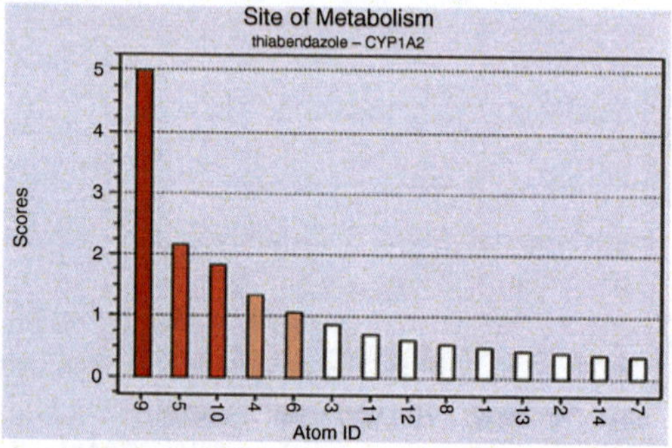

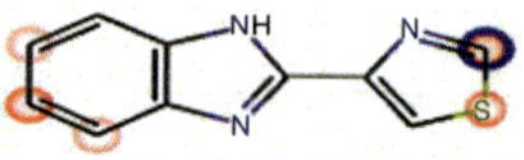

Thiabendazole

Fig. 7.8 *Top panel*, predictions of sites of metabolism of amodiaquine using the docking programme, GOLD [7] and the, *lower panel*, the predictions of sites of metabolism of thiabendazole using the software, Metasite [8]

interpreted in terms of these common reactions such as +16 for hydroxylation or N-oxidation, +32 for dihydroxylation and −14 for demethylation. Some metabolites cannot be easily elucidated by MSMS and might require NMR. In this case, the microsomal incubation is scaled up in order to generate 10–50 μg of metabolite. The ^{1}H-NMR spectrum of the metabolite is obtained, and the changes of resonance are used to determine the site of metabolism, hence the structure of the metabolite.

Knowledge of the metabolite can be used for several purposes in drug discovery. These include guidance in blocking of metabolic hot spots to stabilise the compounds and synthesising the metabolites and testing them for pharmacological and toxicological effects. Figures 7.9 and 7.10 show the metabolite identification in the metabolism of amodiaquine.

7.8.2 *Trapping of Reactive Metabolites*

Some metabolites are very reactive and difficult to isolate and conduct structural studies on since they react with biological components in the incubation. Studies on reactive metabolites are done in two stages: the first is to screen compounds for the generation of reactive metabolites. This is done by screening for the formation of

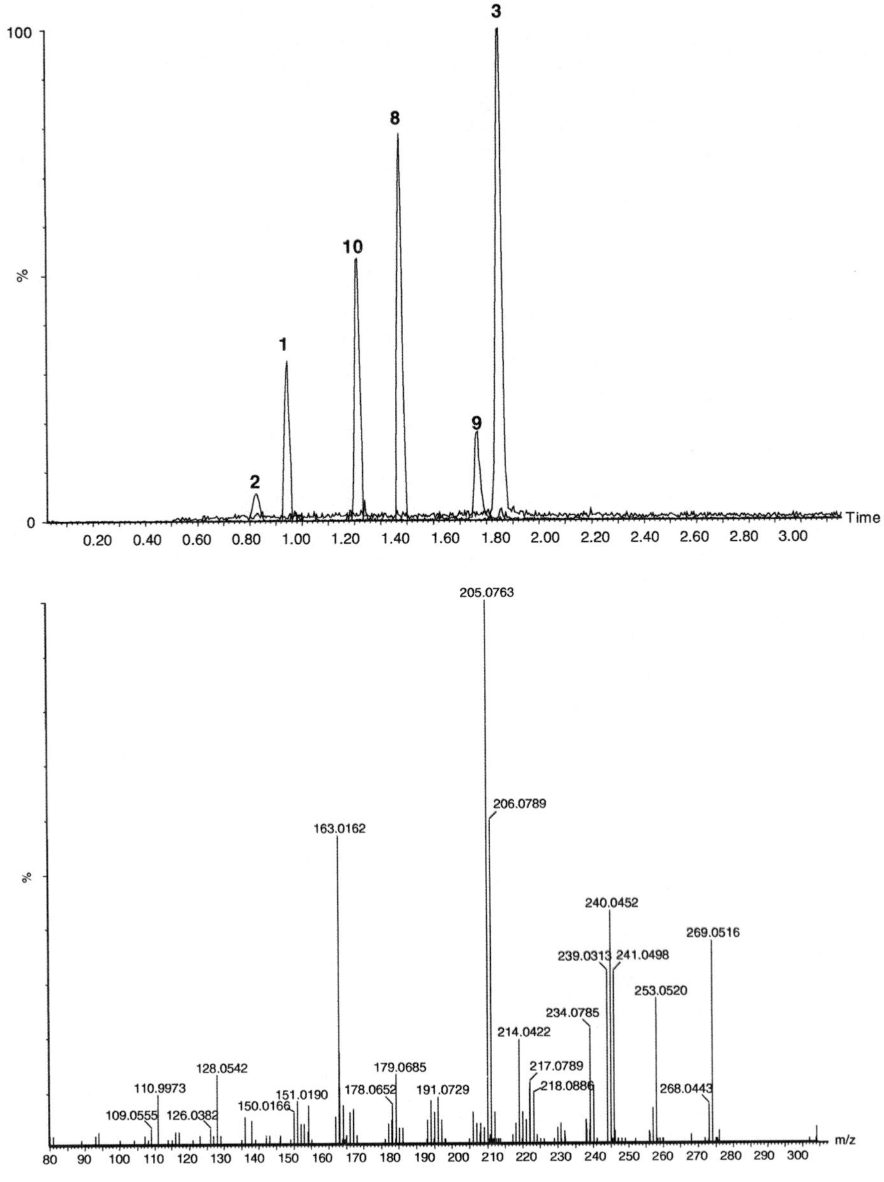

Fig. 7.9 Metabolite identification in the metabolism of (**a**) amodiaquine **1** and its metabolites formed after incubations for 30 min with rCYP1A1. Extracted ion chromatograms of *m*/*z* 356 (**1**), *m*/*z* 328 (**2**), *m*/*z* 317 (**10**), *m*/*z* 315 (**9**), *m*/*z* 301 (**8**) and *m*/*z* 299 (**3**) and (**b**) MS/MS spectra of **3** [9]

glutathione conjugates in human liver microsomes. If iminium ions are likely to be formed, trapping experiments are done using cyanide. Other reactive metabolites are better trapped with methyloxamine or cysteine. For series containing carboxylic

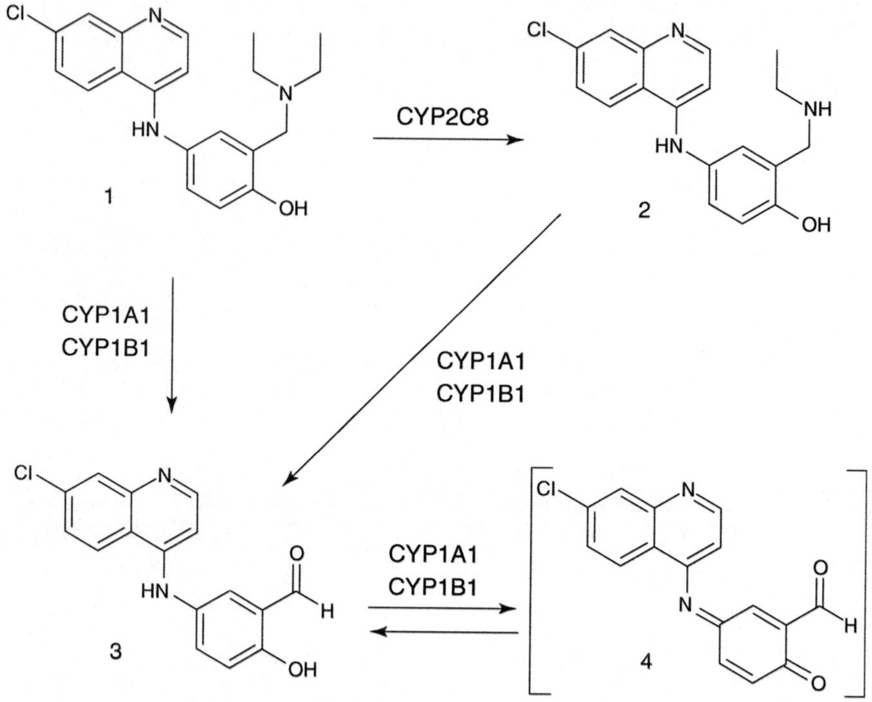

Fig. 7.10 It shows the metabolic scheme in the biotransformation of amodiaquine [9]. **1**—amodiaquine, **2**—desethylamodiaquine, **3**—amodiaquine aldehyde metabolite and **4**—amodiaquine aldehyde quinoniemine metabolite

acid functions, the stability of acyl glucuronides should also be assessed. The isolation and subsequent structural elucidation of such reactive metabolites is a challenge since they usually covalently react with the biological matrix. To overcome this, non-biological systems can be used. The use of electrochemical oxidation is being successfully used to generate most CYP medicated metabolites. Figures 7.10 and 7.11 show how this approach has been used to characterise reactive metabolites of amodiaquine. Both electrochemical oxidation and microsomal incubations where used to characterise the bioactivation of amodiaquine to amodiaquine quinoneimine and the aldehyde metabolite. Trapping experiments with cysteine resulted in four cysteinyl conjugates, and those with glutathione gave four glutathionyl conjugates. MSMS and ^{1}NMR were used to characterise the structures of these metabolites [10]. The generation of the aldehyde metabolite was done by electrochemical oxidation and structural studies done by MSMS and ^{1}NMR. Trapping of the aldehyde metabolite was done with methoxylamine. Trapping experiments with *N*-acetyl cysteine revealed that the aldehyde was further oxidised to an aldehyde quinoneimine species, both in the rCYP incubations and in the electrochemical system (Fig. 7.11).

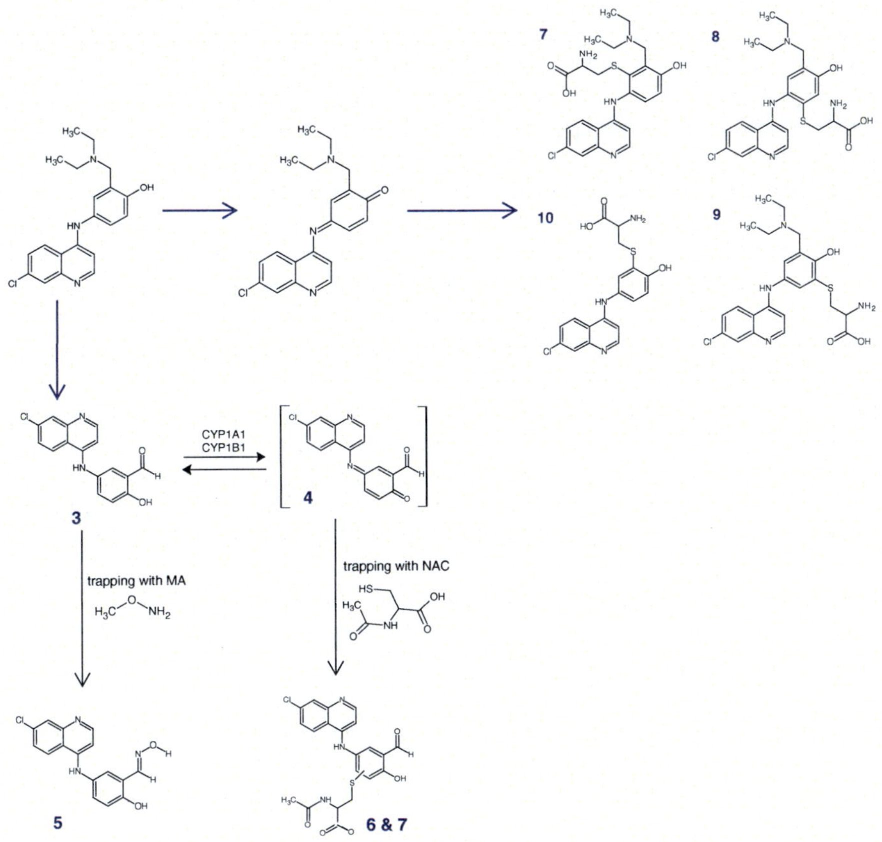

Fig. 7.11 The bioactivation and trapping studies for amodiaquine to amodiaquine quinoneimine and the aldehyde metabolite. The amodiaquine quinoneimine gave 4 cysteinyl conjugates, 7–10. The aldehyde metabolite and its quinoneimine metabolite formed conjugates with methoxyamine (AM) and *N*-acetyl cysteine (NAC) [9, 10]

7.8.3 Enzyme Identification

Towards understanding the basis of metabolic stability and/or bioactivation of new chemical entities, there is a need to identify the enzymes involved in the generation of identified metabolites. This knowledge will be helpful in molecular design towards disrupting the physicochemical determinates of the compound's specific interactions with the enzyme involved. For example, if one shows that a compound is metabolised by CYP2D6, there are well-known qualitative and quantitative structure activity relationships (SAR) associated with substrates of this enzyme that can enable medicinal chemists to reduce metabolic rates by either blocking metabolic hot spots or disrupting the CYP2D6 substrate pharmacophore. This strategy can be used to either solve instability issues or block the generation of reactive metabolites. Enzyme identification is also important towards understanding inter-individual variability in the pharmacokinetics of a drug. If one finds that a drug is solely metabolised by say CYP2D6, then the known genetic and environmental variation of this enzyme in human populations will be expected to reflect the pharmacokinetic variation of the new chemical entity.

Enzyme identification is also used towards predicting likely drug–drug interactions in situations where the test compound is co-administered with a drug that inhibits or induces the enzyme that is mainly responsible for the elimination of the test compound. There are several commonly used methods for enzyme identification (Fig. 7.12a, b, c). It is generally encouraged to use at least two of these methods to arrive at a relatively accurate conclusion on the qualitative and quantitative contributions of various enzymes in the metabolism of a test compound. First, the compound is incubated with sub-cellular fractions known to be involved in the metabolism of drugs. Since most drugs undergo oxidative metabolism by CYPs, identification for these enzymes will be discussed in detail. Data from work on the metabolism of amodiaquine will be used to illustrate the methodology (Fig. 7.12). Data on the identification of CYPs responsible for the metabolism of anti-parasitic drugs will be used to demonstrate the likely general and specific implications of such information in the safe and efficacious use of these drugs in African populations.

(a) Screening of Compound Metabolism Across a Panel of Recombinant Enzymes

In this experimental setup, the compound is incubated with each of a panel of recombinant CYPs such as CYP1A1, 1B1, 2A6, 2B6, 2C8, 2C9, 2C19, 2D6 and 3A4. The incubation in which the test compound is depleted by over 20% is highlighted as a possible contributor to the compound's metabolism. The relative contribution of the highlighted enzymes in the metabolism of the test compound is then determined using the relative activity factor. This involves determining the metabolic clearance of enzyme-specific substrate (e.g. midazolam for CYP3A) in human liver microsomes and also with the recombinant enzyme. The ratio of activity in human liver microsomes/activity in recombinant enzyme would be an expression of nanomoles of recombinant enzyme/mg of liver microsomes. This is referred to as the relative activity factor (RAF) as it estimates the relative amount of CYP3A in the liver microsomes in activity terms. The clearance of the test compound is then determined with recombinant enzyme, the value which will be multiplied by the RAF to give the metabolic activity in the clearance of the test compound due to the specific enzyme. To estimate the relative contribution of various CYPs, the test compound is then incubated with the liver microsomes, which gives the total clearance from the contribution of many other CYPs in the microsomes. For the relative contribution of the specific enzyme(s), the activity obtained from multiplying the RAF with the activity of the specific rCYP is then divided by the compound clearance in the liver microsomes. The fraction of the compound metabolised (f_m) by the specific CYP will then be expressed as a percentage. Table 7.6 shows the application of the above approach in the metabolism of anti-parasitic drugs by CYPs. This knowledge of which CYPs are

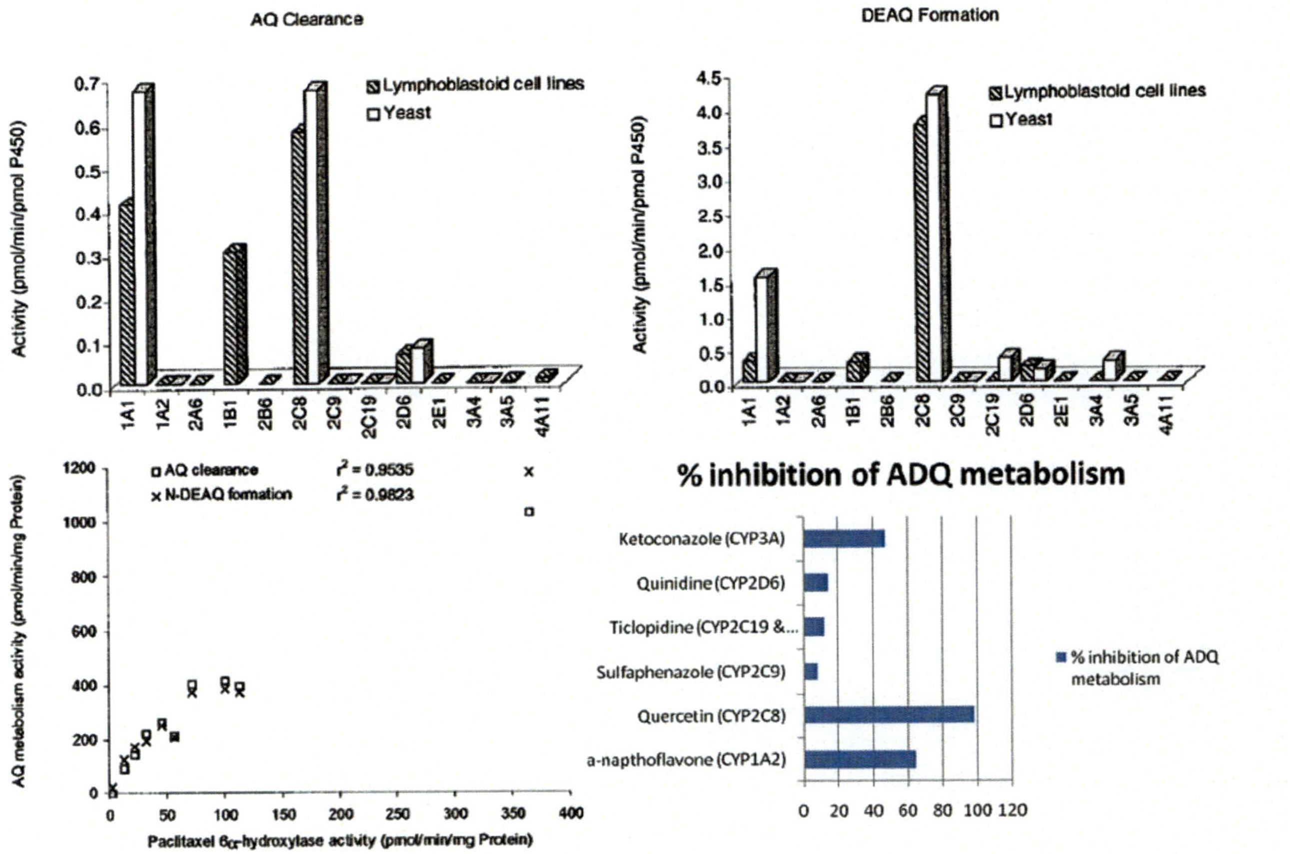

Fig. 7.12 Some methods used in reaction phenotyping as applied to the characterisation of the metabolism of amodiaquine (ADQ) and the formation of desethylamodiaquine. The *top panel* shows the disappearance of amodiaquine in incubations across a panel of rCYPs and the formation of desethylamodiaquine by the same panel of rCYPs. The *lower panel*, *left*, shows the correlation of amodiaquine disappearance and desethylamodiaquine formation in a panel of human liver microsomes characterised for CYP2C8 activity. The *lower panel*, *right*, shows the inhibitory effect of CYP selective diagnostic inhibitors on the disappearance of amodiaquine in incubations with human liver microsomes

Table 7.6 Calculated intrinsic clearance (CL_{int}) of 15 anti-parasitic drugs (APDs) in RLM and HLM, and the relative contributions of CYPs to the metabolism of APDs in HLM using RAFs method [7]

Compound	$CL_{int,RLM}$ (μl/min/mg protein)	$CL_{int,HLM}$ (μl/min/mg protein)	Predicted CL_H (ml/min/kg)	rCYP (% contribution in HLM)								
				1A2	2A6	2B6	2C8	2C9	2C19	2D6	3A4	Sum
Albendazole	43.8	204.4	18.2	53			0.3	3.5	2.2	0.6	5.4	65
Amodiaquine	61.3	440.1	19.1				67			0.2		68
Artemisinin	49.4	19.8	9.8			10				6.5	25	42
Artesunate	113.6	33.2	12.3		120	1.3						121
Chloroquine	3.2	1.5	1.4				54			53	13	120
Dapsone	15.1	6.9	5.0				10	48	12	4.3	31	105
Diethylcarbamazine	8.6	0.7	0.6									ND
Pentamidine	6.9	<0.1	<0.1									ND
Praziquantel	56.3	82.9	16.0	39					14	0.2	30	83
Primaquine	73.9	7.5	5.3	60						23		83
Pyrantel	11.4	4.3	3.4							90		90
Pyrimethamine	<0.9	1.1	1.0									ND
Quinine	37.7	9.1	6.1						3.8		70	74
Thiabendazole	20.1	174.9	17.9	395								395
Tinidazole	8.6	1.2	1.1			12					77	89

responsible for the metabolism of these anti-parasitic drugs enables one to infer pharmacokinetic variability and risk for drug–drug interactions in the clinical use of these drugs.

(b) Correlation Analysis

In this approach, a set of individual human liver microsomes, $n = 5–7$, is chosen which shows a wide spectrum of activities of each of the major human CYPs. This is done using CYP-specific marker substrate reactions such as CYP1A2 (phenacetin to paracetamol), CYP2A6 (coumarin to 7-OH coumarin), CYP2B6 (efavirenz to 8-OH efavirenz), CYP2C9 (diclofenac to 4-OH diclofenac), CYP2C19 (S-mephenytoin to 4-OH mephenytoin), CYP2D6 (bufuralol to 1-OH bufuralol) and CYP3A4 (midazolam to 1-OH midazolam). In addition to the wide range of activity, the activities of each of the CYPs for the various CYPs should not cross-correlate as this will make it difficult to differentiate the roles of such CYPs. The test compound is then incubated with each of these liver microsomes, and the activities are cross-correlated with the activities of the various CYPs. Cross-correlations of >0.7 are indicative of a significant contribution of that enzyme. Figure 7.12 shows such cross-correlations in the identification of the enzymes responsible for the metabolism of amodiaquine.

(c) Use of Potent and Selective Diagnostic Chemical Inhibitors

This approach exploits the knowledge that there are some compounds which demonstrate potent and selective inhibition of some human CYPs. The test compound is therefore incubated with human liver microsomes in the presence and absence of such inhibitors. The extent of inhibition of the test compound's clearance is therefore an estimate of the relative contribution of the inhibited enzyme. The most commonly used potent and relatively selective inhibitors are furafylline and α-naphthoflavone (CYP1A2), sulfaphenazole (CYP2C9), ticlopidine (CYP2C19), quinidine (CYP2D6), quercetin (CYP2C8) and ketoconazole (CYP3A). Figure 7.12 shows how this has been used in estimating the relative contribution of CYPs in the metabolism of amodiaquine. These data conclusively identify CYP2C8 as the major enzyme responsible for the metabolism of amodiaquine to its major metabolite, desethylamodiaquine.

7.8.4 *In Vivo Pharmacokinetics*

Whereas the in silico and in vitro ADMET studies mainly offer mechanistic understanding of pharmacokinetics which is useful in the design of molecules with predicted good PK, in vivo pharmacokinetics are needed to provide a whole system profile of the test compounds. At preclinical stage, various animal models including mice, rat, guinea pig, rabbit, dog and monkey are used to provide pharmacokinetic data on bioavailability, absorption, volume of distribution, half-life and clearance. While these animal models give an integrated outcome of all the

PK processes of absorption, distribution and elimination, their major limitation is that of species differences in such processes between them and humans. This is particularly so for metabolic processes where, for most drugs, there are major interspecies differences. This poses a challenge in making allometric scaling predictions of animal PK to human PK. Other parameters such as volume of distributions and protein binding, however, tend to scale relatively well between species. For drugs which undergo rapid metabolism hence have their clearance limited by blood flow instead of metabolism, they also tend to scale better between species.

Against these limitations, the in vivo preclinical PK is, however, still useful in establishing in vitro–in vivo correlations. This is done against the debateable logic that, if in vitro ADME parameters using sub-cellular fractions from a preclinical animal model can be successfully scaled to the in vivo PK data in that animal, then in vitro ADME parameters derived from human sub-cellular fractions will also be predictive of human PK. A lack of correlation would also be interpreted in a similar way. Most importantly, in vivo PK studies are used to establish plasma concentration–effects relationships in disease animal models. This preclinical proof-of-concept data give projects confidence to take compounds through to human studies.

In routine in vivo PK profiling of compounds the test compound is given to 3 animals i.v and another three animals p.o. Samples are collected over a 24-h period, and PK parameters (Table 7.1) are calculated from the plot of drug concentration versus time using PK software such as Winnonlin (http://www.pharsight.com/products/prod_winnonlin_home.php). If in vitro data show that a test compound inhibits a human CYP whose orthologue can be found in preclinical animal models such as CYP1A2, one can conduct preclinical in vivo drug–drug interaction studies. Figure 7.13 demonstrates such an application of in vivo PK studies in the evaluation of the inhibitory effects of medicinal plant extracts that had been shown to be potent inhibitors of human CYP1A2 in vitro. Using caffeine as a marker for CYP1A2 activity, the effects of each herbal extract were evaluated on its pharmacokinetics in rats. Data from such studies could be used to support recommendations of product label revisions with respect to risk for drug–herb interactions.

7.8.5 In Vitro to In Vivo Correlation

Most preclinical in silico and in vitro data give information that is generally used in a qualitative manner for SAR and sometimes quantitative manner in QSAR modelling to guide the design of molecules. To gain confidence in the lead compounds selected, efforts are then made to make quantitative in vitro ADME to in vivo PK correlations. This is probably the most difficult part in drug discovery pharmacokinetics since many of the in silico and in vitro animal model systems are fraught with many mechanistic and physiological limitations. The ambitious effort of putting together many pieces of ADME information to predict the in vivo outcome has received enormous attention since its success is the key measure of the usefulness of preclinical DMPK.

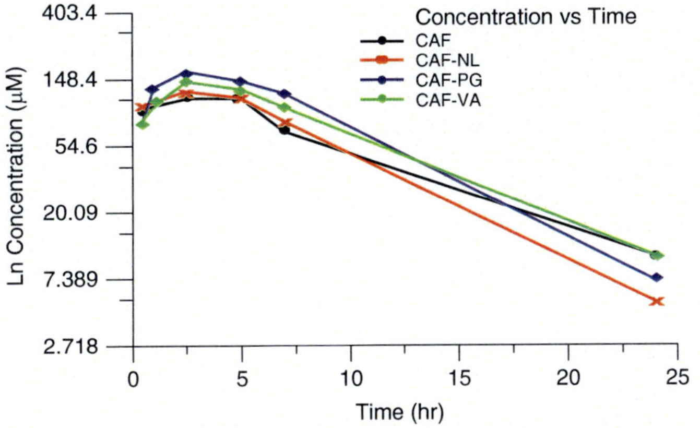

Parameter	Caffeine only	Caffeine and *Piper guineeses*	Caffeine and *Nauclea latifolia*	Caffeine and *Vernonia amygdalina*
T_{max} (Hr)	2.5	2.5	2.5	2.5
C_{max} (μM)	114.73	164.31	120.37	142.65
$T_{½}$ (Hr)	5.7	4.34	4.31	5.22
AUC (hr* μmol)	1372	2005.5	1413.82	1707.9
Cl/F (mg/(hr* μmol)/kg)	0.013	0.009	0.013	0.011
V_d (mg/(μmol)/kg)	0.11	0.06	0.08	0.08

Fig. 7.13 The pharmacokinetics of caffeine, a marker substrate for CYP1A2, in rats, by itself and after co-administration with various plant extracts with reported in vivo antimalarial effects that had been found to inhibit human CYP1A2 in vitro (Moyo et al. unpublished)

7.8.5.1 Absorption (F_a)

The absorption rate constant, K_a, is determined from in vivo animal PK studies by the method of residuals. It is a difficult parameter to predict since it depends on the formulation used. The average K_a from several preclinical animal models is used for humans. K_a has a maximum value of 0.1 min^{-1}. K_a is useful in the estimation of C_{max}, a value important in the predictions of drug–drug interactions and design of toxicological studies. The software, Gastroplus, which takes into account dissolution properties, can be used to give some estimates of this value.

The fraction absorbed, F_a, can be predicted from allometric scaling where rat to human correlations are better than dog to human, with the latter usually over-predicting F_a. For many drugs, the in vitro Caco2 permeability measures, P_{app}, show a positive correlation with F_a. This is particularly so for drugs whose main mechanism of permeability is passive transcellular. The plot of P_{app} versus F_a is, however, very steep in the P_{app} value range of 0.2 and 0.8 $\times$ 10^{-6}. This is why the predictions are generally simplified to $P_{app} > 1.0 \times 10^{-6}$ being associated with F_a of over 80%. Compounds which undergo active transport or are extensively metabolised in the gut are more difficult to predict from permeability studies using Caco2 cells. Better predictions can be made using pieces of intestinal tissues in an Ussing chamber permeability/absorption system.

The software, Gastroplus, can be used to predict F_a where solubility and dissolution rates versus GI transit time are also taken into account. Using this software, more complex modelling that takes into account compound stability in gastrointestinal juice, chemical stability at low pH, efflux, uptake and metabolism in the gut wall can be considered.

The equation used in the prediction of the concentration of drug absorbed is

$$C_{abs} = C_{max} + \frac{k_a \cdot F_a \cdot \text{Dose}}{Q_H}, \tag{7.1}$$

where C_{abs}—concentration of drug absorbed, C_{max}—maximal plasma concentration, K_a—absorption constant, F_a—fraction absorbed and Q_H—liver blood flow.

7.8.5.2 Volume of Distribution (V_d)

Volume of distribution (V_d) can be estimated using the simplified Oie–Tozer equation:

$$V = V_p + f_u \cdot \frac{V_t}{f_{u,t}}, \tag{7.2}$$

where V is volume of distribution, V_p is plasma volume, f_u is unbound fraction in plasma, V_t is volume of tissue and $f_{u,t}$ is the unbound fraction in tissue.

The $f_{u,t}$, V_p and V_t for each species can be obtained from the standard data tables found in the literature. The average $f_{u,t}$ in animals is assumed to be equal to $f_{u,t}$ in humans and is used in Eq. (7.2). The plasma protein binding, f_u, is measured in vitro using human plasma.

Allometric scaling can then be used to predict V_d in humans from animal V_d by plotting the total volume in preclinical species (in units of litres per animal) versus animal body weight (kg) on a log–log scale:

$$\text{Log}_{10}V = a\ \text{Log}_{10}(Bw) + b. \tag{7.3}$$

The free volume of distribution, V_u, can be normalised by taking into account differences in protein binding between species. This normalised value can then be used in the allometric scaling discussed above. Such normalised values tend to give better predictions of V_d in humans with a twofold order of magnitude of the real value:

$$V_{\text{animal,normalized}} = \frac{f_{\text{u,human}} \cdot V_{\text{animal}}}{f_{\text{u,animal}}}. \tag{7.4}$$

Table 7.7 Scaling factors used in the predictions of in vivo clearance from in vitro data

Species	Body weight (kg)	Liver weight (g)	Q_H (ml/min/kg)	MPPGL (microsomes mg prot/g liver)	HPGL (hepatocytes 10^6 cells/g liver)
Mouse	0.025	1.5	152	45	130
Hamster	0.12	6	43	45	120
Rat	0.25	11	80	45	125
Cynomolgus	3	125	44	45	120
Dog	12	384	33	43	120
Human	70	1,680	21	39.8	117.5

7.8.5.3 Hepatic Clearance (CL_H)

Clearance is one of the most important PK parameters because it affects both how much and how often a drug should be administered. When combined with volume of distribution, it gives the half-life, $t_{1/2} = 0.693\ V_d/\mathrm{CL}$, and when combined with fraction absorbed, F_a, it gives the bioavailability (F). The relationship with bioavailability is based on the following relationships: CL = Dose i.v/AUC i.v, and F = (AUC p.o/Dose p.o)/(AUC i.v/Dose i.v). Substitutions with CL will then give F = (AUC p.o/Dose p.o) × CL. Successful prediction of drug clearance from preclinical data is therefore of high priority in drug discovery [11].

Against a background where allometric scaling of animal clearance to humans has not been successful, the development of in vitro metabolism models based on human tissues, microsomes and hepatocytes has resulted in improved in vitro–in vivo correlations. The in vitro–in vivo correlation starts by scaling up the intrinsic in vitro clearance, CL_{int} (μl/min/mg liver microsomes or μl/min/ million hepatocytes), from the test tube experiment to hepatic clearance per kg (ml/kg) animal model or human body weight. This is done using scaling factors for the various species used in drug discovery (Table 7.7).

The scaling equation for in vivo clearance (ml/min/kg body weight) is

$$\text{In Vivo } \mathrm{CL_{int}} = \frac{\text{In Vitro } \mathrm{CL_{int}} \cdot \text{MPPG (or HPGL)} \cdot \text{liver weight}}{1{,}000 \cdot \text{body weight}}. \tag{7.5}$$

To estimate hepatic clearance from this intrinsic metabolic clearance, physiological scaling factors such as liver blood flow (Q_H), plasma protein binding (PPB) and blood plasma partitioning (B/P) are required. These factors are then input into one of several models for the prediction of hepatic clearance (CL_H), the well-stirred model, the dispersion model or the parallel tube model.

In the *"well-stirred" or the "venous equilibrium" model*, the liver is assumed to be a single "well-stirred" compartment, and the unbound concentration in the venous blood leaving the organ is in equilibrium with and equal to the intracellular unbound concentration in the hepatocytes.

The hepatic clearance, CL_H, is expressed as

$$CL_H = \frac{Q_H \cdot f_{uB} \cdot CL_{int}}{Q_H + f_{uB} \cdot CL_{int}}, \tag{7.6}$$

where CL_{int} is the intrinsic clearance [from Eq. (7.5)], reflecting the actual metabolic capacity of the enzyme system when there is free access to substrate, f_{uB} is the free fraction in whole blood and Q_H is the total liver blood flow.

In the *parallel tube model*, the liver is assumed to be composed of a number of parallel tubes, with enzymes uniformly distributed along the tubes. The unbound blood concentration at any point along the tube is assumed to be in equilibrium with the intracellular unbound concentration:

$$CL_H = Q_H\left(1 - e^{\frac{f_{uB} \cdot CL_{int}}{Q_H}}\right). \tag{7.7}$$

The dispersion model is even more physiologically "correct" than the other two models in that it also incorporates axial dispersion of blood caused by the branching and connections of the sinusoids. The degree of dispersion is expressed by the dispersion number, Dn. The model assumes that the diffusion of drug along the sinusoids is much more rapid than the blood transit through the liver. At its two extremes, the dispersion model collapses to either the parallel tube model ($Dn \rightarrow 0$) or the well-stirred model ($Dn \rightarrow \infty$). The elimination capacity is expressed by the efficiency number Rn:

$$Rn = \frac{f_{uB} \cdot CL_{int}}{Q_H}. \tag{7.8}$$

The mathematical expression is more complex than for the other two models:

$$CLH = Q_H\left[1 - \frac{4a}{(1+a)^2 2e^{(a-1)/2Dn)} - (1-a)^2 2e^{(a-1)/2Dn)}}\right], \tag{7.9}$$

where $a = (1 + 4RnDn)^{1/2}$.

The difference between these three liver models is in their description of the concentration profile across the liver. This has most effect for high clearance compounds, which results in the highest difference in concentration across the liver. A comparison of the predictive value of the three models was studied for 28 compounds, and none of the models gave consistently better results than the other models. Since the well-stirred model is the simplest in nature, it has become the one most commonly used.

In general practice in pharmacokinetic studies, we measure plasma clearance and make the assumption that the B/P is 1.0. This might not be true for some drugs such as chloroquine and some 4-aminoquinolines which tend to partition into red blood cells. For these drugs, plasma clearance is not equal to blood clearance. In general, we also assume that binding to microsomes or cellular components is not important, but for some drugs, clearance predictions are improved by taking into account binding to in vitro biological components. Correction of microsomal binding significantly improves the prediction of clearance of acidic compounds. The ideal equation for the prediction of clearance is therefore

$$\mathrm{CL_b} = \frac{Q_\mathrm{H} \times \left(\mathrm{CL_{int}} \times \frac{\mathrm{fu_{blood}}}{\mathrm{fu_{mic}}}\right)}{Q_\mathrm{H} + \left(\mathrm{CL_{int}} \times \frac{\mathrm{fu_{blood}}}{\mathrm{fu_{mic}}}\right)}. \tag{7.10}$$

Using this equation, it is clear that the clearance of a drug cannot be greater than the blood flow through that organ. For compounds with very high intrinsic clearance, the equation reduces to $\mathrm{CL_b} = Q_\mathrm{H}$, and for low instrinsic clerance drugs, the equation reduces to $\mathrm{CL_b} = \mathrm{CL_{int}}$. Compounds are then ranked as low clearance ($<25\%\ Q_\mathrm{H}$) or high clearance ($>75\%\ Q_\mathrm{H}$).

7.8.5.4 Bioavailability (*F*)

Bioavailability is determined by both absorption and hepatic clearance:

$$F = F_\mathrm{a} + f_\mathrm{g} + \left(1 - \frac{\mathrm{CL_h}}{Q_\mathrm{H}}\right), \tag{7.11}$$

where $\mathrm{CL_H}/Q_\mathrm{H}$ = extraction ratio.

In a scenario where there is complete absorption and no gut metabolism, bioavailability will be equal to 1−extraction ratio (*E*). For cases where this is not so, one needs to calculate the concentration of the drug that is absorbed, C_abs, using Eq. (7.1):

$$C_\mathrm{abs} = C_\mathrm{max} + \frac{k_\mathrm{a} \cdot F_\mathrm{a} \cdot \mathrm{Dose}}{Q_\mathrm{H}}.$$

At the preclinical stages, K_a can be calculated from animal model PK studies and assumed to be the same in humans; F_a, fraction absorbed, can be estimated from Caco2 permeability studies. Estimation of human bioavailability from preclinical animal by allometric scaling is generally not good but is reasonable with data from monkey studies.

7.8.5.5 Half-Life ($t_{1/2}$)

The half-life along with the therapeutic index and the PK–PD relationship dictates the dosing frequency. It is also used in attempts to estimate human C_max, hence important for safety and CYP inhibition studies. Combining CL and V_d at steady state (V_{ss}) gives the effective half-life: $t_{1/2} = 0.693\ V_d/CL$. Allometric prediction of half-life from preclinical animals generally under-predict this parameter, with human half-lives being >5-fold higher than those in rats and threefold higher than in dogs and monkeys. Use of in vitro data for CL from human cellular and sub-cellular systems and in vivo V_{ss} from preclinical animal models gives improved results for half-life in humans.

7.9 Drug–Drug Interactions

Metabolism-based drug–drug interaction (DDI) is responsible for clinically important safety and efficacy issues in the use of medicines. Inhibition is generally associated with increased plasma concentration of the victim drug which might result in increased incidences and severities of adverse drug reactions (ADRs). On the other hand, induction generally results in reduced plasma levels of the victim drugs which might result in sub-therapeutic concentration which can promote the emergence of drug resistance in the treatment of infectious diseases. Prediction of the effects of induction is still qualitative while that for inhibition can now be done quantitatively. In this chapter, we will present the approaches used in the prediction of inhibition based drug–drug interactions.

With the data generated from in vitro experiments on the IC_{50}, inhibition constant (K_i), mechanism of inhibition and estimated pharmacologically effective concentration, EC_{50}, one can make predictions of the likelihood of enzyme inhibition based drug–drug interactions in the early phases of drug discovery. The Food and Drug Administration, FDA, has already come up with comprehensive guidelines on the determination of the risk for DDIs (http://www.fda.gov/cder/guidance/index.htm). For general evaluations of DDIs, one can use the assumption that effective plasma concentrations will be around 1.0 μM for the selective and potent new chemical entities. This assumption should be used with caution and re-evaluate compounds for DDI risk as PK/PD studies generate a more accurate estimation of likely plasma concentrations.

The simple prediction equation is therefore $[I]/K_i$, where $[I]$ is the concentration of inhibitory test compound. The value to use for this concentration is a highly debated point as it can be total C_{max}, C_{ss}, unbound C_{max}, C_{ss}, total or unbound I_{max} (the concentration entering the liver from the portal vein). FDA guidelines are based on total plasma concentrations, C_{max}, but most researchers are increasingly using the unbound concentration entering the liver. FDA has ranked the risk for inhibition based on the above assumption as follows:

K_i	I/K_i	In vivo interaction
Reversible mechanism		
<1 μM	>1	Highly likely
1–50 μM	0.1–1	Possible
>50 μM	<0.1	Remote

A more accurate way of quantifying DDI is to determine the extent of inhibition as the reduction in clearance or increase in AUC. This can be achieved using the following equations:

$$\text{Fraction of remaining activity vi/vo} = (V_{\max}S/K_{\mathrm{m}}(1 + I/K_{\mathrm{i}}))/(V_{\max}S/(K_{\mathrm{m}} + S))$$

$$= \frac{(K_{\mathrm{m}} \pm S)}{K_{\mathrm{m}}(1 + I/K_{\mathrm{i}} + \mathrm{S})}$$

$$\text{Fraction of inhibition } (i) = 1 - (\text{vi/vo}) = \frac{1 - -((K_{\mathrm{m}} \pm S))}{K_{\mathrm{m}}(1 + [I]/K_{\mathrm{i}} + S)} = \frac{[I]}{[I] + K_{\mathrm{i}}(1 + S/K_m)}.$$

With the assumption that in vivo, drug concentrations (*S*) will be much lower than K_{m}, the fractional inhibition reduces to $[I]/([I] + K_{\mathrm{i}})$. The decrease in clearance or increase in AUC can then be estimated by the equation:

$$\begin{aligned} \mathrm{CL}_{\text{with inhibitor}}/\mathrm{CL}_{\text{without inhibitor}} &= \mathrm{AUC}_{\text{with inhibitor}}/\mathrm{AUC}_{\text{without inhibitor}} \\ &= 1/(1 - [I]/[I] + K_{\mathrm{i}}) \\ &= 1 + [I]/K_{\mathrm{i}}. \end{aligned}$$

Since most drugs are eliminated by many routes, the equation needs to take into account the fraction of the drug's clearance, f_{m}, which is by the inhibited enzymes. The fraction of the metabolic route (f_{m}) inhibited is very important in estimating the risk for DDI. $f_{\mathrm{m}} > 0.8$ is associated with high risk for DDI since the metabolism of the affected drug significantly depends on the affected enzyme. AUC increases of >2-fold are associated with high risk for DDI.

In the early phases of drug discovery, when IC_{50} only are determined, one can estimate the K_{i} for competitive inhibition as the worst case scenarios where $K_{\mathrm{i}} = IC_{50}/2$. For non-competitive inhibition, $K_{\mathrm{i}} = IC_{50}$.

For the estimation of inlet concentration, the following equation is used:

$$[I]_{\text{inlet,u,max}} = \left([I]_{\max} + \frac{k_{\mathrm{a}} \times F_{\mathrm{a}} \times D}{Q_{\mathrm{H}}}\right) * f_{\mathrm{u}}.$$

Assuming $C_{\max}$ to be the effective concentration interacting with the drug-metabolising enzymes, we have estimated the likelihood of DDI in the use of anti-parasitic drugs. Using the simple model of $C_{\max}$, we have estimated the likelihood of DDI in the use of anti-parasitic drugs (Table 7.8). In vivo studies on the inhibitory effects of thiabendazole and artemisinin on CYP1A2-mediated metabolism confirmed these in vitro data. Thiabendazole and artemisinin inhibited the formation of paraxanthine from caffeine by 92% and 66%, respectively [12].

Table 7.8 Inhibitory effects of some anti-parasitic drugs on drug-metabolising CYPs [13]

CYP/compound	K_i (μM)	Type of inhibition	Plasma C_{max} (μM)	Inhibitory potency ($[I]/K_i$)	Predicted % inhibition $[([I]/([I] + K_i)) \times 100]$
CYP1A2					
Artemisinin	0.43	Competitive	1.38	3.20	76
Niclosamide	2.70	Mixed	Negligible	–	Negligible
Thiabendazole	1.54	Mixed	89	57.80	98
Primaquine	0.22	Competitive	0.44	2.0	67
Dihydroartemisinin	3.67	Competitive	2.50	0.68	41
CYP2D6					
Quinine	15.51	Competitive	15.41	0.99	50
Chloroquine	12.68	Competitive	0.39	0.031	3
Amodiaquine	2.1	Competitive	0.74	0.35	26
Desethylamodiaquine	4.13	Mixed	444	107.51	99
Proguanil	6.76	Mixed	0.76	0.11	10
Cycloguanil	5.97	Competitive	0.21	0.04	3

7.10 Lead Optimisation

The lead optimisation stage utilises all in silico, in vitro and in vivo tools to build knowledge that can guide the medicinal chemists in addressing ADME issues in the selected series and enable PK scientists to make first dose in man predictions of new chemical entities. Before the onset of lead optimisation (LO), the candidate drug target profile (CDTP) must be formulated with respect to oral bioavailability, clearance, effective half-life, therapeutic dose, drug–drug interactions caused by the CD and risk for the generation of reactive metabolites. These can differ significantly depending on the intended use of the drug, e.g. if the drug is aimed at long duration of action, then the design must aim for a drug with a larger half-life ($t_{1/2}$) and a smaller elimination rate constant (K_{el}). If on the other hand a drug is too toxic or is required for a short duration of action, the design should aim for a smaller half-life and larger elimination rate constant. For a typical once-daily oral drug, the CDTP has generally a bioavailability of >25% to reduce the risk of large variability and a dose size of <300 mg to ensure that it fits in average-size tablet. Other CDTP goals are a low risk for reactive metabolites, minimal or reduced enzyme or transporter inhibition or induction to avoid drug–drug interactions.

During lead optimisation, iterations of DMPK and pharmacological evaluations, design of molecules and synthesis of molecules define a high point in the creative process of drug discovery (Fig. 7.14). The depicted movement of compounds and information shows an integrated paradigm that has come to characterise most drug discovery approaches used by leading pharmaceutical companies. This approach requires multidisciplinary teams that are tightly coupled to each other for rapid data turn-around times, design and implementation strategies.

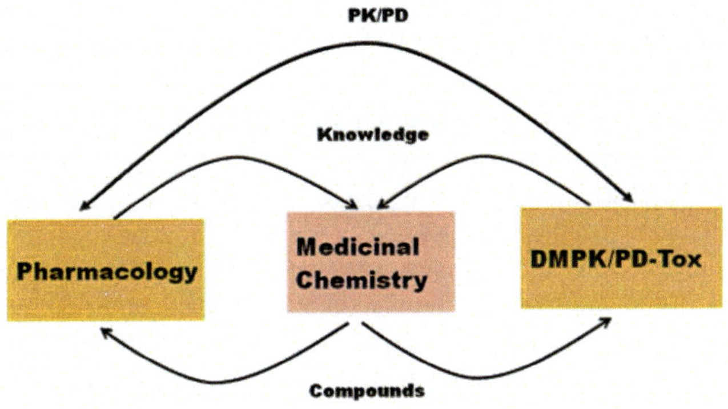

Fig. 7.14 The integrated drug discovery paradigm that is driven by a tight interplay of DMPK, medicinal chemistry and pharmacology [14]

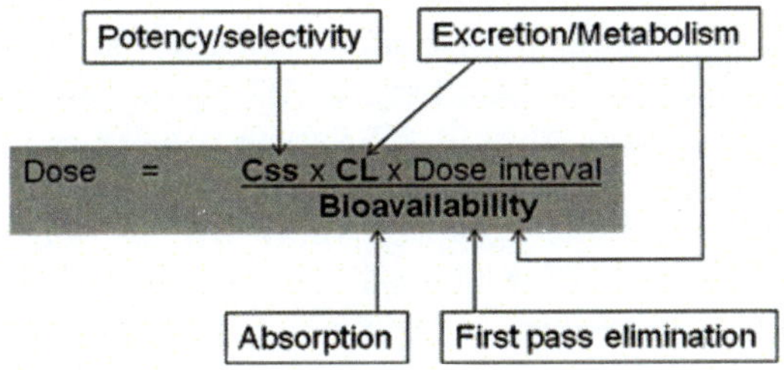

Fig. 7.15 Equation for dose estimation showing the importance of clearance and bioavailability

The variables described in this chapter clearly indicate that the ADME process is dependent on a host of chemical properties. It is also evident that lipophilicity (the tendency of a compound to partition into a non-polar lipid matrix versus an aqueous matrix) has a major impact on all the ADME processes, making it one of the most important parameters determining both drug pharmacokinetics and pharmacodynamics. It has been shown to correlate with compound solubility, permeability, metabolism, protein binding, distribution and pharmacologic potency. This property has resulted in the so-called physicochemical dilemma where it positively correlates with parameters which require opposite optimisation directions. For example, high lipophilicity is predictive of poor solubility but predictive of good permeability, yet one needs both good solubility and permeability for good oral absorption. High lipophilicity positively correlates with metabolic instability, high enzyme inhibitory effects and high drug target potency, making it difficult to independently optimise for these parameters. These are some of challenges lead optimisation scientists will have to solve towards the nomination of a candidate drug with predicted good PK/PD properties for the first time in human studies. The preclinical ADME and PK parameters culminate in dose in man estimation (Fig. 7.15).

7.11 Conclusions

The ADME and PK studies discussed in this chapter are now an integral part of the drug discovery efforts in Africa where they guide medicinal chemistry and pharmacology in the design and eventual selection of candidate drugs predicted to be safe and efficacious in humans. Figure 7.14 shows the new paradigm of drug discovery where ADME and PK have evolved from descriptive sciences to predictive sciences that are contributing to the reduction of attrition rates of new chemical entities. For the pharmaceutical industry, integration of DMPK in drug discovery and development has successfully addressed this parameter as a hurdle in this complex and expensive process. It has left issues of therapeutic efficacy and toxicity as outstanding causes of high attrition rates of new chemical entities. Continued ADMET/PK research is needed to tease out its contribution to the continuing challenge of poor efficacy and idiosyncratic reactions associated with some new chemical entities.

For Africa, there is need to have more laboratories and institutions with knowledge and technical capacity to provide ADME/PK support to drug discovery projects. The successful effort at AiBST can be used as a template for the promotion of ADMET/PK on the continent. Drug discovery activities in Africa are projected to increase in the framework of the African Network for Drug and Diagnostic Innovation (ANDI) which was recently initiated (http://www.andi-africa.org). AiBST will contribute to ANDI projects through support of medicinal chemistry- and pharmacology-driven projects with ADME/PK input and services. New applications of ADME/PK towards the safe and efficacious use of herbal medicines are also under development at AiBST. Preliminary studies have resulted in the setup of in vitro and in vivo platforms for the evaluation of herbal medicines for drug–herb metabolic interactions [15]. In yet another innovative approach, we have initiated collaborations with medicinal chemists in implementing ADMET-guided drug discovery from natural products and through rescuing of old drugs by addressing their ADMET liabilities [16]. Our work on amodiaquine aims at addressing the toxicity issues associated with the antimalarial whilst our work on praziquantel aims at addressing the issue of low oral bioavailability associated with this antischistosomicide. With respect to pharmacogenetics of drug metabolism, our work is showing that for some drugs such as efavirenz, the genetic polymorphism of CYP2B6, the enzyme responsible for the drug's metabolism and elimination, could be a clinically relevant biomarker in personalising the treatment of HIV/AIDS patients. These developments are further expanding the utility of ADMET/PK in drug discovery, development and rational clinical use of medicines.

References

1. Prentis RA, Lis Y, Walker SR (1988) Pharmaceutical innovation by the seven UK-owned pharmaceutical companies (1964–1985). Br J Clin Pharmacol 25:387–396
2. Kola I, Landis J (2004) Can the pharmaceutical industry reduce attrition rates? Nat Rev Drug Discov 3:711–716
3. Kerns EH, Di L (2008) Drug-like properties: concepts, structure design and methods. From ADME to toxicity optimisation. Elsevier, Amsterdam
4. Ana Ruiz-Garcia A, Bermejo M, Moss A et al (2008) Pharmacokinetics in drug discovery. J Pharm Sci 97:654–690
5. Gilbaldi M (1975) Biopharmaceutics and clinical pharmacokinetics. Lea & Febiger, Philadelphia
6. Masimirembwa CM, Otter C, Berg M et al (1999) Heterologous expression and kinetic characterization of human cytochromes P-450: validation of a pharmaceutical tool for drug metabolism research. Drug Metab Dispos 27:1117–1122
7. Li XQ, Björkman A, Andersson TB et al (2003) Identification of human cytochrome P(450) s that metabolise anti-parasitic drugs and predictions of in vivo drug hepatic clearance from in vitro data. Eur J Clin Pharmacol 59:429–442
8. Thelingwani RS, Zvada SP, Hugues D et al (2009) In vitro and in silico identification and characterisation of thiabendazole as a mechanism-based inhibitor of CYP1A2 and simulation of possible pharmacokinetic drug-drug interactions. Drug Metab Dispos 37:1286–1294
9. Johansson T, Jurva U, Grönberg G et al (2009) Novel metabolites of amodiaquine formed by CYP1A1 and CYP1B1: structure elucidation using electrochemistry, mass spectrometry, and NMR. Drug Metab Dispos 37:571–579
10. Jurva U, Holmén A, Grönberg G et al (2008) Electrochemical generation of electrophilic drug metabolites: characterization of amodiaquine quinoneimine and cysteinyl conjugates by MS, IR, and NMR. Chem Res Toxicol 21:928–935
11. Masimirembwa CM, Bredberg U, Andersson TB (2003) Metabolic stability for drug discovery and development: pharmacokinetic and biochemical challenges. Clin Pharmacokinet 42:515–528
12. Bapiro TE, Sayi J, Hasler JA et al (2005) Artemisinin and thiabendazole are potent inhibitors of cytochrome P450 1A2 (CYP1A2) activity in humans. Eur J Clin Pharmacol 61:755–761
13. Bapiro TE, Egnell AC, Hasler JA et al (2001) Application of higher throughput screening (HTS) inhibition assays to evaluate the interaction of antiparasitic drugs with cytochrome P450s. Drug Metab Dispos 29:30–35
14. Masimirembwa CM, Thompson R, Andersson TB (2001) *In vitro* high throughput screening of compounds for favorable metabolic properties in drug discovery. Comb Chem High Throughput Screen 4:245–263
15. Gwaza L, Wolfe AR, Benet LZ et al (2009) *In vitro* inhibitory effects of *Hypoxis obtusa* and *Dicoma anomala* on cyp450 enzymes and pglycoprotein. Afr J Pharm Pharmacol 3:539–546
16. Chibale K, Guantai E, Masimirembwa C (2011) Extracting molecular information from African natural products to facilitate unique African-led drug-discovery efforts. Future Med Chem 3:257–261
17. Lipinski C, Lombardo F, Dominy BW et al (1997) Experimental and computational approaches to estimate solubility and permeability in drug discovery and development settings. Adv Drug Deliv Rev 23:3–25

Chapter 8
Marine Bioprospecting in Southern Africa

Michael T. Davies-Coleman and Suthananda N. Sunassee

Abbreviations

BMS	Bristol-Myers Squibb
HIF-1	Hypoxia-inducible factor-1
NCI	National Cancer Institute
ORI	Oceanographic Research Institute
Rhodes	Rhodes University
SCC	Squamous cell carcinoma
SCOC	Squamous cell oesophageal cancer
SIO	Scripps Institution of Oceanography
SKB	SmithKline Beecham
STS	Soft tissue sarcoma

8.1 Background

Many sessile marine organisms (invertebrates, algae and microorganisms) produce natural products as a chemical defence against predation or in a chemically mediated response to inter-species competition for limited resources (e.g. space on a reef or nutrients). Interestingly, over the last four decades, the isolation and identification of marine natural products by chemists, working in tandem with biochemists and pharmacologists, has provided important lead compounds for the pharmaceutical industry, especially in the field of new anti-cancer drug discovery [1–3]. While an estimated 15 marine natural products are in various phases of clinical development (mostly related to oncology) [4], only one marine secondary metabolite trabectidin (**1**), isolated from the Caribbean marine ascidian

M.T. Davies-Coleman (✉) • S.N. Sunassee
Department of Chemistry, Rhodes University, P.O. Box 94, Grahamstown 6140, South Africa
e-mail: m.davies-coleman@ru.ac.za; snsunassee@gmail.com

K. Chibale et al. (eds.), *Drug Discovery in Africa*,
DOI 10.1007/978-3-642-28175-4_8, © Springer-Verlag Berlin Heidelberg 2012

Ecteinascidia turbinate [5], has reached the market place. Yondelis®, the commercial dosage form of **1**, is prescribed for patients with soft tissue sarcoma (STS) and was the first drug to be prescribed for STS in a quarter of a century, netting 17€ million in sales for the pharmaceutical company PharmaMar in the first quarter of 2010 [6].

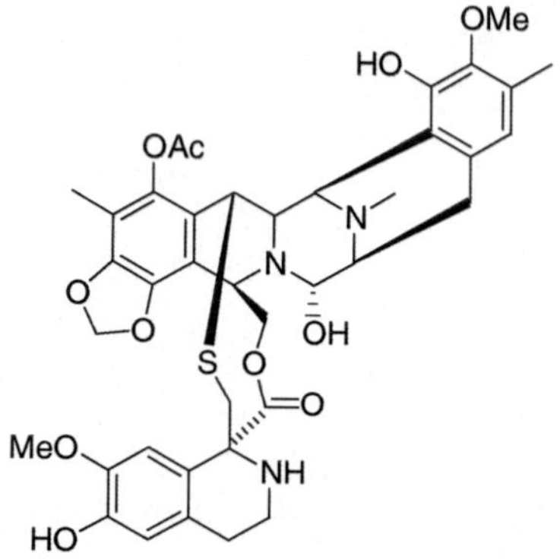

1

Marine biota are a unique reservoir of biomolecular diversity not only because of the large number of species present in the oceans but also because the marine environment presents very different physiological and ecological challenges compared to those experienced by terrestrial species. As a consequence of the differences between terrestrial and marine habitats, the initial expectation that marine organisms utilize unique biosynthetic pathways or exploit variations on established biosynthetic pathways to biosynthesize structurally unusual biomolecules has been validated by the complexity and novelty of the plethora of marine natural products isolated from the world's marine biota over the last 30 years [7].

Marine bioprospecting or biodiscovery is the search for marine natural products and other materials that may be of benefit to society, i.e. for use as medicines, agrochemicals, marine antifouling coatings, etc. Possessing a vast and unique biodiversity of both terrestrial and marine flora and fauna, southern Africa is widely recognized as one of the world's most biodiverse regions [8]. While bioprospecting of southern Africa's terrestrial phytochemical resources for new pharmaceuticals is well established [9], very little is known about the natural product diversity or biomedical potential of the vast and uniquely diverse intertidal and benthic marine invertebrate populations occurring off the coast of southern Africa.

8.1.1 Southern Africa's Unique Marine Environment

Significant variations in marine diversity around the approximately 3,000-km-long southern African coastline, stretching from Namibia in the west to southern Mozambique in the east, are linked directly to the occurrence of three

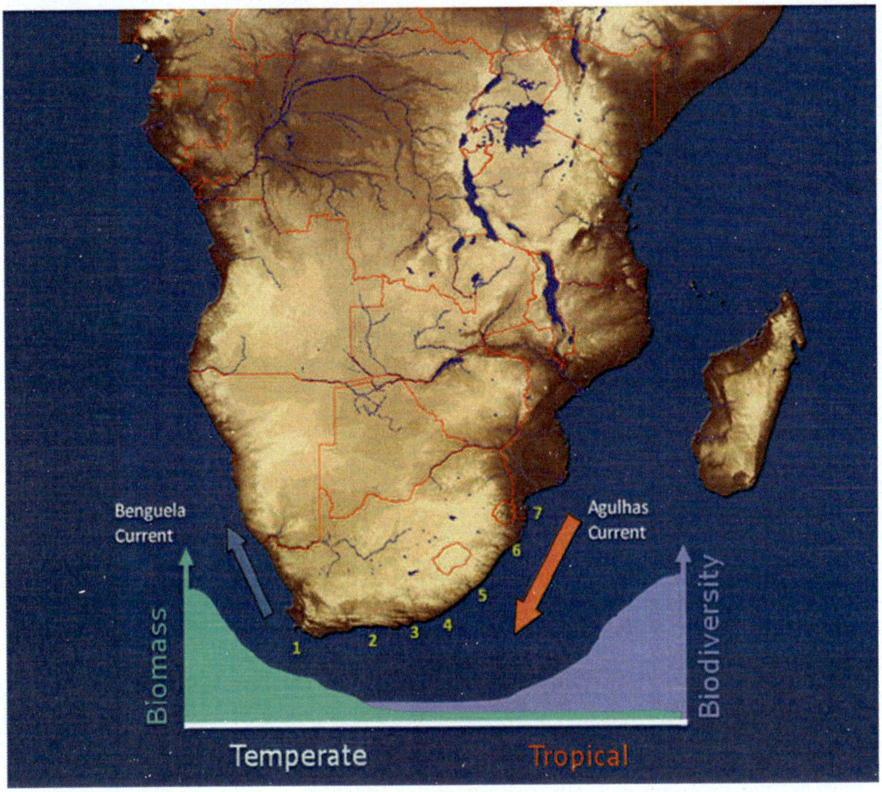

Fig. 8.1 Southern Africa showing the general direction of the two major offshore currents, the overall estimated distribution of biomass and biodiversity in tropical and temperate waters and the seven major marine invertebrate collecting sites (1990–2010): 1. Cape Peninsular, 2. Tsitsikamma National Park, 3. Algoa Bay, 4. Coffee Bay, 5. Aliwal Shoal, 6. Sodwana Bay and 7. Ponto do Ouro

biogeographical zones: the cool temperate west coast, the warm temperate south-east coast and the subtropical east coast [10]. The biogeographical zones are largely the product of the oceanographic conditions surrounding the warm Agulhus and cold Benguela ocean currents that follow the edges of the continental shelf on the east and west side of the subcontinent, respectively (Fig. 8.1).

While the offshore colder waters of the west coast support a larger biomass of marine organisms arising from the upwelling of cold nutrient rich water associated with the Benguela current, centres of greater marine biodiversity, in common with similar areas around the world, are found in the warmer water of the east coast (Fig. 8.1). Although many circumtropical species are found along the coasts of southern Africa's subtropical regions, a significant proportion of the over 10,000 species of marine organisms recorded off southern Africa coast are reported to be endemic [11]. High levels of endemism in chemically defended species are of obvious interest to marine bioprospectors because of the increased potential for the discovery of novel biologically active chemical entities from these species.

Fig. 8.2 The South African marine tube worm *Cephalodiscus gilchristi* (Photo: Louise Lange)

8.2 An Historical Overview of Marine Bioprospecting in South Africa

The first marine bioprospecting forays along the coast of South Africa occurred around the Cape Peninsula (1972–1981) [12] and were facilitated by Pettit from Arizona State University [13, 14]. Although these collections of marine organisms occurred over three decades ago, one compound, cephalostatin 1 (**2**) [13] isolated from an elusive South African hemichordate marine tube worm *Cephalodiscus gilchristi* (Fig. 8.2), continues to attract attention due to its potency and novel mechanism of pharmacological action [15–17]. Cephalostatin 1 and its 18 closely related naturally occurring and semi-synthetic analogues (US Patents 4873245, 5047532, 5583224 and WO 8908655) are widely regarded as some of the most cytotoxic secondary metabolites (**1** has ca. 1 nM activity National Cancer Institute's (NCI) 2 day 60 cell line panel) ever to be screened by the NCI [16], and the recently renewed interest in this compound is discussed in Sect. 8.2.1.

Despite the clear biomedicinal potential of southern African marine organisms, revealed by Pettit and co-workers, marine biodiscovery along the coast of southern Africa waned during the 1980s and only regained momentum in 1992 with the initiation of two parallel collaborative marine bioprospecting programmes between two research institutions based in South Africa, viz. Rhodes University (Rhodes) and the Oceanographic Research Institute (ORI) and their different international academic and pharmaceutical company partners. Some of the products and longer term implications of the collaborations between ORI, the University of Tel Aviv in Israel and the Spanish pharmaceutical company PharmaMar (1992–1998) and between Rhodes, Scripps Institution of Oceanography (SIO) and SmithKline Beecham (SKB, now GlaxoSmithKline) in the USA (1992–1995) are discussed here. These two programmes were followed by a collaborative initiative between Rhodes and the NCI (1998 until present) and between SIO and Bristol-Myers Squibb (BMS, 2000–2003) [18], both of which contributed to the building of resource bases of South African marine invertebrate extracts in the USA and South Africa, respectively, and which are currently available for new screening initiatives. No patentable South African natural products have yet emerged from either of the latter two collaborative programmes.

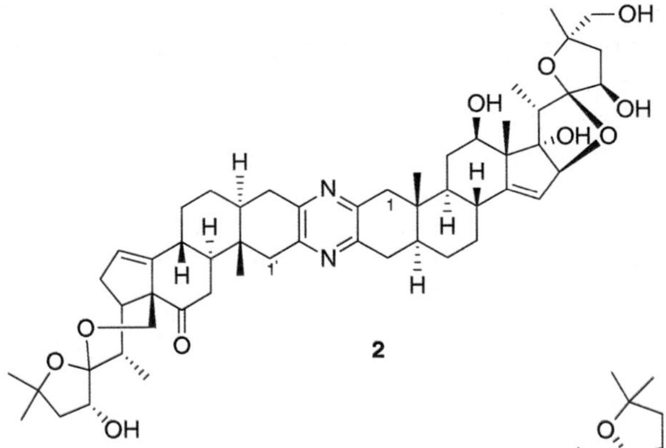

2

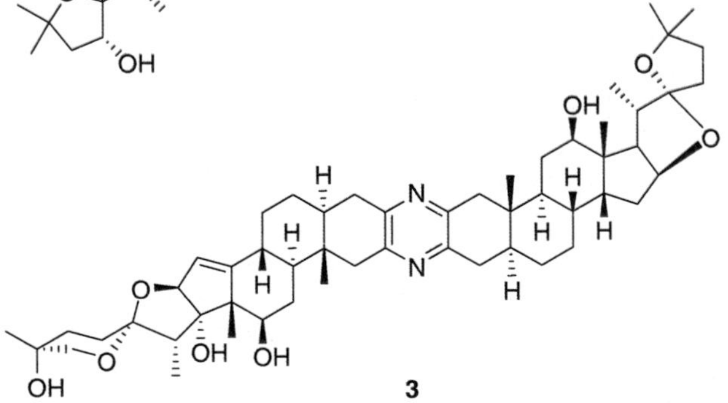

3

8.2.1 Cephalostatin 1: A Potent South African Marine Anti-neoplastic Agent

Aqueous and methanolic extracts of the initial collection of *C. gilchristi* in 1972 were shown 2 years later to induce significant life span extension in the NCI's in vivo murine lymphocytic leukaemia mouse model. However, the chemical structures of the cephalostatin metabolites present in the *C. gilchristi* extracts proved elusive with the structure of **2** only emerging in the chemistry literature some 13 years later [13]. Paradoxically, given the fundamental taxonomic differences between marine worms and ascidians (also known as tunicates or sea squirts) the trisdecacyclic *bis*-steroidal pyrazine structures of the 19 different cephalostatins isolated from *C. gilchristi* are closely related to those of a further 26 members of the ritterazine series, e.g. ritterazine B (**3**) obtained from the Japanese ascidian *Ritterella tokioka*. Fuchs and co-workers have recently compiled a comprehensive review of the structural chemistry, comparative cellular mechanisms of anti-neoplastic activity and different synthetic approaches to the cephalostatins and ritterazines [19, 20]. The clinical development of the most cytotoxic compound, **2**, in the cephalostatin series has been hampered by the problem of supply; a frequent occurrence in marine drug development given the small amounts of secondary metabolites usually produced by marine organisms [21]. A recollection of *C. gilchristi* (450 kg) from South Africa in 1990 afforded a mere 100 mg of **2**, far short of the multi-gram quantities required for clinical development, necessitating the synthesis of **2** as the only viable alternative to resolve the supply problem [20]. Despite the appearance in the literature of several syntheses of either portions of **1** or simplified analogues of **2**, Lacour et al.'s [22] and more recently Shair and co-worker's [17] syntheses of **2** (both groups have utilized the abundant plant-derived steroid hecogenin acetate as a starting point) are enantioselective syntheses of this marine natural product that provide hope for resolution of the preclinical demand for several grams of this compound.

Motivating endeavours towards the successful syntheses of **2** is an increased understanding of the novel cytotoxicity of this compound to different cancer cell lines. The role of **1** in bringing about apoptosis or controlled cell death in cancer cells has been extensively studied. Cephalostatin 1 has been shown to induce an unusual apoptotic pathway in leukaemia cancer cells through activation of one of the major apoptosis enzymes, caspase 9, without the participation of the apoptosome [16]. Recently, a comparative bioinformatics screening of 43,000 small chemical entities against cancer cell lines known to possess the *p16* tumour cell suppression gene revealed that **2** is the most consistently cytotoxic compound to these cell lines [17]. The *p16* gene encodes for a tumour suppressor protein which is able to arrest tumour cell division by binding to two of the cyclin-dependent kinases (CDK4 and CDK6) responsible for ensuring the smooth transition of cancer cells from the G1 to the S phase in the cell cycle. Although the mechanisms by which **2** may be selectively cytotoxic to cancer cell lines possessing the *p16* gene are unknown, further investigation of the mode of action of this South African marine natural product against these specific cancer cell lines may lead to future important advances in cancer chemotherapeutics.

8.2.2 Sodwana Bay: An Epicentre of Marine Invertebrate Biodiversity and Biomedicinal Potential

The warm subtropical waters of Sodwana Bay, situated on the northern coast of KwaZulu Natal, South Africa (Fig. 8.1), support the southernmost coral reefs in Africa. The Indo-Pacific coral reefs, including those found in Sodwana Bay, sustain

Table 8.1 Cytotoxicity of compounds **4–9** and **12–14** against the PharmaMar panel of four cancer cell lines

Compound	Patent number	Cell line IC_{50} (μM)			
		P-388	A-549	HT-29	MEL-28
4	(WO 9701334) [28]	10	1.0	20	–
5	(WO 9701334) [28]	2.0	0.2	2.0	2.0
6	(WO 9701334) [28]	10.5	0.02	10.5	10.5
7	(WO 0020411) [30]	0.04	0.02	0.02	0.04
8	(US 5661175) [32]	0.015	0.015	0.007	0.015
9	(US 5661175) [30]	0.02	–	–	–
12	(WO 9632388) [41]	1.7	1.7	0.20	2.0
13	(WO 9632388) [41]	2.7	11.2	11.2	11.2
14	(WO 9632388) [41]	1.1	1.1	0.28	2.2

the most diverse marine invertebrate and vertebrate fauna in the world [23]. Associated with a proliferation in tropical and subtropical marine invertebrate diversity is a cornucopia of marine natural products, many of which may possess as yet unrealized biomedicinal potential [24]. The 6-year (1992–1998) collaborative bioprospecting programme in Sodwana Bay between Schleyer at ORI, Kashman at the University of Tel Aviv and the pharmaceutical company PharmaMar has been the most productive to date in the southern African region and has resulted in the discovery of 33 new bioactive compounds and three known compounds [12], of which nine have been patented for their anti-cancer properties (Table 8.1).

An analysis of the phyletic distribution of secondary metabolites reported from Sodwana Bay marine invertebrates reveals that only three invertebrate phyla, viz. Porifera (sponges), Tunicata (ascidians) and Cnidaria (soft corals) have provided 70%, 21% and 9%, respectively, of the bioactive secondary metabolites isolated from the invertebrates collected from this region of the South African coast [12]. The chemical structures and bioactivities of marine natural products isolated by Kashman and others from Sodwana Bay marine invertebrates have been reviewed previously [12]. PharmaMar's interest in the discovery of new anti-cancer compounds provided the screening platform for bioassay-guided fractionation and chromatography of the Sodwana Bay marine invertebrate extracts and the screening of extracts and pure compounds against one or more of four cancer cell screens (P-388 murine leukaemia, A-549 human lung carcinoma, HT-29 human colon carcinoma and MEL-28 melanoma) which has dominated Kashman and co-workers' studies of South African bioactive marine natural products. We summarize here the nine anti-tumour marine natural products patented from the Sodwana Bay marine bioprospecting programme, bringing the current total number of patented South African marine natural products to 18 (nine of which are naturally occurring cephalostatins).

Rudi and Kashman [25, 26] initially isolated a group of triterpenoids, the sodwanones A–F, from the marine sponge *Axinella weltneri*. A later recollection of the same sponge furnished the sodwanones G–I [27]. Of these nine compounds, only sodwanones A (**4**), G (**5**), H (**6**) and I were reported to be cytotoxic in four different cancer cell line screens [27, 28]. Rudi et al. [27] subsequently patented **4–6** as anti-tumour compounds, making provision in their patent to cover any compounds that possess the general sodwanone backbone structure [28]. The IC_{50} values of these three compounds against different cancer cell lines are summarized in Table 8.1.

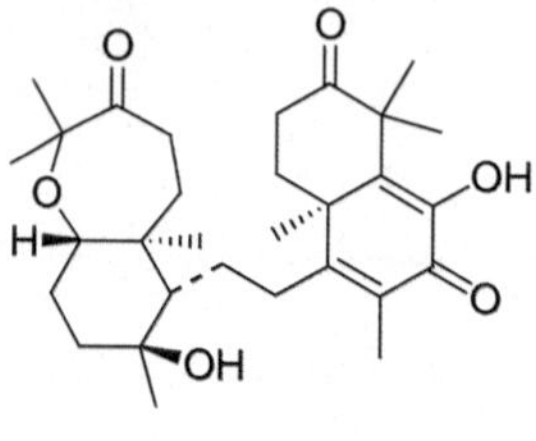

4

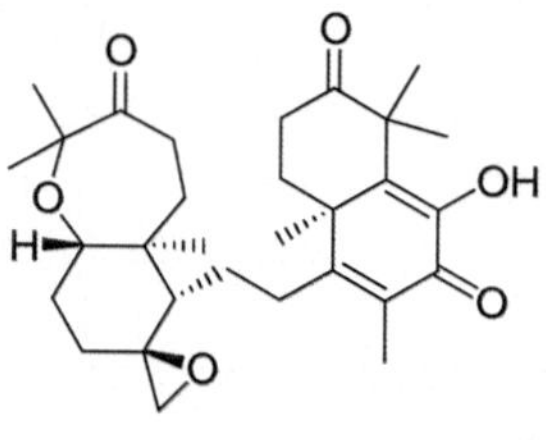

5

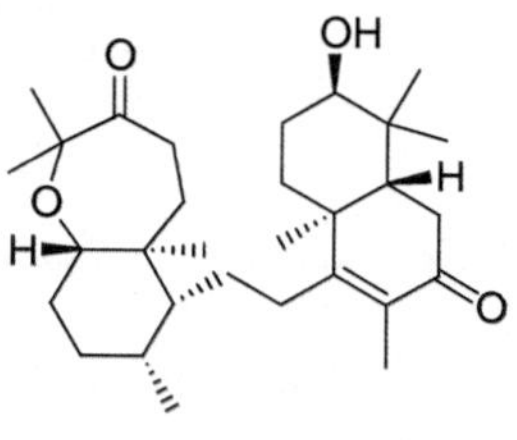

6

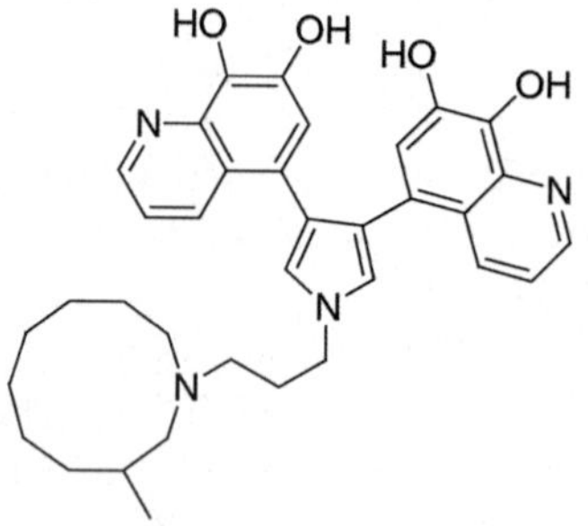

7

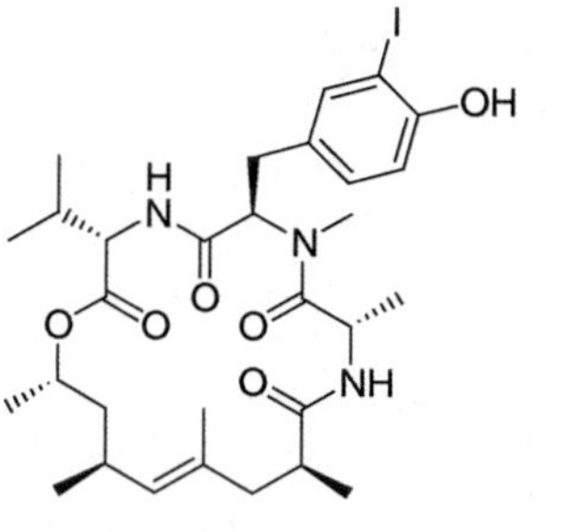

8

The Sodwana Bay marine sponge *Haliclona tulearensis* afforded the cytotoxic alkaloid halitulin (**7**) (Table 8.1) [29]. The potent cytotoxicity of compound **7** prompted Kashman et al. [30] to patent halitulin as a potential anti-tumour agent. Kashman also extended the patent to include compounds containing any substituent attached to the central pyrrole moiety of **7** [30], suggesting that the azamacrocycle at this position is not essential for cytotoxicity.

The cyclic peptide geodiamolide TA (**8**) and tripeptide hemiasterlin (**9**) isolated from the Sodwana Bay marine sponge *Hemiasterella minor* (Kirkpatrick) exhibited strong cytotoxicity against the PharmaMar four cancer cell line panel at relatively low concentrations (Table 8.1) [31, 32]. No details of the absolute configuration of **8** were included in the paper describing the isolation of this compound or in the subsequent patent of geodiamolide TA as an anti-tumour compound. The relative configuration shown here, however, was inferred from congruency of the spectroscopic data with the related compounds geodiamolides A–F [28]. Following

Kashman and co-workers' initial discovery of hemiasterlin, the anti-tubulin activity of this tripeptide **9** and the closely related analogues hemiasterlin A and B were first reported by Anderson and co-workers [33]. The relative structural simplicity of **9** along with its ability to produce abnormal mitotic spindles at nano-molar concentrations and induce microtubule depolymerization at higher concentrations earmarked this compound for further development as a potential anti-cancer agent [34]. Consequently, a synthetic programme to explore structure–activity relationships of hemiasterlin and its bioactive naturally occurring analogues resulted in the discovery of the phenylalanine analogue HTI-286 (**10**) as a synthetically more accessible and more potent tubulin polymerization inhibitor [35]. Both **9** and **10** and a further hemiasterlin analogue E7974 (**11**) [36] were placed into clinical development for the possible treatment of a variety of cancers including hepatic, prostate and bladder cancer [6, 37–39] although currently only the latter is being actively worked on.

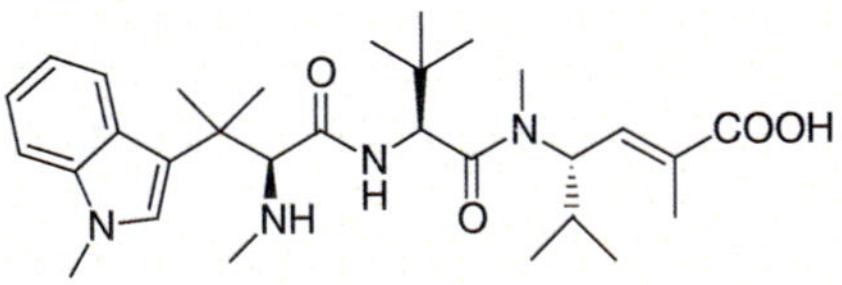

9

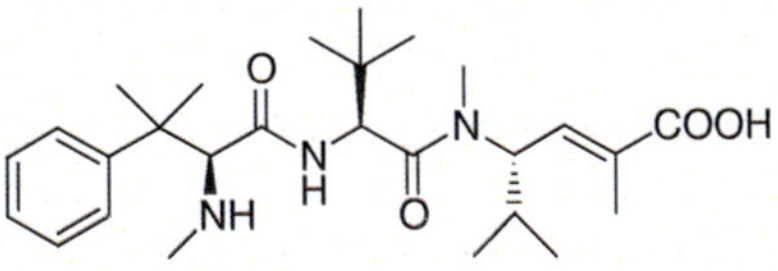

10

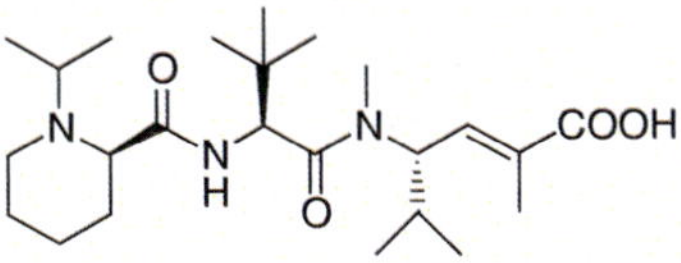

11

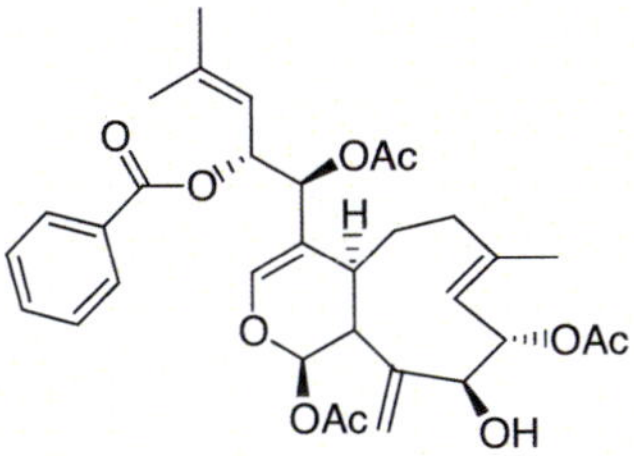

12

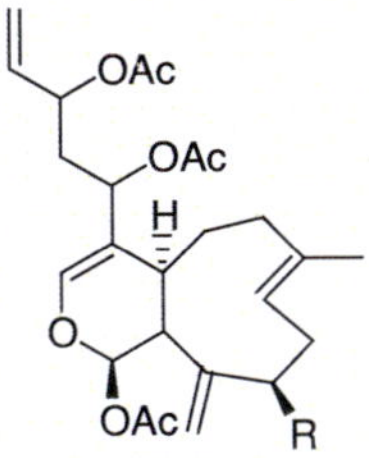

13 R = H
14 R = OH

The xenicane diterpenes, antheliatin and zahavins A and B, (**12–14**) from the Sodwana Bay soft coral *Eleutherobia aurea* were found to be strongly cytotoxic against the PharmaMar panel of four cancer cell lines and were also subsequently patented (Table 8.1) [40, 41].

8.2.3 Marine Bioprospecting Elsewhere Along the Southern African Coast

The first bioprospecting collections of marine invertebrates outside of Sodwana Bay were made from Coffee Bay on the previously named Transkei's Wild Coast (Fig. 8.1) in 1992 and provided ca. 160 extracts that were initially screened in a series of simple antimicrobial in-house assays at SIO. However, no new bioactive compounds were identified, which is not surprising given the limitations of the screening programme. Some 15 years after their original preparation, the resubmission of these same extracts into a series of screens designed to detect modulators of autophagy afforded four new amino steroids, the clionamines A–D (**15–18**), from South African specimens of the ubiquitous sponge *Cliona celata* [42]. This clearly shows that the integrity of marine invertebrate extracts can be sustained for decades provided they are kept in frozen storage. Autophagy is a highly conserved process in which proteins and cell organelles are packaged into double-membraned vesicles known as autophagosomes. While **15–18** all induced autophagosome accumulation, a medicinal application for these compounds is unlikely. They may, however, find a role as chemical biology tools that can be utilized to explore the mechanistic details of autophagy, a process of controlled cell death that is still not well understood [42].

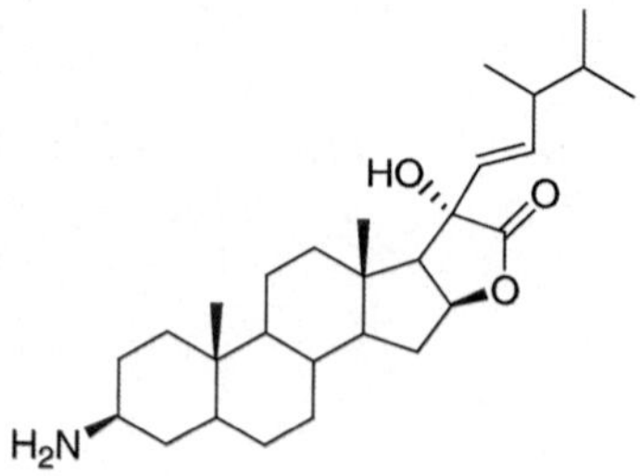

15

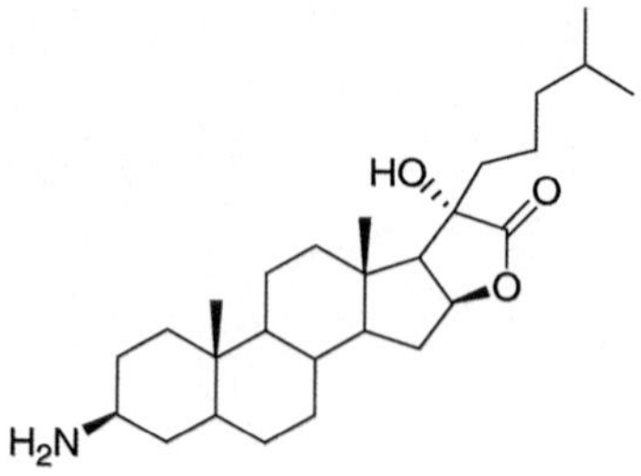

16

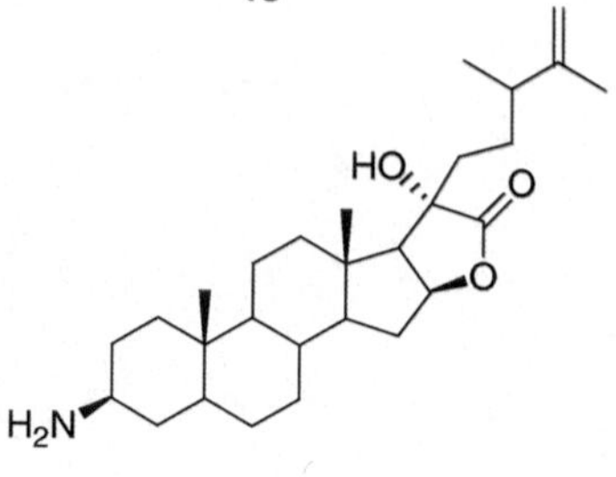

17

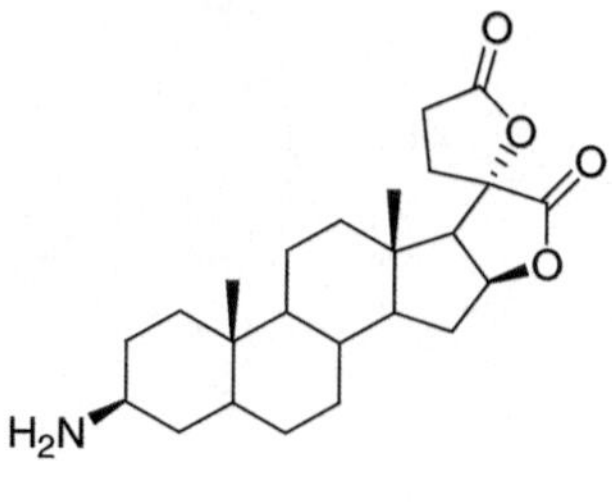

18

Over the period 1994–1995, a large-scale collection of 336 different marine invertebrates was made from three locations along the Southern African coast: Aliwal Shoal, Ponto do Ouro and the Tsitsikamma National Park (Fig. 8.1) as part of a collaborative programme between Rhodes, SIO and SKB. A total of 634 aqueous and various organic soluble extracts were prepared from this collection, and each extract was screened in a battery of biomedicinal screens at SKB. Although an acceptable hit rate of 3.1% was observed in the SKB screening programme, only known compounds, e.g. oroidin (**19**), variabilin (**20**) and avarol (**21**), were found to be potential phosphokinase C, endothelin-converting enzyme and DNA topoisomerase inhibitors, respectively. Despite a significant investment by SKB into the establishment of a sustainable marine biodiscovery research initiative in South Africa, no marine invertebrate-derived products were patented from this 2-year programme.

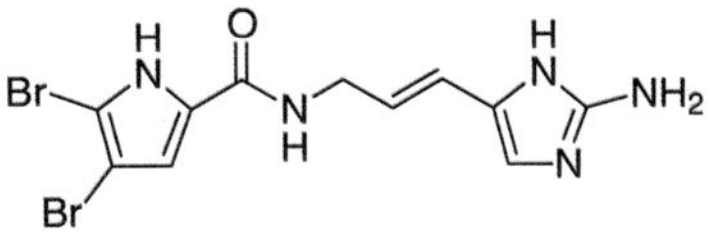

19

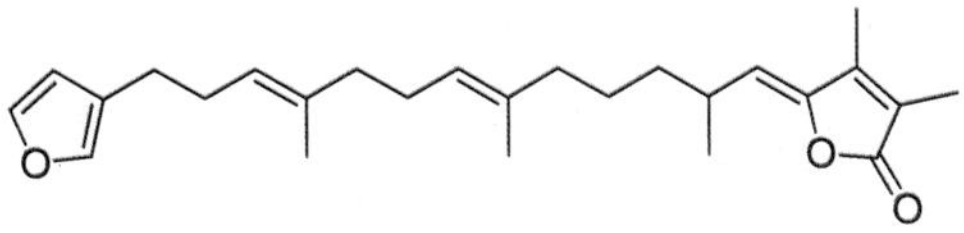

20

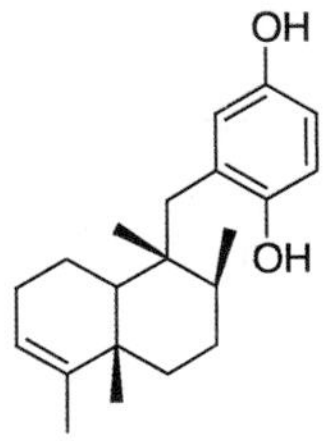

21

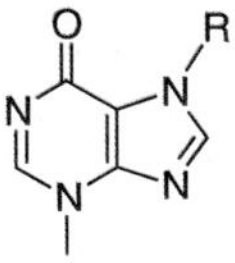

22

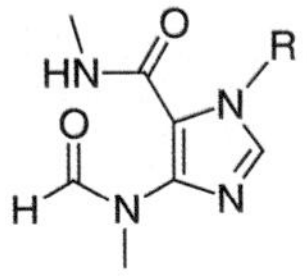

23

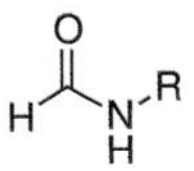

24

R = 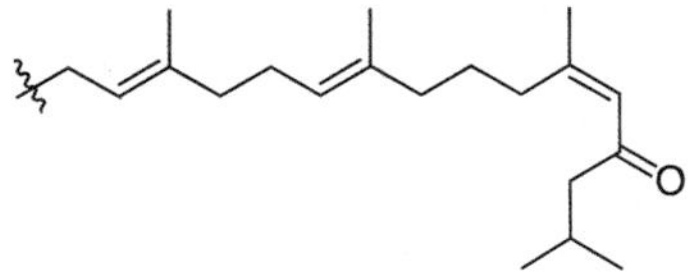

The return of all the extracts from SKB to Rhodes in 1996 has continued to provide a unique source of marine natural products for the ongoing marine biodiscovery research in South Africa. For example, a series of prenylated alkaloids, the malonganenones A–C (**22–24**), originally isolated from the marine gorgonian *Leptogorgia gilchristi* [43] collected off Ponto do Ouro in 1994, as part of the Rhodes/SIO/SKB marine bioprospecting initiative, have recently been shown to inhibit the aggregate suppression activity of the malarial parasite heat shock protein PfHsp70-1 [44].

A concerted effort between Rhodes and the National Cancer Institute (NCI) to find new potential anti-cancer compounds in Algoa Bay, South Africa (Fig. 8.1), between 1998 and 2000 resulted in the collection of over 300 marine invertebrates. Although no new compounds with significant anti-cancer activity were found, the deposition of extracts prepared from these invertebrates in the NCI's natural product repository has provided access (controlled by the terms of the MOU between Rhodes and the NCI) to these extracts by research groups outside of South Africa. As a result, the sodwanone A, **4**, and seven new closely related sodwanones were isolated from an NCI's repository extract prepared from an

Algoa Bay *Axinella* sponge by Nagle and co-workers [45]. A HIF-1 (T47D) inhibition assay was used to initially identify the *Axinella* extract and to guide the subsequent isolation of the sodwanones (IC_{50} of **4** ca. 20 μM) from this extract. Hypoxia-inducible factor-1 (HIF-1), first identified as a hypoxia-activated transcription factor which promotes tumour cell adaption to hypoxic conditions, is a molecular target of increasing importance in anti-cancer drug discovery and development [45].

8.2.4 South African Marine Natural Products and Oesophageal Cancer

Oesophageal cancer is one of the most prevalent forms of cancer worldwide and occurs in two main, but aetiologically unrelated, subtypes: adenocarcinoma and squamous cell carcinoma [46, 47]. Adenocarcinoma occurs mainly in developed countries and is principally associated with gastric reflux and Barrett's oesophagus [46–48]. Barrett's oesophagus is a metaplastic change of the flat squamous cells of the oesophageal mucus layer to columnar epithelium and has been found in patients suffering from this condition to increase the risk of developing adenocarcinoma [48]. Squamous cell carcinoma (SCC) is the most common form of oesophageal cancer and is found predominantly in developing countries such as South Africa [46, 48]. Squamous cell oesophageal cancer (SCOC) is the second most common cancer reported in poor rural and peri-urban populations in South Africa, with an age-standardized incidence rate of 16.22 per hundred thousand people [49]. The relatively high incidence of SCOC in South Africa has been reported to be linked to extraneous factors such as smoking, alcohol consumption, diets poor in fresh fruit and vegetables as well as the consumption of foods contaminated with a *Fusarium* fungus (which produces a carcinogen called fumonisin) [46, 47]. SCOC is a significant health problem due to the late diagnosis of the disease, a direct consequence of the asymptomatic nature of the tumour which only becomes evident once the cancer has metastasized to the extent where it hinders the swallowing of food [46]. Unfortunately, late diagnosis results in only a 10% survival rate in affected patients [46, 49] and the likelihood of low levels of remission in patients after either surgery or chemotherapy [50]. The prevention of SCOC can be achieved through changes in lifestyle, including diversification of diet [47]. Presently, the most common and favourable form of chemotherapeutic intervention for SCOC is the anti-cancer drugs cisplatin and 5-fluorouracil [51, 52], resulting in complete remission in only 20–30% of early diagnosed cases [51]. Combination chemotherapy using a cisplatin-containing treatment, with other active agents such as either irinotecan, epirubicin, docetaxel or paclitaxel, was also attempted in some cases [50], but it has become increasingly apparent in the last few years that the modest success rate of the current conventional chemotherapeutic agents against SCOC have reached a plateau [52, 53].

Table 8.2 Summary of the IC_{50} values of the linear triprenyl toluquinones and toluhydroquinols **29–34** isolated from the *L. millecra* against the oesophageal cancer cell line WHCO1 [54]

	Compounds tested					
	25	**26**	**27**	**28**	**29**	**30**
IC_{50} (μM)	37.9	83.3	9.5	12.9	42.7	32.7
95% CI[a]	37.7–38.0	82.1–84.5	9.4–9.5	12.5–13.3	42.6–42.7	32.5–33.0

[a]Confidence interval

A recent emphasis in oesophageal cancer drug discovery has been placed on the identification of molecular markers, e.g. growth factor receptor pathways and protein kinase pathways such as the c-Jun (AP1), as potential targets in SCOC chemotherapy [47, 52]. Marine prenylated toluquinones and toluhydroquinol agonists (**25–30**) of the c-Jun signalling pathway in WHCO1 (an oesophageal cancer cell line derived from a South African patient) were discovered in extracts of the endemic South African nudibranch *Leminda millecra*, originally collected in Algoa Bay (Fig. 8.1) [54]. Compounds **25–33** have been shown to induce apoptosis in oesophageal cancer cells by triggering the JNK/c-Jun signalling pathway through the generation of reactive oxygen species (ROS) [54]. The toluhydroquinol **27** exhibited the highest cytotoxicity ($IC_{50} = 9.5$ μM) against the WHCO1 oesophageal cancer cell line (Table 8.2), and its total synthesis has been reported by Li et al. [55].

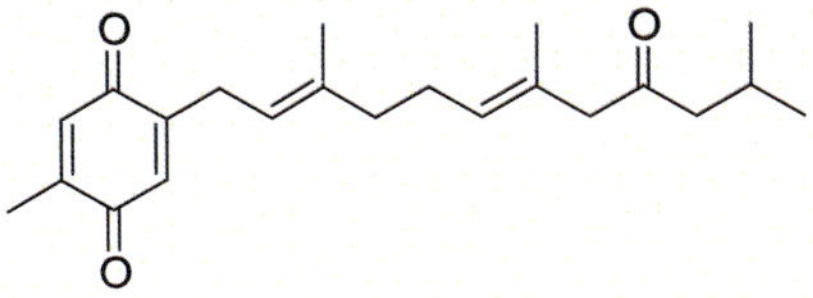

25

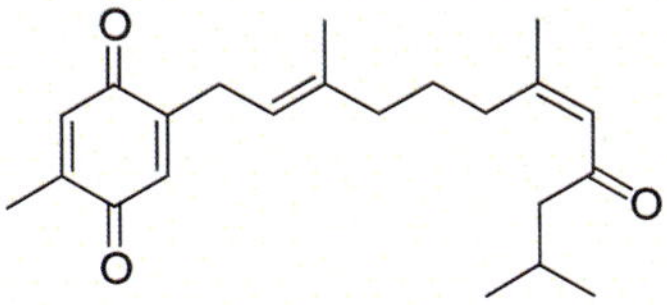

26

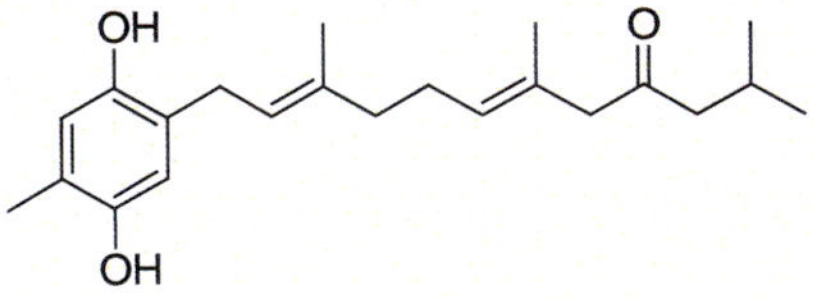

27

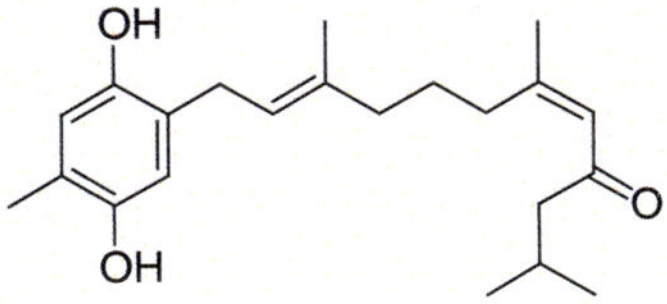

28

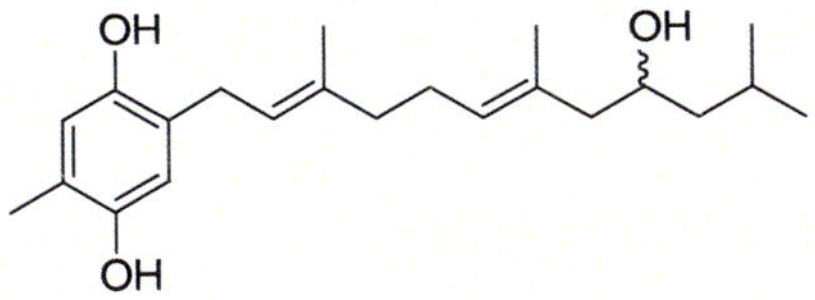

29

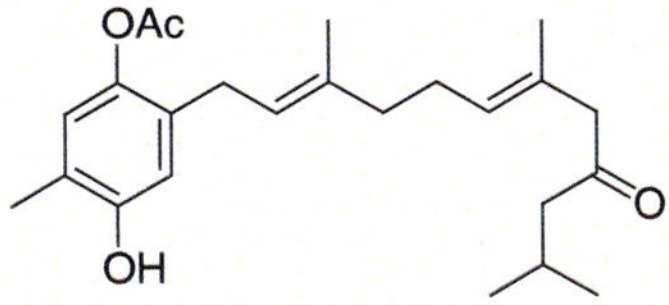

30

Table 8.3 Cytotoxicity of selected South African marine natural products against the oesophageal cancer cell lines WHCO1, WHCO6 and KYSE30 [53, 56, 57]

Compound	Cell line IC_{50} (μM)		
	WHCO1	WHCO6	KYSE30
3	1.2	1.7	1.4
31	3.0		
32	61.0	>200	>200
33	38.0	85.5	66.0
34	1.6	5.7	3.2
35	56.0	78.3	77.8
36	0.7	6.8	3.2
37	3.5	9.6	5.7
38	1.1	1.1	1.5

The search for potential marine chemotherapeutic agents against SCOC forms part of an ongoing collaboration between Davies-Coleman (Marine Biodiscovery Group at Rhodes) and Hendricks (Division of Medical Biochemistry at the Medical School of the University of Cape Town) and has led to the identification of several further South African marine invertebrate-derived natural products (**31–38**) with significant cytotoxicity to oesophageal cancer cells (Table 8.3) [53] .

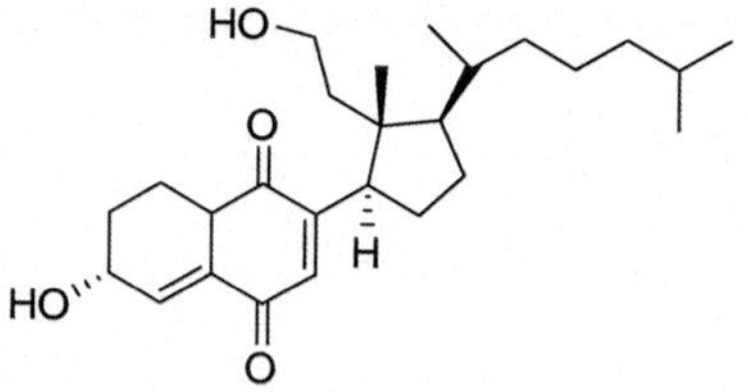

31

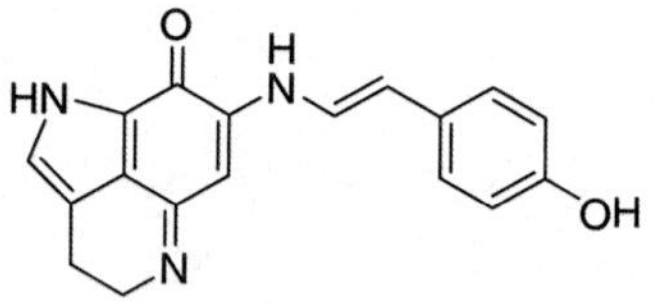

32

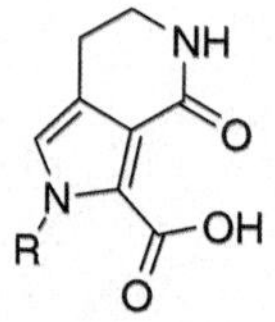

33 R = **a**

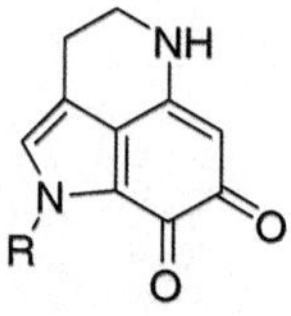

34 R = **a**
35 R = H

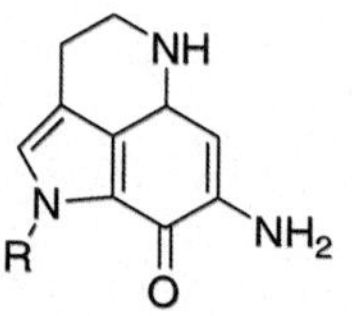

36 R = **a**

a =

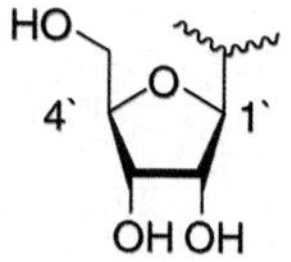

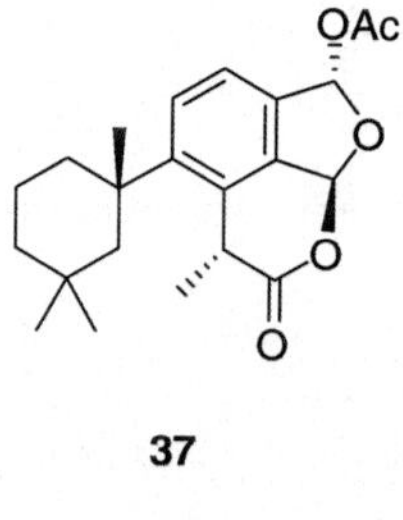

37

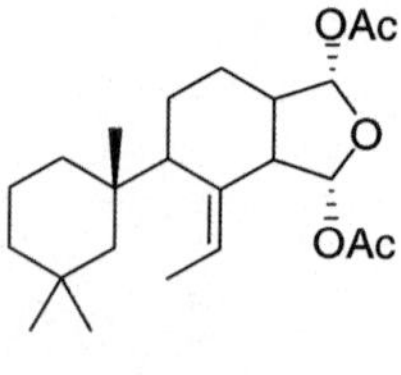

38

8.3 Conclusion

Over four decades of marine bioprospecting in southern Africa has identified the marine invertebrate communities of the south western Indian Ocean as a rich source of new bioactive marine secondary metabolites with biomedicinal potential. However, only a very small fraction of the marine invertebrate population, occurring off the richly biodiverse coast of southern Africa, has thus far been explored for their pharmaceutical potential. Therefore, much still needs to be achieved including the extension of bioprospecting initiatives to incorporate an investigation of the secondary metabolites produced by the symbiotic microbial communities associated with marine invertebrates, e.g. sponges, in addition to the marine microbes (especially actinobacteria) found in marine sediments. The relatively rapid advances made in southern African marine bioprospecting in the last two decades can be attributed, in no small measure, to the involvement of international partners such as SKB, BMS and the NCI. The impact on international partnerships of the recently promulgated National Environmental Laws Amendment Act 14 of 2009 in South Africa, which seeks to monitor and control the discovery phase of bioprospecting both on land and in South Africa's coastal waters, has yet to be determined.

References

1. Cragg GM, Grothaus PG et al (2009) Impact of natural products on developing new anti-cancer agents. Chem Rev 109:3012–3043
2. Fusetani N (2010) Biotechnological potential of marine natural products. Pure Appl Chem 82:17–26
3. Glaser KB, Mayer AMS (2009) A renaissance in marine pharmacology: from preclinical curiosity to clinical reality. Biochem Pharmacol 78:440–448
4. Sashidhara KV, White KN et al (2009) A selective account of effective paradigms and significant outcomes in the discovery of inspirational marine natural products. J Nat Prod 72:588–603
5. Cuevas C, Francesch A (2009) Development of Yondelis (trabectedin, ET-743). A semisynthetic process solves the supply problem. Nat Prod Rep 26:322–337
6. European Science Foundation (2010) Position Paper 15. Marine biotechnology: a new vision and strategy for Europe. http://www.esf.org/index.php?eID=tx_nawsecuredl&u=0&file=fileadmin/be_user/research_areas/marine/pdf/Publications/MBPP15_MarineBiotechnology.pdf&t=1305543601&hash=63880103a556f163639f0be4178d1512. Accessed 23 May 2010
7. Andersen RJ, Williams DE (2000) Pharmaceuticals from the sea. In: Hester R, Harrison R (eds) Chemistry in the marine environment, 13th edn. RSC Press, Cambridge
8. Jarvis LM (2009) Building brick by brick. Chem Eng News 87:13–20
9. Mulholland DA, Drewes SE (2004) Global phytochemistry: indigenous medicinal chemistry on track in southern Africa. Phytochemistry 65:769–782
10. Branch ML, Branch GM (1992) The living shores of Southern Africa. Struik, Cape Town
11. Branch GM, Griffiths CL et al (1994) Two oceans. David Philip, Cape Town
12. Davies-Coleman MT (2005) Bioactive natural products from southern African marine invertebrates. In: Atta-ur-Rahman (ed) Studies in natural products chemistry, 32nd edn. Elsevier, Amsterdam

13. Pettit GR, Inoue M et al (1988) Isolation and structure of the powerful cell-growth inhibitor cephalostatin-1. J Am Chem Soc 110:2006–2007
14. Pettit GR, Herald CL et al (1993) Isolation and structure of the powerful human cancer cell-growth inhibitors spongistatins 4 and 5 from an African *Spirastrella-Spinispirulifera* (Porifera) 1. J Chem Soc Chem Comm 1805–1807
15. Lopez-Anton N, Rudy A et al (2006) The marine product cephalostatin 1 activates an endoplasmic reticulum stress-specific and apoptosome-independent apoptotic signaling pathway. J Biol Chem 281:33078–33086
16. Rudy A, Lopez-Anton N et al (2008) The cephalostatin way of apoptosis. J Nat Prod 71:482–486
17. Fortner KC, Kato D et al (2010) Enantioselective synthesis of (+)-cephalostatin 1. J Am Chem Soc 132:275–280
18. Fenical W, Jensen P et al (2003) New anticancer drugs from cultured and collected marine organisms. Pharm Biol 41:6–14
19. Moser BR (2008) Review of cytotoxic cephalostatins and ritterazines: isolation and synthesis. J Nat Prod 71:487–491
20. Lee S, LaCour TG et al (2009) Chemistry of trisdecacyclic pyrazine antineoplastics: the cephalostatins and ritterazines. Chem Rev 109:2275–2314
21. Proksch P, Edrada-Ebel R et al (2003) Drugs from the sea – opportunities and obstacles. Mar Drugs 1:5–17
22. LaCour TG, Guo C et al (1998) Interphylal product splicing: the first total syntheses of cephalostatin 1, the north hemisphere of ritterazine G, and the highly active hybrid analog, ritterostatin GN1N. J Am Chem Soc 120:692–707
23. Williams GC (1993) Coral reef octocorals – an illustrated guide to the soft corals, sea fans and sea pens inhabiting the coral reefs of northern Natal. Durban Natural Science Museum, Durban
24. Blunt JW, Copp BR et al (2011) Marine natural products. Nat Prod Rep 28:196–268
25. Rudi A, Goldberg I et al (1993) Sodwanones A-C, three new triterpenoids from a marine sponge. Tetrahedron Lett 34:3943–3944
26. Rudi A, Kashman Y et al (1994) Sodwanones A-F, new triterpenoids from the marine sponge *Axinella weltneri*. J Nat Prod 57:1416–1423
27. Rudi A, Goldberg I et al (1995) Sodwanones G, H, and I, new cytotoxic triterpenes from a marine sponge. J Nat Prod 58:1702–1712
28. Rudi A, Kashman Y et al (1996) Cytotoxic triterpenes from a marine sponge. WO Patent 9701334
29. Kashman Y, Koren-Goldshlager G et al (1999) Halitulin, a new cytotoxic alkaloid from the marine sponge *Haliclona tulearensis*. Tetrahedron Lett 40:997–1000
30. Kashman Y, Koren-Goldshlager G et al (1999) Cytotoxic alkaloids including halitulin, isolation from *Haliclona tulearensis*, and antitumor activity. WO Patent 2000020411
31. Talpir R, Benayahu Y et al (1994) Hemiasterlin and geodiamolide TA; two new cytotoxic peptides from the marine sponge *Hemiasterella minor* (Kirkpatrick). Tetrahedron Lett 35:4453–4456
32. Kashman Y, Gravalos DG (1995) Hemiasterlin and geodiamolide TA from the sponge *Hemiasterella minor*, and methods using them for treatment of tumors. US Patent 5661175
33. Anderson HJ, Coleman JE et al (1997) Cytotoxic peptides hemiasterlin, hemiasterlin A, and hemiasterlin B include mitotic arrest and abnormal spindle formation. Cancer Chemother Pharmacol 39:223–226
34. Kingston DGI (2009) Tubulin-interactive natural products as anticancer agents. J Nat Prod 72:507–515
35. Nieman JA, Coleman JE et al (2003) Synthesis and antimitotic/cytotoxic activity of hemiasterlin analogues. J Nat Prod 66:183–199
36. Kuznetsov G, TenDyke K et al (2009) Tubulin-based antimitotic mechanism of E7974, a novel analogue of the marine sponge natural product hemiasterlin. Mol Cancer Ther 8:2852–2860

37. Vashist YK, Tiffon C et al (2006) Inhibition of hepatic tumor cell proliferation in vitro and tumor growth in vivo by taltobulin, a synthetic analogue of the tripeptide hemiasterlin. World J Gastroenterol 12:6771–6778
38. Hadaschik BA, Adomat H et al (2008) Intravesical chemotherapy of high-grade bladder cancer with HTI-286, a synthetic analogue of the marine sponge product hemiasterlin. Clin Cancer Res 14:1510–1518
39. Hadaschik BA, Ettinger S et al (2008) Targeting prostate cancer with HTI-286, a synthetic analog of the marine sponge product hemiasterlin. Int J Cancer 122:2368–2376
40. Rudi A, Ketzinel S et al (1995) Antheliatin and zahavins A and B, three new cytotoxic xenicane diterpenes from two soft corals. J Nat Prod 58:1581–1586
41. Rudi A, Kashman Y et al (1996) Reef-inhabiting soft coral cytotoxic xenicane diterpenes with anti-tumor activities, *Anthelia glauca* antheliatin and *Alcyonium aurea* zahavin a and zahavin b. WO Patent 9632388
42. Keyzers RA, Daoust J et al (2008) Autophagy-modulating aminosteroids isolated from the sponge *Cliona celata*. Org Lett 10:2959–2962
43. Keyzers RA, Gray CA et al (2006) Malonganenones A-C, novel tetraprenylated alkaloids from the Mozambique gorgonian *Leptogorgia gilchristi*. Tetrahedron 62:2200–2206
44. Cockburn IL, Pesce E et al (2011) Screening for small molecule modulators of Hsp70 chaperone activity using protein aggregation suppression assays: inhibition of the plasmodial chaperone PfHsp70-1. Biol Chem 392:431–438
45. Dai J, Fishback JA et al (2006) Sodwanone and yardenone triterpenes from a South African species of the marine sponge *Axinella* inhibit hypoxia-inducible factor-1 (HIF-1) activation in both breast and prostate tumor cells. J Nat Prod 69:1715–1720
46. Hendricks D, Parker MI (2002) Oesophageal cancer in Africa. IUBMB Life 53:263–268
47. Stoner GD, Wang LS et al (2007) Chemoprevention of esophageal squamous cell carcinoma. Toxicol Appl Pharm 224:337–349
48. Pickens A, Orringer MB (2003) Geographical distribution and racial disparity in esophageal cancer. Ann Thorac Surg 76:S1367–S1369
49. Mqoqi N, Keller P et al (2003) Incidence and geographical distribution of histologically diagnosed cancer in South Africa, 1996–1997. National Cancer Registry of South Africa, South African Institute for Medical Research, Johannesburg
50. Tew WP, Kelsen DP et al (2005) Targeted therapies for esophageal cancer. Oncologist 10:590–601
51. Lordick F, Stein HJ et al (2004) Neoadjuvant therapy for oesophagogastric cancer. Brit J Surg 91:540–551
52. Scheithauer W (2004) Esophageal cancer: chemotherapy as palliative therapy. Ann Oncol 15:97–100
53. Whibley CE, Keyzers RA et al (2005) Antiesophageal cancer activity from southern African marine organisms. Ann N Y Acad Sci 1056:405–412
54. Whibley CE, McPhail KL et al (2007) Reactive oxygen species mediated apoptosis of esophageal cancer cells induced by marine triprenyl toluquinones and toluhydroquinones. Mol Cancer Ther 6:2535–2543
55. Li CH, Chen XS et al (2005) The first total synthesis of triprenylquinone and hydroquinones. Chin Chem Lett 16:1024–1026
56. Davies-Coleman MT, Froneman W et al (2005) Anti-oesophageal cancer activity in extracts of deep-water Marion Island sponges. S Afr J Sci 101:489–490
57. van Wyk AWW, Gray CA et al (2008) Bioactive metabolites from the South African marine mollusk *Trimusculus costatus*. J Nat Prod 71:420–425

Chapter 9
Natural Product-Based Drug Discovery Against Neglected Diseases with Special Reference to African Natural Resources

Sami A. Khalid

9.1 Introduction

It seems quite appropriate to emphasize the importance of the discovery of chemotherapeutic agents for the treatment of neglected tropical diseases (NTDs) at a time when billions of people desperately need medications to combat these diseases. Despite this challenge facing us, today's interest in natural products on the part of the pharmaceutical industry has markedly decreased as a result of their market-driven interest and new drug discovery strategies based on high throughput screening of synthetic compounds made by combinatorial chemistry. Nevertheless, it should be emphasized that combinatorial chemistry as practised by nature has yielded a bewildering array of secondary metabolites having exotic structures, rich in complex stereochemistry and possessing a wide spectrum of bioreactive functionalities [1]. Compounds derived from natural products (NPs) have made a great impact on the pharmaceutical industry; of the 1,010 new chemical entities (NCEs) that were approved by the US Food and Drug Administration (FDA) and other relevant agencies between January 1981 and June 2006, 43 were intact natural products (NPs) and a further 232 (23%) were second-generation NP derivatives [2]. The second-generation compounds were primarily semi-synthetic NP analogues created to improve properties such as solubility and pharmacokinetics. The Dictionary of Natural Products Version 16.2 (2008) cited 265,123 entries as natural products and their derivatives [3]. In reality, natural products cover a much wider and larger chemical space than combinatorial and synthetic compounds due to the diversity of natural products in terms of their chiral centres, ring fusion and richness in functional groups which render them viable for a wider ligand affinity and better specificity to biological targets [4]. Interestingly, natural products are

S.A. Khalid (✉)
Faculty of Pharmacy, University of Science & Technology, P.O. Box 11507, Khartoum, Sudan
e-mail: khalidseek@hotmail.com

K. Chibale et al. (eds.), *Drug Discovery in Africa*,
DOI 10.1007/978-3-642-28175-4_9,

further characterized to be often orally bioavailable despite violating many of the parameters stated in Lipinski's "rule of five" [5].

9.2 Neglected Tropical Diseases

NTDs are a group of chronic disabling infections affecting more than a billion people worldwide, mainly in Africa and mostly those living in remote rural areas, urban slums or conflict zones. People suffering from NTDs are predominantly afflicted by poverty, and they constitute an unattractive market to private-sector research and development (R&D) investment. These diseases not only affect our health but also represent a vicious cycle of socioeconomic events which reinforce and feedback on each other, leading towards inescapable poverty of a sizable number of the population (http://www.thelancet.com/series/neglected-tropical-diseases).

The current situation portrays the devastating burden of NTDs in the African context and represents real challenges to African scientists. The global threat posed by NTDs compelled the World Health Organization (WHO) to release its first report on the situation last year [6]. The number of NTDs enumerated in this report reached 17, including three soil-transmitted helminthiases. Although malaria is not mentioned among the NTDs in this report, possibly due to a resurgence at the global level of increased funding by various international bodies (e.g. Roll Back Malaria Initiative, RBM) and philanthropic organizations (e.g. The Bill & Melinda Gates Foundation) during the last decade, malaria still remains one of the NTDs; this is mainly due to the fact that the death toll from this disease in Africa constitutes over 90% of the total number of global deaths and due to its direct/indirect economic impact which results in an estimated US$ 12 billion annual income loss in sub-Saharan Africa, which translates to a 1.3% annual loss in gross domestic product (GDP) in this malaria-endemic African region [7]. Recent publications still include malaria together with other NTDs such as leishmania, trypanosomiasis, schistosomiasis, onchocerciasis, lymphatic filariasis and dengue fever [8]. Each year, there are between 300 and 500 million clinical cases of malaria; estimates of the number who die from the disease range between 1 and 2.5 million annually. A disproportionate number of deaths from malaria occur among the poor, and about half of those who die are children and pregnant women [6]. Currently, there is no vaccine for malaria, and the parasite has already developed resistance against almost all the currently available antimalarials. Unfortunately, new reports indicate the emergence of resistance against artemisinin derivatives as well [9].

9.2.1 Leishmaniasis

Leishmaniases are diverse clinical manifestations that are caused by parasitic protozoa of the genus *Leishmania* and are transmitted to humans (and animals) by the bite of female sand flies belonging to the genera *Phlebotomus* (Old World) and *Lutzomyia* (New World). The disease is endemic in tropical regions of Africa, America and the Indian sub-continent and in the sub-tropics of southwest Asia and the Mediterranean. Some forms of the disease are anthroponotic (transmitted between humans), while others are zoonotic (involving an animal reservoir). There are about 20 known species of *Leishmania* that have been associated with various disease forms in man [6]. Leishmaniases are classified into three groups: cutaneous (CL), mucocutaneous (MCL) and visceral (VL) leishmaniasis on the basis of their clinical symptoms, with severity of the disease ranging from the disfiguring cutaneous type in which nodules form at the site of infection, leaving scars upon healing, to the lethal visceral leishmaniasis. Currently marketed drugs for treatment are antimonials, amphotericin B and pentamidine as well as miltefosine which has recently approved for oral treatment of VL [6].

9.2.2 African Trypanosomiasis (Sleeping Sickness)

The geographical range of human African trypanosomiasis (HAT) is restricted to sub-Saharan Africa, where there are suitable habitats for its vector, the tsetse fly. Together with the animal form of African trypanosomiasis, known as *nagana*, human disease is a major cause of rural underdevelopment in sub-Saharan Africa [6]. The disease mainly affects poor and remote rural regions. Disease transmission occurs in children and adults during activities such as farming, hunting, fishing or washing clothes. The African trypanosome's pathogenic form belongs to the species *Trypanosoma brucei*, which has three subspecies: *T. b. gambiense*, which causes endemic disease in central and West Africa; *T. b. rhodesiense*, which causes more acute disease in east and southern Africa and *T. b. brucei*, which usually infects domestic and wild animals, but not humans. The strict geographical separation between *T. b. gambiense* and *T. b. rhodesiense* could soon change, however, because the continued spread of *T. b. rhodesiense* in Uganda towards the northwest might cause an overlap of the distributions of the two forms of disease [10].

Trypanosomes are transmitted by blood-feeding tsetse flies of the genus *Glossina* from one mammalian host to another. About 30 species and subspecies of tsetse flies exist, and these are separated into three groups that prefer different habitats and show different abilities to transmit *T. b. gambiense* or *T. b. rhodesiense* sleeping sickness. The flies' biology has one major peculiarity—they are viviparous [10].

The disease appears in two stages, the first haemolymphatic stage and the second meningoencephalitic stage, which is characterized by invasion of the central nervous system (CNS). Disease caused by either of the two parasites leads to coma and death if left untreated. *T. b. gambiense* infection is characterized by a chronic progressive course. According to models based on survival analysis, the estimated average duration of such infection is around 3 years, which is evenly divided between the first and second stages. *T. b. rhodesiense* disease is usually acute, and death occurs within weeks or months [6].

Three major epidemics have ravaged the African continent in the past century due to a progressive re-emergence of the disease, which reached its peak in the late 1990s in the Democratic Republic of the Congo, Angola, Central African Republic, southern Sudan and Uganda [10].

Infection is generally fatal if left untreated. *T. b. brucei* is not infectious to man. At present, pentamidine and suramin are prescribed in the early disease state before CNS involvement, while melarsoprol is used against late-stage disease. Eflornithine (DFMO) is only useful against *T. b. gambiense*, but the requirement for large doses limits its use. Diminazene (Berenil) and isometamidium are drugs for veterinary use. The development of new medicines against HAT underwent a serious setback when the new oral first-stage diamidine drug, pafuramidine maleate (DB289), failed almost at the end of the development programme due to its nephrotoxicity [10]. To date, there are no drugs at the clinical stage for treatment of HAT. However, the nitroimidazole, fexinidazole, has been advanced from discovery to the preclinical stage. The present study is designed to obtain safety, tolerability and pharmacokinetic (PK) data after single and multiple oral administration of increasing doses of fexinidazole in healthy male sub-Saharan African subjects. This study will also assess the relative bioavailability of fexinidazole administered as a tablet in comparison with oral suspension and will assess the impact of concomitant food intake on the relative bioavailability of fexinidazole after single oral dose administration [http://clinicaltrialsfeeds.org/clinical-trials/show/NCT00982904].

9.2.3 American Trypanosomiasis (Chagas Disease)

Although Chagas disease, caused by the protozoan parasite *Trypanosoma cruzi*, does not occur in Africa and is primarily found in South and Central America [6], we have included it in our screening battery mainly to find patterns of selectivity and possibly to speculate on the mechanisms of action. It is estimated that 16–18 million people are infected by *T. cruzi*. Infection mainly occurs through contact with the faeces of a triatomine insect (vector), which defecates after sucking blood at night. The insect resides in crevices in the walls and roofs of poorly constructed houses, usually in rural and periurban areas throughout Latin America. The parasite is transmitted when a person inadvertently enables the parasite-contaminated faeces to make contact with any break in the skin (including the bite), the eyes or mouth. Other modes of transmission include transfusion of infected blood, oral transmission through contaminated food, vertical transmission and organ transplantation. Up to 30% of patients will develop heart damage, and up to 10% will develop damage to their oesophagus, colon or autonomic nervous system, or all of these, in the late chronic phase of the disease. Patients ultimately die, usually suddenly, with death caused by arrhythmias, often occurring in early adulthood. Nifurtimox and benznidazole are the available drugs for the chagasic patients; however, because they have known toxicity and limited efficacy, new drugs are urgently needed [6].

9.3 Discovery of Chemotherapeutics for NTDs

We have witnessed a concomitant advance and explosion in the field of discovery of chemotherapeutic agents against neglected disease from natural products during the last three decades. Notwithstanding, it is worth mentioning at this juncture that only 21 (1.3%) out of 1,556 approved drugs were specifically developed against NTDs during the 1975–2004 period, despite the fact that NTDs account for 11.4% of the global diseases burden [11]. The focus on preparations from plants with known medicinal properties goes back to antiquity. The discovery of pure compounds as active principles in plants was first described at the beginning of the nineteenth century, and the art of exploiting natural products has become part of the molecular sciences. The discovery of quinine from *Cinchona succirubra* (Rubiaceae) and its subsequent development as an antimalarial drug represented a milestone in the history of drug discovery from nature. The isolation of quinine from *Cinchona* bark by Pelletier and Caventou (1820) for the treatment of malaria long before the discovery of the *Plasmodium* parasite in a blood smear by Laveran in Algeria (1880), and the subsequent explanation in 1902 by Nobel Prize winner, Sir Ronald Ross, of how the parasite enters our bodies, represents a unique case in the history of drug discovery where the drug (i.e. quinine) was discovered before the differentiation and elaboration of the disease (i.e. malaria) itself.

9.3.1 The African Perspective

Investigating the virtually untapped African natural sources of novel antiprotozoal compounds remains a major challenge and a source of novelty in the era of combinatorial chemistry and genomics.

Although Africa appears culturally heterogeneous with thousands of distinct tribes, dialects and languages, it is united by a common thread of traditional medical systems and spiritual attributes. African biodiversity, coupled with this deeply rooted African ethnobotanical heritage, has already contributed a number of novel chemical entities resulting in potent pharmacotherapeutics, and Africa still remains a promising untapped reservoir for the discovery of more diverse chemical entities. African flora have already demonstrated their chemical diversity by providing a series of novel chemical entities including the anticancer bisindole alkaloids, vincristine and vinblastine of Madagascar periwinkle (*Catharanthus roseus*), xanthine alkaloids from the Abyssinian *Coffea arabica*, cathartic sennosides from the leaves and pods of the Sudanese *Cassia senna* and the South African *Aloe ferox*, besides the cardiotonic glycosides, strophanthin and ouabain from the Mozambican seeds *Strophanthus hispidus* and the closely related *S. gratus*. The discovery of physostigmine from the West African Calabar bean (*Physostigma venosum*) opened up a new chapter in ophthalmology by offering treatment for glaucoma and myasthenia gravis and provided the template for a new generation of synthetic drugs.

Although approximately 30,000 species have been reported to occur in tropical Africa including a total of 2,667 genera in 293 vascular plant families, these do not include the over 269 families, comprising over 1,519 genera and over 12,000 plant species found in Madagascar [12]. The pharmacotherapeutic potential of the African biodiversity has not yet been realized since much of the research carried out is fragmented with little or no focus on drug discovery. The recent slowdown in bioprospecting and discovery of drugs from African biodiversity is mainly due to the fragmented nature of natural products research across Africa. African researchers are not only challenged by NTDs encroaching upon the continent, but they are also confronted with meagre financial resources, poor infrastructure and non-accessibility to modern technological platforms which are crucial elements for drug discovery. While the African continent is virtually considered as the epicentre of pathogens endemicity, the African scientific community is very poorly represented in setting the research agenda and priorities. The recent establishment of the European Foundation Initiative for African Research into Neglected Tropical Diseases (EFINTD), which consists of five European foundations and aims to combat NTDs by offering funding for postdoctoral fellows from sub-Saharan Africa to pursue scientific careers in their home continent, is undoubtedly a move in the right direction (http://www.ntd-africa.net/), provided that concerted and concomitant efforts are made to maintain a sound scientific capacity in Africa. Such efforts are a major requirement for stopping the brain drain and building a critical mass of young African scientists to guarantee ownership and sustainability of long-term control programmes.

As a priority, we definitely need to further consolidate the African ethnobotanical heritage by developing safe, efficacious and affordable phytomedicines to serve the needs of the custodians of this traditional knowledge, with special emphasis on the indigenous diseases. This should be done at an early stage of our endeavour and before we embark on costly and time-consuming research geared only towards bioprospecting. This huge African reservoir of biodiversity should be tapped and made available to the population through a systematic preclinical research based on the development of phytopharmaceutical formulations rather than the isolation of single chemical moieties [9].

There is a pressing need to focus more strongly on the short-term applicability of research and its relevance to national and regional health problems in Africa, as well as its benefits to the local population. Instead of concentrating on pure basic research and scientific impact factors, more research efforts should be dedicated to operational research and the better application of existing tools in the health system. Lack of the appropriate research platforms in Africa, coupled with insufficient funds to enable them to work efficiently in their home countries, is indeed a major factor hindering African scholars trained abroad from returning to their home countries to pursue careers in health research.

The burden of NTDs is very costly in terms of human suffering as well as contributing to poverty and underdevelopment. The lack of effective, safe and affordable chemotherapeutics against these diseases, coupled with the non-availability of vaccines for any of these diseases, represents an insurmountable challenge for African countries [13]. Hence, the majority of currently available drugs have a number of shortcomings and drawbacks due to the development of resistance and toxicity, as well as non-affordability due to their high cost, coupled with limited availability. Meanwhile, there is a severe shortage of novel leads against NTDs that meet the desired target product profiles for these diseases, and there are certain gaps or bottlenecks in R&D pipelines associated with tropical diseases. These bottlenecks can be mainly summarized as follows: (a) preclinical research does not begin despite our current knowledge of drug targets and lead compounds; (b) validated candidate drugs do not enter clinical development because of company strategy and (c) new or existing drugs do not reach patients due to registration problems, lack of production, high prices or not being adapted to the local conditions of use since most of the medications are injectable and costly.

It is envisaged that any new antiprotozoal drug candidate should be safely dosed orally for a relatively short period to induce its curative action at affordable cost to the population in endemic countries, should have good stability in the tropics and should have a low propensity to generate resistance within a short period. Therefore, drug discovery of new antiprotozoal drugs against NTDs is urgently needed to reduce the mortality and morbidity associated with them. An additional criterion desired for drugs against African trypanosomiasis is efficacy against both early and late stage of both major species *T. b. rhodesiense* and *T. b. gambiense*.

Since markets for drugs against NTDs are primarily in poor countries, marketing opportunities are generally considered to be limited. Therefore, it is of prime importance to streamline coordinated approaches involving multidisciplinary networks of academic and government institutions, researchers in several disciplines, as well as partnerships between the pharmaceutical industry and the public sector in both developed and developing countries. Fortunately, during the past decade, there has been a dramatic increase in interest in R&D directed towards producing new drugs for tropical diseases, and various partnerships involving academic consortia, industry, governments and philanthropic organizations are now dedicated to drug discovery, preclinical and late-stage development candidates and drug development in countries where most patients cannot afford to pay prices for the patented drugs they need. Accordingly, a number of strategies have been identified to streamline drug discovery of leads from natural sources against NTDs. These strategies include (a) the identification of antiprotozoal active compounds from natural sources products by evaluation of their biological activity following strictly bioactivity-guided fractionation of plant extracts, followed by the semi-synthesis or total synthesis of the new drugs; (b) the optimization of existing agents to improve their pharmacodynamics and pharmacokinetic profiles and (c) the identification of new chemotherapeutic targets by using the available genome sequences of human pathogens and their vectors and subsequent discovery of compounds that act on these targets [13]. These strategies usually require the evaluation of the activity of a compound in a validated "whole organism" screen against the disease-causing parasite. Reversal of parasite resistance should be considered an alternative approach to be pursued [14].

The aforementioned strategies are being attempted by a number of initiatives and networks in order to promote the identification of "hits" or "leads" and to develop a globally accessible database containing a prioritized list of drug targets across the range of disease-causing parasites within the portfolio of TDR (a programme on Tropical Diseases Research sponsored by WHO). These include the Compound Evaluation and the Natural Products Initiative (CEN) and the Genomics and Discovery Research Committees (GDR) which were initiated a couple of years ago at the Special Programme for Research and Training in TDR in order to evaluate compounds in validated whole organism and molecular target assays, promote the use of natural products as starting points for the discovery of new lead compounds and support development of new tools and methods to accelerate drug discovery [13]. Recently, a few organizations have been established with a special interest in the development of new antiprotozoal drugs, among them are the Tropical Disease Initiative (TDI) and the Institute for OneWorld Health (IOWH). The World Bank, governments and charities are highly encouraged by a number of international agencies to subsidize developing country purchases at a guaranteed price, as well as the creation of non-profit venture-capital firms (virtual pharmas) that look for promising drug candidates before commencing drug development through contracts with corporate partners. The IOWH, a not-for-profit pharmaceutical company, based in San Francisco, is focused on developing safe, effective and affordable new medicines for people with infectious diseases in the developing world. Similarly, the TDR programme is also supporting drug discovery and public–private partnerships (PPPs) in order to address the imbalance in funding between developed and developing world diseases. In these PPPs, the private partner can expand its business opportunities in return for assuming the new or expanded responsibilities and risks. Among these PPPs are the Drugs for Neglected Diseases initiative (DNDi) and Medicines for Malaria Venture (MMV). The DNDi was created in 2003 by the Oswaldo Cruz Foundation (Brazil), the Indian Council of Medical Research, the Kenya Medical Research Institute, the Ministry of Health Malaysia, Pasteur Institute (France), Médecins sans Frontières (MSF) and the UNDP/World Bank/WHO/TDR to improve the quality of life and the health of people suffering from neglected diseases. The purpose of DNDi is to combine the skills and research capacity in academia, the pharmaceutical industry and contract researchers, public–private partnerships or product development partnerships aimed at creating focused research consortia that address all aspects of drug discovery and development. These consortia not only emulate the projects within pharmaceutical and biotechnology industries, such as the identification and screening of natural product libraries, medicinal chemistry, pharmacology and pharmacodynamics, formulation development and manufacturing, but also use and strengthen existing capacity in disease-endemic countries, particularly for the conduct of clinical trials [8]. The Drugs for Neglected Diseases *initiative* (DND*i*) has adopted a model closely related to that of a virtual biotechnology company for the identification and optimization of drug leads.

The MMV was established in 1999 to bring public and private-sector partners together to fund and provide managerial and logistical support for the discovery, development and delivery of new medicines to treat and prevent malaria. Pharmaceutical companies are also increasing their participation in this process. GlaxoSmithKline, for example, is focused on discovering, developing and making new drugs and vaccines available for the treatment or prevention of diseases in the developing world, primarily malaria, but also HIV/AIDS and tuberculosis. The Novartis Institute for Tropical Diseases in Singapore and the Wellcome Trust are focusing on malaria in partnership with MMV. Sanofi-Aventis has established the Impact Malaria Programme and continues collaborating with TDR and DNDi. Pfizer Inc. has also been collaborating with TDR to provide compounds and drug discovery support to identify leads for malaria, African sleeping sickness, Chagas disease, leishmaniasis and other tropical diseases. Novartis has also developed a project for a Fund for R&D in Neglected Diseases (FRIND). The establishment of the African Network for Drugs and Diagnostic Innovation (ANDI) will definitely promote research for drug discovery for NTDs. The concept of ANDI was launched by several African institutions, through TDR, in October 2008, in Abuja, Nigeria.

9.3.2 Protocols for Discovering Antiparasitic Natural Compounds

Undoubtedly the conventional screening systems utilizing the whole organism have considerable advantages over the sub-cellular target-oriented options; they involve all targets and take drug disposition phenomena into consideration. The whole organism in vitro models are generally used for finding "hits". Selected reference standard protozoa of interest are usually cultivated either in cell-free medium or in cell cultures. The in vitro intracellular macrophages of *L. donovani* amastigotes and *T. cruzi*, as well as the axenic culture of blood stage forms of *T. b. rhodesiense* and *L. donovani*, are used in the evaluation of our extracts/compounds. Unfortunately, the screening discussed in the present work did not include *T. b. gambiense*. It should be noted at this juncture that the use of promastigotes for screening should be discouraged, and the activity of the extracts/compounds against any particular parasite should be discriminated from specific cell toxicity. Confirmation of "hits" resulting from in vitro screening in in vivo models such as Swiss mouse is indispensable to determining the pharmacokinetic, metabolic and toxicological parameters. *P. falciparum* culture in human erythrocytes is the standard in vitro model used for malaria, while *P. berghei* is used in the mouse model (4-day test).

Generally, extracts/compounds are dissolved in standard (20 mM or 20 mg/ml) stock solutions in DMSO using fourfold serial dilutions in standard 96-well microplates as a fixed concentration in all screens to facilitate spectrophotometric or visual reading whenever possible. Although it is advisable to express activity for compounds with known molecular weights in μM, it is a very common practice to express the activity for extracts, mixtures and pure compounds in μg/ml. Pure compounds exhibiting IC_{50} values higher than 32 μg/ml (or μM) are usually classified as inactive, while compounds with IC_{50} values lower than 4 μg/ml (or μM) are considered as promising "hits" and should be further evaluated in a secondary in vitro screen applying a full dose titration using twofold serial dilutions. Cytotoxicity was assessed using rat skeletal myoblasts (L6 cells).

Extracts showing IC_{50} values <1 μg/ml and cytotoxicity >10 μg/ml (i.e. selectivity index >10) are usually considered promising candidates for bioactivity-directed fractionation.

Due to the limited space, the discussion in this chapter is restricted to some selected and structurally defined bioactive secondary metabolites that have been tested in standard in vitro and/or in vivo assays.

9.3.3 Antiparasitic Compounds from African Plants

9.3.3.1 Phenolic Compounds

Flavonoids

Flavonoids are widespread in the plant kingdom. In contrast to some ethnomedical reports, up to 1986, there was no scientific evidence of their antiparasitic activity [14]. However, these ubiquitous natural products have attracted renewed interest following the demonstration that some methoxylated flavanones found in *Artemisia annua* (Asteraceae) act synergistically with artemisinin against *P. falciparum* in vitro [15].

The in vitro antiprotozoal screening of 16 Asteraceae species of Sudanese origin revealed the prominent activity of the dichloromethane extract of *Ageratum conyzoides* L. against bloodstream forms of *T. b. rhodesiense* ($IC_{50} = 0.78$ μg/mL) and noticeable activities against *L. donovani* ($IC_{50} = 3.4$ μg/ml) as well as *P. falciparum* ($IC_{50} = 8.0$ μg/ml) [16]. Bioactivity-guided fractionation of this extract yielded five highly methoxylated flavonoids (**1–5**) along with the chromene derivative, encecalol methyl ether (**6**) [16].

The spectroscopic analysis revealed that all the flavonoids isolated are highly methoxylated flavones identified by means of mass- as well as NMR-spectroscopic analyses as 5,6,7,8,5′-pentamethoxy-3′,4′-methylenedioxyflavone (eupalestin, **1**); 5,6,7,5′-tetramethoxy-3′,4′-methylenedioxyflavone (**2**); 5,6,7,8,3′,4′,5′-heptamethoxyflavone (5′-methoxynobiletine, **3**); 5,6,7,3′,4′,5′-hexamethoxyflavone (**4**) and 4′-hydroxy-5,6,7,3′,5′-pentamethoxyflavone (ageconyflavone C, **5**) [16].

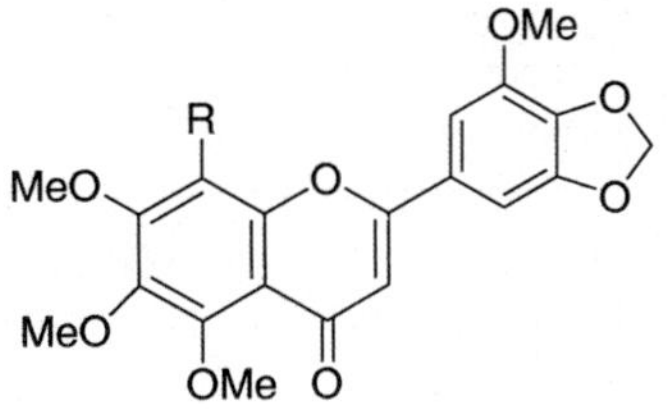

	R
1	OMe
2	H

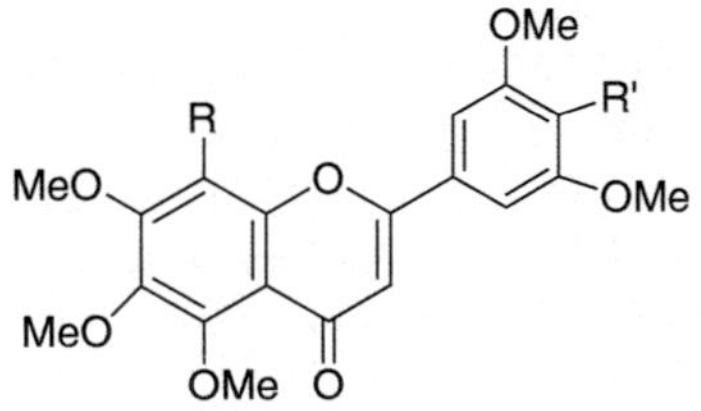

	R	R'
3	OMe	OMe
4	H	OMe
5	H	OH

All the flavonoids exhibited certain antitrypanosomal activity against *T. b. rhodesiense* with IC_{50} values: 6.67, 7.29, 4.76, 8.58 and 3.01 μg/ml, respectively. Compounds, 5,6,7,5′-tetramethoxy-3′,4′-methylenedioxyflavone (**2**) and 5,6,7,8,3′,4′,5′ heptamethoxyflavone (**3**), also showed activity against *T. cruzi*, with IC_{50} values of 19.5 and 26.4 μg/ml, respectively. Three flavones, namely, 5,6,7,8,3′,4′,5′-heptamethoxyflavone (5′-methoxynobiletine, **3**), 5,6,7,3′,4′,5′-hexamethoxyflavone (**4**) and 4′-hydroxy-5,6,7,3′,5′-pentamethoxyflavone (**5**), exhibited activity against *L. donovani* with IC_{50} values of 5.29, 8.61 and 3.56 μg/ml, respectively. Only 5,6,7,8,5′-pentamethoxy-3′,4′-methylenedioxyflavone (eupalestin, **1**), 5,6,7,5′-tetramethoxy-3′,4′-methylenedioxyflavone (**2**) and 5,6,7,3′,4′,5′-hexamethoxyflavone (**4**) showed antiplasmodial activity against the *P. falciparum* chloroquine-resistant strain with IC_{50} values of 4.57, 4.26 and 2.99 μg/ml, respectively. None of the isolated flavonoids exhibited high cytotoxicity against the L6 cell line. The crude fraction of *A. conyzoides* also yielded a chromene derivative identified as encecalol methyl ether (**6**) with very low levels of activity against *T. b. rhodesiense* and *T. cruzi* with IC_{50} values of 78.4 and 28.6 μg/ml, respectively [16].

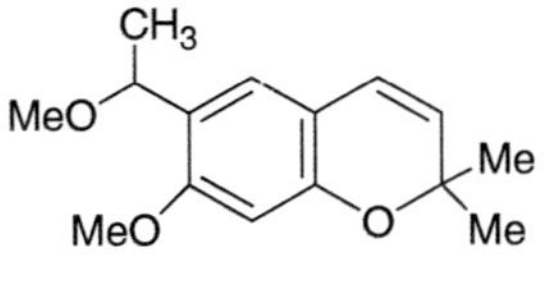

6

Moreover, the antitrypanosomal activity of the *A. conyzoides* extract was found to decrease considerably while stored in solution. An unstable compound was detected in the fresh extract by HPLC, which was converted rapidly into the encecalol methyl ether while stored in methanolic solution. This compound, deemed to represent a constituent with antitrypanosomal activity, could not be isolated from the extract in intact form. It was expected on grounds of its mass spectrum to represent encecalol angelate, which was therefore prepared by total synthesis [17]. The total synthesis of encecalol angelate has been achieved by a linear six step synthesis with overall 15% yield. The synthesized encecalol angelate was chromatographically and spectroscopically identical with the natural product. It degraded in methanol with a half-life of approximately 6 h to yield encecalol methyl ether (**6**). The antiprotozoal activity of the synthetic encecalol angelate against *T. b. rhodesiense* as well as *T. cruzi*, *L. donovani* and *P. falciparum* was investigated. It was found to possess low activity against these pathogens so that it is most likely that this compound is not responsible for the high antitrypanosomal activity of the fresh crude plant extract [16, 17].

In order to reveal the structure–activity relationship among flavonoids displaying antiprotozoal activity, a large set of flavonoid aglycones and glycosides were assessed for their in vitro and in vivo activity against *T. b. rhodesiense*, *T. cruzi* and *L. donovani* as well as their cytotoxicity against mammalian L6 cells which were compared to their antiparasitic activities. The best in vitro trypanocidal activity against *T. b. rhodesiense* was exerted by 7,8- dihydroxyflavone ($IC_{50} = 68$ ng/ml), followed by 3-hydroxyflavone, rhamnetin and 7,8,3′,4′-tetrahydroxyflavone ($IC_{50} = 0.5$ μg/ml) and catechol ($IC_{50} = 0.8$ μg/ml). The activity against *T. cruzi* was moderate, and only chrysin dimethylether and 3-hydroxydaidzein had IC_{50} values less than 5.0 = μg/ml. The majority of the metabolites tested possessed remarkable leishmanicidal potential. Fisetin, 3-hydroxyflavone, luteolin and quercetin were the most potent, giving IC_{50} values of 0.6, 0.7, 0.8 and 1.0 μg/ml, respectively. The flavonoids, 7,8-dihydroxyflavone and quercetin, appeared to ameliorate parasitic infections in mouse models. The tested compounds lacked cytotoxicity in vitro and in vivo. Some general trends with respect to the structure–activity relationships related to the antiparasitic activity of flavonoids were discussed, but it was not possible to draw clear and detailed quantitative structure–activity relationships (QSAR) for any of the bioactivities [18].

Lignans

A variety of types of lignans are encountered among natural products with a broad spectrum of medicinal properties. However, few lignans have been tested against parasitic protozoa. The arylnaphthalene lignan, justicidin A (**7**) which was isolated from the aerial parts of *Haplophyllum tuberculatum* (Rutaceae), exhibited moderate antimalarial activity ($IC_{50} = 1.9$ μg/ml), and it has been hypothesized that justicidin A inhibits the parasite by disturbing the mitochondrial structure and/or function [14].

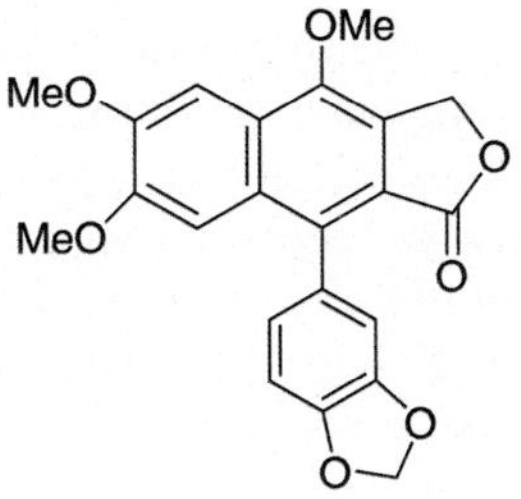

7

9.3.3.2 Acetogenins

The Annonaceous acetogenins are waxy substances composed of C-32 or C-34 long-chain fatty acids that are combined with a 2-propanol unit at C-2 to form a terminal α,β-unsaturated γ-lactone. They usually cyclize to form one, two or three tetrahydrofuran (THF) or tetrahydropyran (THP) rings near the middle of the aliphatic chain. Various hydroxyls, double bounds, carbonyls and acetyls can be located throughout the molecule. These compounds have been associated with a number of biological activities including antiprotozoal, insecticidal, antimitotic, cytotoxic, fungicidal and pesticidal activity [19, 20].

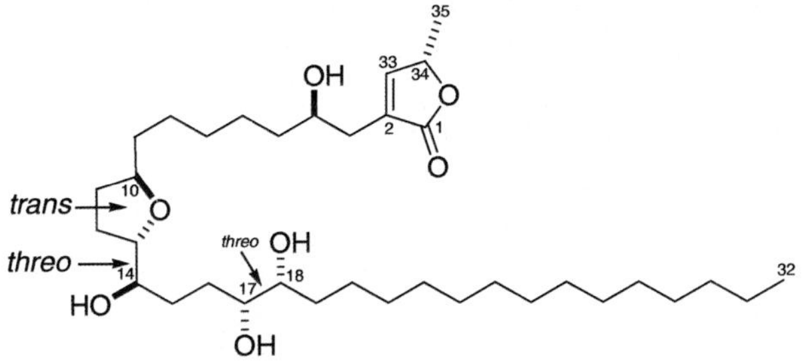

8

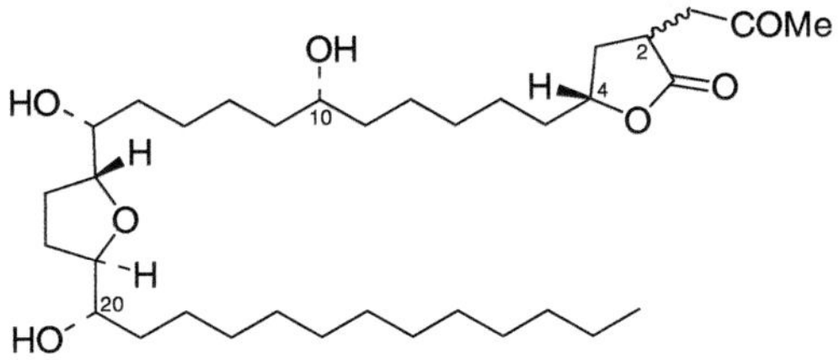

9

Our screening of a dozen of acetogenins isolated from three Annonaceous species resulted in the isolation of 6 weakly active acetogenins with the exception of the mono-THF ring acetogenin, gigantetrocine (**8**), isolated from the African plant *Uvariopsis congolana*. This compound exhibited very significant ($IC_{50} = 0.019$ μg/ml) antitrypanosomal activity specifically against *T. cruzi* and less activity ($IC_{50} = 0.88$ μg/ml) against *T. b. rhodesiense* trypomastigotes with relatively minimal cytotoxicity ($IC_{50} = 5.5$ μg/ml). This may be due to the inhibition of the mitochondrial NADH oxidase activity. Gigantetrocine (**8**), however, exhibited weaker activity against the related kinetoplastid, *L. donovani* ($IC_{50} > 1$), and moderate activity against the chloroquine-resistant *P.* falciparum ($IC_{50} = 6.72$ μg/ml).

Gigantetrocine was originally isolated from the Asian taxon, *Goniothalamus giganteus* [21]. Interestingly, the antiplasmodial activity of *U. congolana* has been associated with the acetogenin-rich fraction of this plant [22]. However, the structurally related isoannonacin (**9**) showed weak activity with $IC_{50} = 19.4$ and 25.7 μg/ml against *T. cruzi* and *T. b. rhodesiense*, respectively.

9.3.3.3 Terpenoids

Triterpenoids

The triterpenoid quinone methides comprise a relatively small group of unsaturated and oxygenated D:A-*friedo-nor*-oleananes. Due to the restricted taxonomic distribution of these nor-triterpenoids to the family Celastraceae, they have been coined the general name celastroides [23]. These compounds are characterized by their extended chromophore over the A and B rings of the triterpenoid skeleton, as exemplified by celastrol (**10**), pristimerin (**11**) and isoiguesterol (**12**) isolated by our group from the root bark of *Maytenus senegalensis*. Preliminary screening of a number of Sudanese plant extracts showed prominent antiplasmodial activity of the root bark of *Maytenus senegalensis* (*Celastraceae*).

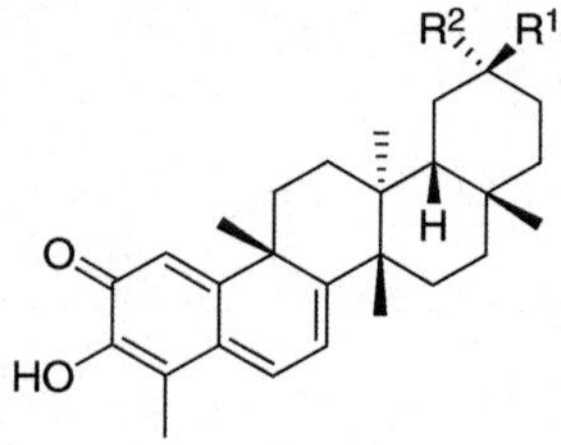

	R_1	R2
10	Me	COOH
11	Me	COOMe
12	H	CH_2OH

The crude dichloromethane extract possessed IC_{50} values of 3.9 and 10 μg/ml on chloroquine-sensitive *P. falciparum* (3D7) and resistant (Dd2) strains with no effect on lymphocyte proliferation even at the highest tested concentration; the IC_{50} was greater than 100 μg/ml. Liquid–liquid separation of the methanolic extract of *M. senegalensis* revealed that the dichloromethane extract possessed an IC_{50} value of only 2.1 μg/ml. Column fractionation of dichloromethane extract gave one fraction with IC_{50} value of 0.5 μg/ml [24].

Bioassay-guided fractionation of the dichloromethane extract resulted in the isolation of (20α)-3-hydroxy-2-oxo-24-nor-friedela-1(10),3,5,7-tetraen-carboxylic acid-(29)-methylester, pristimerin (**11**). The in vitro antiplasmodial activity of the isolated compound against the chloroquine-resistant strain (Dd2) of *P. falciparum* was $IC_{50} = 0.5$ μg/ml, and its in vitro antileishmanial activity performed on promastigotes of *L. major* was $IC_{50} = 6.8$ μg/ml [25]. Further investigation of the extract revealed the presence of celastrol (**10**) and isoiguesterol (**12**) with $IC_{50} = 0.18$ and 0.03 μg/ml against the *P. falciparum* chloroquine-resistant strain, respectively.

Limonoids

The triterpenoid skeletons might undergo a variety of structural modifications which could lead to loss of four skeletal carbon atoms to form mainly C_{26}-based tetranor-triterpenoids known as the limonoids. These compounds are further characterized by a β-substituted furan ring at C-17. The modifications of the tetracyclic triterpene skeleton in plants have reached their zenith in four closely related families of the order Rutales: Rutaceae, Meliaceae, Cneoraceae and Simaroubaceae. Limonoids are frequently present in a number of plants reputed to exert antipyretic and/or antimalarial effects in African traditional medicine. *Melia azedarach* L., *Azadirachta indica* A. Juss, *Khaya senegalensis* (Desr.) A. Jus and *Citrus aurantium* feature very prominently among the medicinal plants used in the African traditional medicine [26, 27]. However, our previous investigation of various limonoids showed that A-*seco*-limonoids derived from the Rutaceae such as limonin, nomilin and rutaevin exhibited no appreciable antiplasmodial activity [14].

One well-known representative from the family Meliaceae is *Azadirachta indica*, the neem tree, which is widely used in the African traditional medicine as a decoction for the treatment of malaria. Results obtained from our early studies indicate that gedunin (**13**) has an IC_{50} value of 0.8 μg after a 48-h exposure (0.3 μg after 96 h), roughly equivalent to quinine under the same experimental conditions [14, 28]. However, our subsequent investigation in mice infected with *P. berghei* showed negligible suppression of parasitaemia, and no clear dose-response effect has been observed, while concurrent oral administration of gedunin (**13**) with a mixture of volatile oil of dill (*Anethum graveolens* L.) resulted in total clearance of parasitaemia in a dose-response manner. It was eventually revealed that this significant increase of in vivo activity was due to the presence of 6-allyl-4,5-dimethoxy-1,3-benzodioxol (dillapiole) in dill oil [9].

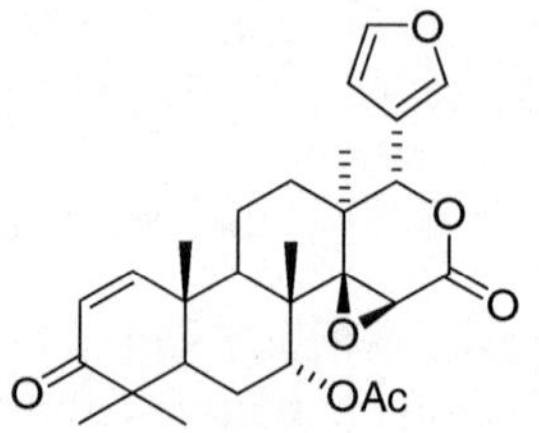

13

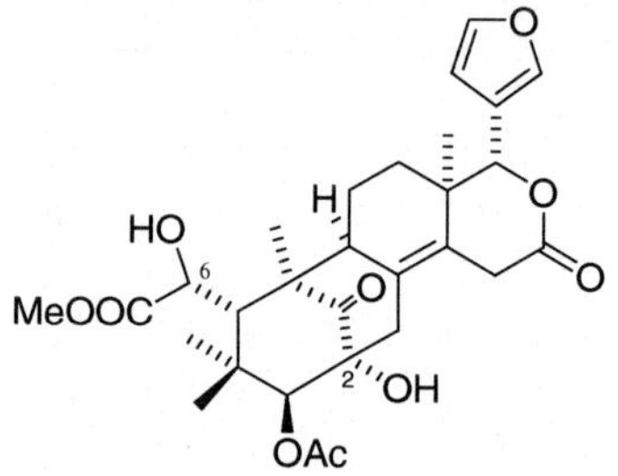

14

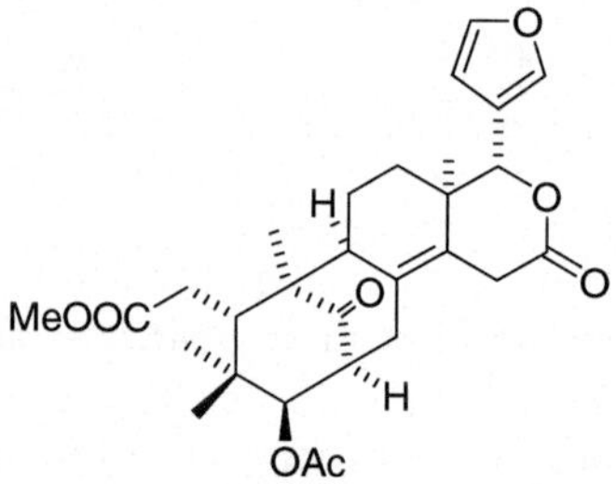

15

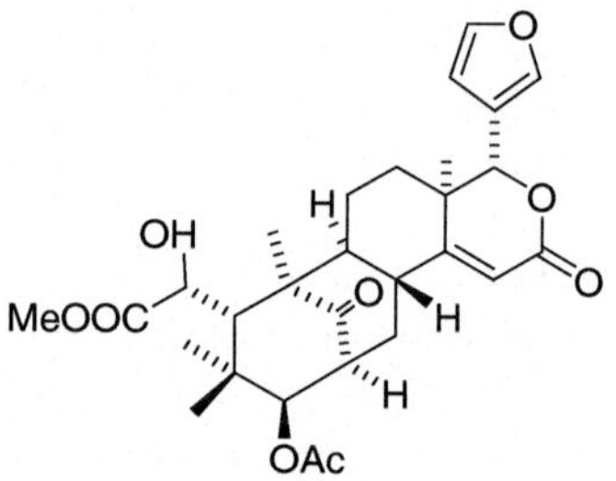

16

Dillapiole is a potent cytochrome CYP450 inhibitor with broad substrate specificity. Application of the structurally related inhibitors, apiol and myristicin, resulted in varying degrees of synergism [9]. This example gives credence to the importance of in vivo studies and the biotransformation mediated by CYP450 subfamily enzymes in modulating the therapeutic effects of certain chemical compounds. This is one example among the frequently encountered in vitro active compounds that ultimately fail to reproduce their bioactivity when tested against a whole organism due to, among other reasons, their poor bioavailability profiles. Therefore, one should be extremely careful in extrapolating in vitro data to in vivo situations, as has been demonstrated by our previous research on a number of occasions.

In view of our interest in the antiprotozoal activity of members of the Meliaceae family, we have conducted a bioactivity-guided fractionation of the chloroform extract of the bark of *Khaya senegalensis*. Repeated chromatographical separation of a very complex mixture resulted in the isolation of three tetranortriterpenoids of the mexicanolide group [29]. Among them, a new compound has been characterized as 2,6-dihydroxyfissinolide (**14**). The two others compounds were characterized as fissinolide (**15**) and methyl 3β-acetoxy-6-hydroxy-1-oxomeliac-14-enoate (**16**). The latter compound (**16**) has been poorly described in the literature, and the assignments of some signals in the 13 C NMR spectrum of fissinolide have been reversed, but the signals of the 1 H NMR spectrum of methyl 3β-acetoxy-6-hydroxy-1-oxomeliac-14-enoate have been fully assigned [28]. Fissinolide (**15**) was slightly active in vitro against chloroquine-resistant strains of *P. falciparum* ($IC_{50} = 48 \pm 3$ mM) and promastigotes of *L. major* ($IC_{50} = 69 \pm 13$ mM). The 2,6-dihydroxy derivative **14** exerted $IC_{50} = 0.1220.08$ mM against *P. falciparum* and $IC_{50} > 0.20$ mM against *L. major*. The methyl 3β-acetoxy-6-hydroxy-1-oxomeliac-14-enoate (**16**) exhibited no significant antiprotozoal activity ($IC_{50} > 0.20$ mM) against either *P. falciparum* or *L. major*. None of these mexicanolide (**14-16**) inhibited the phytohemagglutinin-A-induced proliferation of lymphocytes [29].

Diterpenes

Among the 10 phorbol esters we isolated from *Jatropha curcas* and croton oil (*Croton tiglium*, Euphorbiaceae), the nitrogen-containing phorbol ester, sapintoxin D (**17**), exhibited the highest antitrypanosomal activity against *T. cruzi* ($IC_{50} = 1.90$) and *T. b. rhodesiense* ($IC_{50} = 7.40$ μg/ml) trypomastigote, respectively, with unexpectedly low cytotoxicity ($IC_{50} > 10$) [unpublished results].

Although most of phorbol esters possess both high antiparasitic activity as well as high cytotoxicity to mammalian cells, this compound seems a particularly attractive candidate for further development due to the good selectivity index.

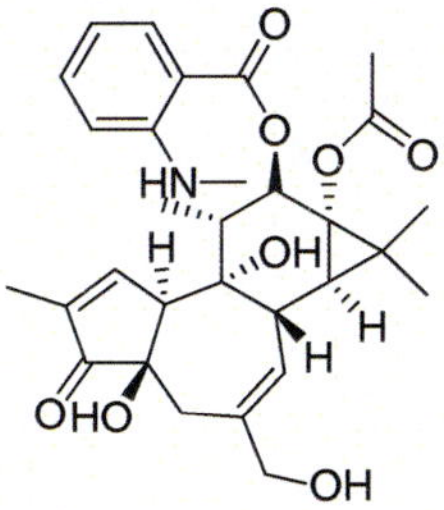

17

However, this compound showed much less toxicity to the taxonomically related *Leishmania* spp. It is pertinent to mention that the chemically related compound, 12-*O*-tetradecanoyl phorbol-13-acetate, which has been previously isolated from croton oil (*Croton tiglium*), *Sapium* and *Euphorbia* species exhibited significant toxicity towards *T. cruzi* at a concentration as low as 0.1 μg/ml, but this compound is considered to be one of the best known protein kinase C-activating diterpene phorbol esters which proved to have high cytotoxicity at therapeutic doses.

Sesquiterpene Lactones

In vitro screening of the dichloromethane extract of the aerial part of *Xanthium brasilicum* Vell. [syn. *X. strumarium* var. *brasilicum* (Vell.) Baker in Mart.] revealed its promising antiparasitic activity against *T. b. rhodesiense*, *T. cruzi* and *L. donovani* as well as *P. falciparum*. Bioactivity-guided fractionation resulted in the isolation of six bioactive sesquiterpenes [30]. Among them are four sesquiterpene lactones (STL) of the xanthanolide series (4,5-seco-guaianolide-type). They were identified by spectroscopic means as 8-epixanthatin (**18**), 8-epixanthatin 1β and 5β-epoxide (**19**) and as the dimers pungiolide A (**20**) and pungiolide B (**21**).

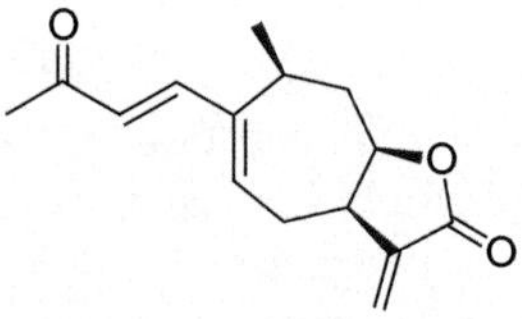

18

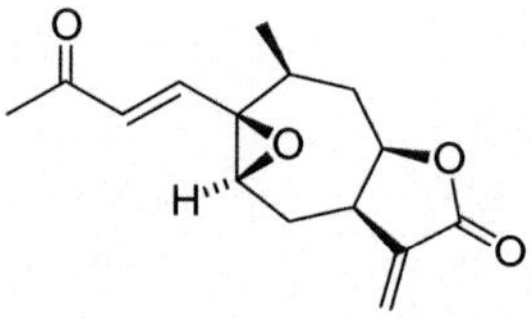

19

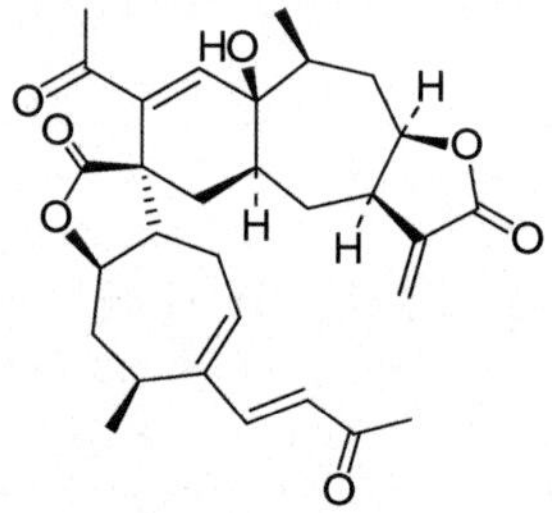

20

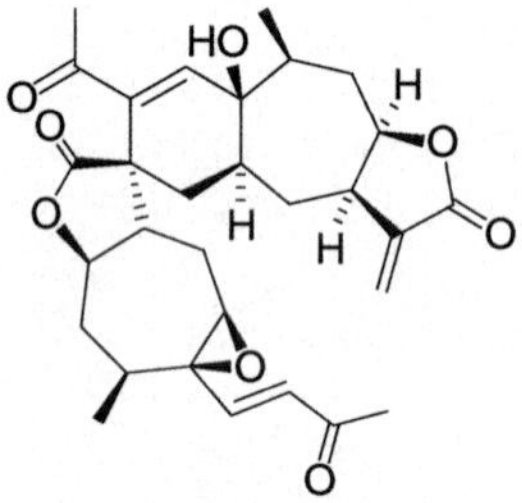

21

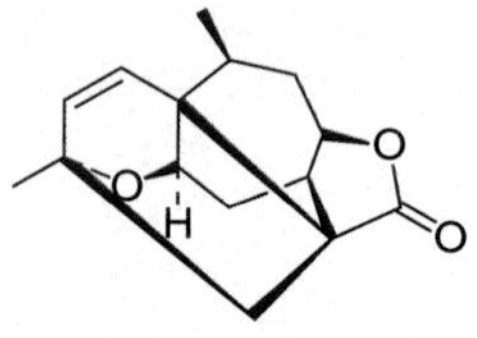

22

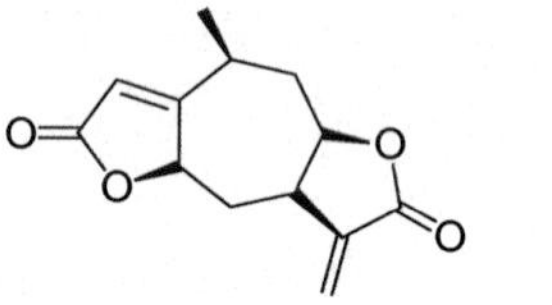

23

The other two isolated compounds were spectroscopically identified as the xanthanolide sesquiterpene lactones, xanthipungolide (**22**) and 4,15-dinor-1,11(13)-xanthadiene-3,5β:12,8β-diolide (**23**). While xanthipungolide (**22**) turned out to be inactive against the tested parasites, the dinor-xanthanlide (**23**) showed significant activity against *T. b. rhodesiense* (IC_{50} = 1.92 μg/ml) and *L. donovani* (IC_{50} = 6.33 μg/ml). As the most active single compound from this extract, 8-epixanthatin 1β,5β-epoxide (**19**) showed IC_{50} values of 0.09, 2.95, 0.16 and 1.71 μg/ml (0.33, 11.3, 0.6 and 6.5 μM) against *T. b. rhodesiense*, *T. cruzi*, L. *donovani* and *P. falciparum*, respectively, while its cytotoxicity against rat myoblast cells used as a control was determined to be 5.8 μg/ml (22.1 μM). Besides assessment of their antiprotozoal activity, the structural assignments for the dimeric xanthanolides pungiolides A (**20**) and B (**21**) were reinvestigated and their stereochemistry fully established [30].

Prompted by results of our previous studies on the antitrypanosomal activity of the sesquiterpene lactone, helenalin (**24**) ($IC_{50} = 0.051$ μg/ml) and some of its congeners against *T. b. rhodesiense* [31], a set of 40 STLs compounds were subjected to structure–activity (in vitro) studies against *T. b. rhodesiense*, *T. cruzi*, *L. donovani* and *P. falciparum*.

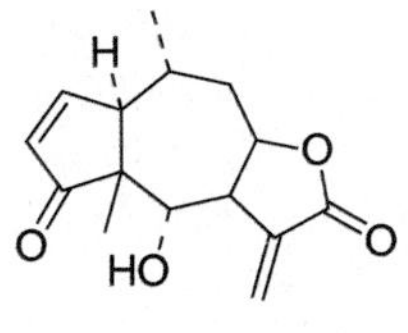

24

Furthermore, cytotoxic activity against L6 rat skeletal myoblast cells was assessed. Some of the compounds possess high activity, especially against *T. b. rhodesiense* (e.g. helenalin and some of its esters with IC_{50} values of 0.05–0.1 μM, which is about 10 times lower than their cytotoxic activities). All investigated antiprotozoal activities are significantly correlated with cytotoxicity, and the major determinants for their activity are α,β-unsaturated structural elements, which is known to be essential for other biological activities of STLs. It was observed, however, that certain compounds are considerably more toxic against protozoa than against mammalian cells, while others exhibit more cytotoxicity than activity against the protozoa.

This comparative quantitative structure–activity relationships (QSAR) analysis was therefore undertaken in order to discern the antiparasitic activity of STLs against *T. b. rhodesiense* versus their cytotoxicity [32]. Both activities were found to depend to a large extent on the same structural features and molecular properties, and any observed differential activity was attributed to subtle structural differences which would be difficult to exploit in terms of lead optimization. It is clear, however, that QSAR studies may play an important role in the discovery of antiparasitic drugs and maybe are used as a predictive tool for the molecular development. As a result, there is an increasing interest on the development of rational approaches for antiparasitic drugs discovery. In this sense, a very important role may be played by computer-aided drug design techniques based on multi-target quantitative structure–activity relationships (mt-QSAR) studies [32].

9.3.3.4 Alkaloids

Benzo[c]Phenanthridine Alkaloids

Benzo[*c*]phenanthridine alkaloids, although known for many years, have recently been the subject of increasing interest because of their wide range of pharmacological activities, including antiviral, antitumor, antimicrobial, antifungal and anti-inflammatory properties. The decoction of the leaves of *Argemone mexicana* L.

(Papaveraceae) is widely used in Sudanese traditional medicine for the treatment of malaria and early-stage trypanosomiasis. In vitro evaluation of the aqueous extract displayed the most prominent antitrypanosomal activity against *T. b. rhodesiense* (IC_{50} = 0.09 μg/ml). Bioactivity-directed fractionation of this extract resulted in the isolation of the quaternary benzo[*c*]phenanthridine alkaloid, sanguinarine (IC_{50} = 0.05 μg/ml). The in vitro cytotoxicity of sanguinarine (**25**) on three cell lines (HT-29, MRC-5 and NBZa) was less than antitrypanosomal activity by factor of 60, 17 and 70, respectively (IC_{50} values, 3.0, 0.85 and 3.5 μg/ml). Thus, sanguinarine (**25**) appears to show selective toxicity towards *T. b. rhodesiense* [33]. Based on its quaternary nature and polarity it is anticipated, however, that sanguinarine would not be able to treat the late stage of the disease due to its inability to cross the blood–brain barrier.

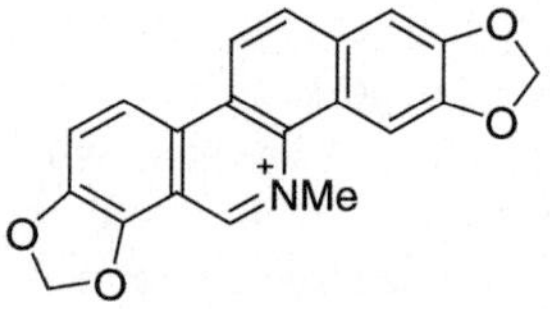

25

Our investigation of the antiplasmodial activity revealed that the activity is entirely confined to the aqueous extract (IC_{50} = 4.5 μg/ml) when tested against chloroquine-resistant strains of *P. falciparum*. Sanguinarine (**25**), however, exhibited weak antiplasmodial activity (IC_{50} = 2.65 μg/ml) [33].

Sesquiterpene Pyridine Alkaloids

Sesquiterpene pyridine alkaloids, based on a highly oxygenated dihydro-β-agarofuran core on a sesquiterpene moiety, constitute a relatively small group of natural products frequently found in plants of the family Celastraceae and, recently, have also been described in plants of the Hippocrateaceae. This observation reinforces the recent botanical classification in which the two families, Celastraceae and Hippocrateaceae, appear to be lumped together into the Celastraceous stock.

Sesquiterpene pyridine alkaloids have been of interest due to their cytotoxicity against several human tumour cell lines and their insect antifeedant and insecticidal activities. Our continuing investigation of the root bark of *Maytenus senegalensis* (Lam.) Excell resulted in the isolation of wilforine (**26**) with prominent in vitro activity (IC_{50} = 0.003 μg/ml) against *P. falciparum* chloroquine-resistant strain, but also showing very high toxicity against human cell lines (3 pg/ml) [unpublished results].

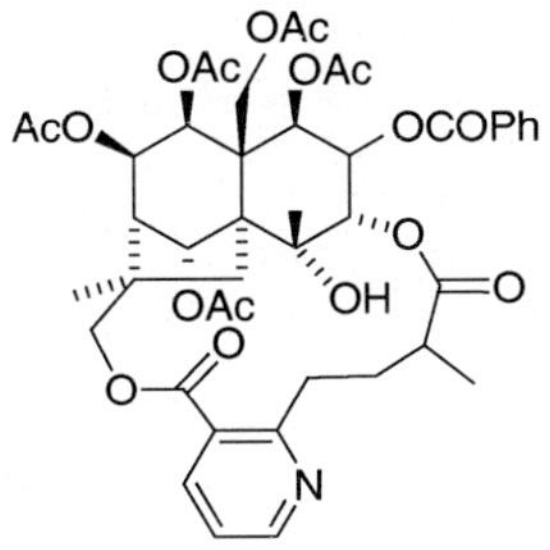

26

Amaryllidaceae Alkaloids

The family Amaryllidaceae is well known to produce biosynthetically related alkaloids based on the isoquinoline nucleus, although their structures differ considerably. Their range of activity includes antiviral, cytotoxic, antitumor, immunostimulatory, analgesic, anti-inflammatory and acetylcholinesterase inhibitory. To date, more than a dozen Amaryllidaceous alkaloids are reported to exhibit promising antiprotozoal activity [34].

These reports encouraged us to investigate the antiparasitic activity of the bulbs of *Pamianthe peruviana* [35]. One novel and three known alkaloids were obtained. The novel alkaloid, referred to here as pamianthine (**27**), co-occurs along with two lycorine-type alkaloids, namely, hippadine (**28**) and lycorine (**29**), as well as a crinane-type alkaloid, epihaemanthidine (**30**). These compounds displayed antitrypanosomal activity against *T. b. rhodesiense* with IC_{50} values of 34.9, 10.6 and 0.1 μg/ml, respectively, with no cytotoxicity ($IC_{50} \geq 90$ μg/ml). Lycorine (**29**) was also active against both *P. falciparum* chloroquine-sensitive (K1, $IC_{50} = 0.85$) and susceptible strains (NF54, $IC_{50} = 1.75$ μg/ml), while epihaemanthidine (**30**) showed significant activity against *T. cruzi* and the *P. falciparum* chloroquine-resistant K1 strain with $IC_{50} = 1.3$ and 074 μg/ml, respectively. The cytotoxicity, of the latter compound, however, was pronounced when tested in L6 (IC50 = 0.8 μg/ml). Pamianthine (**27**) displayed no activity against *P. falciparum* and *T. cruzi*, while it expressed a high selective toxicity against *T. b. rhodesiense* ($IC_{50} = 9.8$ μg/ml) with practically no cytotoxicity ($IC_{50} \geq 90$ μg/ml).

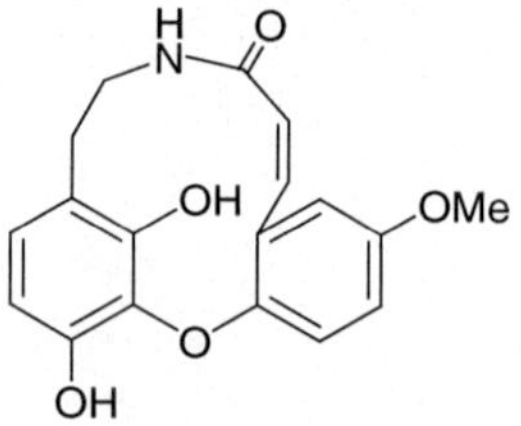

27

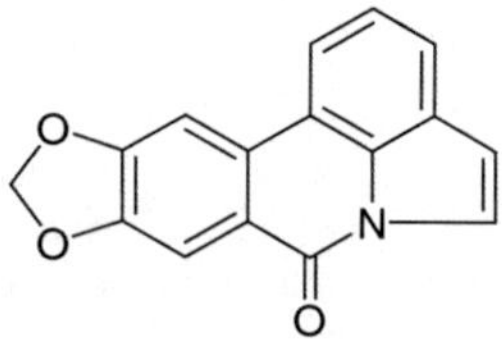

28

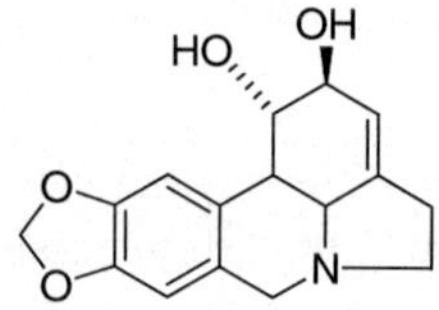

29

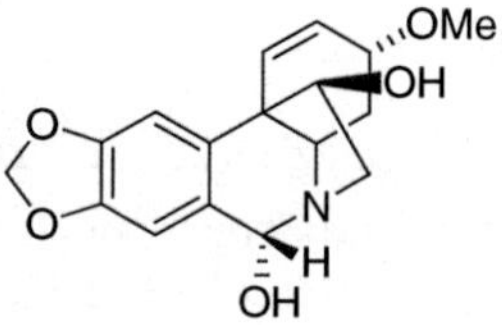

30

Previous phytochemical work on *Crinum amabile* Donn resulted in the isolation of lycorine (**29**) and two other Amaryllidaceous alkaloids as the principal antimalarials of this taxon [34].

Indole Alkaloids

In the course of our phytochemical investigations of African medicinal plants used for treatment of malaria, the chloroform extract of the root bark of *Nauclea latifolia* (Rubiaceae) was subjected to bioactivity-directed fractionation to yield the glucosidic monoterpene indole alkaloid, strictosamide (**31**). This alkaloid exhibited relatively weak antiparasitic activities with IC_{50} of 54.1, 55.7, 29.7 and 18.5 μg/ml against the trypomastigotes of *T. b. rhodesiense*, *T. cruzi*, *L. donovani* amastigotes and the *P. falciparum* chloroquine-resistant (K1 strain), respectively. The cytotoxicity of strictosamide, however, was 51.9 μg/ml, which provides a safe therapeutic window for the treatment of the last two parasites [unpublished results].

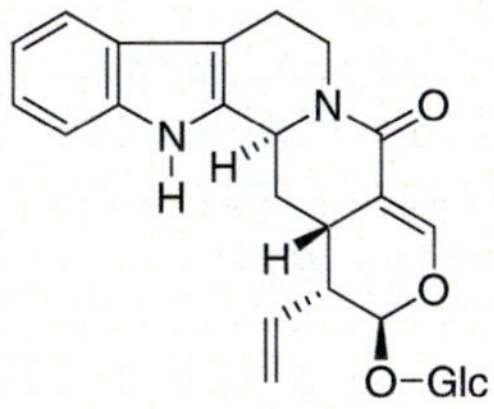

31

It is of interest to note that a number of other species of this genus (e.g. *N. orientalis, N. diderrichii* and *N. pobeguinii*) are frequently used in the African traditional medicine as antimalarials, and some commercial formulations based on *N. latifolia* are currently marketed in the Democratic Republic of the Congo for uncomplicated malaria [36]. The antiplasmodial activity of these extracts has been attributed mainly to strictosamide (strictosidine lactam) and other related indole alkaloids. There is a real controversy, however, regarding the reported in vitro IC_{50} values of the strictosamide against the chloroquine- resistant strain (K1) with extreme values ranging between 0.37 μg/ml [37] and 547 μg/ml [38]. Furthermore, there is some degree of scepticism about the ability of the strictosamide to enter the parasite and reach the intracellular target due to its glycosidic nature. Nevertheless, the in vivo results give credence for the oral use of the decoction in the African traditional medicine considering the envisaged metabolic activation of strictosamide as a result of its de-glucosylation to yield mainly its respective parent alkaloid (i.e. aglycone) which is capable to penetrate the parasite and reach its intracellular target. However, strictosamide rapid recrudescence is considered as one of the main drawbacks of this alkaloid [36]. The metabolic fate of strictosamide has recently been demonstrated in rats' bile using ion trap–TOF mass spectrometer and mass defect filter technique [39]. The presence of other metabolites bearing 4, 9-dihydro-3 H-β-carboline moiety resulted from the metabolic transformation of strictosamide, or its metabolic derivatives may partly contribute to our understanding of the antiparasitic activity associated with the traditional use of strictosamide-bearing plants. β-carbolines have attracted attention due to the variety of their biological activities as a result of their intercalating into DNA, inhibition of topoisomerase and monoamine oxidase besides their antiparasitic activities [34].

9.4 Conclusions and Future Directions

A large number of antiprotozoal secondary metabolites with immense structural variety have been isolated from higher plants. However, despite all the analytical techniques available, considerable work is still needed to determine their mechanisms of action. Furthermore, many of these compounds have only been subjected to in vitro testing, and in vivo results are still lacking. Plant-derived secondary metabolites have provided many medicinal drugs in the past. Even though there exist an important number of natural compounds that have demonstrated potential as possible antiprotozoal agents, most of them do not meet all the criteria required to be viable for their commercialization. Notwithstanding these problems, natural compounds will continue to play an important role in the development of a new generation of antiprotozoal drugs.

Considering the current interest in screening plants and/or marine organisms for antiprotozoal activity, and our incomplete knowledge of their mechanism of action, potentially useful species and/or compounds should be tested principally in animals in order to determine their effectiveness in whole organism systems, with particular emphasis on toxicity. Furthermore, both the preclinical and clinical protocols used need to be standardized to enable a more systematic search and to facilitate interpretation and comparison of the results. Major changes are occurring in R&D for drugs to treat tropical and neglected diseases, with the emergence of many organizations involved in product development. The pressing challenge at the moment is to ensure that efficacious and safe medications are delivered to the people who need them and to ensure that scientists in endemic countries are involved in the whole process.

A host of public–private partnerships (PPPs) have emerged to address these challenges. Their approach is to stimulate R&D for neglected diseases while minimizing business risk. This R&D is typically done as not-for-profit or no-profit-no-loss, by partnerships involving public, multilateral and bilateral agencies, pharmaceutical companies, NGOs and philanthropies. These partnerships, especially those developing new medical products based on the needs identified by the most disadvantaged countries—known as product development public–private partnerships (PDPPPs)—are seen as a positive force. They have raised great expectations for expanding the pool of products available for improving the health status of the most deprived populations [7].

To guarantee the long-term sustainability of these programmes, greater involvement of disease-endemic countries has to be built into the PPP organizational model. A prompt, more focused and result-oriented technology transfer and capacity building will support a future role of disease-endemic countries in discovering and developing the drugs they need. The Pool for Open Innovation against Neglected Tropical Diseases, administered by BIO Ventures for Global Health (http://www.bvgh.org/), seeks to make the patents and, at the discretion of a pool contributor, know-how of companies and organizations more widely available for the development of therapeutics for "NTDs". The Pool contributors recognize that there are no or few commercial returns for therapeutics for NTDs, but the social returns will be enormous. Therefore, the main objective of the Pool is to catalyse and incentivize research into NTDs by making patents and know-how ("IP") more widely available, on terms that facilitate the development of new therapeutics, and to make the process efficient and cost-effective.

Despite the numerous hurdles confronting science in sub-Saharan Africa, African scientists have made notable achievements in some institutions, and the forecast for future progress of science in Africa seems to be promising despite enduring long civil wars and economic crises. Apparently, many countries have now entered a period of rapid growth, and the output of publications is now rising rapidly; sub-

Saharan researchers produced approximately 0.8% of the total papers in the Scopus database in 1996 and 1% by 2009. African scientists need to build on this momentum in order to combat poverty and rampant infectious diseases. Nevertheless, there remains an alarming disparity among the African nations in terms of their scientific activity [40].

Acknowledgement The author would like to thank all previous and present collaborators for their inspiration over the last several years. In particular, I want to acknowledge the productive collaboration with Prof. Timothy Geary at Institute of Parasitology, McGill University, Canada; Prof. Thomas Schmidt at Institute of Pharmaceutical Biology and Phytochemistry, University of Münster, Germany; and Prof. Reto Brun and his group of Medical Parasitology and Infection Biology, Swiss Tropical and Public Health Institute, Basel, Switzerland.

References

1. Verdine GL (1996) The combinatorial chemistry of nature. Nature 384(Suppl):11–13
2. Newman DJ, Cragg GM (2007) Natural products as sources of new drugs over the last 25 years. J Nat Prod 70:461–477
3. Buckingham J (2008) Dictionary of natural products. Chapman and Hall, Oxfordshire
4. Singh SB, Culberson JC (2010) Chemical space and the difference between natural products and synthetics. In: Buss AD, Butler MS (eds) Natural product chemistry for drug discovery. RSC publishing, Cambridge
5. Lipinski CA (2003) Chris Lipinski discusses life and chemistry after the Rule of Five. Drug Discov Today 8:12–16
6. ***First WHO report on neglected tropical diseases: working to overcome the global impact of neglected tropical diseases (2010) Crompton and Peters (eds). WHO Press, Geneva
7. Berger M, Murugi J, Buch E et al. (2010) Strengthening pharmaceutical innovation in Africa. Council on health research for development (COHRED); new partnership for Africa's development (NEPAD)
8. Chatelaine E, Loset J-R (2011) Drug discovery and development for neglected diseases: the DND*i* model. Drug Des Devel Ther 5:175–181
9. Khalid SA (2011) Towards efficacious and safe African phytopharmaceuticals of pharmacopeial quality, Natural Products Communications, submitted for publication
10. Brun R, Blum J, Chappuis F et al (2010) Human African trypanosomiasis. Lancet 375 (9709):148–159
11. Chirac P, Torreele E (2006) Global framework on essential health R&D. Lancet 367 (9522):1560–1
12. Miller JS, Gereau RE (2000) Therapeutic potential of plant-derived compounds: Realizing the potential. In: Cutler SJ, Cutler HG (eds) Biologically active natural compounds: pharmaceuticals. CRC Press, Boca Raton, FL
13. Nwaka S, Ramirez B, Brun R et al (2009) Advancing drug innovation for neglected diseases-criteria for lead progression. PLoS Negl Trop Dis 3(8):e440. doi:10.1371/journal.pntd.0000440
14. Khalid SA, Farouk A, Geary TG et al (1986) Potential antimalarial candidates from African plants: an *in vitro* approach using *Plasmodium falciparum*. J Ethnopharmacol 15:201–209
15. Elford BC, Roberts MF, Phillipson JD et al (1987) Potentiation of the antimalarial activity of qinghaosu by methoxylated flavones. Trans R Soc Trop Med Hyg 81(3):434–436
16. Nour AMM, Khalid SA, Kaiser M et al (2010) The antiprotozoal activity of methylated flavonoids from Ageratum conyzoides L. J Ethnopharmacol 129:127–130

17. Harel D, Khalid SA, Kaiser M et al (2011) Encecalol angelate, an unstable chromene from Ageratum conyzoides L. Total synthesis and investigation of its antiprotozoal activity. J Ethnopharmacol 137:620–625
18. Tasdemir D, Kaiser M, Brun R et al (2006) Antitrypanosomal and antileishmanial activities of flavonoids and their analogues: *in vitro*, *in vivo*, structure-activity relationship, and quantitative structure-activity relationship studies. Antimicrob Agents Chemother 50(4):1352–1364
19. Rosa Martha Perez Gutierrez (2007) Handbook of compounds with antiprotozoal activity isolated from plants. Nova Science Publishers, New York
20. Johnson HA, Oberlies NH, Alali FQ et al (2000) Thwarting resistance: annonaceous acetogenins as new pesticidal and antitumor agents. In: Cutler SJ, Cutler HG (eds) Biologically active natural compounds: pharmaceuticals. CRC Press, Boca Raton, FL
21. Zeng L, Zhang Y, Ye Q et al (1996) *cis*-Gigantrionenin and 4-acetyl gigantetrocin A, two new bioactive annonaceous acetogenins from *Goniothalamus giganteus*, and the stereochemistries of acetogenin 1,2,5-triols. Bioorg Med Chem 4:1271–1279
22. Boyoma FF, Kemgnea EM, Tepongning R et al (2009) Antiplasmodial activity of extracts from seven medicinal plants used in malaria treatment in Cameroon. J Ethnopharmacol 123:483–488
23. Gunatilaka AAL (1996) Triterpenoid quinonemethides and related compounds (Celastroloids). In: Herz W, Kirby GW, Moore RE et al (eds) Progress in the chemistry of organic natural products. Springer, New York
24. El Tahir A, Satti GM, Khalid SA (1999) Antiplasmodial activity of selected Sudanese medicinal plants with emphasis on *Maytenus senegalensis* (Lam.) Exell. J Ethnopharmacol 64(3):227–233
25. Khalid SA, Friedrichsen G M, Christensen SB et al. (2007) Isolation and characterization of pristimerin as the antiplasmodial and antileishmanial agent of *Maytenus senegalensis* (Lam.) Exell. ARKIVOC (ix): 129-134
26. Watt SM, Breyer-Brandwijk MG (1962) The medicinal and poisonous plants of South and Eastern Africa. E. and S. Livingstone, Edinburgh
27. Ade-Serano MA (1982) Antimalarial and antilymphocytotoxic properties of *Azadirachta indica* (Dongo Yaro). Journal of African Medicinal Plants 5:85–93
28. Khalid SA, Duddeck H, Gonzalez-Sierra M (1989) Isolation and characterization of an antimalarial agent of the neem tree *Azadirachta indica*. J Nat Prod 52(5):922–926
29. Khalid SA, Friedrchsen GM, Kharazmi A et al (1998) Limonoids from Khaya senegalensis. Phytochemistry 49:1769–1772
30. Nour AM, Khalid SA, Kaiser M et al (2009) The antiprotozoal activity of sixteen Asteraceae species native to Sudan and bioactivity-guided isolation of xanthanolides from *Xanthium brasilicum*. Planta Med 75:1363–1368
31. Schmidt TJ, Brun R, Willluhn G et al (2002) Anti-trypanosomal activity of helenalin and some structurally related sesqiterpene lactones. Planta Med 68:750–751
32. Schmidt TJ, Nour AM, Khalid SA et al (2009) Quantitative structure – antiprotozoal activity relationships of sesquiterpene lactones. Molecule 14:2062–2076
33. Khalid SA, Schmidt TJ, Brun R (2001) Bioactivity directed isolation of sanguinarine as the main trypanocidal alkaloid of Argemone mexicana. Int. Congress of the Society for Medicinal Plant Research, Abstract book, Erlangen, Germany
34. Osorio EJ, Robledo SM, Bastida J (2008) Alkaloids with antiprotozoal activity. In: Cordell GA (ed) The alkaloids. Academic, Oxford
35. Ali H (2002) Isolation and evaluation of biologically active secondary metabolites from selected medicinal plants of Sudan, PhD, University of Bonn, Bonn
36. Mesiaa K, Cimangaa RK, Dhooghe L et al (2010) Antimalarial activity and toxicity evaluation of a quantified Nauclea pobeguinii extract. J Ethnopharmacol 131:10–16
37. Abreu P, Pereira A (2001) New indole alkaloids from *Sarcocephalus latifolius*. Nat Prod Lett 15:43–48

38. He ZD, Ma CY, Zhang HJ et al (2005) Antimalarial constituents from *Nauclea orientalis* (L.). Chem Biodivers 2:1378–1386
39. Liang Y, Xiao W, Dai C et al (2011) Structural identification of the metabolites for strictosamide in rats bile by an ion trap-TOF mass spectrometer and mass defect filter technique. J Chromatogr B 879:1819–1822
40. Irikefe V, Vaidyanathan G, Nordling L et al (2008) Science in Africa: the view from the front line. Nature 474:556–559

Chapter 10
"Now I Heal with Pride"—The Application of Screens-to-Nature Technology to Indigenous Knowledge Systems Research in Botswana: Implications for Drug Discovery

Kerstin Andrae-Marobela, Aku N. Ntumy, Masego Mokobela, Mthandazo Dube, Angelina Sosome, Mbaki Muzila, Bongani Sethebe, Keitseng N. Monyatsi, and Barbara N. Ngwenya

Abbreviations

CBD	Convention on Biological Diversity
IK	Indigenous knowledge
IKS	Indigenous knowledge systems
NME	New molecular entities
R&D	Research and development
STI	Sexually transmitted infections
STN	Screens-to-Nature

K. Andrae-Marobela (✉)
Department of Biological Sciences, University of Botswana, P. Bag UB 00704, Gaborone, Botswana

Center for Scientific Research, Indigenous Knowledge and Innovation (CesrIKi), P.O. Box 70237, Gaborone, Botswana
e-mail: marobelak@mopipi.ub.bw

A.N. Ntumy • M. Mokobela • M. Dube • A. Sosome • M. Muzila • B. Sethebe
Department of Biological Sciences, University of Botswana, P. Bag UB 00704, Gaborone, Botswana

K.N. Monyatsi
African Regional Intellectual Property Organisation (ARIPO), P.O. Box 4228, Natal Road, Harare, Zimbabwe

B.N. Ngwenya
Okavango Research Institute (ORI), University of Botswana Maun Campus, P. Bag 285, Maun, Botswana

Center for Scientific Research, Indigenous Knowledge and Innovation (CesrIKi), P.O. Box 70237, Gaborone, Botswana

K. Chibale et al. (eds.), *Drug Discovery in Africa*,
DOI 10.1007/978-3-642-28175-4_10, © Springer-Verlag Berlin Heidelberg 2012

10.1 Introduction

The innovation potential of the pharmaceutical industry has come under considerable criticism. A recent analysis of research and development (R&D) and approval of new molecular entities (NMEs) including new biologics by the US Food and Drug Administration (FDA) between 1950 and 2008 revealed that the rate of production of new drugs had been stagnant [1]. The number of new drugs that are approved annually is not greater today than it was 50 years ago [1]. This is astonishing. Neither the exponential growth of investment in pharmaceutical research and development, nor the emergence of new biological technologies, nor the organizational restructuring of the pharmaceutical sector has significantly contributed to a higher rate of production of new drugs. Consequently, the study concluded that the innovative capacity of the established R&D model might have reached its limits, and the industry's efforts to embrace new approaches to innovation are of particular importance [1].

The chief of the European Medicines Agency (EMA), Thomas Lönngren, criticized the unsustainable R&D model of the drug industry by stating that of the estimated US $85 billion spent globally each year on R&D, around US $60 billion can be considered as wasted if one takes into account how few NMEs were produced [2]. One study, for example, reported that despite a 70% increase in drug research investment and related activities during the period of 1994 to 2004, there was an actual decline in launching NMEs [3]. One factor that limits successful approval is the high attrition rate during drug development. Only 1 out of 12 drugs entering clinical trials successfully emerges as a new drug. This is mainly due to the lack of appropriate bioavailability, poor pharmacokinetics, and significant adverse effects [2, 4].

The safety of new drug candidates is the defining issue in drug development, but it is usually addressed relatively late in the process. Rather, overemphasis has been on nonclinical aspects of drug discovery, and this has caused many drug candidates that have traveled far in the "pipeline" to fail drug approval [2]. "Never before did scientists have to consider (. . .) safety (. . .) aspects so early in R&D as they must today," admit research strategists in the pharmaceutical industry [5]. In line with this statement, the question is how can pharmaceutical research and classical drug discovery processes gain new dynamics of producing safer drug candidates in a less wasteful way?

One attractive strategy emerges from a new school of thinking in pharmaceutical research termed "reverse pharmacology." In this approach, documented clinical experiences are integrated with experimental whole-systems observations at the beginning of a drug discovery process. The purpose is to identify leads which are subsequently developed into drug candidates by detailed laboratory preclinical and clinical studies [6]. This approach reverses the classical practice of "laboratory to clinic" to a "clinic to laboratory" process [7]. That way, safety issues receive attention in the initial phases of the drug discovery processes and not at the end.

Subsequently, researchers have called for indigenous knowledge[1]-based platforms in combination with systems biology to guide drug discovery [8–11]. This approach considers the richness and diversity of natural product-based traditional medicines with its documented long-term human use as a strength. This is not only more effective to screen for bioactivities [12] and to (re)discover new molecular entities from natural products [13–16], but such an approach serves to address safety issues early on in the drug discovery process. Therefore, it is estimated that the drug discovery and development phase can be reduced significantly in both time and costs [6]. For instance, the reverse pharmacology approach used in the development of an antimalarial phytomedicine in Mali resulted in the introduction of a new effective standardized herbal antimalarial remedy after 6 years of research at a cost of only 400,000€ [17] compared to estimated costs of US $1–1.5 billion using the classical path of drug discovery [6].

However, efforts to develop indigenous knowledge-guided platforms for drug discovery bring in their own dynamics and challenges. Particularly, an inadequate intellectual property rights regime in relation to indigenous knowledge systems is an issue of major importance. Unevenly developed access and benefit-sharing mechanisms for indigenous knowledge holders and their communities have led to widening gaps between indigenous knowledge holders and researchers, culminating in mutual mistrust. Unethical practices of exploiting natural resources for commercial purposes by researchers and companies without the consent of and benefit for involved indigenous knowledge holders and their communities have contributed to an atmosphere in which it is difficult to establish collaborations.

Even though there have been attempts in recent years to protect indigenous knowledge holders through important international agreements, such as the Convention on Biological Diversity (CBD), there are challenges in implementing the legal frameworks under these initiatives. For example, a recent compilation of case studies on how the CBD provision on access and benefit sharing is implemented in relation to the exploration of biodiversity for commercially valuable genetic and biochemical resources ("Bioprospecting") revealed that

---

[1] We define indigenous knowledge in a broad sense referring to established knowledge originating from or adapted to a local context. Indigenous knowledge systems refer to harnessing of indigenous knowledge to produce a particular product or result. However, the term "indigenous knowledge" does not exist as such in Setswana and other local languages in Botswana. The term which is used by communities is "kitso ya setso," and traditional healers in Botswana call themselves "dingaka tsa setso." "Setso" implies a relation to history and culture and can be translated into "traditional" if one takes into account that history and culture are dynamic. Therefore, we use the term indigenous knowledge in a more general context and the term "traditional" in a more specific context. We use the term "traditional healer" not "traditional doctor" to recognize the diversity of types of healers, which include herbalists, faith healers, sangomas, bone setters, traditional birth attendants, and other healing professions.

benefit sharing remains largely unresolved [18]. The study noted that there is a tendency to use secondary sources, such as databases, as surrogates for indigenous knowledge and that there is a trend to declare indigenous knowledge systems in general as a public domain for which it is perceived that no benefit-sharing schemes need to be established [18].

Some ethnopharmacologists had to admit that technological advances in drug discovery have only marginally contributed to a beneficial development of indigenous knowledge systems that would result in meaningful returns to indigenous knowledge holders. Particularly, the lack of transparency in the research process, the lack of access of indigenous knowledge holders to research results, and the absence of a continuous dialogue between indigenous knowledge holders and scientists were criticized [19]. Furthermore, the fact that many countries have not yet established policies [20] and comprehensive implementation mechanisms to promote, protect, regulate, and develop indigenous knowledge systems tends to encourage the view that indigenous knowledge systems are part of an informal sector, which does not match the standard of institutionalized science. For example, biomedical practitioners often state that they are in favor of collaborations with traditional healers, but prefer collaboration to be on their terms, such as limited to referral of patients from healers to biomedical practitioners [21, 22]. Similarly, biomedical scientists often use an extractive approach when interacting with indigenous knowledge holders. In many cases, researchers obtain plant material and information on its medical uses from traditional healers, but never share research results with them. Indigenous knowledge holders—usually well-respected individuals in their communities—are reduced to anonymous "informants" or "respondents" in research publications and remain completely alienated from the research process [23]. Furthermore, in the health sector where indigenous knowledge systems (IKS) are particularly attractive for drug discovery, the often unclear legislative status of traditional healers contributes to an atmosphere where they hesitate to enter into collaborations which would reveal healing practices, which are not necessarily covered by the law.

Therefore, in this chapter, we describe how we developed a relatively simple set of field-suitable screening methods to detect bioactivities in plants, the Screens-to-Nature (STN) Technology, into a participatory tool to initiate and consolidate collaborations between researchers and traditional healers, and evaluate the potential of this approach for drug discovery from natural products. The first section of the chapter gives a brief overview of indigenous knowledge systems research in Botswana. The following sections explain and illustrate our approach using the STN system in two regions of Botswana, assess whether the STN technology is a suitable tool for drug discovery, evaluate whether the STN system is beneficial to traditional healers, illuminate interfaces and innovation processes as they became visible during the application of the STN technology, and conclude by outlining challenges and perspectives as experienced by our approach.

10.2 Indigenous Knowledge Research in Botswana

Indigenous knowledge systems are rich and diverse in Botswana. This is reflected in the Botswana National Vision 2016 document, which states that "while much can be borrowed from other countries we will need to look within our resources and culture to find the sources of innovation that will allow us to shape our own future."[2] A number of policies and corresponding strategic plans describe the various domains of indigenous knowledge and recognize their potential for sustainable economic development, such as biodiversity and natural resources management, pharmaceuticals, agriculture, breeding, and animal management. For example, the recent draft policy by the Ministry of Infrastructure, Science and Technology recommends the "infusion of indigenous knowledge in the national R&D agenda" and favors "partnerships (that) should bring about integration of the indigenous knowledge and modern science in order to deliver culturally acceptable solutions for local problems" [24].

However, comprehensive implementation strategies and monitoring/evaluation plans are not established to translate the commitment to the development of indigenous knowledge systems into tangible results. Interfaces between research and indigenous knowledge, which would be important for a focused, competitive R&D strategy, such as drug discovery from natural products using an indigenous knowledge-guided platform are not materializing to full potential. This is so because regulations to address access and benefit sharing of natural resources, suitable intellectual property regimes, and guidelines on setting up ethically sound collaborations are largely absent. There is currently no policy in place that clarifies the legal status, rights, and needs of traditional healers and traditional medicine in general. As a consequence, indigenous knowledge systems research in Botswana remains largely fragmented and uncoordinated.

As of now, the Department of Chemistry at the University of Botswana has a long-standing and strong record of natural product and phytochemistry research, which has accumulated an in-depth understanding of a broad range of natural product compound classes, their bioactivities, and their applications in diverse areas including catalysis, synthesis, medicine, and agriculture [25–29]. The same applies to the Botswana College of Agriculture that has established research groups on pharmacognosy and agroforestry with emphasis on medicinally important trees and food plants [30–32]. However, the establishment of concrete research partnerships with indigenous knowledge holders and communities remains subjected to individual initiatives that have never been comprehensively evaluated.

To address some of these gaps, the Center for Scientific Research, Indigenous Knowledge and Innovation (CesrIKi) was founded at the University of Botswana in

[2] Vision 2016 (1997) A long term vision for Botswana—towards prosperity for all. Government of Botswana, Gaborone, Botswana.

2006 and became operational in 2007. The main objectives of this interdisciplinary center are:

1. To contribute to national economic development and poverty alleviation by exploring Botswana's comparative advantage in natural resources and indigenous knowledge systems
2. To translate IKS in innovative processes to the benefit of communities
3. To develop IKS conscious scientists who will have reciprocal relationships with both rural communities and the formal sector
4. To support a paradigm shift in pedagogical approaches to ensure that the local communities endowed with natural resources will actively participate in applied and basic research initiatives for the present and future development of generations of scientists

A first project funded by the United Nations Development Program Global Environment Facility Unit (UNDP-GEF) to document IKS in relation to human health and food systems sought to explore appropriate modalities for interaction with indigenous knowledge holders and their communities and to develop a solid base for long-term and mutually beneficial relationships that are transparent and respectful. In this framework, which spanned the period of 2008 to 2010, we developed and applied the STN systems as a participatory tool.

10.3 A New Approach: Participatory Research with the "Screens-to-Nature" (STN) System

The STN system is comprised of a set of field-suitable qualitative bioassays which detect various bioactivities of plant extracts. The screening system was developed through a collaboration between Rutgers University and the University of Illinois who jointly founded the Global Institute for Bioexploration (GIBEX) in 2003 with "a mission to empower scientists from the developing world to carry out their own therapeutic lead discovery and to promote sustainable exploration of local biodiversity for products related to human health" (for a more detailed account on GIBEX, see [33]). The STN bioassays focus on health targets which are relevant to the disease burden in developing countries, such as infectious agents (bacteria, fungi, protozoa), as well as enzymes and enzyme inhibitors associated with metabolic and infectious diseases (glucosidase-inhibitors, protease inhibitors) and on general health protection via antioxidant properties of phytochemical constituents [33, 34]. The STN bioassays are designed for screening of ethanol extracts which can be prepared in the field from 2 g plant material, and the assays use nonpathogenic model organisms for a first assessment of bioactivities. Some of the assays can be performed in a relatively short time and produce visible results within 10 to 30 min. This was attractive for us, as screening experiments could be easily accommodated within a workshop format.

We decided to approach communities and traditional healers from two different areas in Botswana. The first site was the village of Mmankgodi, situated approximately 35 km from the capital Gaborone in the Kweneng District (Map 10.1a). The Mmankgodi community consists of close to 5,000 inhabitants and unites the two major ethnic groups, the Bakwena and the Bahurutshe. This village was also chosen because it possesses an interesting blend of urban and rural influences. We first consulted the community through traditional customary structures called "dikgotla" which are headed by a local chief, the "kgosi," and subchiefs, "dikgosana." The project was introduced to representatives from each ward of the village, and ample space was given for expressing views and concerns. After obtaining permission from the "kgosi" and the community, one plant-testing workshop was conducted on May 30, 2008, with participation of the cultural cooperative of Bahurutshe Cultural Village in Mmankgodi. The cooperative was founded by women in the village mainly for the generation of small income out of the Cultural Village's activities, which attracts local and foreign tourists. Some of the women of the cultural cooperative are very knowledgeable in medicinal plants and were interested in small economic activities related to selling of herbs or simple herbal products. The second workshop was organized on March 18, 2010, with traditional healers from Mmankgodi. The women of the cultural cooperative contributed a selection of popular medicinal plants or their parts, and traditional healers collected together with us some plant samples on the hill where Bahurutshe Cultural Village is located. During plant collection, our botanist explained the importance of proper identification of plant species and illustrated ways of botanical preservation of the morphology of plants. Participants helped to grind the plant material and to prepare an ethanolic extract from selected medicinal plants.

These extracts were then used to perform three experiments in smaller groups detecting antioxidant activities, as well as protease and glucosidase inhibitory properties from plant extracts [34; see also Sect. 10.4.3]. Working in small groups allowed active participation in the experimental procedure (Fig. 10.1a, b) and gave ample space to explain the experiment, share the results, and discuss their implications and meaning (Fig. 10.1c). This approach is in striking contrast to previous experiences where scientists never gave feedback on laboratory test results. Our results were shared during the workshop and discussed. Examples included the fact that antioxidant activities might be useful in preventing strokes and that natural product-based protease and glycosidase inhibitors are potentially interesting in the area of drug development for diseases like HIV/AIDS, malaria, and diabetes. This created much excitement. Participants felt that they have actively learned something and, perhaps more importantly, felt that indeed their knowledge and "modern" scientific knowledge can work hand in hand resulting in new discoveries.

Encouraged by this positive experience, we conducted a workshop in a similar manner with traditional healers in Ngamiland District in Northwestern Botswana (Map 10.1b). Ngamiland District is the third largest in the country and the most ethnically diverse. The district possesses a unique ecosystem, the Okavango Delta, the largest inland freshwater ecosystem in the world. The oldest inhabitants of the

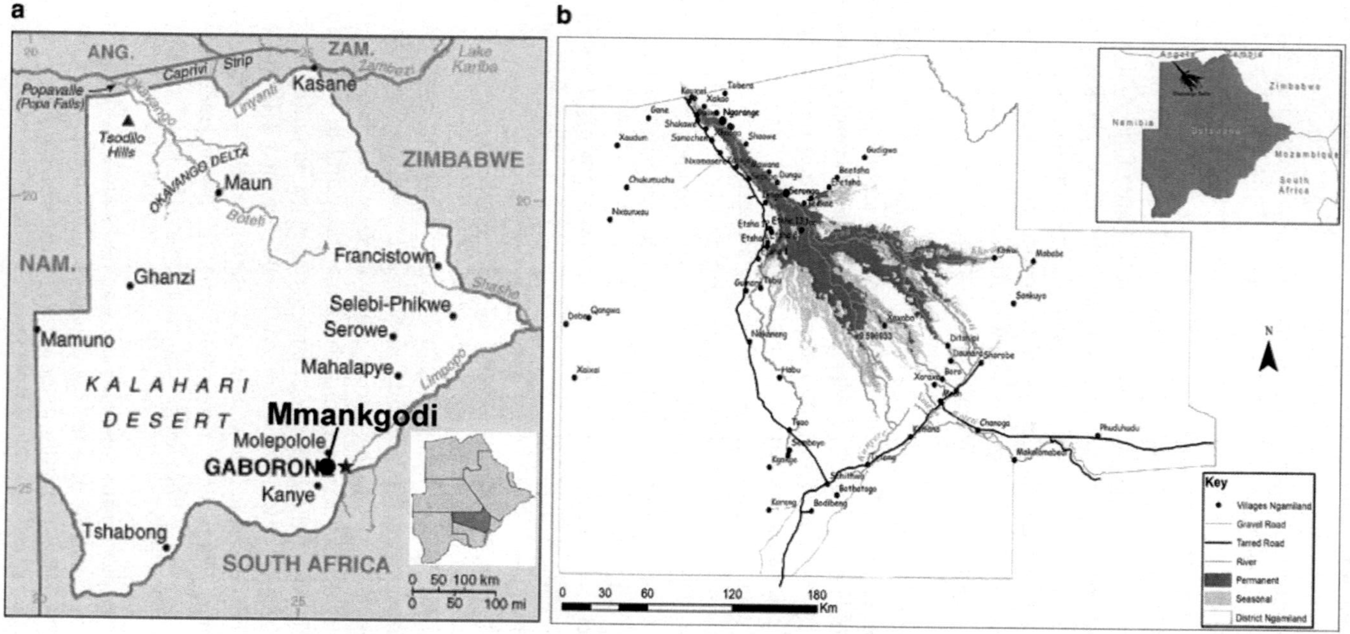

Map 10.1 Study sites. (**a**) Location of the village of Mmankgodi in Kweneng District and (**b**) Ngamiland District. Note villages along the main road south and southeast of the Okavango Delta (Kareng, Bodibeng, Sehitwa, Toteng, Maun, Chanoga, Makalamabedi, and Shorobe)

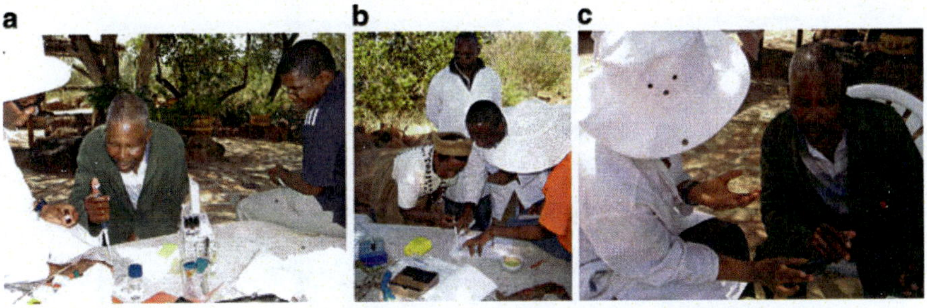

Fig. 10.1 Participation in and discussion of STN assays. A traditional healer (**a**) and a woman of the Bahurutshe cultural cooperative (**b**) assist in a STN assay whose results are discussed (**c**)

Okavango basin are the Basarwa or San, a collective name used for several groups of Khoisan-speaking people. Ethnic groups in the western part of the Okavango are HamBukushu, Bayei, Bakgalakgadi, and BaSubiya. Over 95% of people in the Okavango Delta directly or indirectly depend on natural resources found in the wetland to sustain their livelihoods [35]. Many of the close to 3,000 plant species in Botswana are found in this area.

Fifty-five traditional healers and public health workers originating from eight different villages in Ngamiland, including Kareng, Bodibeng, Sehitwa, Toteng, Maun, Chanoga, Makalamabedi, and Shorobe, participated in the plant-testing workshop on September 23, 2008, in Maun, the capital of Ngamiland. Traditional healers brought their medicines and prepared extracts. Public health workers helped with pipetting. In the same manner as described above, three groups conducted three different experiments and shared results, which were discussed collectively after displaying them on posters. Again, traditional healers appreciated that researchers were not hiding results, but shared them and that researchers took time to explain. As this workshop included public health workers, mainly nurses and health education officers from local clinics, a new dynamic was noticeable. For the first time, public health workers could see that indeed traditionally used medicinal plant extracts do have effects, which can be scientifically detected. As the workshop concluded, traditional healers contributed 55 plant samples for further testing, for antibacterial, antiprotozoal, and other STN assays, which required more time and could not be performed during the workshop.

Our concept was to discard the notion that indigenous knowledge holders are reduced to "informants" and work towards an approach that sees traditional healers as true research partners, who have the right to transparency and access and benefit sharing. Accordingly, after having performed all STN assays available at that time, we organized a "report-back" workshop on the April 23, 2009, with 61 participants (Fig. 10.2). Recognizing that not all healers would feel comfortable discussing their therapeutics in public, healers received individual reports with test results for the plant samples they had submitted, and they were awarded certificates of appreciation for their contribution to IKS research in Botswana. Public health workers

Fig. 10.2 Report-back workshop shares STN results with traditional healers and public health workers in Maun, Ngamiland District

received a summary report that showed percentages of plant samples displaying antibacterial, antifungal, antiprotozoal, enzymatic, and enzyme inhibitory activities. Then, a discussion that led into charting a way forward followed. Hence, follow-up activities and future collaboration on more specific questions were agreed upon.

Since then, CesrIKi has managed to extend and maintain a network of traditional healers who consistently submit plant samples for STN testing. This is also supported by the Botswana traditional healer umbrella organization "Baitseanape ba setso." At a general dissemination workshop that took place on October 7, 2010, the president of the organization, Mr. Setilo, publicly stated the full support of his organization for CesrIKi's efforts to build research partnerships with traditional healers.

10.4 STN: A Suitable Primary Screening Tool for Drug Discovery

How relevant are these primary screening assays for drug discovery processes? To answer this question, we summarize in this section some results we obtained during the 2-year STN program at the University of Botswana. We made use of the above-described established network of traditional healers and community members who were knowledgeable in medicinal plants and encouraged them to submit plant samples. We emphasized the importance of correct plant identification and asked healers to provide a sample for botanical authentication, or, alternately, STN team members went with healers into the field and directly collected plant samples for proper botanical identification. Voucher specimens were deposited in the University of Botswana Herbarium.

10.4.1 Investigated Plant Species: A Snapshot of Botswana's Diversity

In total to date, 621 plant samples have been investigated using the STN system. Close to half of the samples (47%) were provided by traditional healers. Sampling of approximately half of the remaining 53% was mainly guided by available literature on medicinal uses of plants and can be considered as indirect ethnobotanical sampling, while the rest of plant samples were collected randomly. The overall number of plant samples included samples that represent different plant parts of the same plant species, as well as "repetitions" of the same plant part from the same species, but collected at different times of the year and at different geographical locations. Of these, 515 plant samples (83%) have been botanically authenticated, while the identity of 106 plants (17%) remains unknown, though in most cases their vernacular name could be documented. The total number of botanically authenticated plant samples represents 214 different plant species of which 178 (83%) are indigenous to Botswana, 33 species (15%) are exotic species (based on classification according to [36]), and three plant species (2%) are of unknown provenance.

The plants investigated encompassed 71 of the 182 families found in Botswana [36] (Table 10.1). The most frequently investigated families were Fabaceae and Asteraceae, which contributed 33 and 20 different species and are known to be large and diverse families. Lamiaceae, Euphorbiaceae, Asphodelaceae, and Apocynaceae were also investigated with 10, 10, 9, and 8 species, respectively. Many of these families are well documented as possessing medicinal properties [32, 36–38].

A wide range of plant parts were investigated, including underground organs (roots, tubers, bulbs, and corms), flowers, fruits, shoots, stems and twigs, seeds/pods, and mixtures of two or more different plant parts. Most of the plant samples tested with the STN system were extracts from underground organs (47% of all samples), followed by leaf (27%) and bark extracts (4.8%) (Fig. 10.3). The prominence of underground organs of plants concurs with previous findings that the plant part which is mostly used in traditional medicines in Botswana is the root [23]. This differs from some reports about medicinal plant uses in other countries. In South America leaves principally contribute to traditional medicinal preparations [39], possibly explained by the differences in the countries' climates (dry versus tropical respectively).

Table 10.1 Plant families of species screened by the STN system

Plant family	No. of species	Plant family	No. of species
Asteraceae	**20**[a]	Malpighiaceae	1
Alliaceae	1	Malvaceae	6
Amaranthaceae	5	Meliaceae	1
Anacardiaceae	7	Moraceae	1
Apocynaceae	**8**	Moringaceae	2
Araceae	1	Myrothamnaceae	1
Asclepiadaceae	1	Myrtaceae	1
Asparagaceae	6	Ochnaceae	1
Asphodelaceae	**8**	Oleaceae	2
Bignoniaceae	1	Orchidaceae	1
Bombacaceae	1	Papaveraceae	1
Boraginaceae	2	Pedaliaceae	4
Burseraceae	2	Plumbaginaceae	1
Cannabaceae	1	Poaceae	2
Capparaceae	7	Polygalaceae	1
Celastraceae	2	Polygonaceae	2
Clusiaceae	1	Pteridaceae	1
Combretaceae	5	Rhamnaceae	1
Commelinaceae	1	Rubiaceae	2
Convulvulaceae	3	Rutaceae	2
Crassulaceae	2	Salvadoraceae	1
Cucurbitaceae	4	Sapindaceae	2
Cyperaceae	2	Selaginellaceae	1
Dracaenaceae	1	Solanaceae	7
Ebenaceae	4	Sterculiaceae	2
Euphorbiaceae	**9**	Thymelaeceae	1
Fabaceae	**33**	Tiliaceae	2
Flacourtiaceae	1	Verbenaceae	5
Hyacinthaceae	3	Vitaceae	2
Hypoxidaceae	1	Welwitschiaceae	1
Lamiaceae	**10**	Zygophyllaceae	2
Lauraceae	1	Unknown	4

Total number of families: 71

[a]Plant families contributing the largest number of species are indicated in bold

10.4.2 The Screening Platform

The high HIV/AIDS prevalence rate of 15–28% in Botswana [40] has reconfigured the pattern of antimicrobial infections in the country. Opportunistic infections, such as tuberculosis and sexually transmitted infections (STIs), as well as emerging multidrug (MDR) and extensively drug-resistant (XDR) mycobacterial strains and resistant causative agents for STIs, have become a major public health threat. Therefore, we modified the original STN set of bioassays by including pathogenic bacterial strains, which enabled us to obtain more specific screening data on

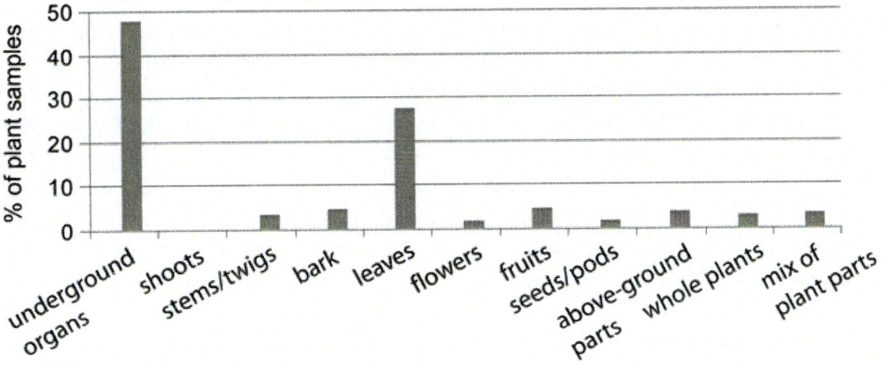

Fig. 10.3 Plant parts investigated with the STN system

clinically relevant pathogens or pathogen model organisms. In addition to using nonpathogenic bacterial strains found in saliva as simple model organisms to detect general antibacterial activities when exposed to plant extracts [34], extracts were screened for inhibitory properties against *Mycobacterium aurum*. *M. aurum* is a fast-growing, nonpathogenic mycobacterium, which has very similar drug susceptibility characteristics as *Mycobacterium tuberculosis*. Because of the structural similarities of mycolates, which are responsible for the permeability of the cell envelope for antimicrobials, *M. aurum* has been recommended as the most suitable model organism to identify new potential therapeutics against *M. tuberculosis*, particularly cell wall inhibitors [41, 42]. Furthermore, we included *Neisseria gonorrhoeae*, the causative agent of gonorrhea, in the STN-screening program based on the observation that treatment and/or management of STIs is one of the major areas of expertise of traditional healers in Botswana [23]. Yeast infections, such as those caused by *Candida albicans*, constitute one of the major opportunistic infections in immune-compromised patients. Up to 90% of HIV-infected individuals suffer from at least one episode of candidiasis, which is commonly characterized by the development of oral thrush [43]. The STN system uses simple baker yeast (*Saccharomyces cerevisiae*) as a model organism to screen for antifungal bioactivities. The STN bioassay panel comprised further the detection of α-glucosidase and α-amylase inhibitors, which have attracted interest as potential therapeutic agents as antidiabetics [44] or as suppressors of the HIV replication cycle mainly at the entry stage [45], and trypsin protease inhibitors which have been implicated with anticarcinogenic [46] and anticoagulant activities [47].

Finally, two types of protozoa viability assays were performed using *Bodo caudatus* and *Acanthamoeba castellanii*. *B. caudatus* is a nonpathogenic member of the family Bodonidae, which is related to the genera *Leishmania* and *Trypanosoma* (Trypanosomatidae) by belonging to the same order of Kinetoplastida [48]. *Leishmania* and *Trypanosoma* include the pathogenic human parasites, such as

T. cruzi and *T. brucei*. The *B. caudatus* viability assay can therefore be considered as a preliminary screening system for anti-trypanosomal activity. Amoebae are a very diverse group of protozoans of which at least six forms are parasitic in humans. Most relevant of these is *Entamoeba histolytica*, which causes amebiasis and dysentery and remains a challenging health problem in developing countries with insufficient sanitation infrastructure [49]. *Acanthamoeba* is a genus common in soil and freshwater and is often found associated with water equipment, contact lenses, medicinal pools, ventilation, etc. *Acanthamoeba* have gained increasing clinical relevance mainly as causative agents of an often seriously progressing keratitis among contact lenses users [50] and are implicated in causing encephalitis particularly in an immune-compromised health context [51]. The STN system uses the nonpathogenic *A. castellanii* (Neff strain) to test anti-amoebic activity of plant extracts.

10.4.3 Methodology of STN-Screening Assays

The screening of plant extracts using the above-mentioned organisms and enzymatic systems was performed as previously described [34]. Briefly, nonpathogenic bacterial strains from saliva, as well as *N. gonorrhoeae* and *M. aurum* ($OD_{550} = 0.5$, equivalent to 600×10^6 c.f.u./ml according to McFarland standard), were cultivated on LB agar, 7H8 agar supplemented with 10% Oleic/Albumin/Dextrose/Catalase (OADC), and Thayer-Martin agar, respectively, in 48-well plates in the presence or absence of ethanolic plant extracts and incubated overnight or for 48 h. The presence or absence of bacterial colonies was observed from duplicate assays classifying the complete absence of colonies as antibacterial activity. Penicillin, streptomycin, and rifampicin were used as controls.

The viability of *S. cerevisiae* and the protozoa *B. caudatus* and *A. castellanii* after exposure to ethanolic plant extracts for 24 h in 96-well plates was detected using 3-[4,5-dimethylthiazol-2-yl]-2,5-diphenyltetrazolium bromide (MTT) as viability indicator. Econazole and $CuSO_4$ were used as controls for *S. cerevisiae* and protozoa, respectively. Antifungal and antiprotozoal activities of plant extracts were detected from duplicate assays by the lack of violet color formation from MTT (violet color indicates a metabolically intact organism).

α-Glucosidase (isolated from pea shoots) and α-amylase (present in saliva) inhibitory properties of plant extracts were qualitatively evaluated by exposing enzymes to solidified starch agar in the presence or absence of plant extracts. The intact starch surface was visualized with aqueous iodine-solution which results in a dark blue pigment formation.

Lastly, protease inhibitory plant extracts were identified using gelatin-coated X-ray films. The gelatin top layer on an X-ray film strip was exposed to trypsin alone as control or to trypsin in the presence of plant extracts.

Table 10.2 Bioactivities determined by the Screens-to-Nature (STN) system

Assay	No of samples tested	Frequency	
		No of positives	%
Anti-bacterial			
General (Saliva)	254	58	23
Mycobacterium aurum	373	42	11
Neisseria gonorrhoea	240	80	33
Anti-fungal			
Saccharomyces cerevisiae	592	95	16
Glucosidase/amylase inhibition			
α-amylase (saliva)	431	66	15
α-glucosidase (pea shoots)	467	41	9
Trypsin inhibition	599	199	33
Anti-protozoa			
Bodo caudatus	325	16	5
Acanthamoeba castellanii	128	10	8

10.4.4 The STN-Screening Results

In total, nine different STN bioassays were performed (Table 10.2). The most prominent bioactivities detected in plant extracts were anti-*N. gonorrhoeae* and trypsin inhibitory properties with 33% of all plant samples displaying positive results. In comparison, bioactivities of plant extracts inhibiting growth of *M. aurum* were detected with 11%, considerably lower than anti-*N. gonorrhoeae* activities, which can most likely be explained by the unique cell wall composition of mycobacteria which effectively prevents penetration of a wide range of compounds [52]. Alpha-glucosidase inhibitor and antiprotozoal assays appeared much more discriminatory, detecting α-glucosidase inhibitors in 9% of extracts tested, anti-*B. caudatus* activities in 5%, and anti-*A. castellanii* activities in 8% of all tested samples. The low percentage of antiprotozoal activities might reflect the fact that plants do have numerous antibacterial and antifungal defense mechanisms, but possess lesser secondary metabolites active against protozoa of this type. Inhibitory activities against bacterial strains from saliva, α-amylase inhibitors, and antifungal activities were detected in a similar range of 23%, 15%, and 16%, respectively.

10.4.5 Correlation of Bioactivities with Plant Families

Although the broad distribution of plant samples over a high number of plant families resulted in a relatively low sample size for evaluating a correlation between bioactivities and specific plant families, we observed some preliminary trends. For example, a total of nine species of Euphorbiaceae have been screened, and in five or more species antimicrobial activities could be detected. In fact, 11% of species showing bioactivity in general antibacterial (saliva), anti-*S. cerevisiae*, and anti-*N. gonorrhoeae* assays were from Euphorbiaceae. This supports recent literature documenting the use of Euphorbiaceae plants against sexually transmitted diseases [37]. Species from the Fabaceae showed significant trypsin inhibitory activities. Out of 33 species investigated, 20 inhibited trypsin in the STN assay. Thus, 19% of the total plant species inhibiting trypsin (106 species) were from the Fabaceae. Extracts from the species of the Solanaceae showed remarkable α-glucosidase inhibitory properties. Five out of seven species inhibited α-glucosidase, attributing 17% of plant species with this bioactivity to the Solanaceae. All four species of the Verbenaceae displayed anti-yeast activity, but no significant other properties as detected by STN assays. This finding correlates with reports of anti-yeast activity (*C. albicans*) of certain species within this plant family [53].

10.4.6 Correlation of Bioactivities with Different Plant Parts

We then chose three examples of bioassays to evaluate whether bioactivities correlated with a specific plant part from which the extract was prepared. For example, in 43% of underground parts extracts and in 30% of leaf extracts, anti-*N. gonorrhoeae* activities could be detected, while the remaining 27% of positive bioactivities were distributed over a wide range of plant parts, including bark, flowers, stems/twigs, and above-ground parts. A similar distribution was seen when investigating anti-*M. aurum* activities. 52% of active extracts were prepared from underground organs, while 19% active extracts were obtained from leaves. α-Amylase inhibitors were found to be present in 52% of root/tuber extracts and in 32% of tested leaf extracts. These findings correlate reasonably with traditional medicinal uses attributed to specific plant parts where roots are mostly used in medicinal preparations, followed by leaves [23]. A slightly different picture was obtained when evaluating antiprotozoal activities. Underground organs dominated significantly where 80% of extracts displayed anti-*A. castellanii* and 69% anti-*B. caudatus* activities; no leaf extracts had anti-*A. castellanii* and 12% had anti-*B. caudatus* activities. A possible explanation is that the two types of protozoa are predominantly found in soil and groundwater and the root system would be the most suitable part of the plant to generate secondary metabolites as defense mechanism against parasites.

10.4.7 The Reliability of the STN Assays

All STN assays as described above generate qualitative results. In a first effort to establish whether the STN bioassays are suitable as a primary screening tool for drug discovery, we conducted quantitative follow-up assays using *C. albicans*, a clinically relevant fungal organism, and *M. aurum* for rescreening the qualitative anti-*S. cerevisiae* and anti-*M. aurum* plant extracts. Based on the STN prescreening, a number of plant extracts with favorable minimal inhibitory concentration (MIC) and minimal fungicidal and bactericidal concentrations (MFC and MBC), respectively, could be identified [54]. Quantitative MTT viability assays confirmed 60% of detected anti-*B. caudatus* activities using the STN protozoa assay and 98% of anti-*N. gonorrhoeae* properties detected in plant samples screened by the STN system (unpublished results). We wish to note that we observed false-negative results, for example, when matching STN results for *S. cerevisiae* with low MICs obtained for anti-*C. albicans* plant extracts, as well as false-positive STN results when comparing with quantitative growth inhibition of *B. caudatus* mediated by plant extracts, or both when using *M. aurum* as screening organism for qualitative (STN) and quantitative viability assays. Furthermore, the reproducibility of STN results using plant samples of the same species and plant part varied. Antimicrobial assays showed a better reproducibility than enzymatic STN assays. One has, however, to keep in mind that the same plant species may vary in their secondary metabolite profile depending on different ecological parameters, such as soil and rainfall patterns, and on different seasons of the year. Therefore, the variability in reproducing qualitative results may not be a consequence of the unreliability of the STN system, but the tested plant material. After having established mutually trustful relations with traditional healers as described in Sect. 10.5, we are now able to collect plant samples *with* traditional healers and take—with their consent—GPS data of plant locations. More detailed studies to correlate bioactivities with natural plant habitats are therefore more feasible in order to comprehensively conclude the reliability of the STN assays. In conclusion, we found that the STN assays are useful as a prescreening tool to detect particularly antimicrobial activities from plant extracts.

10.5 Is the STN System Beneficial to Indigenous Knowledge Holders?

When we introduced the STN system as a participatory research tool earlier in this chapter (Sect. 10.3), we described the excitement of community members and traditional healers with participating in scientific experiments characterizing "their" medicinal plants and with sharing of the results. This, however, might only be the perception of the researchers. In order to evaluate how indigenous knowledge holders perceive the research partnership in the framework of the STN system, we found it more appropriate to ask them directly. Therefore, we conducted a survey soliciting the opinions of 28 traditional healers with whom we worked closely in the Ngamiland District.

The participants in this survey covered a wide range of different types of healers including herbalists and bone setters (herbalists who use a set of bones "ditaola" to

Table 10.3 What interested you most about plant testing?

	Frequency	Percentage (%)
Learn more about plant activities	8	31
Some form of validation	8	31
Understand plants to heal patients better	4	15
Cooperation with scientists	4	15
Learn about contents and toxicity	1	4
Conserve knowledge for further generations	1	4

$N = 26$

Table 10.4 What did you learn from plant testing

	Frequency
Better understanding on how plants work	10
Plants are useful in medical treatments	6
Scientific validation is possible	5
Plants have value and should be preserved	4
Will better attend to patient confidentiality	3

$N = 26$

communicate with ancestors) as the majority (9/28 each), herbalists who are also faith healers (6/28), faith healers only (2/28), and traditional birth attendants (2/28). 82% of healers were male and 18% were female, which reflects the general male dominance of the healing profession in Botswana [23, 55]. Most of the traditional healers (22/28, 81%) first learned that medicinal plants can be tested for bioactivities when they were approached by us. Only 2 of 28 healers were in contact with researchers before. When we asked the participants whether they found the STN plant-testing workshops useful, all of the healers ($N = 28$) answered in the affirmative. Table 10.3 summarizes their answers when we inquired in more detail what interested the healers most about plant testing. Approximately a third of healers (31%) wanted to learn more about plant activities, and another third (31%) were interested in some form of validation of their plant uses. Some healers (15%) related plant testing directly to a healing context, and equally 15% were mainly interested in promoting the collaboration with scientists. Learning about contents and toxicity, and the conservation of knowledge for future generations were issues of interest to two healers. We also inquired whether healer's expectations had been met when the plant-testing workshops were conducted, and we asked the participants what they had learned from performing the STN experiments. Healers mostly answered that they have a better understanding on how medicinal plants work, that plants are indeed useful in medical treatments, and that scientific validation is possible. Interestingly, four healers stated that plant testing had encouraged them to protect and conserve the plants which they are frequently using and expressed their desire to protect their intellectual property in relation to plant uses, as well as in relation to their general knowledge about plants and their medicinal applications (Table 10.4). When

Table 10.5 In what way has the research partnership benefitted you as traditional healers?

	Frequency
Encouraged to share knowledge and work together	10
Better understanding of plant activities and a form of validation	6
"Now, I heal with pride"/Work with confidence/Researchers protect us from being undermined	5
General acknowledgement/appreciation	4
General increase of knowledge	3
More communication and publicity	2
Preservation of traditional knowledge	1

$N = 26$

asked ($N = 26$) whether the plant testing has motivated healers to learn more about "their" plants, all of them stated that it had indeed increased their motivation to understand medicinal plants better.

In enquiring how traditional healers perceive their role in the research partnership with scientists, the majority of all respondents (11/28, 42%) described their part in the research partnership as being the provider of traditional medical knowledge and of medicinal plants. However, over a third of healers (9/28, 35%) emphasized that their role is mainly to contribute to collaboration efforts. One healer summarized, "Collaboration is important for us to be respected, not hidden" (T.G., 23.4.2009). Several healers emphasized that collaborative efforts should be based on elements of equality:

> "We are both learning. We as traditional healers contribute what we know and you as scientists take it further" (M.T. 23.4.2009)
>
> "We are giving what we know and you are giving what you know" (G.C. 23.4.2009)

Three participants (12%) saw their role as generally important, and the remaining respondents felt that they were not yet able to describe their role as they had joined the network only recently.

We then wanted to know whether the partnership with scientists had actually benefitted traditional healers. 93% of respondents (26/28) were of the opinion that it had, while two healers did not yet comment on it, as they were still new in the partnership. Table 10.5 summarizes in what way traditional healers felt they have benefitted from the research partnership. Though healers initially expected to learn more about their plants (Table 10.3), they saw the main benefit in the way the research partnership encouraged the sharing of knowledge and working together. One healer explained:

> "The main benefit comes from combining ideas and knowledge from different people with different ideas to give our healing respect" (M.M., 23.4.2009)

Another participant put the benefit in much more concrete terms:

> "I benefit a lot. I can get information, I get a certificate of appreciation, my work might be published. Through that, people will appreciate that my plants have been tested and do have value." (L.T. 23.4.2009)

Table 10.6 What should be improved in the research partnership on plant testing

	Frequency
Better cooperation with medical doctors, nurses, government and amongst healer's associations themselves	8
More and more regular workshops	5
It is fine as it is	4
Simpler language/more explanation	2
Better logistics	2
More healers should be encouraged to submit plants	1
Plant testing should take place closer to healers	1
There should be more work on conservation issues	1
Consequent follow-up on suggestions	1

$N = 25$

Having gained a better understanding of plant activities and having seen some form of scientific validation was the second most important benefit for healers. Significantly, several healers expressed the feeling that they now "heal with pride" and that researchers helped them "from being undermined." One healer characterized how the STN research partnership was perceived:

> "There is equality. You researchers do not belittle us or look down on us" (M.N., 23.04.2009)

To us, this is one of the most important achievements of the application of the STN system to indigenous knowledge research. It shows that scientists can play a positive role in building trustful collaborations, which is—as in any research partnership—a precondition for establishing successful research platforms. According to the healers, other benefits included a general acknowledgement of their healing profession through science, a general increase of knowledge, as well as more communication and publicity. The last statements do reflect an increased interaction between healers themselves, as they were brought together by workshops, and the fact that the workshops were covered by local print media and radio stations, as well as in the national television channel (BTV).

Consequently, all healers ($N = 28$) stated that they would like to continue with the research partnership. When asked what kind of research should be conducted in the near future, the majority of healers suggested that generally more plants should be tested (4/26), but also more specific plants which are used for specific conditions (13/26). Here, traditional healers recommended primarily medicinal plants they use according to their specific areas of expertise, such as the treatment of STIs and other conditions. Interestingly, a considerable number of healers (7/26) suggested more research on medicinal plants potentially useful for the treatment/management of HIV/AIDS which indicates a sincere concern of healers to combine all efforts in combating the pandemic. Other suggestions included more research on the conservation of medicinal plants, an issue which emerged particularly after individual reports with testing results were given to the healers.

Finally, we wanted to know what should be improved in future collaborations. The majority of healers pointed toward one limitation of the current collaborative efforts, namely, the lack of inclusion of biomedical doctors, government representatives, and representatives of healer's associations themselves (Table 10.6). Healers found it important to initiate closer collaboration and dialogue with sectors that still have reservations about traditional healing. Healers emphasized that collaboration has to have continuity and requested more regular workshops together. Though there were many suggestions for improvement, such as using simpler language, better logistics, having plant-testing facilities nearer to healers, as well as follow-up on suggestions made and encouraging more healers to submit plants, a number of healers were also of the opinion that the current model of collaboration was suitable as it stands.

10.6 New Dynamics, Interfaces, and Innovation

As much as interdisciplinary research platforms give rise to issues emerging from various interfaces between disciplines and in interchange within the socioeconomic and legislative science context, it is important to understand that the STN-research partnership with indigenous knowledge holders has also triggered much more than an increased interest in scientific evaluation of medicinal plants.

"Now that I know that they have activities and indeed are useful, this partnership has motivated me to conserve my plants," summarized one healer (N.M., 23.4.2009) reflecting the concern about a sustainable use of natural resources, which was expressed by many of his colleagues.

The issue of cultivation of medicinal plants has been a subject of debate among healers in our workshops. There seems to be an agreement that some medicinal plants can be cultivated and that it would be desirable to do so, while others, particularly medicinal plants which are associated with "bad luck" or plants whose prescription and use in treatment requires close monitoring by the healer, should not.

Other healers pointed directly to the issue of intellectual property rights, which is a legitimate issue on the interface of natural resources, bioactivities, and drug discovery: "The plant testing has motivated me to preserve and protect my intellectual property" (T.P., 23.4.2009). It is important to note that the question of intellectual property rights was placed here in a positive context. This might have been a consequence of the fact that we respected intellectual property as well. Each healer obtained an individual report summarizing the screening results, and results were discussed confidentially. It remained up to the healer whether and with whom results are shared. Protection of rights emerged as an issue in the framework of a transparent research partnership of mutual trust, instead of being used as a justification to maintain secrecy and a general unwillingness to engage in dialogue. One healer explained:

> "This partnership has shown me that knowledge should not be kept to ourselves. We should try to teach others so that we can unite" (T.T., 23.4.2009)

We noted that the STN testing of plants led to new views on how scientific results can be incorporated into healing. Though we explained in report-back

Fig. 10.4 Labeling and vouchering of plant samples by traditional healers. (**a**) Various medicinal plants and (**b**) plant press invented by P. Magakwa (traditional healer)

workshops the limits of in vitro bioactivity tests and emphasized that it would be appropriate not to change an established treatment regime, healers quickly recognized new potentials: "Plant testing has shown us that our plants might be useful for other conditions which we have not treated before" (N.M., 23.4.2009). Though this might be seen as the most controversial aspect of the application of the STN system, since this type of screening was intended first and foremost as a noninterventionist research tool, healer's conclusions nevertheless reflect the fact that knowledge never stands still. Indigenous knowledge systems were never static, but developed and adapted and continue to do so over time under the influence of many factors [56]. Traditional healers have always experimented with combining different medicinal plants. They have reinvented their healing expertise and have adapted to "new" diseases, such as tuberculosis and sexually transmitted infections during colonial times [57] and HIV/AIDS in this century. Scientific findings will therefore contribute to the dynamics of traditional healing. However, it would be important to investigate whether, and to what extent, the STN system has influenced certain established healing practices or has triggered new treatment regimes to ensure that no bodily harm is caused on the basis of qualitative STN in vitro screening results.

Lastly, during our interaction with indigenous knowledge holders in the framework of the STN program, we emphasized the need to properly identify plant species. Healers quickly understood how different names in different languages for the same plant or the same name for different plants can lead to mistakes in plant-testing experiments. Figure 10.4 illustrates how healers reacted to the challenge of correct botanical authentication and how they tried to assist researchers in that endeavor. Figure 10.4a shows how two healers in Maun started to properly label and arrange their medicinal plants in their consultation hut. Figure 10.4b displays an innovative method of vouchering invented by a traditional healer using simple material, such as cardboard and wire, to preserve the morphology of plant samples for botanical identification. The plant name and its use are recorded on the cardboard. Both examples illustrate that contrary to widespread assumptions that traditional healers are somewhat disorganized and have difficulties in grasping systematic scientific approaches, they emerged as reliable research partners, who adapted quickly to new methods and transformed some of them according to their abilities and means.

10.7 Conclusion

The Screens-to-Nature (STN) approach described in this chapter showed that a cooperation based on mutual trust between scientists and indigenous knowledge holders is possible, even in a setting where legislative regulations on indigenous knowledge systems, on bioprospecting, and on access- and benefit-sharing mechanisms are largely absent.

The STN approach positively contributed to collaborative efforts in a number of ways. Firstly, sharing of results throughout the screening process of medicinal plants was in stark contrast to extractive research methods and equipped traditional healers to assess the usefulness of the research partnership. Secondly, plant testing as a participatory research approach served to demystify science in a way that traditional healers saw opportunities to gain knowledge instead of "protecting" their knowledge from science. Thirdly, the STN screening contributed to developing consciousness of traditional healers in other areas, such as the protection and conservation of natural resources and intellectual property rights. This is important, as informed decisions on just access and benefit-sharing systems and an adequate intellectual property rights regime can only be made if indigenous knowledge holders are actively participating in the research process.

The STN-screening platform has yielded results that are acceptable to both traditional healers and scientists. The STN system emerged as a useful prescreening method particularly with respect to the identification of antimicrobials from natural products and has created an important interface between IKS and drug discovery research. This interface has been perceived as of strategic importance for efforts to build a pan-African Natural Products Library (pANPL), which is described elsewhere in this book, and in whose framework the STN screening is combined with mechanism-based screening assays for antiparasitic drug discovery.

The success of the STN program in bridging the gap between indigenous knowledge holders and scientists assisted in prompting the Ministry of Infrastructure, Science and Technology to commission CesrIKi with the development of a national Indigenous Knowledge Policy, that is, to address adequate access and benefit mechanisms, and is combined with respective implementation and monitoring plans. Taken together, we have come a step further toward an indigenous knowledge-based research platform for drug discovery in Botswana.

Acknowledgments This study was sponsored by a grant of the Office of Research and Development (ORD) of the University of Botswana to KAM (No. R835). We are grateful to the directors of the Global Institute for Bioexploration (Gibex), Dr. I. Raskin and Dr. M.-A. Lila, whose consistent and generous material support in form of consumables and organisms kept the STN screening going. We are indebted to G. Joseph, who established most of the original STN assays. We wish to express our sincere gratitude to B. Abegaz, who initiated the STN/University of Botswana partnership, and to the traditional healers in the Ngamiland District and the community of Mmankgodi with whom we embarked on an exciting journey of participatory natural product research. Finally, we thank N. Makate for critically reading the manuscript.

References

1. Munos B (2009) Lessons from 60 years of pharmaceutical innovation. Nat Rev Drug Discov 8:959–968
2. Lancet T (2011) Where will new drugs come from? Lancet 377:97
3. Mahajan R, Gupta K (2010) Food and drug administration's critical path initiative and innovations in drug development paradigm: challenges, progress and controversies. J Pharm Bioallied Sci 2(4):307–313
4. Bharath EN, Manjula SN, Vijaychand A (2011) *In silico* drug design-tool for overcoming the innovation deficit in the drug discovery process. Int J Pharm Pharm Sci 3(2):8–12
5. Schmid EF, Smith DA (2005) Is declining innovation in the pharmaceutical industry a myth? Drug Discov Today 10(15):1031–1039
6. Patwardhan B, Vaidya ADB (2010) Natural products drug discovery: accelerating the clinical candidate development using reverse pharmacology approaches. Indian J Exp Biol 48:220–227
7. Vaidya ADB (2006) Reverse pharmacological correlates of ayurvedic drug actions. Indian J Pharm 38:4
8. Fabricant DS, Farnsworth NR (2001) The value of plants used in traditional medicine for drug discovery. Environ Health Perspect 109(1):69–75
9. Verpoorte R, Choi YH, Kim HK (2005) Ethnopharmacology and systems biology: a perfect holistic match. J Ethnopharmacol 100:53–56
10. Patwardhan B, Vaidya ADB, Chorghade M et al (2008) Reverse pharmacology and systems approaches for drug discovery and development. Curr Bioact Compd 4:1–12
11. Kong D-L, Li X-J, Zhang H-Y (2009) Where is the hope for drug discovery? Let history tell the future. Drug Discov Today 14(3/4):115–119
12. Svetaz L, Zuljan F, Derita M et al (2010) Value of ethnomedical information for the discovery of plants with antifungal properties. A survey among seven Latin American countries. J Ethnopharmacol 127:137–158
13. Newman DJ, Cragg GM (2007) Natural products as sources of new drugs over the last 25 years. J Nat Prod 70(3):461–477
14. Wang Y (2008) Needs for new plant-derived pharamaceuticals in the post-genome era: an industrial view in drug research and development. Phytochem Rev 7:395–406
15. Harvey AL (2008) Natural products in drug discovery. Drug Discov Today 13(19/20):894–901
16. Ji H-F, Li X-J, Zhang H-Y (2009) Natural products and drug discovery. EMBO Rep 10 (3):194–200
17. Willcox ML, Graz B, Falquet J et al (2011) A 'reverse pharmacology' approach for developing an anti-malarial phytomedicine. Malar J 10(Suppl 1):S8
18. Aguilar-Støen M, Dhillion SS, Rosendal GK (2006) Bioprospecting under different technological, biological and regulatory settings: trends and challenges. Environ Sci Policy. doi:10.1016/j.envsci.2006.07.003

19. Jäger AK (2005) Is traditional medicine better off 25 years later? J Ethnopharmacol 100:3–4
20. WHO (2005) National policies on traditional medicine and regulation of herbal medicines – report of a WHO global survey. World Health Organization, Geneva
21. Barbee E (1986) Biomedical resistance to ethnomedicine in Botswana. Soc Sci Med 22(1):75–80
22. Madiba SE (2010) Are biomedicine health practitioners ready to collaborate with traditional health practitioners in HIV/AIDS care in Tutume subdistrict of Botswana. Afr J Tradit Complement Altern Med 7(3):219–224
23. Andrae-Marobela K, Ngwenya BN, Monyatsi KN et al (2010) Documentation and promotion of indigenous knowledge – based solutions for Botswana – an ethnosurvey. Research report. Center for Scientific Research, Indigenous Knowledge & Innovation (CESRIKI), Gaborone
24. Ministry of Infrastructure, Science & Technology (MIST) (2010) Draft national policy on research, science, technology and innovation. Republic of Botswana, Gaborone
25. Majinda RRT, Abegaz BM, Bezabih M et al (2001) Recent results from natural product research at the University of Botswana. Pure Appl Chem 73(7):1197–1208
26. Abegaz BM, Davies-Coleman MT (2009) A brief history of natural products research in Africa, pharmacognosy in action. In: Cragg GM, Beutler JA, Jones WP (eds) The American Society of Pharmacognosy – 50 years of progress in natural product research 1959–2009. The American Society of Pharmacognosy, Omnipress, Madison, WI, pp 149–150. http://www.pharmacognosy.us/wordpress/wp-content/uploads/ASP-History-Chapter-5.pdf
27. Bwire RN, Majinda RR, Masesane IB et al (2009) From nature through chemical synthesis towards use in agriculture: Oryzoxymycin case study. Pure Appl Chem 81(1):105–112
28. Mihigo SO, Mammo W, Bezabih M et al (2010) Total synthesis, antiprotozoal and cytotoxicity activities of rhuschalcone VI and analogs. Bioorg Med Chem 18:2464–2473
29. Chingwaru W, Majinda RR, Yeboah SO et al (2011) *Tylosema esculentum* (Marama) tuber and bean extracts are strong antiviral agents against rotavirus infection. Evid Based Complement Alternat Med. doi:10.1155/2011/284795
30. Mojeremane W, Legwaila GM, Mogotsi KK et al (2005) Monepenepe (*Cassia abbreviata*): a medicinal plant in Botswana. Int J Bot 1(2):108–110
31. Motlhanka DMT, Habtemariam S, Houghton P (2008) Free radical scavenging activity of crude extracts and 4′-O-Methylepigallocatechin isolated from roots of *Cassine Transvaalensis* Burtt-Davy from Botswana. Afr J Biomed Res 11:55–63
32. Motlhanka DM, Makhabu SW (2010) Medicinal and edible wild fruit plants of Botswana as emerging new crop opportunities. J Med Plants Res 5(10):1836–1842
33. Kellogg J, Joseph G, Andrae-Marobela K et al (2010) Screens-to-nature: opening doors to traditional knowledge and hands-on science education. NACTA J 54(3):41–48
34. Joseph G (2008) Gibex 'Screens-to-Nature' Manual. Ruttgers University, New Brunswick
35. Kgathi DL, Ngwenya BN (2005) Community-based natural resource management and social sustainability in Ngamiland. Botswana notes and records – special edition in human interactions and natural resource dynamics in the Okavango Delta and Ngamiland 37:40–60
36. Setshogo MP (2005) Preliminary checklist of the plants of Botswana. Southern African Botanical Diversity Network Report 37. University of Botswana Herbarium, Gaborone
37. Setshogo MP, Mbereki CM (2011) Floristic diversity and uses of medicinal plants sold by street vendors in Gaborone, Botswana. Afr J Plant Sci Biotech 5(1):69–74
38. Grace OM, Simmonds MSJ, Smith GF et al (2008) Therapeutic uses of *Aloe* L. (Asphodelaceae) in Southern Africa. J Ethnopharmacol 119:604–614
39. Sanz-Biset J, Campos-d-la-Cruz J, Epiquien-Rivera E et al (2009) A first survey on the medicinal plants of the Chazuta valley (Peruvian Amazon). J Ethnopharmacol 122(2):333–362
40. WHO/UNAids (2010) 2010: A global view of HIV infection. World Health Organisation. (WHO)/UNAids, Geneva
41. Chung GAC, Aktar Z, Jackson S et al (1995) High-throughput screen for detecting antimycobacterial agents. Antimicrob Agents Chemother 39:2235–2238

42. Gupta A, Bhakta S, Kundu S et al (2009) Fast-growing, non-infectious and intracellularly surviving drug-resistant *Mycobacterium aurum*: a model for high-throughput anti-tuberculosis drug screening. J Antimicrob Chemother 64:774–781
43. Runyoro DKB, Ngassapa OD, Matee MIN et al (2006) Medicinal plants used by Tanzanian traditional healers in the management of *Candida* infections. J Ethnopharmacol 106:158–165
44. Tundis R, Loizzo MR, Menichini F (2010) Natural products as alpha-amylase and alpha-glucosidase inhibitors and their hypoglycaemic potential in the treatment of diabetes: an update. Mini Rev Med Chem 10(4):315–331
45. Robina I, Moreno-Vargas A, Carmona AT et al (2004) Glycosidase inhibitors as potential HIV entry inhibitors. Curr Drug Metab 5(4):329–361
46. Waladkhani AR, Clemens MR (2008) Dietary phytochemicals and cancer. In: Watson RR, Preedy VR (eds) Botanical medicine in clinical practice. CAB International, Wallingford
47. Prezelj A, Anderluh PS, Peternel L et al (2007) Recent advances in serine protease inhibitors as anti-coagulant agents. Curr Pharm Des 13(3):287–312
48. Hughes AL, Piontkivska H (2003) Phylogeny of Trypanosomatidae and Bodonidae (Kinetoplastida) based on 18S rRNA: evidence for paraphyly of *Trypanosoma* and six other genera. Mol Biol Evol 20:644–650
49. Ximenez C, Moran P, Rojas L et al (2009) Reassessment of the epidemiology of amebiasis: state of the art, infection. Genet Evol 9:1023–1032
50. Lindsay RG, Watters G, Johnson R et al (2007) *Acanthamoeba keratitis* and contact lens wear. Clin Exp Optom 90 (5):351–360 [see p 11]
51. Schuster FL, Visvesvara GS (2009) Free-living amoebae as opportunistic and non-opportunistic pathogens of humans and animals. Int J Parasitol 34:1001–1027
52. Niederweis M, Danilchanka O, Huff J et al (2010) Mycobacterial outer membranes: in search of proteins. Trends Microbiol 18(3):109–116
53. Parekh J, Chanda S (2008) *In vitro* antifungal activity of methanol extracts of some Indian plants against pathogenic yeast and moulds. Afr J Biotech 7(23):4349–4353
54. Mukanganyama S, Ntumy AN, Maher F et al (2011) Screening for anti-infective properties of selected medicinal plants from Botswana. Afr J Plant Sci Biotech 5 (1-Special Edition): 1–7
55. Ntloedibe-Kuswani GS (1999) Bongaka, women and witchcraft, Women's Worlds 99. The 7th international interdisciplinary congress on women, Session VII, Gendering the Past, Tromso, Norway
56. Sillitoe P (1998) The development of IK: a new applied anthropology. Curr Anthropol 39(2):223–252
57. Molefi RK (1996) A medical history of Botswana 1885–1966. The Botswana Society, Gaborone

Chapter 11
Innovative Approaches to Exploiting Traditional Medicines in Malaria

Philippe Rasoanaivo and Solofoniaina Razafimahefa

11.1 Introduction

There are approximately 60,000 plant species in African Union countries, which represent roughly a quarter of the world plants. Unfortunately, despite the wealth and endemicity of the African plant biodiversity and associated cultures, the region has only contributed 83 of the world's 1,100 leading commercial medicinal plants.

Investigation of medicinal plants for drug discovery has mainly followed the Western approach of single active constituents using standard pharmacological methods. Unfortunately, there are several examples of herbal medicines with reputedly excellent therapeutic effects, which subsequently produce disappointing results when evaluated in the laboratory with the standard biological screenings. Conversely, there are several bioactive compounds isolated from plants with excellent biological activity in the laboratory, which are ineffective or too toxic to use in human patients.

The advent of modern techniques, namely, combinatorial chemistry, high throughput screening (HTS), bioinformatics, "omics" methods, etc., has revolutionized drug discovery. Unfortunately, biopharmaceutical companies attempting to increase productivity through these novel technologies have fallen short of achieving the desired results [1]. Following the trend of these novel technologies, there has also been a paradigm shift in the investigation of plants, focusing more on chemical diversity than

P. Rasoanaivo (✉)
Ecole Supérieure Polytechnique, Université d'Antananarivo, BP 1500 Antananarivo, Madagascar

Institut Malgache de Recherches Appliquées, Avarabohitra Itaosy, lot AVB 77, BP 3833 Antananarivo, Madagascar
e-mail: rafita@moov.mg

S. Razafimahefa
Institut Malgache de Recherches Appliquées, Avarabohitra Itaosy, lot AVB 77, BP 3833 Antananarivo, Madagascar

K. Chibale et al. (eds.), *Drug Discovery in Africa*,
DOI 10.1007/978-3-642-28175-4_11,

traditional uses. The exploration of biodiversity for new sources of biologically active natural products termed bioprospecting is based on massive random collecting, systematic extraction and medium-to-high throughput biological screening. Disappointingly, the success in terms of new medicines reaching the market has not increased with the application of the modern technologies [2].

The World Health Organization estimates that up to 80% of population in Africa depends on traditional medicine for their primary health care requirements [3]. This is attributed to cultural acceptability, efficacy against certain types of diseases and ailments, physical accessibility and economic affordability as compared to modern medicine. Focusing on malaria, 1,277 medicinal plants from 160 families were reported and classified according to their importance, for the treatment of this tropical disease which afflicts mainly the poor and underprivileged populations of sub-Saharan Africa [4]. Based on our continuing ethnobotanical fieldwork for the past seven years, in which we have recorded nearly one hundred new antimalarial plants in Madagascar, it is likely that many other unpublished species are used to treat this disease. African scientists should therefore get "outside the box" and think otherwise on how to better harness traditional knowledge to drug discovery and formulation. Research on antimalarial drugs has been mainly focused on killing the parasites in the blood stage and has rarely considered other mechanisms. To this end, it may be important to go beyond the boundary of classical test protocols, and devise modified protocols to demonstrate biological activities. After a brief review of traditional medicine and malaria cases in Madagascar, we will give an overview of relevant plant-based malaria research with emphasis on research in Madagascar, and we will present an innovative approach to exploiting traditional medicines for malaria.

11.2 Outline of Traditional Medicine and Related Matters in Madagascar

11.2.1 Traditional Medicine

Traditional medicine has an important place in health care delivery in Madagascar. In most cases, it is a first port of call before orthodox medicine and a last resort when all orthodox efforts fail. Two main features characterize the traditional medicine in Madagascar: the oral tradition for communication and the secrecy of treatments with plants. To the best of our knowledge, there is only one document entitled (free translation) "*List of plants with healing properties written by one Malagasy medicine-men in the year 1880*" written by a healer in our native language [5]. A few Malagasy healers have preserved their knowledge in written form, but it has been practically impossible to have access to these documents.

11.2.2 Medical Beliefs

It is believed that a gift of healing is bestowed on practitioners by his ancestors and gods. Malagasy traditional medicine is holistic in approach, using mainly herbs combined with some aspects of spirituality. Health problems are not only attributed to pathological explanations but also to supernatural forces. It appears that traditional healers utilize some astonishing magical or religious practices to get rid of the evil spell before administration of a plant decoction to cure the disease. Furthermore, a patient is not only seen as a physical being but also as a body with soul and spirit living in equilibrium. Diseases are interpreted as imbalance between these three components. It is believed that this natural equilibrium can be regained by treatment with herbs because herbs are natural. Following this view, only nature can restore nature. However, the Malagasy people also believe that there is a balance between the harmful and the curative powers of medicinal plants, and they pray to gods and the ancestors to move this balance towards healing. Patients taking herbal infusions expect benefits from the ingredients, and the power of ancestors and gods is invoked. Spiritual significance is more important to them than biological properties of the remedy. Another characteristic of Malagasy traditional medicine is its strong connection with the observation of what occurs in the universe. Finally, beliefs of the Madagascar population with regard to illness are also dictated by the materialization of the physiological disturbances. To cure a disease, its cause and effects must be physically removed, either via the gastrointestinal tract, vomiting and/or by bleeding.

11.2.3 Inventory of Ethnomedical Knowledge

Madagascar has developed a wealth of ethnomedical knowledge over many centuries. The first written ethnomedical knowledge in Madagascar document dates back to the 1600s [6]. The first medical doctorate thesis defended by a Malagasy citizen concerned the investigation of a plant used as an ordeal poison [7], and the next two were devoted to the inventory of indigenous medical practices and beliefs, including malaria [8, 9]. Overall, hundreds of papers, books and dissertations have been published, many of them in the realm of "grey" literature. A systematic inventory, compilation and computerized processing of all published and unpublished ethnomedical work was done by the *Institut Malgache de Recherches Appliquées* (IMRA). Nearly 4,500 plants have been reported to have medicinal virtues. The database named PLANTMED is regularly updated with information published locally or obtained from our own ethnobotanical fieldwork investigations. It has been very helpful in the selection of plants in drug discovery programmes. Particularly, data on 239 medicinal plants used to treat malaria in Madagascar were published [10], and based on updated data retrieved from PLANTMED, nearly 300 medicinal plants are now used to treat this tropical disease in Madagascar.

11.3 Malaria Cases in Madagascar

The native word "*tazo*" (which generally means fever) was the first term used for malaria, and later, the term *tazomoka* (fever of anopheles origin) was commonly used for this disease. Malaria appears to have existed in Madagascar for several centuries [6].

The first written document on the subject was a tragic story presented as a doctoral thesis at the Faculty of Medicine in Paris in the year 1827 [11]. A relevant characteristic of malaria in Madagascar is its sudden and unpredictable resurgence. The first reported malaria epidemic occurred in 1882 in the highlands and was surprisingly accompanied by a massive epizootic that killed half of the zebus in Madagascar [12]. Ten years later, local populations and European residents began to hear rumours of a new disease called *ramanenjana*, a strange and inexplicable epidemic of dancing mania, which was first experienced in the southern part of the island in February 1963 and reached the capital where it started becoming common in March. It appeared in the form of choreomania, in which dancing and singing are performed to an extraordinary and abnormal extent [13]. In a work presented as a medical doctoral thesis in France by a Malagasy citizen, this epidemic was always reported to occur at the time when the rice crop was ripening and hence was claimed to be a choreomania of malarial origin [14].

Another malaria epidemic occurred in 1878 in Ankazobe, approximately 100 km from Antananarivo under the name *tazobe* (great malaria), *tazon'avaradrano* (malaria of the northern part of Antananarivo) and *aretin'olona* (someone's disease), killing thousands of people [15]. One year, later, another severe malaria epidemic took place in Fianarantsoa, South Highlands, under the local names *rapo-rapo* and *safotany* [16]. In 1903, a concomitant epidemic of malaria and influenza in Antananarivo killed thousands of people [17]. But the most severe malaria epidemic probably occurred in 1905–1906, again in the Central Highlands, and was reported to have killed several thousands of people [18, 19]. Another known tragic epidemic took place in the 1980s, once more in the Central Highlands as one of the country's most devastating tropical diseases [20]. The next malaria epidemic was reported to have appeared in 1994 in the southern part of Madagascar [21]. Since then, several restricted malaria epidemics were reported to have taken place in some part of the Island, some of them co-occurring with chikungunya disease.

11.4 Antimalarials from Plants of Madagascar

Natural compounds of plant origin have provided the best antimalarials known to date. The alkaloid quinine was isolated in 1820 from the South American *Cinchona* species traditionally used as an antimalarial remedy by the Incas in Peru. It was the first drug to be introduced for malaria chemotherapy and served as a template for the synthesis of chloroquine that has been used since the 1940s. On the other side of the globe in the 1970s, Chinese scientists discovered the remarkable antimalarial activity of artemisinin derived from *Artemisia annua*. Since then, many plant species used traditionally for the treatment of malaria have been evaluated for antimalarial potential. In the last three decades, several compounds of various chemical classes have been isolated from plant species, of which some fulfil the main prerequisites needed for new lead antimalarial agents [reviewed in 22–25]. An informative book dealing with the entire field of antimalarial and antivectoral natural products has been published [26]. And recently, a special issue of *Malaria Journal* was devoted to the uses of natural compounds in the development of new antimalarial treatments: (http://www.malariajournal.com/supplements/10/S1).

Despite great efforts to control it, malaria is still a serious problem in Madagascar, causing about 10,000 deaths and one million cases each year according to the statistics published by the Ministry of Health. It is likely that malaria deaths are much higher than the official data since those occurring in remote areas are not adequately recorded. Research on malaria chemotherapy began in 1985 following the sudden resurgence of this disease during the 1980s. The severity of the infection is such that local populations now recognize the hitherto unknown condition to which they have given the name *bemangovitra* (disease of great shivering). Shortage of appropriate drugs due to the serious deterioration of the economic situation in Madagascar at that time led the population back to the massive use of traditional herbal remedies. We took this unique opportunity to carry out a massive ethnobotanical fieldwork in different regions of Madagascar [10]. In the course of these ethnobotanical investigations, we particularly learned that rural populations in Madagascar treat *bemangovitra* by means of self-medication with 1–2 tablets of chloroquine (a dose thought to promote chloroquine resistance), together with a decoction made from various plant species termed chloroquine adjuvants. We assumed that some of these species may enhance chloroquine efficiency and might therefore reverse chloroquine resistance on the part of malaria parasites.

11.4.1 Preliminary Screening of Plants for Antiplasmodial Activity

Six medicinal plants commonly used to treat malaria in Madagascar were first evaluated for their antiplasmodial activity in collaboration with researchers at the *Institut de Médecines et d'Epidémiologie Tropicales, Hôpital Bichat-Claude Bernard, Paris*, who have facilities to conduct antiplasmodial tests [27–29]. However, active constituents were not isolated. We learned from this collaborative work that the availability of the biological tests situated right in our institute is of paramount importance for a successful natural product research involving a bioassay-guided fractionation. Through adequate training activities and procurement of basic equipment, our institute has been able to perform a battery of biological screening. Unfortunately, for 4 years, our research was discontinued due to equipment problems, resuming again in 2009, thanks to the IFS/PRISM project which donated a Hidex Chameleon Muliplate Reader with fluorescence, luminescence, absorption and LCS detection.

11.4.2 Search for Naturally Occurring Chemosensitizers

Synthetic drugs that reverse chloroquine resistance have attracted much attention since the pioneering work of Martin et al. in the 1990s [30]. Thus, it was demonstrated that several calcium antagonists, various tricyclic antidepressants and tricyclic antihistaminics restored in vitro and/or in vivo chloroquine sensitivity to resistant strains of *Plasmodium* malaria [31].

At that time, very little was known about natural products that could enhance chloroquine effectiveness, although relevant data exist in ethnomedical knowledge. A research project was therefore initiated to evaluate this potential for treating malaria. Nearly ten chloroquine-adjuvant plants were recorded and five of them investigated in our laboratory. Two of them, *Hazomalania voyroni* and *Strychnos myrtoides*, gave promising results. All bioassay-directed fractionation has been carried on in our institute, while structure elucidation of the bioactive compounds has been done in collaboration with western laboratories.

11.4.2.1 Investigation of *Hazomalania (= Hernandia) voyroni* (Hernandiaceae)

Hazomalania (= Hernandia) voyroni, endemic to Madagascar at the genus level, grows in the southwest of the island. Preliminary, bioassay-directed fractionation of the crude extract from stem barks showed that the bioactive constituents were located in the polar alkaloid fraction [32]. Bioactivity-guided fractionation led to the isolation of a new group of *bis*-isoquinoline alkaloids, which we termed hervelines A, B, C and D (Fig. 11.1), and which were composed of pavine and benzyltetrahydroisoquinoline moieties [33, 34]. Hervelines A, B, C and D are closely related alkaloid dimers, which differ only by the substitution at positions 7 and 12. They were isolated by countercurrent distribution using buffer solutions at decreasing pH as a mobile phase, strongly indicating that they differed in their relative basicity and polarity. The hervelines A–D have moderate antiplasmodial activity. But, strikingly, the structural difference led to a drastic variation in the chloroquine-enhancing activity ranging from synergism for herveline C and herveline B to simple additive effect for herveline A and antagonism for herveline D. In the less polar fractions, two known dehydroaporphines, dehydrodicentrine and dehydrocrebanine, were isolated (Fig. 11.2). They showed in vitro antiplasmodial strong effects on the chloroquine-resistant *Plasmodium falciparum* FCM29 strain, with median inhibitory concentrations (IC_{50}) of 80 ng/ml and 72 ng/ml, respectively. They may consequently serve as lead compounds for the design of more effective antimalarial drugs.

Unfortunately, *H. voyroni* is classified as an endangered species of tree because its trunk is extensively used for manufacturing water-resistant coffins, dugouts and wooden buildings. It was not possible to continue further investigations on the alkaloids of this plant to bring a standardized extract into therapeutic use. However,

Fig. 11.1 Structure of hervelines A–D

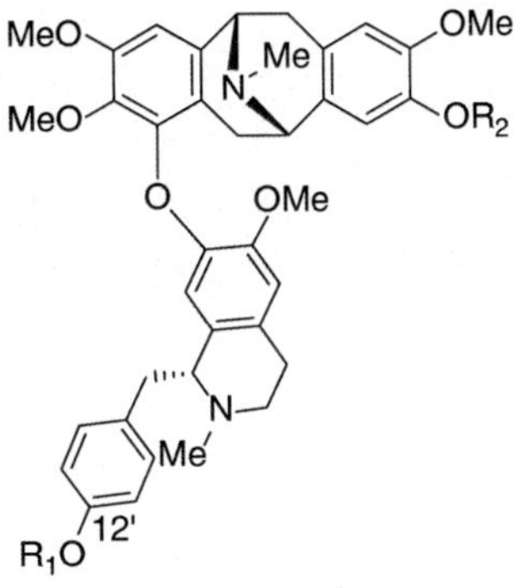

R_1= H; R_2= Me: herveline A

R_1= Me; R_2= H: herveline B

R_1= Me; R_2= Me: herveline C

R_1= H; R_2= H: herveline D

Fig. 11.2 Structure of dehydroaporphines isolated from *Hazomalania voyroni*

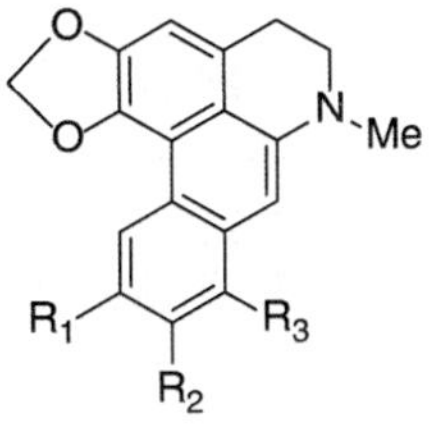

R_1 = R_2 = OMe; R_3 = H : dehydrodecentrine

R_1 = H; R_2 = R_3 = OMe : dehydrocrebanine

the crude alkaloid extract of *H. voyroni* has some constituents acting as antimalarials and others as drug-activity enhancers, which is a relevant example of the way in which nature can provide clues for a chemotherapy strategy in the treatment of malaria.

11.4.2.2 Investigation of Madagascar *Strychnos*

Our work on Madagascar *Strychnos* started first with *S. mostueoides* whose decoction is used to treat malaria. Antiplasmodial evaluation of the crude alkaloids showed a moderate activity (IC_{50} = 8.5 μg/ml). Nevertheless, we investigated this plant phytochemically for academic purposes resulting in the isolation of two new and six known alkaloids [35]. Two of them, the new alkaloid named malagashanine and the known strychnobrasiline (Fig. 11.3), will be discovered to be the relevant constituents for chloroquine-potentiating effect in Madagascar *Strychnos* alkaloids.

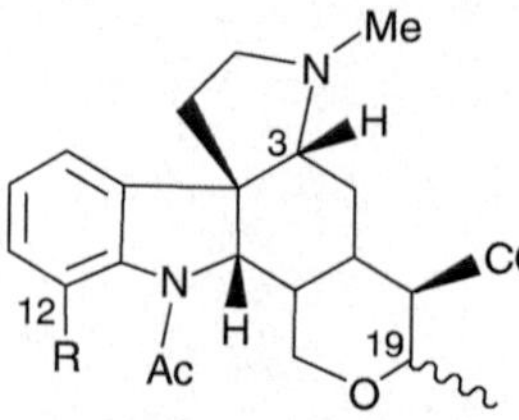

R= H; 19S: malagashanine
R= OH, 19S: 12-hydroxy-malagashanine
R= OH, 19R: 12-hydroxy-19-*epi*-malagashanine

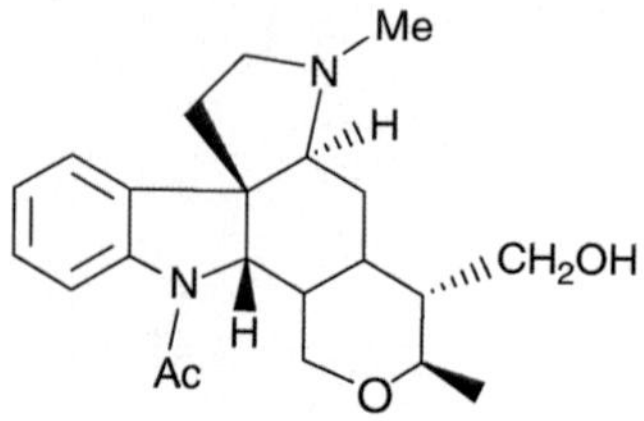

Malagashanol

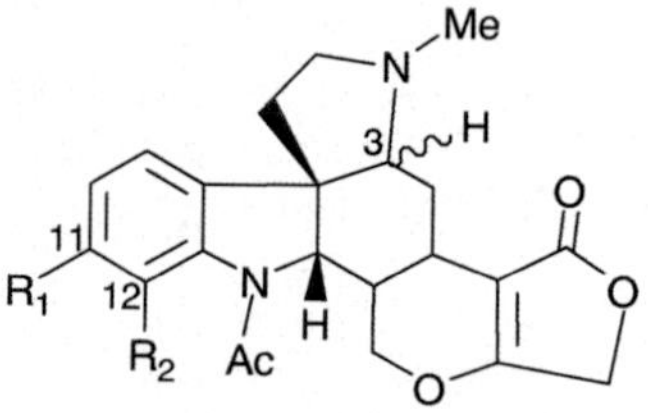

Myrtoidine series

R_1= OMe, R_2= H, 3βH: myrtoidine
R_1= OMe, R_2= H, 3αH: 3-*epi*-myrtoidine
R_1= R_2= H, 3βH: 11 -demethoxy-myrtoidine
R_1= R_2= H, 3αH: 3-*epi*-11-demethoxy-myrtoidine
R_1= H, R_2= OH, 3βH: 11-demethoxy-12-hydroxy-myrtoidine
R_1= H, R_2= OH, 3αH: 3-*epi*-11-demethoxy-12-hydroxy-myrtoidine

Fig. 11.3 Structure of alkaloids isolated from *Strychnos myrtoides* and *S. diplotricha*

It is important to stress that most of the information that we obtained on the traditional uses of the Malagasy *Strychnos* come from J. P. Abrahama, a highly experienced person in the field of taxonomic identification of plants and ethnobotany. He tragically died in May 1996 as a consequence of an insect bite during a botanical fieldwork [36]. During a series of ethnobotanical fieldworks conducted in the 1990s with him in the Eastern rain forests, we observed that he discretely used an infusion of a scraped stem of a plant every time we went to the forests. We asked him about the purpose of this practice. He told us evasively that the infusion of this plant, known under the vernacular name *retendrike*, protects him against malaria. This prompted us to evaluate the antiplasmodial activity of this plant, but the results were rather disappointing since the crude alkaloids had a low activity (IC_{50} ~100 μg/ml). Many months later, we tactfully made inquiries into Abrahama's practice, and we learned that, in fact, the infusion is used in combination with 1–2 tablets of chloroquine. This guided our work into drug combination assessment, which resulted in the isolation of the two major bioactive constituents, malagashanine and strychnobrasiline. They were devoid of any antiplasmodial action, but when combined with chloroquine at doses much lower than required for antimalarial effect, they enhanced in vitro chloroquine action against the chloroquine-resistant strain FCM29. The results were confirmed in vivo with the crude alkaloid extract [37]. The structure of malagashanine was revised as a new subtype of *Strychnos* alkaloids, the N_b-C(21)-secocuran type with an unusual 3βH configuration, and confirmed by X-ray analysis [38]. Six minor alkaloids (Fig. 11.3) of *S. myrtoides* structurally related to malagashanine were isolated by the countercurrent distribution technique in a Craig apparatus [39]. It is worthwhile underlining the existence of two configurations for C-3 and C-20 in this series of alkaloids.

Another *Strychnos* species, *S. diplotricha*, is widely used to treat malaria in Antalaha (north-east Madagascar). This prompted us to phytochemically investigate the crude alkaloid, leading to the isolation of malagashanine and strychnobrasiline as major compounds, together with minor constituents also related to the parent compound but possessing both 3αH and 3βH configurations [40].

Up to now, malagashanine and structurally related alkaloids have only been isolated from Madagascar *Strychnos* and constitute a novel series of N_b-C(21)-secocuran type of *Strychnos* alkaloids endowed with chloroquine-enhancing activities. It is also important to underline the positive correlation between the ethnomedical use of the three Madagascar *Strychnos* species as chloroquine adjuvants in different regions of Madagascar by culturally different ethnic groups and the knowledge of the chloroquine-enhancing effect of the two common constituents, malagashanine and strychnobrasiline.

11.4.2.3 Clinical Observational Studies with *Strychnos myrtoides*

The potential of the tricyclic antidepressant desipramine and the tricyclic antihistaminic cyproheptadine to improve the efficacy of chloroquine against chloroquine-resistant *P. falciparum* in humans was investigated [41]. However, these clinical studies provided no evidence for enhanced chloroquine efficacy in vivo through the use of desipramine and cyproheptadine in doses corresponding to the usual therapeutic range .Thus, synthetic compounds that have a good pharmacological profile in experimental models may lack useful application in humans. After the traditional recipe and basic toxicity study were confirmed, we submitted the infusion of *S. myrtoides* to a clinical observational study, following approval by the ethical committee of the Ministry of Health. The method we used was to select patients for whom antimalarial treatment failure was observed upon their admission to hospital and to treat them with an infusion of 0.5 g of powdered stem bark or, alternatively, chloroquine together with the infusion if the quantity of antimalarial uptake was judged insufficient. Twenty cases of antimalarial drug failure were recorded in the study. The most remarkable results were the complete elimination of parasitaemia in two patients for whom treatment failure had been observed with chloroquine and in one patient with anaphylactic shock following an injection of quinine salts. Four patients with quinine intolerance were also successfully treated with the same therapeutic regimen [42].

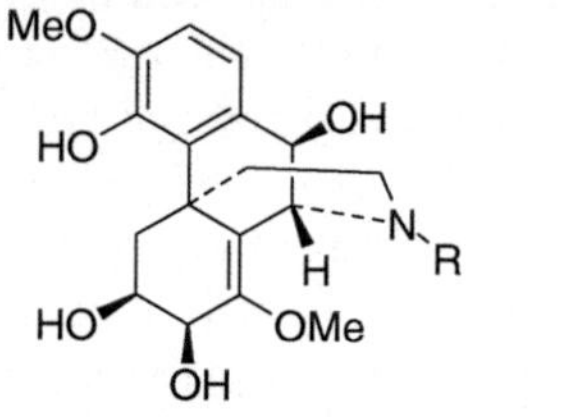

Fig. 11.4 Structure of tazopsine and NCP-tazopsine

11.4.3 Lead Alkaloid for the Hepatic Stage of Plasmodium *Malaria*

Global research directed at antimalarial drug discovery has principally focused on the asexual blood stages of malaria parasites. Only two 8-aminoquinoline drugs, primaquine and tafenoquine, kill the liver stage of *Plasmodium*. A serious disadvantage of the 8-aminoquinolines is the toxicity caused by haemolysis of red blood cells in patients with some types of G6PD deficiency, a frequent condition in African populations. Following the inaugural meeting of the Research Initiative on Traditional Antimalarial Methods held in Moshi (Tanzania) in 1999, the Pitié-Salpêtrière Hospital, Paris, France, and IMRA decided to implement a screening programme of Malagasy antimalarial plants targeting the *Plasmodium* hepatic stage. Pharmacological models are available at the French laboratory, particularly the culture of the liver stage of the parasite in vitro, which is technically more challenging than that of the erythrocytic stages.

The most prominent results were the discovery of a new morphinan alkaloid named tazopsine (Fig. 11.4) isolated from the Malagasy species *Strychnopsis thouarsii* endemic to Madagascar at genus level, which specifically inhibited the development of *P. yoelii* and *P. falciparum* hepatic parasites in cultured primary hepatocytes, with inhibition being most pronounced during the early developmental stages [43]. Other structurally related derivatives were also isolated [44]. Among several semi-synthetic derivatives, N-cyclopentyl-tazopsine (NCP-tazopsine) was found to be specifically active in vitro against the liver stage, but inactive against the blood forms of the malaria parasite. Furthermore, oral administration of NCP-tazopsine completely protected mice from a sporozoite infection.

Other morphinans of the morphine series, which differ from tazopsine and NCP-tazopsine mainly by the inverse absolute configuration of asymmetric carbons on the piperidine ring, were not found to be inhibitory when tested against the *P. yoelii* liver stages. The discovery of this new class of molecules could lead to the development of a true causal prophylactic drug and may prevent the propagation of the parasites and the pathology. The treatment targets the early stages of malaria infection in which liver-stage parasites are far less numerous than blood-stage parasites. This would make it more difficult for the parasite to develop the drug resistance that hampers conventional malaria treatment programmes. NCP-tazopsine has undergone further pharmacological investigations in primate models.

11.4.4 Multi-use of Bioactive Natural Products

In addition to their useful bioactivities, natural products can be used for many purposes, as biochemical tools to probe the mechanism of action, as starting materials for the semi-synthesis of derivatives, as source of inspiration for the identification of useful pharmacophores and many others.

11.4.4.1 Malagashanine as a Biochemical Tool

Natural products have been used extensively to elucidate complex cellular mechanisms, leading to the identification of important targets for therapeutic intervention. In the area of malaria, the elucidation of the mechanism of chloroquine resistance and its reversal has generated much interest and stimulated intense debate. It is generally accepted that chloroquine-resistant strains of *Plasmodium falciparum* accumulate far less drug than chloroquine-sensitive strains. However, the mechanism involved in the acquisition of the resistance phenotype remains a matter of controversy and ongoing investigation. The synthetic compound verapamil has been extensively used as a biochemical tool to probe the mechanism of chloroquine resistance and its reversal. In the area of natural products, the herveline series and malagashanine, possessing unique structures found hitherto in the Madagascar plants, may also be useful biochemical tools for understanding the above mechanism.

Malagashanine was active in vivo, and selectively enhanced in vitro the action of quinolines (chloroquine, quinine and mefloquine), aminoacridines (quinacrine and pyronaridine) and halofantrine, all of which are believed to exert their antimalarial effect by binding to haematin [45]. Furthermore, using tritiated chloroquine, we carried out several kinetic experiments and concluded that malagashanine prevents chloroquine efflux from, and stimulates chloroquine influx into, drug-resistant *P. falciparum* [46]. We then postulated for the first time the existence of a bifunctional chloroquine transporter that is involved in these processes. In this context, resistant parasites modify the chloroquine importer into a chloroquine exporter, and reversing chloroquine resistance would also mean restoring the initial importing role of the presumed bifunctional chloroquine transporter. With the availability of the Hidex Chameleon Multiplate Reader in 2009, further investigations have been under way to clarify the exact nature of this putative bifunctional transporter. Particularly, malagashanine was also shown to reduce the activity of the glutathione transferase (PfGST) to 80% but showed a significant time-dependent inactivation of this enzyme with respect to control [47].

One relevant unpublished result in the “ex vivo” test was the observation of antiplasmodial activity in the blood serum after 8 h treatment. This may be explained either by reabsorption of malagashanine or formation of active metabolite(s).

11.4.4.2 Strychnobrasiline as a Starting Material for Semi-synthesis

Many complex bioactive natural products can be isolated in significant quantities without great difficulty. As a result, these readily available compounds provide valuable starting materials for the rational generation of useful derivatives or libraries of compounds prepared through semi-synthesis and biocatalysis.

Strychnobrasiline was by far the major constituent of *Strychnos myrtoides* with a 1.5% yield. As it was devoid of in vivo chemosensitizing activity, we used this compound as a starting material for chemical derivatizations. Several chemical reactions were used to generate various derivatives. Two unexpected molecular rearrangements deserve attention (Fig. 11.5). First, reduction of N_a-deacetyl-strychnobrasiline with lithium aluminium hydride gave a rearranged derivative [48]. Secondly, oxidation of alkaloids may also afford unexpected rearrangement reactions. At this point, treatment of N_a-deacetyl-strychnobrasiline with 3-chloroperoxybenzoic acid led to the formation of a nitrone with a new indole skeleton obtained by the Baeyer-Villiger rearrangement [49]. Strychnobrasiline was then used for the semi-synthesis of malagashanine derivatives. The strategy consisted of opening the C ring by hydrolysis of the iminium ion form followed by the Schiff reaction [50].

11.4.4.3 Design and Synthesis of Chemosensitizing Pharmacophores

Surprisingly, we found that the hexahydro derivative of malagashanine reverses chloroquine resistance with comparable potency to the parent compound. Another unrelated result was the discovery of new related tropane alkaloid aromatic esters named pervilleines A–F (Fig. 11.6) isolated from *E. pervillei* endemic to Madagascar, found to be excellent modulators of the multidrug resistance phenotype in cancer, with activity comparable to the standard MDR modulators verapamil and cyclosporine A [51]. These results prompted us to investigate the role of natural products as sources of inspiration or messages to be decrypted for the design and synthesis of novel chemosensitizers. To this end, we observed that the three series of naturally occurring chemosensitizers (namely, hervelines, malagashanine and pervilleines) have unrelated structures, and this is also true for the synthetic chemosensitizers in malaria, the calcium entry blockers, some tricyclic antidepressants and some tricyclic antihistaminics. This raises the problem of functional versus chemical diversity.

In its simplest structural form, the hexahydro derivative of malagashanine is a 1,4-diamine. We therefore assumed that a basic chemosensitizing pharmacophore might be present in most malaria reversers. At this point, careful inspection of the

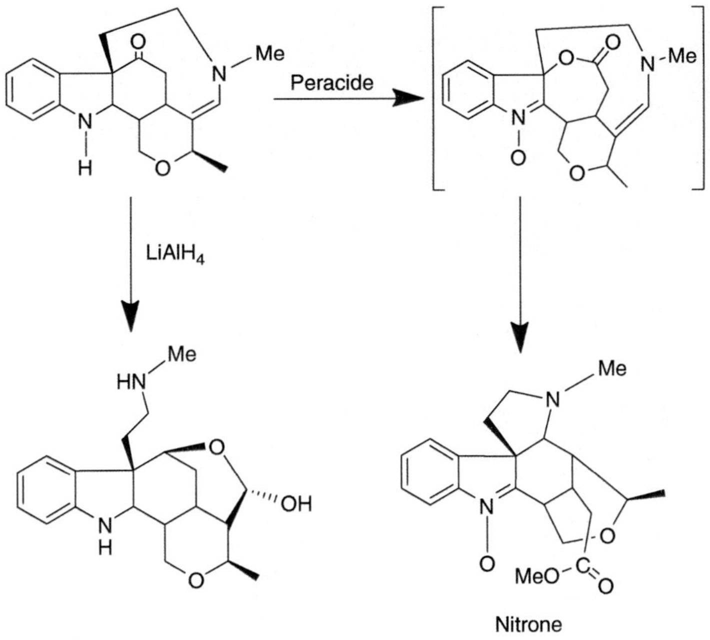

Fig. 11.5 Rearrangements of strychnobrasiline in oxidative and reductive conditions

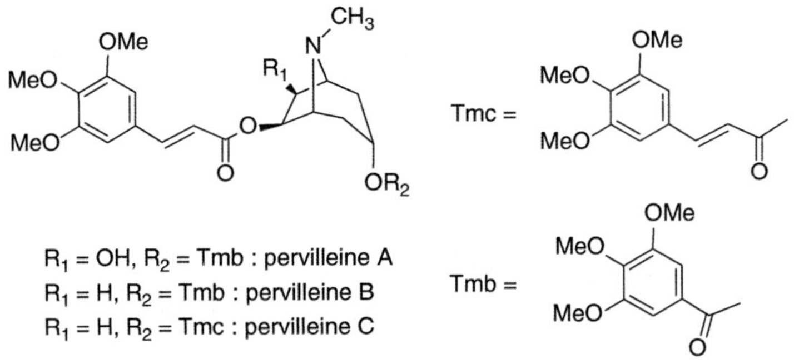

Fig. 11.6 Structures of pervilleine A, B and C

structures of synthetic chemosensitizers surprisingly showed that, although they appear to possess unrelated structures, most of them have in common a basic fragment, namely, the *N*-phenyl-1,3-diamino-propane and 1-phenyl-4-amino unit. A similar fragment is found in the pervilleine series. These observations led us to postulate a unifying chemosensitizing pharmacophore in malaria (Fig. 11.7) [52]. We performed a preliminary confirmation of this hypothesis by synthesizing and pharmacologically evaluating several related structures [53].

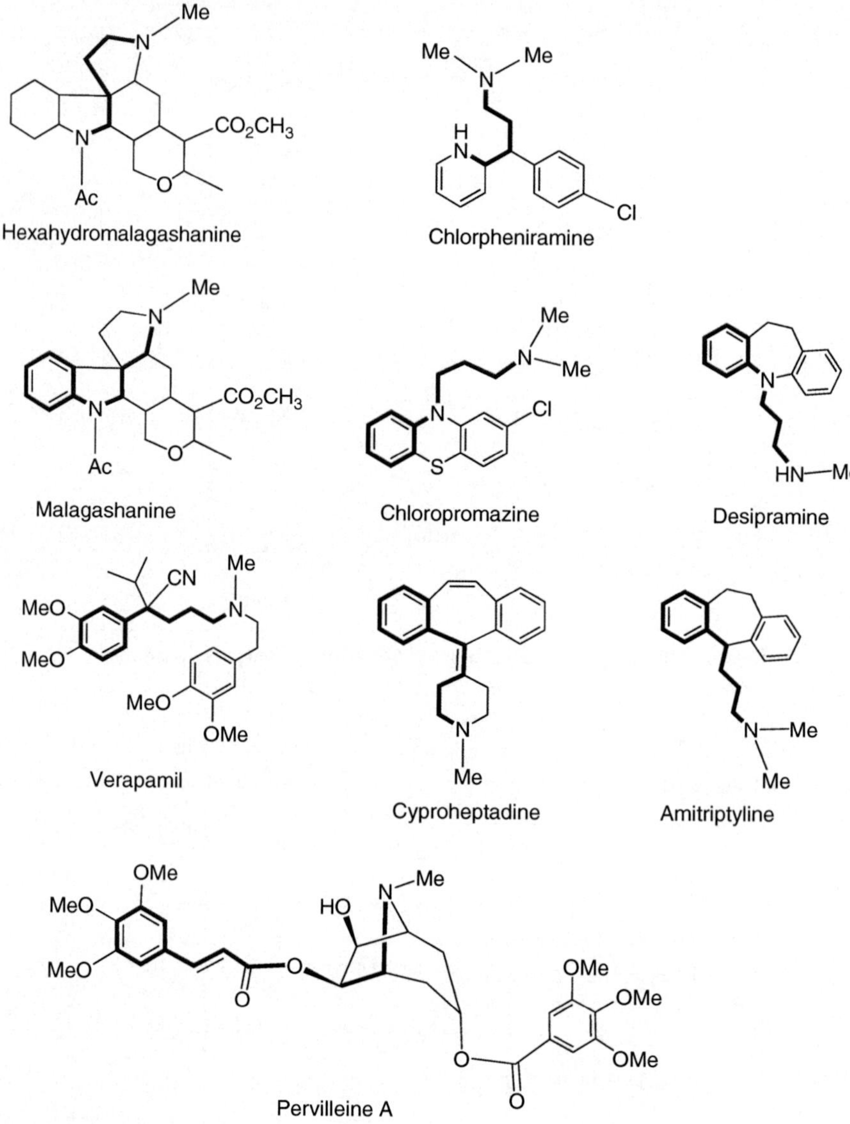

Fig. 11.7 Putative chemosensitizing pharmacophores (highlighted in *bold*) in naturally occurring and synthetic antimalarial reversers

The 1,4-diamino structure is reminiscent of that of polyamines, namely, putrescine, spermidine and spermine, which play a central role in cellular growth, differentiation and neoplastic transformation. Polyamine-based structures may thus offer a wide range of structural possibilities, for example, by replacing some of the nitrogen atoms by a phenyl group, for designing new chemosensitizers with useful clinical relevance in the treatment of malaria and cancer, rescuing previously highly successful treatment regimens for future use. Furthermore, chloroquine has a 1,4-diamino fragment in its structure. This begs the question of whether this is a coincidence in the context of our proposed pharmacophore and whether the data are relevant to understanding chloroquine resistance. It was reported that chloroquine derivatives with an appropriately modified side chain retained activity against chloroquine-resistant strains of *P. falciparum*, strongly suggesting that the diamino fragment may play an important role in the mechanism of chloroquine resistance.

We further exploited our hypothesis by synthesizing nearly 70 derivatives of di- and triamines within a training programme. One relevant result was the observation that chemosensitizing activity was inversely proportional to antiplasmodial effect [54]. Another noteworthy result (to which we have already referred) was the observation that herveline D, which had the strongest antiplasmodial activity, lacks chloroquine-enhancing effects while herveline C, which had the strongest chemosensitizing activity, possesses the lowest antiplasmodial activity. These results are also in line with the observation that chemosensitizing phenothiazines transformed them into antimalarial drugs by increasing the number of basic groups in the alkylamino side chain [55]. All these studies point to the role of the chemosensitizing pharmacophore hypothesis as defined above in antimalarial activity, chloroquine resistance and its reversal.

Following the reported chemosensitizing activities of voacamine in cancer [56], we evaluated its reversing effects in malaria. It was shown to lack chloroquine-enhancing activity, and its in vitro antiplasmodial effect was not potentiated by malagashanine [57].

11.4.4.4 Gallic Acid as an Antiplasmodial Enhancer

The number of natural products is limited, whereas millions of hybrids as combinations of parts of different natural products can be prepared. This new approach seems promising in the development of leads for medicinal applications, as the biological activity of several new hybrids exceeds that of the parent compounds [58].

Trimethoxy derivatives of gallic and coumaric acids play an essential role in the chemosensitizing activities in the pervilleine series. In our antimalarial screening programme detailed below, we found that an ethyl acetate extract of *Piptadenia pervillei* displayed strong antiplasmodial activities [59]. Bioassay-guided fractionation afforded two active compounds, namely, (+)-catechin-3-*O*-gallate and (+)-catechin-5-*O*-gallate, while the parent compound (+)-catechin had low antiplasmodial activities [60]. Independently of this work, we found that the introduction of gallic acid in a synthetic compound significantly enhanced its antiplasmodial effect [61]. Catechin-3-*O*-gallate was reported to inhibit the biosynthesis of type II fatty acid [62]. Careful examination of these structures led to the identification of a putative unit composed of two polyhydroxyphenol rings, joined by an ester linkage that might be required for their inhibitory activity against type II fatty acid biosynthesis (Fig. 11.8).

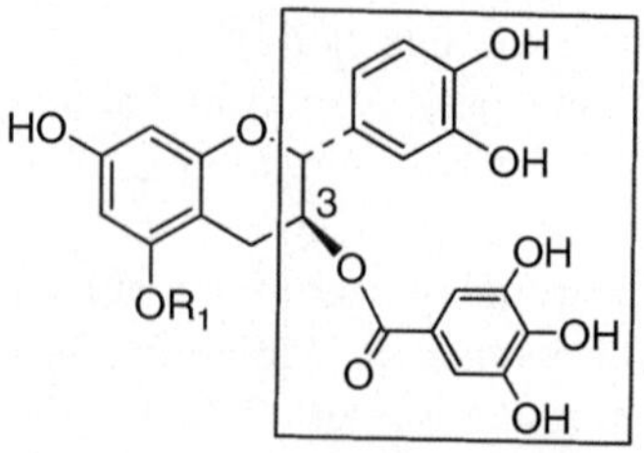

(+)-Catechin-3-O-gallate

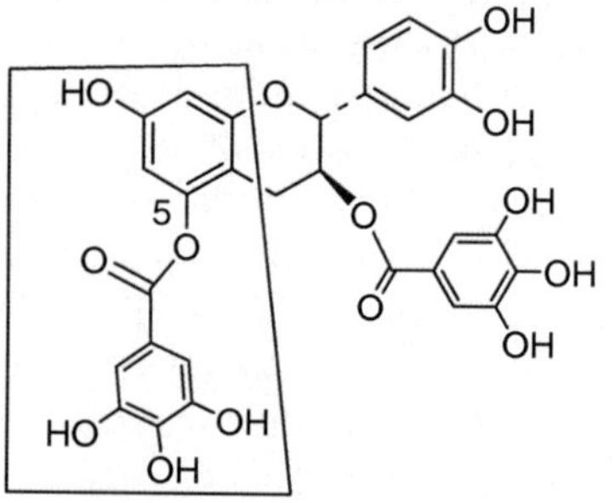

(+)-Catechin-5-O-gallate

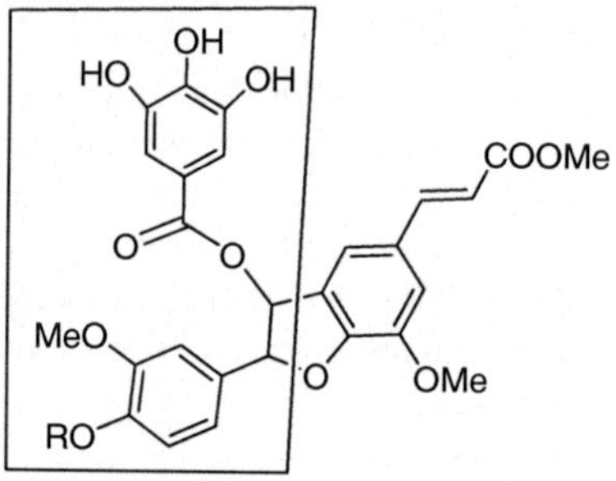

Synthetic compound showing antiplasmodial activity

Fig. 11.8 Putative chromophores (framed) that may inhibit the biosynthesis of fatty acid of type II

The fatty acid biosynthesis type II pathway has received considerable interest as a candidate therapeutic target in *Plasmodium falciparum* as shown by the various works on triclosan and derivatives. The putative unit might therefore give clues for the design and synthesis of inhibitors of the biosynthesis of type II fatty acid.

11.5 Lessons Learned from Milestones of Antimalarial Chemotherapy

11.5.1 Evolving Feature of Drug Discovery

In the historical point of view outlined in the Fig. 11.9, drug discovery has followed the evolution of science and technology, moving from a holistic empirical approach to a reductionist computational approach. Originally, it is most likely that ancestral people tested medicines on themselves and later on patients. With the development of pharmacology, a shift to animal laboratory models occurred. This was then followed by the use of isolated organs and later on the use of microorganisms. And with the identification of receptors, drug development focused on molecular targets to further speed up the screening of compounds at the nanomolar scale for pharmacological

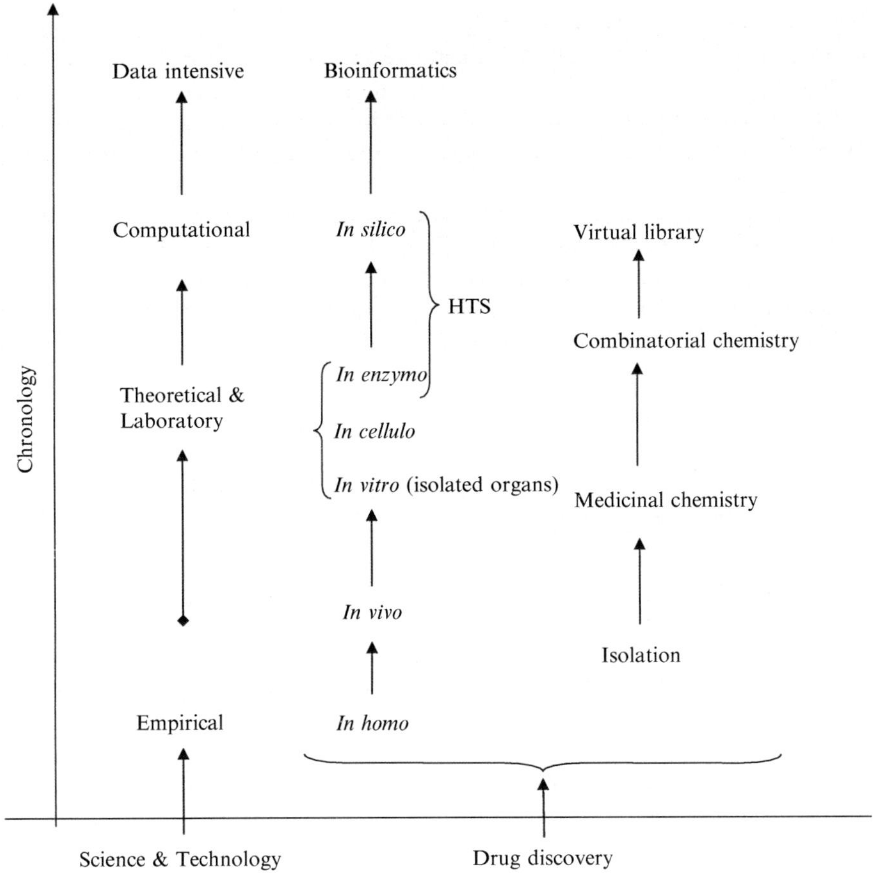

Fig. 11.9 Evolving feature of drug discovery

activity. This has resulted in the present-day high throughput screening (HTS) methods, which, through miniaturization and robotization, can screen one million samples in a short time. In parallel, the development of computer-based science has led to virtual screening. This evolving feature of drug discovery is outlined in Fig. 11.9.

11.5.2 The First Antimalarial Drug from Traditional Knowledge

The history of antimalarial drug discovery provides one excellent example of the evolving feature of drug discovery. One legend reported that a feverish Indian, desperate with thirst, came across a pool of stagnant water and drank from it. He noticed it was very bitter and worried that he had been poisoned by the quina-quina (bark of barks) trees surrounding the pool. As it turned out, his fever quickly disappeared. Since then, South American Indians have been using *cinchona* growing on the eastern slopes of the Andes Mountains from Venezuela to Bolivia to treat fevers for many centuries. The Jesuits learned of the "quina-quina bark" miracle drug in Peru and subsequently introduced *cinchona* bark into medical use in Europe around 1630. In 1631, it was used in Rome to cure the city which was a malarious region due to the marshes and swamps that surrounded it. In the 1680s, the *cinchona* bark was widely accepted as a malarial remedy. When Charles II was cured of malaria at the end of the seventeenth century, it became popular in London. In 1820, the key curative principle of the *cinchona* bark was isolated from the powdered bark by Caventou and Pelletier who named it "quinine".

11.5.3 The Discovery of New Powerful Antimalarials Stimulated by Wars

Medicinal chemistry on antimalarial agents was not developed until the twentieth century, stimulated by serious shortages of the supply of quinine during the two world wars. In fact, it was these wars that mainly stimulated the compelling search for new antimalarial agents for military needs. In the interwar period, German scientists conducted intensive research to develop synthetic alternatives to quinine for preventing and treating malaria. In 1933, their efforts were rewarded with the discovery of quinacrine (Atabrine®). When the United States and its allies lost access to natural sources of quinine at the beginning of World War II, quinacrine became, and remained throughout the war, the mainstay of Allied troops' malaria prevention and treatment.

Nevertheless, a huge programme of screening and clinical testing of potential antimalarials was begun in the USA in 1941 and independently in Great Britain. More than 16,000 candidate antimalarial compounds were tested by US scientists. This resulted in several drugs, the most useful of which was chloroquine. In fact chloroquine had already been discovered by German scientists in 1934, but was rejected because of toxicity concerns [63]. Meanwhile in the UK, scientists had developed a completely different antimalarial drug based on inhibition of folic acid synthesis, leading to the discovery of proguanil followed 6 years later by pyrimethamine. The combination of these two antimalarial drugs known under the name Fansidar® became widely used in both prophylaxis and treatment.

The next episode in antimalarial drug discovery was the Vietnam War, malaria being a prime problem for all armed forces. Military research institutions on both sides of the Pacific started to screen substances for their antimalarial properties. In the enormous US Army drug discovery programme, 237,056 synthetic compounds were screened, and the programme came up with two drugs: mefloquine and halofantrine, products of what has been called "the military-industrial complex" [64]. However, serious adverse effects have now limited their therapeutic uses [65]. On the other side of the globe, the Leadership of Vietnam asked for help from the Leadership of China who assured him that China's scientists could develop a new malaria cure. To this end, the Chinese Academy of Traditional Chinese Medicine was requested to undertake the programme. The Academy employed not only traditional Chinese practitioners but also research scientists. Following this programme, Chinese scientists rediscovered the remarkable antimalarial activity of qinghaosu (artemisinin) derived from the ubiquitous shrub *Artemisia annua* [66]. Artemisinin and its derivatives form the basis of ACT strategy.

11.5.4 In Vivo Tests First Used for Evaluating Antimalarial Activities

The in vitro antiplasmodial test was first available in 1979 [67]. Until the 1950s, screening of antimalarial drugs was carried out on avian malaria. This was less than satisfactory as the avian malaria species differ in a number of ways from those that infect humans. The discovery in the 1950s of wild rodent malaria parasites, and later on other rodent species that could infect laboratory rats, transformed the tests used for drug development. And it is in that context of in vivo tests that the useful antimalarial drugs used today, excepting the antifolate derivatives, were discovered. One point that deserves attention was the 60-day observation of treated animals in the US Army antimalarial drug development programme. Survivors at the end of this period were considered to be cured.

11.5.5 Antimalarials in Modern Drug Discovery

With the advent of the modern drug discovery approach based on random, in vitro, mechanism-based, high throughput screening in the initial phases of plant drug discovery, nearly two million compounds in GlaxoSmithKline's chemical library were screened for inhibition of *Plasmodium falciparum* using the LDH method [68]. Useful hits were discovered, and the results were made publicly available to encourage drug lead identification efforts and further research into this disease. Similarly, 309,474 chemicals were screened using a chemical genetic approach, and this screening programme led to the discovery of relevant hits [69].

11.5.6 Lessons Learned from Screening Madagascar Plant Extracts for Antiplasmodial Activity

In the 2000s, we started two screening programmes for antimalarial drug discovery from Madagascar plants using the in vitro isotopic microtest. Plants were selected on the basis of either their traditional uses as antimalarials or their endemicity. In the first programme, 50 extracts were screened [59]. In the second programme, one hundred and ninety plants belonging to 67 families and 144 genera were collected in five different ecosystems of Madagascar to encompass reasonable chemical structure diversity [70]. In these programmes, dehydroaporphines were shown to be the most active in vitro with $IC_{50} = 80$ nM. Extracts of *Piptadenia pervillei* were found to be the most active with $IC_{50} = 0.5$ µg/ml. As already mentioned, bioassay-guided fractionation led to the identification of catechin and its gallate derivatives as the active constituents [60]. In view of the overall paucity of the results, and taking into consideration the wealth of ethnomedical knowledge in Madagascar, we thought we needed to return to the historical methods of drug discovery [71]. These methods involve a holistic approach with patient and in vivo animal experiments in the centre, using mixtures and appropriate pharmacological methods to prove activity and detect possible synergy among plant extract constituents [72] and the occurrence of prodrugs. To this end, instead of the linear method of modern drug discovery, we need to explore an innovative approach using a flexible strategy, moving beyond the boundary of pharmacological standards. This method has a close similarity to what was termed "reverse pharmacology" [73]. Reverse pharmacology is defined as the combination of traditional knowledge and the application of modern technology and processes to provide valuable drugs. Its aim is to reverse the routine "laboratory-to-clinic" pathway to "clinics-to-laboratories".

11.6 Exploring Innovative Possibilities

Before exploring innovative possibilities, we need to make a critical analysis of previous investigations

11.6.1 Critical Analysis of Ethnobotany-Based Drug Discovery in Malaria

There are two opposite Greek philosophies dealing with two distinct approaches: the first philosophy related to a single drug is that "Every healing has its quintessence"; the second related to multi-component drugs, "Nature does nothing without purpose or uselessly". The former approach forms the basis of the "single target-single compound" paradigm of drug discovery in pharmaceutical companies. In the latter approach, it is assumed that complex, multifactorial diseases may require multi-component, multifunctional therapies, and that complex molecular interactions produce effects that may not be achieved by a single component.

Research on antimalarial drugs has been mainly focused on killing the parasites but rarely considers other mechanisms. These studies are often limited to the evaluation of crude plant extracts against malaria parasites in vitro. In some cases, active compounds have been isolated, and in relatively few studies, compounds or extracts have been assessed in vivo for their activities against mice infected with malaria parasites. In vitro activities do not always predict in vivo effects, and the lack of in vitro activities does not preclude in vivo effects [74]. Furthermore, a common assumption in the current search for active compounds is that those with higher in vitro potency at their target, hopefully at the nanomolar level, have greater potential to translate into successful, low-dose therapeutic applications. This great reliance on simplistic in vitro assays does not take into account absorption, distribution, metabolism, excretion and toxicity (ADMET) characteristics [75]. It took some researchers in the industry many years to realize this mistake and begin focusing more on network pharmacology, considering multi-target strategies over single-target approaches [76].

One characteristic of pure natural products is that many of them do not comply with Lipinski's rule of five. During evolution and natural selection, humans have developed effluxing mechanisms for detoxification to prevent pure naturally occurring compounds being absorbed [77]. This can be exemplified by the poor bioavailability of artemisinin, quinine and curcumin.

Furthermore, many antimalarial herbal remedies may exert their anti-infective effects not only by directly affecting the parasite but also by stimulating natural and adaptive defence mechanisms of the host. The immune system is the first line of host defence, and it is always associated with a complex inflammatory process and the production of reactive oxygen species. But their activity can also induce severe immuno-pathology which is responsible for the disease symptoms and harmful effects, especially in cerebral malaria. However, the antimalarial effects, the immune system activities, the inflammatory activities and the antioxidant effects of plant extracts have been studied in isolation from each other, not in a holistic approach. We think that the complexity of the malaria disease and the holistic traditional approach also requires a holistic way of studying this tropical disease.

Regarding clinical trials of antimalarial plants, successful results were obtained with crude extracts as infusions or decoctions, but it should be noted that parasite recrudescence was experienced by a high proportion of patients, and it is this recrudescence which likely causes drug resistance. A significant challenge is therefore how to avoid recrudescence in humans.

11.6.2 Host-Parasite Co-evolution

All living organisms evolve from birth to extinction, and their continuity is sustained by the reproduction system. Any factors that interrupt this process will unavoidably lead to adaptive survival mechanisms. Furthermore, living organisms are not separate from the rest of nature. They live in an evolving equilibrium with all components of the ecosystem in which they stay, and they have the ability to adapt to any change in the ecosystem. Medicinal plants are not therefore an immutable reservoir of natural products. They are part of complex and evolving ecosystems characterized by the disappearance/appearance phenomenon also called natural selection.

Symbiosis refers to relationships between organisms that show an intimate relation with each other. Three types of symbiosis have been recognized depending on the nature of the relationships: parasitism, commensalism and mutualism. Over half of all living species of plants and animals are parasitic, which by definition involves intimate association with unfavourable impact on hosts. In human malaria, the *Plasmodium* parasites derive nourishment from the host, and this is detrimental to the host. To counteract this, the host has developed defence mechanisms to prevent the parasite from causing any harm if it has entered the body. But the parasite has also developed survival mechanisms. Therefore, host and *Plasmodium* are engaged in antagonistic co-evolution, where evolution in the host defences leads to counter adaptations in the parasite, which selects for improved resistance in the host, and so on, leading to endless non-equilibrium evolutionary dynamics. But neither have malaria parasites succeeded in eliminating humans nor have humans succeeded in eradicating malaria parasites.

The first line of host defence against an intrusion into the body is the immune system which involves mechanisms that directly attack the pathogen to block invasion or eliminate the invading microorganism by producing effectors such as cytokines. This is known as a resistance mechanism. However, excess immune response leads to immune-mediated diseases called immuno-pathology. This is reminiscent of the Yin and Yang Chinese philosophy which is based on complementary opposites that are interconnected and interdependent in the natural world as part of a dynamic system and may ebb and flow over time [78]. Consequently, if one effect reaches its peak, it will naturally begin to transform into the opposite effect. To illustrate this fundamental principle of life, IFN-γ and TNF-α responses are associated with protective immunity to *P. falciparum*, but severe *P. falciparum* malaria is accompanied by high levels of circulating cytokines, including IFN-γ and TNF-α [79]. Tolerance is defined as all of the mechanisms that regulate the self-harm that can be caused by the immune response. The phenomenon of natural antimalarial immunity known as premunition, in which infected individuals can withstand the presence of parasites in their blood at levels that would elicit sickness in unprotected individuals, has been described in Madagascar [80] and in Tanzania [81]. This may be explained in terms of the body tolerating the *Plasmodium* spp. rather than inducing a large immune response that would cause pathology. An important body of literature has been generated regarding this subject [82]. The sum of resistance and tolerance defines a host's defensive capacity. While most emphasis have been directed to eliminating the invading malaria parasites or blocking its invasion, very little has been done on the tolerance mechanisms. We suggest these may be the most useful in terms of medical intervention because they are less likely to affect resistance to the malaria parasites.

11.6.3 Would Traditional Medicine Give Answers to Malaria Tolerance and Protection?

11.6.3.1 Going Beyond the Standard Antimalarial Tests

In our continuing investigation of medicinal plants traditionally used to treat malaria, we learned during fieldwork that infusion of the scratched barks of a medicinal plant is claimed to cure malaria if the treatment is taken at an early stage of the disease. Standard in vitro and in vivo antimalarial tests carried out on the lyophilized aqueous

Table 11.1 Considerations towards cross-disciplinary thinking with malaria infections

Parasite side	Host side
Hydrolysis of haemoglobin makes amino acids available for parasite development, but this process also releases haeme, which is extremely toxic to the parasite. Parasite transforms haeme into an "*inert crystal" named malarial pigment or haemozoin by forming unique dimers that then crystalize* [83]	But haemozoin is not "inert" in the area of immune system. It has profound suppressing effects on dendritic cell function, central player in the generation of immune responses. Large excess of haemozoin enhances immuno-pathology or facilitates the parasite's survival by depressing beneficial immune system [84]
Chloroquine primarily kills the parasites by binding to free haeme, which prevents its detoxification into haemozoin [83]	Natural malaria infection starts in the skin, but paradoxically, most experimental infections bypass the skin entirely. Chloroquine displays also other mechanisms of action, including immuno-modulatory and anti-inflammatory effects. Co-administered with some vaccines in skin immunization, it confers protection in some way [85 and references cited in]
Drug *resistance* is defined as the ability of parasite to counteract drug predators	*Resistance* is defined as the ability of the host to block invasion or eliminate the invading parasite
Parasite *immuno-tolerance* (unresponsiveness of the adaptive immune system to parasite, ensuring infection and tolerance to reinfection) [86]	*Host tolerance* (ability of the host to regulate the self-harm that can be caused by the immune response) [87]
Delicate balance between pro-inflammatory and anti-inflammatory responses is a major determinant of the clinical outcome of a malaria infection	

extract failed to demonstrate antimalarial effects activities in the malaria blood stage. We went beyond the protocol of the standard in vivo 4-day suppressive tests and fortuitously found, at day 8, a drastic drop in parasitaemia was observed in treated mice, together with a proliferation of lymphocytes, whereas the control mice died. The results were reproducible. It was a starting point for a new cross-disciplinary thinking in malaria outlined in Table 11.1.

Overall, little attempt has been made to "think on both sides (parasite and host) of the coin". The balance between two complementary opposites (pro-inflammatory versus anti-inflammatory, protective immunity versus immune pathology) in the natural world as part of a dynamic system, and which may ebb and flow over time should be drawn into our approaches to malarial treatments.

11.6.3.2 New Approach by Combining Two Elements

Malaria drug combination has been based on a "curative + curative" approach. Based on our cross-disciplinary thinking, we use the "curative + protective" approach. Curcumin may be a good candidate [88], but the concentration of several biological activities in one compound, such as immuno-stimulating, anti-inflammatory and antioxidant, together with its poor bioavailability, is not ideal for the approach.

11.6.3.3 Shifting the Paradigm of Drug Efficacy

In drug discovery strategies, a common underlying assumption is that compounds with higher in vitro potency and selectivity at their target(s) have greater potential to translate into successful, low-dose therapeutics. In a multi-component phytomedicine, several moderate biological activities may be more useful than one single strong activity to ensure adequate balance, as with a thermostat, on complementary opposite parameters that are interconnected and interdependent in complex diseases such as malaria. Malaria is a complex, multifactorial disease which probably requires multi-component, multifunctional therapies, and the complex molecular interactions produce effects that may not be achieved by single component. To exploit fully the potential of traditional medicine, we need to consider its holistic approach and the necessity of having a multidisciplinary team.

11.7 Concluding Remarks

A potentially effective strategy to achieve complete malaria protection is to modulate the immune response with drugs that neutralize suppressive functions and potentiate protective responses [89]. Our ongoing research on a Madagascar antimalarial plant using adequate drug/extract combination has given promising results in malaria treatment, tolerance and protection, as well as passive transfer immunity in mouse models. Our results to date pave the way for new studies and may open the door to a new avenue of investigation into malaria immunity.

African traditional medicine is an untapped reservoir of remedies for the treatment of tropical disease in Africa. Investigations should however be taken within the context of the African culture, taking into consideration the holistic feature of this medicine. Where there are no modern drugs, there could be a safe and effective local treatment.

Acknowledgements We thank warmly the CNRS (France), the IFS/PRISM, TWAS, and CYROI (La Réunion) for various supports.

References

1. Lahana R (2003) Who wants to be irrational? Drug Discov Today 8:655–656
2. Newman DJ, Cragg GM (2007) Natural products as sources of new drugs over the last 25 years. J Nat Prod 70:461–477
3. Kasilo OMJ, Trapsida JM, Mwikisa CN et al (2010) An overview of the traditional medicine situation in the African Region. WHO Afr Health Mon 14:7–15
4. Willcox ML, Bodeker G (2004) Traditional herbal medicines for malaria. BMJ 329:1156–1159
5. Anonymous (1978) Zava-maniry mahasitrana nosoratan'ny Ombiasa iray tamin'ny taona 1880. Alimanaka Dina 1:12–17

6. De Flacourt E (1642) Histoire de la Grande Isle de Madagascar. In: Grandidier A, Froidevaux H, Grandidier G (eds) Collection des Ouvrages Anciens Concernant Madagascar. Union Coloniale, Paris
7. Rasamimanana J (1891) Contribution à l'étude de l'action physiologique de la tanghinine cristallisée extraite de *Tanghinia venenifera*. Thèse de Doctorat Médecine, Lyon
8. Ramisiray G (1901) Croyances et pratiques médicales des Malgaches. Thèse de Doctorat Médecine, Paris
9. Ranaivo C (1902) Pratiques et croyances médicales des Malgaches relatives aux accouchements et à la médecine infantile. Thèse de Doctorat Médecine, Paris
10. Rasoanaivo P, Petitjean A, Ratsimamanga-Urverg S et al (1992) Medicinal plants used to treat malaria in Madagascar. J Ethnopharmacol 37:117–127
11. Havet EN (1827) Dissertation sur une maladie qui règne à l'Ile de Madagascar: conseils hygiéniques à suivre pour l'éviter. Thèse de Doctorat Médecine, Paris
12. Panou de Faymoreau A (1860) Nossi-Bé: fièvres intermittentes. Thèse de Doctorat Médecine, Paris
13. Sibree J (1889) The ramanenjana or dancing mania of Madagascar. The Antananarivo Annual XIII: 19–27
14. Andrianjafy (1902) Le ramanenjana à Madagascar : choréomanie d'origine palustre. Thèse de Doctorat Médecine, Montpellier
15. Beauprez (1901) Informations: chronique du mois de juin (le froid et la mortalité des indigènes). Rev Mad 8:593–599
16. Pearse J (1897) A modern epidemic in the Betsileo province: the '*safo-tany*' or '*rapo-rapo*'. The Antananarivo Annual XXI: 32–33
17. Fontoynont M (1903) Grippe et paludisme à Madagascar. La Presse Médicale 72:637–638
18. Fontoynont M (1905) La nouvelle épidémie palustre de Tananarive. Bul Acad Malg 1:334–335
19. Fontoynont M (1905) A propos des épidémies palustres actuelles sur les Hauts-Plateaux de Madagascar. Rev Med Hyg Trop 1:64–68
20. Rabarijaona LP, Rabe T, Ranaivo LH et al (2006) Paludisme sur les Hautes Terres Centrales de Madagascar: stratégies de lutte. Med Trop 66:504–512
21. Champetier de Ribes G, Ranaivoson G, Rakotoherisoa E et al (1994) Une épidémie de paludisme dans le Sud de Madagascar. Arch Inst Pasteur Madagascar 61:66–69
22. Kaur K, Jain M, Kaur T et al (2009) Antimalarials from nature. Bioorg Med Chem 17:3229–3256
23. Oliveira AB, Dolabela MF, Braga FC et al (2009) Plant-derived antimalarial agents: new leads and efficient phytomedicines. Part I. Alkaloids. An Acad Bras Cience 81:715–740
24. Batista R, Júnior AJS, Oliveira AB (2009) Plant-derived antimalarial agents: new leads and efficient phytomedicines. Part II. Non-alkaloidal natural products. Molecules 14:3037–3072
25. Wright CW (2010) Recent developments in research on terrestrial plants used for the treatment of malaria. Nat Prod Rep 27:961–968
26. Willcox M, Bodeker G, Rasoanaivo P (2004) Traditional antimalarial plants and malaria. CRC, Boca Raton, FL
27. Ratsimamanga-Urverg S, Rasoanaivo P, Rakoto-Ratsimamanga A et al (1991) Antimalarial activity and cytotoxicity of *Ficus pyrifolia* and *Rhus (=Baronia) taratana* leaf extracts. Phytother Res 5:32–34
28. Ratsimamanga-Urverg S, Rasoanaivo P, Rakoto-Ratsimamanga A et al (1991) Antimalarial activity and cytotoxicity of *Evodia fatraina* stem bark extracts. J Ethnopharmacol 33:231–236
29. Ratsimamanga-Urverg S, Rasoanaivo P, Le Bras J et al (1991) *In vitro* antimalarial activity and cytotoxicity of *Strychnos mostueoides* (Loganiaceae), *Avicennia marina* (Avicenniaceae) and *Urophyllum lyallii* (Rubiaceae). Discov Innovat 3:81–83
30. Martin SK, Oduola AMJ, Milhous WK (1987) Reversal of chloroquine resistance in *Plasmodium falciparum* by verapamil. Science 235:899–901
31. Rasoanaivo P, Ratsimamanga-Urverg S, Frappier F (1996) Reversing agents in the treatment of drug-resistant malaria. Curr Med Chem 3:1–10

32. Ratsimamanga-Urverg S, Rasoanaivo P, Milijaona R et al (1994) *In vitro* antimalarial activity, chloroquine potentiating effect and cytotoxicity of alkaloids of *Hernandia voyroni*. Phytother Res 8:18–31
33. Rasoanaivo P, Ratsimamanga-Urverg S, Galeffi C et al (1995) A new group of isoquinoline dimers from *Hernandia voyroni*. Tetrahedron 51:1221–1228
34. Rasoanaivo P, Ratsimamanga-Urverg S, Rafatro H et al (1998) Alkaloids of *Hernandia voyroni*: chloroquine-potentiating activity and structure elucidation of herveline D. Planta Med 64:58–62
35. Rasoanaivo P, Galeffi C, De Vicente Y et al (1991) Malagashanine and malagashine, two alkaloids of *Strychnos mostueoides*. Rev Latinoam Quim 22:32–34
36. Rasoanaivo P (1999) Obituary: the Malagasy forester Abrahama. J Ethnopharmacol 68:1–2
37. Rasoanaivo P, Ratsimamanga-Urverg S, Milijaona R et al (1994) *In vitro* and *in vivo* chloroquine potentiating action of *Strychnos myrtoides* alkaloids against chloroquine-resistant strain of *Plasmodium* malaria. Planta Med 60:13–16
38. Caira MR, Rasoanaivo P (1995) Structure of malagashanine, a new alkaloid with chloroquine-potentiating action. J Chem Crystallogr 25:725–729
39. Martin MT, Rasoanaivo P, Palazzino G et al (1999) Minor Nb-(C21) secocuran alkaloids of *Strychnos myrtoides*. Phytochemistry 51:479–486
40. Rasoanaivo P, Palazzino G, Galeffi C et al (2001) The co-occurrence of C(3) epimer Nb,C(21)-secocuran alkaloids in *Strychnos diplotricha* and *S. Myrtoides*. Phytochemistry 56:863–867
41. Warsame M, Wernsdorfer WH, Björkman A (1992) Lack of effect of desipramine on the response to chloroquine of patients with chloroquine-resistant falciparum malaria. Trans R Soc Trop Med Hyg 86:235–236
42. Ramialiharisoa A, Ranaivoravo J, Ratsimamanga-Urverg S et al (1994) Evaluation en clinique humaine de l'action potentialisatrice d'une infusion de *Strychnos myrtoides* vis-à-vis d'antipaludéens. Rev Med Pharm Afr 8:123–131
43. Carraz M, Jossang A, Franetich JF et al (2006) A plant-derived morphinan as a novel lead compound active against malaria liver stages. PLoS Med 3:2392–2402
44. Carraz M, Jossang A, Rasoanaivo P et al (2008) Isolation and anti-malarial activity of new morphinan alkaloids on *Plasmodium yoelii* liver stage. Bioorg Med Chem 16:6186–6192
45. Rafatro H, Ramanitrahasimbola D, Rasoanaivo P et al (2000) Reversal activity of the naturally-occurring chemosensitizer malagashanine in *Plasmodium* malaria. Biochem Pharmacol 59:1053–1061
46. Ramanitrahasimbola D, Rasoanaivo P, Ratsimamanga S et al (2006) Malagashanine potentiates chloroquine anti-malarial activity in drug resistant *Plasmodium* malaria by modifying both its efflux and influx. Mol Biochem Parasitol 146:58–67
47. Mangoyi R, Hayeshi R, Ngadjui B et al (2010) Glutathione transferase from *Plasmodium falciparum* – interaction with malagashanine and selected plant natural products. J Enzyme Inhib Med Chem 25:854–862
48. Trigalo F, Martin MT, Blond A et al (1999) New indolines derivatives from strychnobrasiline, modulators of chloroquine resistance in *Plasmodium falciparum*. Tetrahedron 55:6139–6146
49. Trigalo F, Martin MT, Rasolondratovo B et al (2002) Oxidation of indolines to nitrones and new rearrangement in seco-curane type indoline alkaloids. Tetrahedron 58:4555–4558
50. Trigalo F, Joyeau R, Pham VC et al (2004) Synthesis of modulators of chloroquine resistance in *Plasmodium falciparum*: analogues of malagashanine from strychnobrasiline. Tetrahedron 60:5471–5474
51. Silva GL, Cui B, Chavez D et al (2001) Modulation of the multidrug-resistance phenotypes by new tropane alkaloid aromatic esters from *Erythroxylum pervillei*. J Nat Prod 64:1514–1520
52. Rasoanaivo P, Ramanitrahasimbola D, Rakotonandrasana OL et al (2005) Traditional medicine and resistance modulators. Ethnopharmacologia 35:33–38
53. Chouteau F, Ramanitrahasimbola D, Rasoanaivo P et al (2005) Exploiting a basic chemosensitizing pharmacophore hypothesis. Part 1: design, synthesis and biological evaluation of novel

bicyclic chemosensitizers against drug-resistant malaria parasites. Bioorg Med Chem Lett 15:3024–3028
54. Rakotonandrasana OL (2005) Synthèse de molécules à visée antipaludique mimant la structure des produits naturels. Diplôme d'Etudes Approfondies, Ecole Supérieure Polytechnique, Université d'Antananarivo
55. Kalkanidis M, Klonis N, Tilley L et al (2002) Novel phenothiazine antimalarials: synthesis, antimalarial activity, and inhibition of the formation of beta-haematin. Biochem Pharmacol 63:833–842
56. You M, Ma X, Mukherjee R et al (1994) Indole alkaloids from *Peschiera laeta* that enhance vinblastine-mediated cytotoxicity with multidrug-resistant cells. J Nat Prod 57:1517–1522
57. Ramanitrahasimbola D, Rasoanaivo P, Ratsimamanga-Urverg S et al (2001) Biological activities of the plant-derived bisindole voacamine with reference to malaria. Phytother Res 15:30–33
58. Tietze LF, Bell HP, Chandrasekhar S (2003) Natural product hybrids as new leads for drug discovery. Angew Chem Int Ed Engl 42:3996–4028
59. Rasoanaivo P, Ramanitrahasimbola D, Rafatro H et al (2004) Screening plant extracts of Madagascar for the search of antiplasmodial compounds. Phytother Res 18:742–747
60. Ramanandraibe V, Grellier P, Martin MT et al (2008) Antiplasmodial phenolic compounds from *Piptadenia pervillei*. Planta Med 74:417–421
61. Rakotondramanana DLA, Delomenède M, Baltas M et al (2007) Synthesis of ferulic ester dimers, functionalisation and biological evaluation as potential antiatherogenic and antiplasmodial agents. Bioorg Med Chem 15:6018–6026
62. Zhang YM, Rock CO (2004) Evaluation of epigallocatechin gallate and related polyphenols as inhibitors of the FABG and FABI reductases of bacterial type II fatty acid synthesis. J Biol Chem 279:30994–31001
63. Jensen M, Mehlhorn H (2009) Seventy-five years of Resochin® in the fight against malaria. Parasitol Res 105:609–627
64. Kinnamon KE, Rothe WE (1975) Biological screening in the US Army antimalarial drug development program. Am J Trop Med Hyg 24:174–178
65. Croft AM (2007) A lesson learnt: the rise and fall of Lariam and Halfan. J R Soc Med 100:170–174
66. Liao F (2009) Discovery of Artemisinin (Qinghaosu). Molecules 14:5362–5366
67. Desjardins RE, Canfield CJ, Haynes JD et al (1979) Quantitative assessment of antimalarial activity *in vitro* by a semiautomated microdilution technique. Antimicrob Agents Chemother 16:710–718
68. Gamo FJ, Sanz LM, Vidal J et al (2010) Thousands of chemical starting points for antimalarial lead identification. Nature 465:305–310
69. Guiguemde WA, Shelat AA, Bouck D et al (2010) Chemical genetics of *Plasmodium falciparum*. Nature 465:311–315
70. Rasoanaivo P, Ratsimamanga-Urverg S, Ramanitrahasimbola D et al (1999) Criblage d'extraits de plantes de Madagascar pour recherche d'activite´ antipaludique et d'effet potentialisateur de la chloroquine. J Ethnopharmacol 64:117–126
71. Bourdy G, Willcox ML, Ginsburg H et al (2008) Ethnopharmacology and malaria: new hypothetical leads or old efficient antimalarials? Int J Parasitol 38:33–41
72. Rasoanaivo P, Wright CW, Willcox ML et al. (2011) Whole plant extracts versus single compounds for the treatment of malaria: synergy and positive interactions. Malar J. doi:10.1186/1475-2875-10-S1-S4
73. Patwardhan B, Mashelkar RA (2009) Traditional medicine-inspired approaches to drug discovery: can Ayurveda show the way forward? Drug Discov Today 14:804–811
74. Houghton PJ, Howes MJ, Lee CC et al (2007) Uses and abuses of *in vitro* tests in ethnopharmacology: visualizing an elephant. J Ethnopharmacol 110:391–400
75. Gleeson MP, Hersey A, Montanari D et al (2011) Probing the links between *in vitro* potency, ADMET and physicochemical parameters. Nat Rev Drug Discov 10:197–208

76. Hopkins AL (2008) Network pharmacology: the next paradigm in drug discovery. Nat Chem Biol 4:682–690
77. Owens J (2003) Chris Lipinski discusses life and chemistry after the Rule of Five. Drug Discov Today 8:12–16
78. Cheng JT (2000) Review: drug therapy in Chinese traditional medicine. J Clin Pharmacol 40:445–450
79. Omer FM, Kurtzhals JAL, Riley EM (2000) Maintaining the immunological balance in parasitic infections: a role for TGF-β? Parasitol Today 16:18–23
80. Rasamoel P, Jambou R, Ralamboranto L et al (1998) Portage asymptomatique et accès palustre: un équilibre complexe. Arch Inst Pasteur Madagascar 64:45–47
81. Vounatsou P, Smith T, Kitua AY et al (2000) Apparent tolerance of *Plasmodium falciparum* in infants in a highly endemic area. Parasitology 120:1–9
82. Langhorne J, Ndungu FM, Sponaas AM et al (2008) Immunity to malaria: more questions than answers. Nat Immunol 9:725–732
83. Sullivan DJ (2002) Theories on malarial pigment formation and quinoline action. Int J Parasitol 32:1645–1653
84. Urban BC, Todryk S (2006) Malaria pigment paralyzes dendritic cells. J Biol. doi:10.1186/jbiol37
85. Guilbride DL, Gawlinski P, Guilbride PDL (2010) Why functional pre-erythrocytic and bloodstage malaria vaccines fail: a meta-analysis of fully protective immunizations and novel immunological model. PLoS ONE 5(5):e10685. http://www.plosone.org/article/info%3Adoi%2F10.1371%2Fjournal.pone.0010685. Accessed 19 May 2010
86. Sakaguchi S, Yamaguchi T, Nomura T et al (2008) Regulatory T cells and immune tolerance. Cell 133:775–787
87. Schneider DS, Ayres JS (2008) Two ways to survive infection: what resistance and tolerance can teach us about treating infectious diseases? Nat Rev Immunol 8:889–895
88. Mimche PN, Taramelli D, Vivas L (2011) The plant-based immunomodulator curcumin as a potential candidate for the development of an adjunctive therapy for cerebral malaria. Malar J. doi:10.1186/1475-2875-10-S1-S10
89. Sauerwein RW, Bijker EM, Richie TL (2010) Empowering malaria vaccination by drug administration. Curr Opin Immunol 22:367–373

Chapter 12
Anticancer Drug Repositioning Against Tropical Diseases: The Example of Methotrexate in the Treatment of Malaria

Alexis Nzila and Kelly Chibale

12.1 Introduction

Chemotherapy remains one of the most important tools for the management and control of malaria. However, the rapid selection of malaria parasite resistant to antimalarial drugs can hamper this strategy. To slow the pace of selection of resistance, the World Health Organization has recommended the use of artemisinin combination therapies (ACTs) as first-line treatments of uncomplicated malaria [1]. In Africa, the combinations of lumefantrine/artemether (Coartem®) and amodiaquine/artesunate are currently the two main ACTs used to treat uncomplicated malaria [1–4]. However, recent reports indicate that resistance to artemisinin is emerging in South East Asia, and thus there is concern that this resistance could spread to Africa [5], a scenario which will not only render the current ACTs ineffective, but will also compromise the efficacy of pyronaridine/artesunate and piperaquine/dihydroartemisinin, the two next ACTs that are now in phase III/IV clinical development [6]. Intravenous artesunate has recently become the drug of choice in the treatment of severe malaria [7]. Unfortunately, the spread of artemisinin resistance will also affect the management of severe malaria. Thus, to counterbalance this burgeoning drug resistance problem, new drugs are urgently needed.

One of the strategies to discover new drugs is to reposition, repurpose or find new uses for drugs that have already been used for other indications. This approach, which has the advantage of reducing the cost and shortening the time of drug development, has become an important area of research by the pharmaceutical industry [8, 9]. For instance, in 2004, almost 40% of drugs registered by the Food and Drug Administration (FDA) found new uses in the treatment of various conditions in humans [8]. It is interesting to note, among the few available

A. Nzila (✉) • K. Chibale
Departments of Chemistry and Clinical Pharmacology, and Institute of Infectious Disease and Molecular Medicine, University of Cape Town, Rondebosch 7701, Cape Town, South Africa
e-mail: alexisnzila@yahoo.co.uk; Kelly.Chibale@uct.ac.za

K. Chibale et al. (eds.), *Drug Discovery in Africa*,
DOI 10.1007/978-3-642-28175-4_12,

antimalarials, that some of them have been repositioned. For instance, quinine is used to treat muscle cramps [10], and chloroquine (CQ) (preferably hydroxylchloroquine) is used for the management of rheumatoid arthritis and systemic lupus erythematosus [11]. Artemisinin is being investigated for the treatment of schistosomiasis and cancer [12–14].

12.2 In Vitro Antimalarial Activities of MTX

Several reports indicate that MTX is potent against *Plasmodium falciparum*, including those resistant to the antifolate pyrimethamine (PM), which carry the Ileu-164-Leu *dhfr* codon (dihydrofolate reductase), the mutation associated with higher levels of antifolate resistance, with MTX $IC_{50} < 80$ nM (inhibitory concentration that kills 50% of parasitaemia) [15–18], and $IC_{90/99}$ (inhibitory concentration that kills 90–99% of parasitaemia) falling between 150 and 350 nM. Thus, if such concentration can be achieved in vivo with an acceptable toxicity profile, MTX could potentially become a useful antimalarial drug [18]. However, anticancers are perceived to be toxic, thus not suitable for the malaria treatment. Yet the literature is replete with examples of new uses of anticancers in the treatment of non-neoplastic diseases.

12.3 Anticancer May Be Safe at Low Dose

Most anticancers are used at high dose to block tumour cells, leading to the inhibition of the growth of normal cells. Affected most specifically are cells that multiply actively such as bone marrow cells, e.g. leukocytes; cells of the gastrointestinal mucosa; and hair follicle cells, explaining why bone marrow suppression, mucositis and alopecia (hair loss) are among the most salient side effects of anticancers. However, according to Paracelsus' Law, it is well known that for any drug (including anticancers), there is always a dose range at which the drug is safe.

Paracelsus' Law states "*Sola dosis facit venenum* (only dose makes the poison)," meaning that all substances are poisons, and there are none which are not. The right dose differentiates a poison from a remedy; this principle is also known as the "dose–response effect" [19, 20]. For any drug, there is always a dose range (concentration) that is without any effect, one with a pharmacological effect but with minimal toxicity (or acceptable safety profile) and another with pharmacological and toxic effects. Thus, a molecule becomes a drug if the dose required to treat a complication falls in the second range: pharmacologically active with minimal toxicity. Most drugs used in the treatment of human diseases fall in this group [21].

In this regard, the example of the antimalarial drug chloroquine (CQ) is noteworthy. CQ is used at 10 mg base/kg on days 1 and 2, and 5 mg/kg on day 3 [22]. At this dose, CQ has an acceptable safety profile. However, a dose of 20 mg/kg is considered toxic [22], and fatal cases have been reported from doses as low as 30 mg/kg, only three times higher than the normal dose [23, 24]. This indicates that a slight dose increase shifts CQ's effect from the second range (acceptable safety profile with a pharmacological effect) to the third (life-threatening toxicity). Thus, CQ has a very low safety margin, and yet it has been used widely and is considered to be one of the safest antimalarials.

12.4 Safety of Low Dose of MTX in Human

MTX is one interesting example that vindicates Paracelsus' Law. MTX is used at high dose, up to 5,000–12,000 mg/m^2 per week (130–300 mg/kg) for several weeks for the treatment of cancer, and this dose can yield serum concentrations >1,000 μM, the concentration range that is associated with MTX life-threatening toxicity [25–27]. On the other hand, a 1,000-fold lower dose of MTX (LD-MTX) [0.1–0.35 mg/kg (7.5–25 mg per adult)] is used once weekly in the treatment of rheumatoid arthritis (RA), and sometimes, for many years. At this dose, MTX is safe and a mainstay in the treatment of RA in the Western world [28]. MTX is also the drug of choice for the treatment of juvenile idiopathic arthritis in children (including infants of less than 1 year old), a common rheumatic disease in the Western world [29].

Worldwide, it is estimated that 0.5–1 million adults and 50,000–100,000 children receive LD-MTX weekly for the treatment of rheumatoid and juvenile idiopathic arthritis, respectively. The drug is now being used in the African population as well, and its safety profile is similar to what has been reported in the Western world [30, 31]. LD-MTX is now considered to be one of the safest drugs used in the treatment of RA, and its safety profile has led to its new repositioning in the treatment of various conditions including multiple sclerosis [32], inflammatory bowel disease [33], urticaria [34], chronic cholestatic disorder [35], Wegener's granulomatosis [36], primary biliary cirrhosis [36] and systemic lupus erythematosus [37], among others.

12.5 Proof of Concept of MTX as an Antimalarial in Humans

The antimalarial potential of MTX has been established for almost 40 years now. Two relatively small clinical trials, involving seven patients, have demonstrated that doses as low as 2.5 mg/per day for 3–5 days are effective to treat malaria infection in humans (*Plasmodium falciparum* and/or *P. vivax*) [38, 39]. However, MTX has not come into widespread use because of concerns over toxicity [40, 41]. At the time of the aforementioned clinical trials (which were in the 1970s), no information was available on the safety of LD-MTX. Indeed, LD-MTX only started to be used for the treatment of arthritis from the 1980s, and before then, it was only used at high doses in cancer treatment, doses associated with toxicity. Thus, its use as an antimalarial was abandoned. At present, we have 30 years' experience on the safety of LD-MTX in adults and children; thus, this drug could now be reconsidered and evaluated as a potential antimalarial. Unlike its use in immune diseases, in malaria, MTX will not be used on a chronic basis. It will be a "one-off treatment"; as a result, the risk of toxicity should be even lower.

This information on the in vitro activity of MTX, the safety and possible in vivo efficacy of LD-MTX has led us to re-evaluate the potential of MTX as an antimalarial.

12.6 Re-evaluation of the Antimalarial Potential of MTX in Humans

A phase I evaluation of MTX was conducted to assess the safety and pharmacokinetic profile of this drug in healthy adult male Kenyan volunteers [42]. Twenty five adult volunteers were recruited, admitted and received a 5 mg dose of MTX/day/5 days. Pharmacokinetics (PK) blood sampling was carried out at 2, 4, 6, 12 and 24 h following each dose, and patients were followed up to 42 days. The most important known side effects associated with the chronic use of LD-MTX were solicited. These included nausea, vomiting and oral ulcers, among others. The result clearly indicated the absence of drug-associated side effect during the 42-day follow-up [42].

Based on in vitro data, MTX $>$ 350 nM in blood will be required to clear malaria infection in humans [18]. However, PK analysis of the 25 volunteers showed that the achieved maximum concentration (C_{max}) fell between 160 and 200 nM only, and after 6 h, effective concentration (C_{eff}) was $<$ 150 nM. Thus, at a dose of 5 mg/day/5 days, the achieved MTX blood levels would not be high enough to clear malaria infection. This implies that further studies should be carried out to evaluate the dose range that will yield adequate plasma concentrations and still be well tolerated. One of the approaches would be to reduce the days of treatment from 5 to 3 only and increase the dose to 7.5, 10 or 12.5 mg per day. The total dose for each treatment course should, however, remain around 22.5–37 mg, which is close to the range of doses administered weekly in the treatment of inflammatory diseases (7.5–35 mg per adult). Thus, further studies are warranted to explore the antimalarial potential of MTX.

12.7 Other Anticancers with Antimalarial Potency

Our group has also shown that the anticancer antifolates aminopterin, trimetrexate and pemetrexed are potent against *P. falciparum*, with activity in the nanomolar range (IC_{50} $<$ 50 nM) ([18] and unpublished work). Since all these anticancers are inhibitors of DHFR enzymes, we hypothesise that other anticancer inhibitors of DHFR, such as edatrexate, talotrexin, pralatrexate and piritrexim [43], would also be active against *P. falciparum*. In addition, several other anticancers but non-antifolates have proved potent against *P. falciparum*, and among these are the inhibitors of the microtubulin assembly tubulozole, vinblastine, docetaxel, paclitaxel and dolastatin [44–48]; the DNA cross-linking agent cisplatin [49]; and the proteasome inhibitor Bortezomib [50, 51]. If these drugs were active in vivo at low and safe doses, they could potentially become antimalarials.

12.8 Anticancer Drugs in the Treatment of Non-neoplastic Diseases

The use of MTX in the treatment of non-neoplastic diseases would not be unusual. Indeed, there are many anticancers that are used in the treatment of various complications other than cancer. For instance, cyclophosphamide is used in Behcet's syndrome, idiopathic pulmonary fibrosis, idiopathic thrombocytopenic purpura (ITP), nephritic syndrome, systemic lupus erythematosus (SLE), multiple sclerosis and Wegener's granulomatosis, among others [52, 53]; 6-mercapto-purine is used in Crohn's disease and ulcerative colitis [54]; thalidomide is becoming an important drug in Behcet's syndrome, Crohn's disease and SLE [52, 55]; and vincristine is used in ITP and thrombotic thrombocytopenic purpura [52, 56]. In relation to tropical diseases, the two anticancer agents miltefosine and eflornithine have been repositioned as drugs of choice in the treatment of leishmaniasis and trypanosomiasis, respectively [57, 58]. The common point in all these repositioned drugs is that they are used at doses lower that those used in the treatment of neoplasma; thus, they have better toxicity profiles. Therefore, the use of anticancers could be extended in the treatment of malarial diseases.

12.9 Conclusion

The pharmaceutical industry have now acknowledged the central importance of drug repositioning in the discovery of new active agents, even against diseases for which there are relatively sizeable human and financial resources such as cancer [59]. The rationale is that the large numbers of available drugs should permit, within a shorter time frame and a smaller financial cost, the discovery and development of new uses for drugs. This rationale is in fact more relevant with poverty related diseases such as malaria.

Among drugs that could be repositioned for malaria treatment are anticancer agents. For instance, many anticancers, including MTX, were shown to be potent against malaria parasite several decades ago; however, their perceived toxicity has prevented their development as antimalarials. Yet, as reported by Paracelsus more than 400 years ago, it is "the dose that makes a drug"; this principle has been exploited in the use of several anticancers (including MTX) at low doses in the treatment of various non-neoplastic diseases. This concept could be extended to anticancer bearing antimalarial activity, such as MTX, as discussed in this chapter. Thus, anticancer pharmacopoeia could provide a good platform for the discovery of new antimalarials.

Acknowledgement Financial support from the following sources is gratefully acknowledged: the South African National Research Foundation (NRF), the South African Research Chairs Initiative (SARChI) of the Department of Science and Technology (DST), the South African Medical Research Council (MRC) and the University of Cape Town (UCT).

References

1. Nosten F, White NJ (2007) Artemisinin-based combination treatment of falciparum malaria. Am J Trop Med Hyg 77:181–192
2. Adjei GO, Goka BQ, Binka F et al (2009) Artemether-lumefantrine: an oral antimalarial for uncomplicated malaria in children. Expert Rev Anti Infect Ther 7:669–681
3. Kokwaro G, Mwai L, Nzila A (2007) Artemether/lumefantrine in the treatment of uncomplicated falciparum malaria. Expert Opin Pharmacother 8:75–94
4. Sirima SB, Gansane A (2007) Artesunate-amodiaquine for the treatment of uncomplicated malaria. Expert Opin Investig Drugs 16:1079–1085
5. Dondorp AM, Yeung S, White L et al (2010) Artemisinin resistance: current status and scenarios for containment. Nat Rev Microbiol 8:272–280
6. Anonymous. http://www.mmv.org/sites/default/files/uploads/docs/essential_info_for_-scientists/3Q_Global_Malaria_Portfolio_Slide_by_therapeutic_type.ppt, entry of 15 Oct 2010
7. Sinclair D, Donegan S, Lalloo DG (2011) Artesunate versus quinine for treating severe malaria. Cochrane Database Syst Rev 3:CD005967
8. Ashburn TT, Thor KB (2004) Drug repositioning: identifying and developing new uses for existing drugs. Nat Rev Drug Discov 3:673–683
9. Campas C (2009) Drug repositioning summit: finding new routes to success. Drug News Perspect 22:126–128
10. Miller TM, Layzer RB (2005) Muscle cramps. Muscle Nerve 32:431–442
11. Sibilia J, Pasquali JL (2008) Systemic lupus erythematosus: news and therapeutic perspectives. Presse Med 37:444–459
12. Efferth T (2007) Willmar Schwabe Award 2006: antiplasmodial and antitumor activity of artemisinin–from bench to bedside. Planta Med 73:299–309
13. Krishna S, Bustamante L, Haynes RK et al (2008) Artemisinins: their growing importance in medicine. Trends Pharmacol Sci 29:520–527
14. Utzinger J, Xiao SH, Tanner M et al (2007) Artemisinins for schistosomiasis and beyond. Curr Opin Investig Drugs 8:105–116
15. Dar O, Khan MS, Adagu I (2008) The potential use of methotrexate in the treatment of falciparum malaria: in vitro assays against sensitive and multidrug-resistant falciparum strains. Jpn J Infect Dis 61:210–211
16. Fidock DA, Nomura T, Wellems TE (1998) Cycloguanil and its parent compound proguanil demonstrate distinct activities against *Plasmodium falciparum* malaria parasites transformed with human dihydrofolate reductase. Mol Pharmacol 54:1140–1147

17. Kiara MS, Okombo J, Masseno V et al (2009) In vitro activity of antifolate and polymorphism in dihydrofolate reductase of *Plasmodium falciparum* isolates from Kenyan coast: emergence of parasites with Ile-164-Leu mutation. Antimicrob Agents Chemother 53:3793–3798
18. Nduati E, Diriye A, Ommeh S et al (2008) Effect of folate derivatives on the activity of antifolate drugs used against malaria and cancer. Parasitol Res 102:1227–1234
19. Langman LJ, Kapur BM (2006) Toxicology: then and now. Clin Biochem 39:498–510
20. Rozman KK, Doull J (2001) Paracelsus, Haber and Arndt. Toxicology 160:191–196
21. Nzila A, Okombo J, Becker RP et al (2010) Anticancer agents against malaria: time to revisit? Trends Parasitol 26:125–129
22. Taylor WR, White NJ (2004) Antimalarial drug toxicity: a review. Drug Saf 27:25–61
23. Cann HM, Verhulst HL (1961) Fatal acute chloroquine poisoning in children. Pediatrics 27:95–102
24. Riou B, Barriot P, Rimailho A et al (1988) Treatment of severe chloroquine poisoning. N Engl J Med 318:1–6
25. Fahey JB, DiMaggio C (2007) High-dose methotrexate and primary central nervous system lymphoma. J Neurosci Nurs 39:83–88
26. Fong CM, Lee AC (2006) High-dose methotrexate-associated acute renal failure may be an avoidable complication. Pediatr Hematol Oncol 23:51–57
27. Mantadakis E, Cole PD, Kamen BA (2005) High-dose methotrexate in acute lymphoblastic leukemia: where is the evidence for its continued use? Pharmacotherapy 25:748–755
28. Swierkot J, Szechinski J (2006) Methotrexate in rheumatoid arthritis. Pharmacol Rep 58:473–492
29. Niehues T, Lankisch P (2006) Recommendations for the use of methotrexate in juvenile idiopathic arthritis. Paediatr Drugs 8:347–356
30. Diouf ML, Diallo S, Mbengue M et al (2001) Methotrexate, liver and rheumatoid arthritis in tropical areas. Sante 11:195–200
31. Tahiri L, Allali F, Jroundi I et al (2006) Therapeutic maintenance level of methotrexate in rheumatoid arthritis. Sante 16:167–172
32. Gray OM, McDonnell GV, Forbes RB (2006) A systematic review of oral methotrexate for multiple sclerosis. Multiple Scler 12:507–510
33. Domenech E, Manosa M, Navarro M et al (2008) Long-term methotrexate for Crohn's disease: safety and efficacy in clinical practice. J Clin Gastroenterol 42:395–399
34. Montero Mora P, Gonzalez Perez Mdel C, Almeida Arvizu V et al (2004) Autoimmune urticaria. Treatment with methotrexate. Rev Alerg Mex 51:167–172
35. Novak K, Swain MG (2008) Role of methotrexate in the treatment of chronic cholestatic disorders. Clin Liver Dis 12:81–96, viii
36. Specks U (2005) Methotrexate for Wegener's granulomatosis: what is the evidence? Arthr Rheum 52:2237–2242
37. Wong JM, Esdaile JM (2005) Methotrexate in systemic lupus erythematosus. Lupus 14:101–105
38. Sheehy TW, Dempsey H (1970) Methotrexate therapy for *Plasmodium vivax* malaria. JAMA 214:109–114
39. Wildbolz A (1973) Methotrexate in the therapy of malaria. Ther Umsch 30:218–222
40. Ferone R (1971) Methotrexate therapy for *P. vivax* malaria. JAMA 215:117
41. Laing AB (1972) Methotrexate in malaria. Trans R Soc Trop Med Hyg 66:518–519
42. Chilengi R, Juma R, Abdallah AM et al (2011) A phase I trial to evaluate the safety and pharmacokinetics of low-dose methotrexate as an anti-malarial drug in Kenyan adult healthy volunteers. Malar J 10:63
43. Hagner N, Joerger M (2010) Cancer chemotherapy: targeting folic acid synthesis. Cancer Manag Res 2:293–301
44. Fennell BJ, Carolan S, Pettit GR et al (2003) Effects of the antimitotic natural product dolastatin 10, and related peptides, on the human malarial parasite *Plasmodium falciparum*. J Antimicrob Chemother 51:833–841

45. Koka S, Bobbala D, Lang C et al (2009) Influence of paclitaxel on parasitemia and survival of *Plasmodium berghei* infected mice. Cell Physiol Biochem 23:191–198
46. Schrevel J, Sinou V, Grellier P et al (1994) Interactions between docetaxel (Taxotere) and *Plasmodium falciparum*-infected erythrocytes. Proc Natl Acad Sci USA 91:8472–8476
47. Sinou V, Grellier P, Schrevel J (1996) *In vitro* and *in vivo* inhibition of erythrocytic development of malarial parasites by docetaxel. Antimicrob Agents Chemother 40:358–361
48. Usanga EA, O'Brien E, Luzzato L (1986) Mitotic inhibitors arrest the growth of *Plasmodium falciparum*. FEBS Lett 209:23–27
49. Nair L, Bhasin VK (1994) Cure with cisplatin (II) or murine malaria infection and *in vitro* inhibition of a chloroquine-resistant *Plasmodium falciparum* isolate. Jpn J Med Sci Biol 47:241–252
50. Kreidenweiss A, Kremsner PG, Mordmuller B (2008) Comprehensive study of proteasome inhibitors against *Plasmodium falciparum* laboratory strains and field isolates from Gabon. Malar J 7:187
51. Reynolds JM, El Bissati K, Brandenburg J et al (2007) Antimalarial activity of the anticancer and proteasome inhibitor bortezomib and its analog ZL3B. BMC Clin Pharmacol 7:13
52. Chabner B, Amrein P, Drucker B et al (2006) Chemotherapy of neoplastic diseases. In: Brunton L (ed) The pharmacological basis of therapeutics. McGraw-Hill, Washington, USA, pp 1345–1403
53. Nannini C, West CP, Erwin PJ et al (2008) Effects of cyclophosphamide on pulmonary function in patients with scleroderma and interstitial lung disease: a systematic review and meta-analysis of randomized controlled trials and observational prospective cohort studies. Arthritis Res Ther 10:R124
54. Prefontaine E, Sutherland LR, Macdonald JK et al (2009) Azathioprine or 6-mercaptopurine for maintenance of remission in Crohn's disease. Cochrane Database Syst Rev CD000067
55. Wu JJ, Huang DB, Pang KR et al (2005) Thalidomide: dermatological indications, mechanisms of action and side-effects. Br J Dermatol 153:254–273
56. Mateos J, Perez-Simon JA, Caballero D et al (2006) Vincristine is an effective therapeutic approach for transplantation-associated thrombotic microangiopathy. Bone Marrow Transplant 37:337–338
57. Berman JJ (2008) Treatment of leishmaniasis with miltefosine: 2008 status. Expert Opin Drug Metab Toxicol 4:1209–1216
58. Burri C (2010) Chemotherapy against human African trypanosomiasis: is there a road to success? Parasitology 137:1987–1994
59. Duenas-Gonzalez A, Garcia-Lopez P, Herrera LA et al (2008) The prince and the pauper. A tale of anticancer targeted agents. Mol Cancer 7:82

Chapter 13
Tackling the Problem of Antimalarial Resistance

John Okombo, Leah Mwai, and Alexis Nzila

Abbreviations

ACT	Artemisinin Combination Therapy
ATM	Artemether
AQ	Amodiaquine
Art-D	Artemisinin derivatives
ART	Artesunate
CQ	Chloroquine
CVIET	Cysteine at position 72, valine at position 73, isoleucine at 74, glutamic acid at 75 and threonine at 76
CVIDT	Cysteine at position 72, valine at position 73, isoleucine at 74, aspartic acid at 75 and threonine at 76
CVIKT	Cysteine at position 72, valine at position 73, isoleucine at 74, lysine at 75 and threonine at 76
DEAQ	Desethyl-amodiaquine
DHA	Dihydroartemisinin
DHPS	Dihydropteroate synthase
DHFR	Dihydrofolate reductase
DV	Digestive vacuole
dUMP	Deoxyuridine monophosphate
DDS	Dapsone
GTP-CH	Guanosine triphosphate cyclohydrolase

J. Okombo • L. Mwai
Kenya Medical Research Institute (KEMRI)/Wellcome Trust Collaborative Research Program, P.O. Box 230, 80108 Kilifi, Kenya

A. Nzila (✉)
Departments of Chemistry and Clinical Pharmacology, University of Cape Town, Rondebosch 7701, Cape Town, South Africa

K. Chibale et al. (eds.), *Drug Discovery in Africa*,
DOI 10.1007/978-3-642-28175-4_13,

HFT	Halofantrine
ITP	Preventive treatment in pregnancy
PM/SD	Pyrimethamine/sulfadoxine
LM	Lumefantrine
MFQ	Mefloquine
QN	Quinine
PQ	Piperaquine
pABA	Paraminobenzoic acid
Pfcrt	*Plasmodium falciparum* chloroquine resistance transporter
Pfmdr1	*Plasmodium falciparum* multidrug resistance gene 1
QTL	Quantitative trait loci
SD	Sulphadoxine
SNP	Single nucleotide polymorphism
SERCA	Sarco/endoplasmic reticulum Ca2+ ATPases (SERCAs)
SVMNT	Serine at position 72, valine at position 73, methionine at 74, asparagine at 75 and threonine at 76
VPM	Verapamil

13.1 Background

Malaria is a global public health priority. The control of malaria is hampered by the ability of the parasite to select quickly for resistance against antimalarials. Indeed, there is no single antimalarial in clinical use against which the parasite has not yet developed resistance [1]. Current international strategies for treatment depend on the use of combinations of drugs that include artemisinin, and these strategies are known as artemisinin combination therapy (ACT). Although this strategy is designed to reduce the chance of resistance emerging, there is considerable concern that this will nonetheless happen.

For instance, the combination of lumefantrine (LM) and artemether (ATM), known as Coartem^{TM}, has become the first line of treatment of malaria in many African countries [2–4], but there is concern that resistance to LM could be selected relatively quickly [5–8]. The antimalarial amodiaquine (AQ), whose active metabolite is desethyl-amodiaquine (DEAQ), a close analogue of chloroquine (CQ), is increasingly being used in combination with artemisinin derivatives as first-line treatments in many Western African countries [9, 10]. However, resistance to AQ is common, especially in areas of high CQ resistance [11, 12]. Quinine (QN), one of the oldest drugs, has been the mainstay for the treatment of severe malaria in Africa until 2010, when it was replaced by artesunate [13], and this drug is still recommended after ACT failure for the treatment of uncomplicated malaria [14]. However, in vivo and in vitro QN resistance have been reported in South East Asia and South America, and decrease in vitro activity is now common in Africa [15].

Two other combinations, piperaquine (PQ)/dihydroartemisinin (DHA), and the combination of pyronaridine (PRN)/artesunate (ART), have now reached phase III/IV clinical evaluation as alternative treatment for malaria in Africa [16]. However, when PQ and PRN were used as monotherapy in China, cases of resistance were reported [1]. As discussed earlier, artemisinin derivatives (Art-D) [ATM, DHA and ART] are now the backbone of drug combination. However, resistance to this family of drug is now emerging, threatening the concept of the ACT [17].

Thus, drug resistance in malaria remains a burgeoning problem, and any strategy to control and contain the spread of resistance requires an understanding of the mechanisms of drug resistance. Indeed, understanding the mechanisms of resistance could lead to identifying simple ways, by the use of single markers, to map and monitor the emergence and spread of drug resistance in endemic areas. Such studies could also contribute to clarifying the mode of drug action, which eventually could lead to the discovery of new targets or new drugs.

In this chapter, we have summarized our current knowledge and have presented the existing gaps and challenges in our understanding of the mechanisms of resistance to the main clinically important antimalarial drugs chloroquine (CQ), amodiaquine (AQ), lumefantrine (LM), quinine (QN), mefloquine (MFQ), halofantrine (HFT), artemisinin derivatives and pyrimethamine/sulfadoxine. We have focused on *Plasmodium falciparum* only (we have not discussed information pertaining to *P. vivax*).

13.2 Aminoquinolines (CQ and AQ)

13.2.1 Role of **Pfcrt** *and* **Pfmdr1** *in CQ Resistance*

CQ has been the drug of choice in the treatment of uncomplicated malaria in Africa until the late 1990s, when resistance to this drug reached unacceptable levels [18], leading to its replacement by the antifolate pyrimethamine/sulfadoxine (see Sect. 13.4). Extensive work has been carried out to define the mechanisms of resistance to this drug. The benchmark work was carried out by Wellems's group, using quantitative trait loci (QTL) on the genetic cross of HB3 and Dd2 strains and in vitro activity of CQ. Segments in chromosome 7 were associated with CQ reduced susceptibility [19], and further studies identified *Pfcrt*, as a drug resistance candidate gene [20].

Pfcrt is a 45-kDa protein containing 10 transmembrane domains and is located on the membrane of the digestive vacuole (DV). DV is an acidic, lysosome-like compartment in which haemoglobin is degraded to haematin, a toxic agent resulting from a dimeric form of oxidized haem, and this agent is detoxified by conversion to hemozoin. CQ is a weak base and concentrates in its diprotonated form in DV, where it prevents hemozoin synthesis, leading to parasite death [21–23].

The analysis of the malaria parasite from different endemic settings shows a wide sequence diversity of *Pfcrt* alleles. Single nucleotide polymorphism

(SNPs) or point mutations have been detected at 15 residues, located between codons (amino acid) 72 and 371. The combination of 4 SNPs, at codons 72, 74, 75 and 76 defined the most important haplotypes, which have distinct geographic distributions. The haplotype CVIET (cysteine at position 72, valine at position 73 [which remained unchanged], isoleucine at 74, glutamic acid at 75 and threonine at 76) is found in Asia and Africa; CVIDT and CVIKT haplotype are predominantly found in South East Asia and Pacific regions, respectively; SVMNT in South America and the Pacific region [23, 24].

Of all the mutations, the *Pfcrt* K76T (from lysine to threonine at position 76) is the most critical determinant of CQ resistance in vitro and in vivo, making it a useful marker of CQ resistance. This mutation has been used extensively to map and monitor the emergence and spread of CQ resistance in many malaria-endemic areas [25].

Another gene, the *P. falciparum* multidrug resistance transporter, *Pfmdr1*, an ortholog of mammalian P glycoproteins (also referred to as *pgh1)*, was initially identified as a marker of CQ resistance. This gene is located on chromosome 5 and encodes for a 162-kDa ABC-type transporter, expressed on the membrane of the digestive vacuole. It consists of two homologous halves, each with six predicted transmembrane domains and a conserved nucleotide-binding domain [23].

Five key polymorphisms have been identified in this gene, at codons N86Y, Y184F, S1034C, N1042D and D1246Y [23, 26]. However, studies have shown that the occurrence of *Pfmdr1* mutations alone does not confer a decrease in CQ activity and that this gene plays an ancillary role in CQ resistance. CQ resistance occurs with the selection of *Pfcrt*-N76T mutation, and the additional mutations in *Pfmdr1* (mainly at N86Y) increase the resistance level [23, 26].

13.2.2 Reversal of CQ Resistance

13.2.2.1 Reversal of Resistance with the Use of Chemosensitizing Agents

The observation that resistance in multidrug-resistant tumour cell lines can be reversed by agents that interact with the P-glycoprotein, such as verapamil (VPM) and VPM-related compounds, prompted studies on the use of such agents to reverse chloroquine (CQ) resistance in *P. falciparum*. The first modulation of CQ resistance was reported with the calcium channel blockers VPM and desipramine and was extended to antipsychotic drugs, calmodulin inhibitors, histamine receptor antagonists and antidepressant agents among others [27–29].

This information suggests that antimalarials might be combined with a second agent to restore their effectiveness, and clinical evaluation of some of these agents has been carried out in West Africa, in areas of moderate CQ resistance [30–32]. However, these agents have the disadvantage of being pharmacologically active with systemic effects that may result in a variety of side effects. In addition, the minimum concentrations of these agents needed to chemosensitize CQ resistance (usually more than 1 μM of free drug) [33–36] is not achievable in vivo when normal doses are used. Therefore, high doses have to be used, with all of the attendant risks of toxicity. These limitations may explain why the reversal of CQ resistance has never attained widespread application.

13.2.2.2 Naturally Occurring Reversal of Resistance

Reports have indicated that, for some drugs, mutations that render an organism drug resistant are associated with a loss of fitness. As a result, organisms carrying these mutations will be outgrown by their drug-sensitive counterparts when drug pressure is removed or reduced [37, 38]. The same feature has been reported in malaria with CQ resistance. Indeed, early studies in Gabon, Vietnam and China indicated the possibility of the return of CQ sensitivity following CQ withdrawal [39–41]. CQ reversal has been shown within a decade, in Malawi [42, 43], based on (1) the re-emergence of parasites carrying wild-type codon 76 of *Pfcrt* and, to a lesser extent, wild-type codon 86 of *Pfmdr1* [44]; and (2) the return of CQ in vitro activity [45]. Our group has recently shown that the same trend is occurring in Kenya [46]. This reversal indicates that *Pfcrt* mutant parasites may be loaded with a substantial fitness cost, thus leading to their decline in frequency once drug pressure is removed. This raises the possibility of reintroducing this safe and affordable drug for the treatment of malaria in Africa. However, there is concern that CQ resistance may be rapidly reselected in vivo once the drug comes into widespread use, and consequently, it has been proposed that CQ be used in combination with other antimalarials [43]. A clinical trial of combinations of CQ with ART, atovaquone/ proguanil or azithromycin has been conducted in Malawi. Although the results are still pending (http://www.clinicaltrials.gov, NCT00379821), the possibility of reintroducing CQ to treat uncomplicated malaria exists.

13.2.3 Amodiaquine

AQ, a 4-aminoquinoline closely related to CQ, is rapidly converted to DEAQ in vivo. The half-life of AQ is less than 5 h [47], and DEAQ is slowly eliminated from the body with a half-life >200 h [47, 48]. Thus, though AQ is more active than DEAQ, in vivo, the treatment efficacy is mainly borne by DEAQ [47].

AQ has been combined with ART, and this combination is now the drug of choice in the treatment of uncomplicated malaria in many parts of Western Africa and is currently the world's second most widely used ACT [10]. Early in vitro studies have demonstrated that parasite resistant to CQ are less susceptible to DEAQ, and to a lesser extent to AQ, highlighting that some of the mechanisms of resistance to these drugs may be common [49, 50]. In addition, the efficacy of AQ is reduced in areas of high CQ resistance [12, 51].

This cross-resistance has been confirmed using genetic analyses. Parasites with *Pfcrt*-K76T mutation were less susceptible to DEAQ than wild types [52]. The presence of mutations *Pfmdr1*-N86Y *or Pfmdr1*-D1246Y was also associated with decrease in DEAQ activity [52, 53]. However, since most of these mutations occur on a backdrop of *Pfcrt*-K76T, it is difficult to tease out the impact of *Pfmdr*1 mutations alone on DEAQ activity. On the other hand, in vivo, the presence of *Pfcrt*-K76T has been associated with a decrease in AQ efficacy [54–56].

A recent meta-analysis of AQ efficacy has shown that the presence of *Pfmdr1*-N86Y mutation was associated with a fivefold increase in odd ratios of AQ-treatment failure [25].As it is the case with the in vitro data, this *Pfmdr1* mutation is however likely to occur on a backdrop of *Pfcrt*-K76T. Overall, though CQ resistance is associated with reduced AQ efficacy, AQ retains susceptibility in areas of moderate CQ resistance, hence its use in combination with ART in western parts of Africa [10].

Recent studies have also shed more light on the mechanisms of resistance to AQ. Genetic cross 7G8×GB4, between a CQ-resistant clone from South America carrying the *Pfcrt* SVMNT haplotype and an African clone carrying the CVIET haplotype, has showed that DEAQ-resistant parasites have predominantly an SVMNT haplotype, an indication that genetic patterns of the CQ and AQ resistance might be different [57]. Analyses of field isolates have confirmed the finding that the prevalence of SVMNT haplotype is associated with the use of AQ in many endemic areas [57, 58]. Recently, this AQ-resistant haplotype has been reported in Angola and Tanzania, the two countries where AQ has been widely used [59]. These investigations show that the selection of *Pfcrt*-K76T may not exclusively be the result of the use of CQ. In areas where AQ has been used prior to, or concurrently with, CQ, AQ may be driving the rise of *Pfcrt*-K76T, thus the selection of CQ resistance, an important paradigm in our understanding of the spread of the resistance of CQ [57, 59].

As stated earlier, AQ with ART is used in western parts of Africa, where most parasites are either CVIET or CVMNT [59]. Thus, the relative efficacy of AQ may be the result of the low level of SVMNT parasite populations. Thus, with increased AQ use (as part of the combination with ART) [10], the rate of SVMNT haplotype could become predominant in the population. Monitoring *Pfcrt* polymorphism in these areas would shed more light on molecular mechanisms of resistance to AQ.

13.3 Aryl Amino Alcohol

13.3.1 Lumefantrine

Lumefantrine (LM), also known as benflumetol, is an aryl amino alcohol derivative like QN, MFQ and HFT. These drugs, along with CQ and AQ, are thought to interfere with haem detoxification (Sect. 13.2.1). LM has been combined with the artemisinin derivative artemether, under the name of Coartem™, and this combination is now the most widely used ACT for the treatment of uncomplicated malaria in African countries [2–4].

Previous reports have indicated that the use of Coartem™ is associated with selection of parasites with wild-type *Pfmdr1* (at codon 86) and that such parasites are tolerant to low LM concentrations [6, 7, 60–62]. A recent study has extended the same observation on parasites with wild-type *Pfcrt* (at codon 76) [5]. Using in vitro investigation, we and others have provided evidence that LM in vitro activity is substantially reduced in parasites harbouring the wild type of *Pfmdr1* at position 86 [5, 8], although no difference was found in another study [63]. In addition, our data demonstrated that the decrease in LM in vitro activity is more pronounced in parasites with wild-type genotype at codon 76 of *Pfcrt* [8], an observation that was also reported in strains transfected with *Pfcrt* [5]. All this aforementioned research reveals that the use of Coartem™ selects for wild-type parasite at *Pfmdr1 and Pfcrt*, and in vitro, these parasites are less susceptible to LM and more susceptible to CQ. This inverse relationship in LM and CQ activity raises the possibility of reintroducing CQ once LM resistance is common.

In South East Asia, reduced LM susceptibility following the use of Coartem™ has been reported and has been associated with an increase in *Pfmdr1* copy number (but not mutations) [64]. An increase in the *Pfmdr1* copy number, which is rare in Africa [8], is dominant in South East Asia as a result of the use of MFQ [65]. Thus, LM would have distinct mechanisms of resistance in Africa compared to South East Asia. It is suggested that under LM pressure, African parasites would select for wild-type *Pfcrt* at codon 76 and *Pfmdr1* at codon 86, while in South East Asia, parasites with a high *Pmdr1* copy number would be selected. Nevertheless, it is proposed that the continuous use of LM (as part of Coartem™) in Africa could lead to an increase of not only wild-type parasite populations but also those with multiple copies of the *Pfmdr1* [8], a situation that would be similar to that in South East Asia.

In summary, the use of the Coartem™ in clinics has led to a rapid emergence of LM-tolerant parasites. However, it should be borne in mind that Coartem™ is effective in treating malaria, with high success rates (>90–95%) in many malaria-endemic areas [2]. Thus, the mechanisms of Coartem™ resistance in vivo will likely involve other genes. Nevertheless, we propose that the increase in wild-type *Pfcrt* and *Pfmdr1* parasites (and probably *Pfmdr1* copy number) in the population would be the first step in the selection of resistance to LM. These parasites will form the backdrop of Coartem™ resistance. As a result, an increase in wild-type parasites in the population could be used as an indicator of the emergence of LM resistance.

13.3.2 Quinine

13.3.2.1 Role of *Pfcrt* and *Pmdr1*

For centuries, quinine (QN) has been used to treat malaria and for several decades has been the drug of choice in the treatment of severe malaria [66]. It has recently been chosen, in combination with antibiotics, as a second line in the treatment of uncomplicated malaria after ACT [14].

QN resistance has been reported both in Africa and South East Asia [66], and this led to the investigation of the artemisinin derivative ART as a possible alternative for the treatment of severe malaria [67, 68], and intravenous ART has recently been recommended as the drug choice in severe malaria [13]. However, this strategy could be compromised with the emergence to artemisinin resistance (Sect. 13.3.4) [17].

Though several reports have indicated the emergence of quinine resistance in vivo, there is no evidence of widespread resistance worldwide despite the fact that this drug has been in use for more than 400 years. A review of clinical trials of its antimalarial effect showed that, over a period of 30 years, from the 1970s up to early 2000, the efficacy of QN has not changed in most endemic areas, yet during the same period, the efficacy of chloroquine, pyrimethamine and mefloquine (MFQ) had substantially decreased [69]. In Africa, a reduced cure rate from 80% to 60% has been reported, indicating the emergence of resistance [70–72]. However, detailed observations of these data show that the assessment of resistance was carried out using effectiveness rather than efficacy studies. QN is administered three times per day for 7 days, posing serious limitations on compliance; thus, effectiveness studies could overestimate the rate of treatment failure [66]. So far, cases of bona fide in vivo resistance, based on efficacy studies, have only been reported in South Asia [73, 74].

A detailed review of the literature indicates that QN is relatively efficacious in vivo, but an in vitro decrease in activity is now common in Africa [66]. Several investigations have explored the mechanisms of QN in vitro resistance. Early reports have demonstrated the existence of in vitro cross-resistance between QN and its structural analogues such as CQ, LM, MFQ and HFT [75–79], leading to the hypothesis that these drugs may have some common mechanisms of resistance [80]. This led to the exploration of the role of *Pfcrt* and *Pfmdr1* in relation with QN in vitro activity.

Conflicting results were observed on the association of *Pfcrt*-K76T and QN reduced susceptibility. Correlations were found in some studies, but not in all [22, 23, 63, 81, 82], while transfection studies indicated an increase in QN susceptibility in *Pfcrt*-K76T mutant strains [83, 84]. In South East Asia, QN in vitro activity tends to be inversely related to CQ activity, while worldwide, their activities appear to be positively correlated [85]. These conflicting observations explain the multifactorial nature of QN resistance.

Concerning *Pfmdr1* gene, transfection studies showed that polymorphisms were associated with increased resistance to QN [83, 86–88], supported in field isolates [63, 81], though some studies have shown no association [64, 65, 89]. Increased *Pfmdr1* copy number reduces the activity of QN and those of other arylaminoquinolines such as MFQ and LM [64, 65, 82, 90, 91]. Thus, *Pfmdr1* polymorphism has a bearing on the activity of QN. Further investigations have shown that other transporters modulate QN activity, with the most important one being the Na^+/H^+ exchanger transporter (*Pfnhe*) [86, 92].

13.3.2.2 Role of *Pfnhe*

Using quantitative trait loci (QTL) on the genetic cross of HB3 and Dd2 strains, and in vitro activity of QN, Ferdig et al. [86] identified candidate genes on segments of chromosomes (Chr) 13, 7 and 5 associated with QN reduced susceptibility. The mapped segments of Chr 7 and 5 contain *Pfcrt* and *Pfmdr1*, respectively, while the segment of Chr 13 was narrowed down to *Pfnhe* gene, encoding a putative Na^{+}/H^{+} exchanger [86].

Pfnhe is a 43-kDa protein of 1,930 amino acids and localized on parasite plasma membrane. It contains 12 transmembrane domains and 3 microsatellite regions, msR1, ms3580 and ms4760 [66]. The microsatellite ms4760 is the most variable region of the gene. It is produced by the insertion or deletion of five blocks: DNND (block I), DNNND (block II), NHND (block III), DKNNKND (block IV) and DDNNNDNHNDD (block VI). The number of repeats of each block determines the ms4760 sequence, and so far, over 33 different sequences have been reported. The block II repeats (DNNND) vary between 1 and 4, though less than 5% of isolates analysed so far have 4 repeats [66, 93].

To date, seven studies have addressed the associated between QN activity and *Pfnhe* polymorphism [15, 81, 86, 94–97], mostly summarized in our recent review manuscript [15]. Overall QN activity is significantly reduced in parasites harbouring 2 or more DNNND compared to those with 1 copy number only, though no significant association was reported in 3 studies [81, 94, 95].

The analysis of DDNNNDNHNDD repeat did not yield clear-cut result. Some studies showed a decrease in QN, and other studies had no decrease [15]. No association was found between the combinations of DNNND/DDNNNDNHNDD number and QN in vitro activity.

A recent analysis using allelic exchange (of truncated *Pfnhe* gene) and transfection in parasite with different genetic background (CQ sensitive and CQ resistant) has confirmed the role of *Pfnhe* in conferring QN resistance; however, this was dependent upon strains. Indeed, truncated *Pfnhe* gene in CQ-sensitive strains did not alter their susceptibility to QN, while the opposite was observed with CQ-resistant strain [98], a clear indication that *Pfcrt* (and probably *Pfmdr1*, as discussed earlier) could contribute to QN susceptibility.

13.3.2.3 Association with *Pfcrt* and *Pfmdr*

As discussed earlier, *Pfcrt* and *Pfmdr1* proteins can modulate QN in vitro activity. Thus, it is important to establish the impact of these markers in relation with *Pfnhe* and QN in vitro activity. However, this has not been fully explored yet. Indeed, so far, only two studies have investigated *Pfcrt* and *Pfmdr1* genotypes in relation to *Pfnhe* polymorphisms and QN activity [15, 94]. In one study, the presence of mutant *Pfmdr1*-N86Y (in combination with *Pfcrt*-K76T) rendered parasite harbouring 2 copy of DNNND more resistant to QN than wild-type parasite with the same number of DNNND repeats [15]. The same trend was observed in another study, though the difference was not significant [94].

13.3.2.4 Summary

These early investigations indicate that parasites with 2 copies of DNNND repeats in the ms4067 microsatellite of the *Pfnhe* gene are less susceptible to QN in vitro, and the presence of mutation in *Pfmdr1* may decrease further QN activity. Though QN is still efficacious in Africa, it is on the backdrop of in vitro reduced susceptibility parasites that in vivo QN resistance will emerge. Thus, polymorphism of microsatellite 4760 (in combination with *Pfmdr1* mutations) could be used to monitor the selection and spread of QN reduced in vitro activity.

13.3.3 Mefloquine, Halofantrine

MFQ has been extensively used in South East Asia, by itself or in combination with ART, for the treatment of uncomplicated malaria [99]. This drug is relatively expensive and can cause neurotoxic side effects, limiting its use for mass treatment in Africa. However, it has been used as a prophylactic agent by visiting travellers [100–102].

Though not widely used in Africa, the in vitro activity of MFQ has been investigated against African *P. falciparum* isolates. Overall, the results have shown a high susceptibility of African parasites against this drug, with concentrations that inhibit 50% of parasitaemia growth (IC_{50}s) <30 nM [103–106], though cases of in vitro reduced susceptibility have also been reported [107, 108]. This overall in vitro susceptibility of African parasites to MFQ has been supported by in vivo efficacy of MFQ alone [109, 110] or in combination with Art-D [111–114], though one study showed limited efficacy of MFQ alone [115].

However, in South East Asia, resistance to MFQ has been well established both in vitro [116–118] and in vivo [99]. The mechanisms of resistance to MFQ have been clarified more than 20 years ago, almost a decade before the understanding of the mechanisms of CQ resistance. In 1988, Oduola et al. selected the MFQ-resistant strain, W2mef, by continuous in vitro drug pressure of the parent strain W2.21 [119]. The analysis of P-glycoprotein *Pfmdr1* between W2 and Wmef led to the discovery of the association between amplification of this gene (or increased copy number) and MFQ resistance [120], and a year later, this finding was confirmed in a different study [121]. *Pfmdr1* is the first identified malaria transporter gene to be involved in antimalarial resistance.

The analysis of the field isolates in South East Asia, where MFQ resistance is common, has shown that *Pfmdr1* copy number varies from 1 to 4 and MFQ-resistant parasite (both in vitro and in vivo) has higher copy number of *Pfmdr1* [26, 65, 122].

Interestingly, point mutations or single nucleotide polymorphism (SNPs) also contribute to MFQ resistance. *Pfmdr1* mutant parasites, mainly those at codon 86 (*Pfmdr1*-Y86N), are more susceptible to MFQ than wild-type parasites, and interestingly, the increase in copy number primarily occurs in *Pfmdr1* wild-type parasite [64, 65, 88, 89]. Thus, the use of MFQ selects for *Pfmdr1* wild-type parasite, and it is on this backdrop that the increase in *Pfmdr1* copy number occurs, a situation that is similar with LM resistance (Sect. 13.3.1). *Pfmdr1* copy variation is used as marker of MFQ resistance in South East Asia [64, 65, 88, 89].

Halofantrine (HFT) is another aryl amino alcohol antimalarial. This drug is potent, with $IC_{50} < 10$ nM on average, and is efficacious in humans [123]. However, this drug has not been widely used because of its cardiotoxicity side effects [124]. One study reported the emergence of HFT resistance in central Africa [125], but this finding has been challenged [126, 127]. Overall bona fide in vivo resistance has not been established yet. Nevertheless, the mechanisms of resistance to this drug have been studied in the context of cross-resistance with other aryl amino alcohol, mainly MFQ. The activity of HFT is commonly found to be correlated with MFQ's, that MFQ resistance is associated with decreased HFT activity, and this is primarily associated with an increase in *Pfmdr1* copy number [91, 128, 129].

Recently, using genome-wide association based on microarray analysis of field isolates, a novel locus, PF10_0355, has been found to be associated with HFT reduced in vitro susceptibility. This has been confirmed by functional overexpression that the increase in PF10_0355 copy number decreases HFT activity (along with the activity of MFQ an LM), raising the possibility of the involvement of this gene in the modulation of these aryl amino alcohol antimalarials [130]. However, the exact role of this gene is still speculative since this gene is known to encode for an erythrocyte membrane protein putative, thus a protein involved in invasion. However, it has been proposed that this protein may interact with transporters [130], a type of interaction that has been shown in cancer, yeast and plant cells [131–133]. More studies are needed to clearly define the role of this new drug resistance candidate protein [130].

13.3.4 Artemisinin Derivative

Art-D, found in Qinghao ("blue-green herb"), the *Artemisia annua* or sweet wormwood, has been used as a remedy by Chinese herbalists for more than 2,000 years [134]. Chemically, artemisinin is a sesquiterpene lactone with a peroxide bond that is required for the antimalarial effect. The most commonly used Art-D in clinics are dihydroartemisinin (DHA), ART and ATM [135]. The concern about the emergence and spread of antimalarial resistance led WHO to recommend that antimalarial drug be used as combination therapy, and one of such combinations are artemisinin-based combinations (ACT) [136]. As discussed earlier, Coartem™ and amodiaquine/artesunate are currently the first lines of malaria treatment in

Africa. Two other combinations, piperaquine/dihydroartemisinin and pyronaridine/ artesunate, have now reached phase III/IV clinical evaluation.

Art-D have short half-lives (less than 1 h) but have a rapid killing rate, leading to a parasite reduction biomass by a rate $>10^{10}$ within the first 24 h [137]. Until recently, there has been no evidence for a significant reduction in artemisinin efficacy at either clinical or in vitro levels. In broad terms, the rapid antimalarial action of artemisinins, combined with their pharmacokinetic properties (with terminal elimination half-lives in the order of a few hours), led to the belief that resistance to artemisinins would be very slow to develop since exposure of parasites to sub-therapeutic levels of drug would be very brief. However, these predictions have been contradicted by recent reports on the selection of artemisinin resistance in South East Asia. Indeed, clinical studies have clearly demonstrated a markedly prolonged time to parasite clearance in patients treated with ART, as a result of the emergence of a parasite with reduced in vivo susceptibly to ART [17]. This development is a matter of great concern, and currently, strategies have been proposed to control and contain the spread of resistance, with the specific aim of preventing this resistance to reach the African continent [138, 139].

Several studies have been dedicated to studying the mechanisms of artemisinin resistance. The first studies have focused on the investigation of the role of *Pfcrt* and *Pfmdr1*, the two genes associated with resistance to quinoline-based drugs. Mutations in *Pfcrt* seem not to have a bearing on artemisinin activity. However, polymorphisms in *Pfmdr1* at codons 86, 184, 1034, 1042 and 1246, alone or in combination, have been associated with reduced artemisinin derivatives in vitro by twofold [140]. The increase in *Pfmdr1* copy number was also associated with decrease activity by a factor <2 [140]. Though *Pfmdr1* polymorphism has been proven to have an impact on Art-D in vitro activity, this effect is relatively small and is not associated with reduced in vivo susceptibility [17].

An early study has shown that artemisinin derivatives could target the plasmodium ortholog *PfATPase6* [141]. These studies involved the use of thapsigargin, an analogue of artemisinin derivatives which lacks a peroxide bond and is a known inhibitor of mammalian sarco/endoplasmic reticulum Ca^{2+} ATPases (SERCAs). Heterologous expression using xenopus oocytes and transfection in *P. falciparum* have identified a mutation in a single residue (L263E) in *PfATPases6* that modulates sensitivity to artemisinin [142, 143], suggesting that these gene could contribute to artemisinin resistance.

An analysis of field isolates from Latin America and West Africa has shown that mutations (other than L263E) are associated with reduced in vitro activity to artemether and artesunate [144], and it has been proffered that the intense use of ACT in Niger could have led to the selection of mutation in *PfATPAse-6* [145]. However, it is now clear that polymorphism in *PfATPAse-6* is not associated with the observed decrease in artemisinin efficacy in South East Asia [17].

Other genes have been investigated, including *Pfctp*, *ubp-1* (a gene initially found to be associated with artemisinin resistance in the murine malaria *Plasmodium chabaudi*) and *G7* protein, but none of them has been implicated in artemisinin resistance in *P. falciparum* [140]. Recently, a body of evidence, based on modelling and biological investigations, indicates that artemisinin drugs induce parasite quiescence or dormancy where parasites are protected against drug lethal effect but recover later on and resume a normal growth when drug levels are sub-therapeutic or even cleared from the body. Thus, it is proposed that in vivo resistance could be partly explained by the ability of the artemisinin to induce dormancy [146–148]. Overall, the mechanisms of artemisinin still remain elusive, and further studies are warranted to identify genetic markers associated with resistance to this important antimalarial.

13.4 Antifolate Resistance

13.4.1 Background

Malaria parasites, unlike mammalian cells, lack the necessary machinery to salvage exogenous pyrimidine (uridine, thymidine and cytosine) for RNA and DNA, thus rely on de novo synthesis of these molecules, and this synthesis is dependent upon the availability of folate. Thus, the inhibition of the folate pathway is a seemingly good strategy to block the parasite growth.

Two enzymes, dihydropteroate synthase (DHPS) and dihydrofolate reductase (DHFR), have respectively proven to be critical in the synthesis and the conversion of folate. DHPS mediates the condensation of paraminobenzoic acid (pABA) with hydroxymethyl-dihydropterin pyrophosphate to generate dihydropteroate, and DHFR reduces dihydrofolate to tetrahydrofolate, a form of folate used downstream in the synthesis of deoxyuridine monophosphate (dUMP) which will be converted to other pyrimidine molecules [149].

13.4.2 Antifolate Drugs

Inhibitors of DHPS are sulpha-based analogues of pABA. They include sulphonamide (sulphadoxine, SD) and sulfone (dapsone, DDS). Inhibitors of DHFR compete with dihydrofolate in the active site of the enzyme. The commonly used inhibitors are pyrimethamine, proguanil and chlorproguanil. Proguanil and chlorproguanil are prodrugs converted in vivo to the inhibitors of DHFR cycloguanil and chlorcycloguanil, respectively [149]. Inhibitors of DHPS and DHFR are synergistic, leading to their use in vivo in combination [1].

Chlorproguanil and dapsone were developed under the name of Lapdap™. Combination with artesunate has also been investigated recently, but these combinations have been withdrawn because of DDS toxicity [150, 151]. Proguanil has also been combined with atovaquone (a non-antifolate agent), an inhibitor of electron transport to the cytochrome bc1 complex (coenzyme Q) under the name of Malarone®, and this drug is used as a prophylactic agent against malaria [152]. In this combination, atovaquone synergizes with proguanil, making it a potent combination [153].

The combination of pyrimethamine/sulfadoxine (PM/SD), known as Fansidar^{TM}, has been the most widely used antifolate drug for the treatment of uncomplicated malaria in Africa. This drug was used in Africa in the 1990s for mass treatment, in replacement of the chloroquine. However, resistance to this combination emerged within few years [154, 155], and by the mid-2000s, this drug was withdrawn and replaced by ACT [156].

Despite the emergence of resistance, this drug remains an important drug for the treatment of uncomplicated malaria in pregnancy and for the intermittent preventive treatment in pregnancy and in infancy (ITP) [157]. However, in the context of high PM/SD resistance, its efficacy is greatly compromised [158]. Alternative drugs are being evaluated, including the combination of PM/SD with other antimalarials such as piperaquine and MFQ [159, 160].

13.4.3 Mechanism of Resistance

PM/SD resistance is attributable to parasites that carry point mutations at codons 108 (Ser to Asn), 51 (Asn to Ile) and 59 (Cys to Arg) of *dhfr* gene (these are triple mutant parasites). Resistance is augmented by the selection point mutations at codons 437 (Ala to Gly) and/or 540 (Lys to Glu) or 437 and/or 581 (Ala to Lys) of the *dhps* gene. High levels of in vivo PM/SD resistance are associated with the selection of the mutation at codon 164 (Ile to Leu) of *dhfr* [149, 154, 155]. Interestingly, the existence of this mutation in Africa has been a matter of debate [161, 162]. However, there is now compelling evidence that this mutation does occur in African isolates [163]. Genotyping of *dhfr* and *dhps* has been used widely in Africa to monitor the emergence and selection of resistance to the PM/SD in various malaria-endemic areas [163].

GTP-CH enzyme catalyses the first step of folate biosynthesis. Investigations have demonstrated that parasites highly resistant to the antifolate PM, mainly those with the 164-Leu mutation, have an increased copy number of *gtp-ch* gene, and up to 11 copies have been reported [164]. The presence of mutations in DHFR could be associated with reduced enzyme kinetic properties leading to a fitness cost [162, 164], though in vitro, these reduced enzyme kinetic proprieties have so far yielded contradictory results [165, 166]. Thus, the increase in copy number, which is an adaptive phenomenon, could reflect compensatory mechanisms to maintain sufficient folate product (tetrahydrofolate). This increase in copy number in association with the presence of the 164-Leu mutation has been shown in parasites from South East Asia (an area where parasites with 164-Leu mutation are common), a clear indication that GTP-CH modulates antifolate activity [164], along with changes of amino acids in key positions in DHFR and DHPS enzymes. However, in Africa, more data are still needed to establish whether the increase in *gtp-ch* copy number contributes to antifolate resistance.

13.4.4 Origin of Antifolate-Resistant Parasites

The origin of the African drug-resistant parasites has been a matter of interest. South East Asia has been the focus of the emergence of the multidrug-resistant parasites, and it was proffered that such parasites could have migrated to other malaria-endemic regions including Africa. In the case of CQ, this has been proven with evidence that CQ-resistant parasites originated in South East Asia and spread to Africa [167]. Interestingly, investigations have shown that the same happened with antifolate-resistant parasites. Indeed, African *dhfr* triple mutant parasites (resistant to PM/SD) originated mainly from South East Asia [168–172], though an indigenous origin of some triple mutants, and a few parasites also carrying 164-Leu, was suggested [169, 171, 173]. *dhfr* double mutants (parasite susceptible to PM/SD) are however predominantly of indigenous origin [169–171, 173, 174]. Thus, highly antifolate-resistant parasites may have originated from outside Africa. A similar study was carried out to map the origin of the *dhps* mutants. African *dhps* mutants arose from five geographical foci within Africa, an indication that mutations in this gene may not have been imported from outside Africa [175].

13.5 Concluding Remarks

Currently, in Africa, the artemisinin combinations of Coartem™ AQ/ART are the most widely used drug for the treatment of uncomplicated malaria. Parasites with reduced susceptibly to LM and AQ are now common in Africa. Polymorphisms in *Pfcrt* and *Pfmdr1* have been associated with LM and AQ reduced susceptibility, but these genes alone do not account for resistance to these drugs (at least for LM). Thus, studies are still needed to define the mechanisms of resistance to these drugs.

Most worrying is the emergence of resistance to artemisinin derivatives in South East Asia. If this resistance spreads in Africa, as it has been the case with CQ and antifolates, this will not only render the existing drugs (Coartem™ and AQ/ART) ineffective but also compromise the efficacy of the newly developed combinations of piperaquine/artesunate and pyronaridine/artesunate. Strategies to control and contain the spread of artemisinin resistance have been proposed, especially in preventing this resistance spreading to Africa. However, in the absence of the genetic markers for artemisinin resistance, the goal of such strategies could be difficult to achieve. Thus, it is of paramount importance to define the mechanisms of resistance to artemisinin.

Acknowledgement The authors thank the support of the University of Cape Town.

References

1. Nzila A (2006) Inhibitors of de novo folate enzymes in *Plasmodium falciparum*. Drug Discov Today 11:939–944
2. Makanga M, Krudsood S (2009) The clinical efficacy of artemether/lumefantrine (Coartem™). Malar J 8(Suppl 1):S5
3. Adjei GO, Goka BQ, Binka F et al (2009) Artemether-lumefantrine: an oral antimalarial for uncomplicated malaria in children. Expert Rev Anti Infect Ther 7:669–681
4. Kokwaro G, Mwai L, Nzila A (2007) Artemether/lumefantrine in the treatment of uncomplicated falciparum malaria. Expert Opin Pharmacother 8:75–94
5. Sisowath C, Petersen I, Veiga MI et al (2009) *In vivo* selection of *Plasmodium falciparum* parasites carrying the chloroquine-susceptible pfcrt K76 allele after treatment with artemether-lumefantrine in Africa. J Infect Dis 199:750–757
6. Dokomajilar C, Nsobya SL, Greenhouse B et al (2006) Selection of *Plasmodium falciparum* pfmdr1 alleles following therapy with artemether-lumefantrine in an area of Uganda where malaria is highly endemic. Antimicrob Agents Chemother 50:1893–1895
7. Some AF, Sere YY, Dokomajilar C et al (2010) Selection of known *Plasmodium falciparum* resistance-mediating polymorphisms by artemether-lumefantrine and amodiaquine-sulfadoxine-pyrimethamine but not dihydroartemisinin-piperaquine in Burkina Faso. Antimicrob Agents Chemother 54:1949–1954
8. Mwai L, Kiara SM, Abdirahman A et al (2009) *In vitro* activities of piperaquine, lumefantrine, and dihydroartemisinin in Kenyan *Plasmodium falciparum* isolates and polymorphisms in pfcrt and pfmdr1. Antimicrob Agents Chemother 53:5069–5073
9. Sirima SB, Gansane A (2007) Artesunate-amodiaquine for the treatment of uncomplicated malaria. Expert Opin Investig Drugs 16:1079–1085
10. Zwang J, Olliaro P, Barennes H et al (2009) Efficacy of artesunate-amodiaquine for treating uncomplicated falciparum malaria in sub-Saharan Africa: a multi-centre analysis. Malar J 8:203
11. Mutabingwa TK, Anthony D, Heller A et al (2005) Amodiaquine alone, amodiaquine+sulfadoxine-pyrimethamine, amodiaquine+artesunate, and artemether-lumefantrine for outpatient treatment of malaria in Tanzanian children: a four-arm randomised effectiveness trial. Lancet 365:1474–1480
12. Sasi P, Abdulrahaman A, Mwai L et al (2009) *In vivo* and *in vitro* efficacy of amodiaquine against *Plasmodium falciparum* in an area of continued use of 4-aminoquinolines in East Africa. J Infect Dis 199:1575–1582
13. Dondorp AM, Fanello CI, Hendriksen IC et al (2010) Artesunate versus quinine in the treatment of severe falciparum malaria in African children (AQUAMAT): an open-label, randomised trial. Lancet 376:1647–1657
14. Anonymous (2010) WHO. Guidelines for the treatment of malaria, second edition. http://www.who.int/malaria/publications/atoz/9789241547925/en/index. WHO, Accessed 12 Jan 2011, p 18
15. Okombo J, Kiara SM, Rono J et al (2010) *In vitro* activities of quinine and other antimalarials and pfnhe polymorphisms in *Plasmodium* isolates from Kenya. Antimicrob Agents Chemother 54:3302–3307
16. Anonymous. http://www.mmv.org/sites/default/files/uploads/docs/essential_info_for_-scientists/3Q_Global_Malaria_Portfolio_Slide_by_therapeutic_type.ppt. Accessed 15 Oct 2010
17. Dondorp AM, Nosten F, Yi P et al (2009) Artemisinin resistance in *Plasmodium falciparum* malaria. N Engl J Med 361:455–467
18. Nuwaha F (2001) The challenge of chloroquine-resistant malaria in sub-Saharan Africa. Health Policy Plan 16:1–12
19. Su X, Kirkman LA, Fujioka H et al (1997) Complex polymorphisms in an approximately 330 kDa protein are linked to chloroquine-resistant *P. falciparum* in Southeast Asia and Africa. Cell 91:593–603

20. Djimde A, Doumbo OK, Cortese JF et al (2001) A molecular marker for chloroquine-resistant falciparum malaria. N Engl J Med 344:257–263
21. Sanchez CP, Dave A, Stein WD et al (2010) Transporters as mediators of drug resistance in *Plasmodium falciparum*. Int J Parasitol 40:1109–1118
22. Sanchez CP, Stein WD, Lanzer M (2007) Is PfCRT a channel or a carrier? Two competing models explaining chloroquine resistance in *Plasmodium falciparum*. Trends Parasitol 23:332–339
23. Valderramos SG, Fidock DA (2006) Transporters involved in resistance to antimalarial drugs. Trends Pharmacol Sci 27:594–601
24. Cooper RA, Hartwig CL, Ferdig MT (2005) pfcrt is more than the *Plasmodium falciparum* chloroquine resistance gene: a functional and evolutionary perspective. Acta Trop 94:170–180
25. Picot S, Olliaro P, de Monbrison F et al (2009) A systematic review and meta-analysis of evidence for correlation between molecular markers of parasite resistance and treatment outcome in falciparum malaria. Malar J 8:89
26. Duraisingh MT, Cowman AF (2005) Contribution of the pfmdr1 gene to antimalarial drug-resistance. Acta Trop 94:181–190
27. Egan TJ, Kaschula CH (2007) Strategies to reverse drug resistance in malaria. Curr Opin Infect Dis 20:598–604
28. Henry M, Alibert S, Orlandi-Pradines E et al (2006) Chloroquine resistance reversal agents as promising antimalarial drugs. Curr Drug Targets 7:935–948
29. van Schalkwyk DA, Walden JC, Smith PJ (2001) Reversal of chloroquine resistance in *Plasmodium falciparum* using combinations of chemosensitizers. Antimicrob Agents Chemother 45:3171–3174
30. Gbotosho GO, Happi CT, Sijuade A et al (2008) Comparative study of interactions between chloroquine and chlorpheniramine or promethazine in healthy volunteers: a potential combination-therapy phenomenon for resuscitating chloroquine for malaria treatment in Africa. Ann Trop Med Parasitol 102:3–9
31. Sowunmi A, Adedeji AA, Gbotosho GO et al (2006) Effects of pyrimethamine-sulphadoxine, chloroquine plus chlorpheniramine, and amodiaquine plus pyrimethamine-sulphadoxine on gametocytes during and after treatment of acute, uncomplicated malaria in children. Mem Inst Oswaldo Cruz 101:887–893
32. Sowunmi A, Fateye BA, Adedeji AA et al (2005) Predictors of the failure of treatment with chloroquine in children with acute, uncomplicated, *Plasmodium falciparum* malaria, in an area with high and increasing incidences of chloroquine resistance. Ann Trop Med Parasitol 99:535–544
33. Ciach M, Zong K, Kain KC et al (2003) Reversal of mefloquine and quinine resistance in *Plasmodium falciparum* with NP30. Antimicrob Agents Chemother 47:2393–2396
34. Lehane AM, Hayward R, Saliba KJ et al (2008) A verapamil-sensitive chloroquine-associated H^+ leak from the digestive vacuole in chloroquine-resistant malaria parasites. J Cell Sci 121:1624–1632
35. Martin SK, Oduola AM, Milhous WK (1987) Reversal of chloroquine resistance in *Plasmodium falciparum* by verapamil. Science 235:899–901
36. Martiney JA, Cerami A, Slater AF (1995) Verapamil reversal of chloroquine resistance in the malaria parasite *Plasmodium falciparum* is specific for resistant parasites and independent of the weak base effect. J Biol Chem 270:22393–22398
37. Austin DJ, Anderson RM (1999) Studies of antibiotic resistance within the patient, hospitals and the community using simple mathematical models. Philos Trans R Soc Lond B Biol Sci 354:721–738
38. Lenski RE (1998) Bacterial evolution and the cost of antibiotic resistance. Int Microbiol 1:265–270
39. Legrand E, Volney B, Meynard JB et al (2008) *In vitro* monitoring of *Plasmodium falciparum* drug resistance in French Guiana: a synopsis of continuous assessment from 1994 to 2005. Antimicrob Agents Chemother 52:288–298

40. Schwenke A, Brandts C, Philipps J et al (2001) Declining chloroquine resistance of *Plasmodium falciparum* in Lambarene, Gabon from 1992 to 1998. Wien Klin Wochenschr 113:63–64
41. Liu DQ, Liu RJ, Ren DX et al (1995) Changes in the resistance of *Plasmodium falciparum* to chloroquine in Hainan, China. Bull World Health Organ 73:483–486
42. Laufer MK, Takala-Harrison S, Dzinjalamala FK et al (2010) Return of chloroquine-susceptible falciparum malaria in Malawi was a reexpansion of diverse susceptible parasites. J Infect Dis 202:801–808
43. Laufer MK, Thesing PC, Eddington ND et al (2006) Return of chloroquine antimalarial efficacy in Malawi. N Engl J Med 355:1959–1966
44. Kublin JG, Cortese JF, Njunju EM et al (2003) Reemergence of chloroquine-sensitive *Plasmodium falciparum* malaria after cessation of chloroquine use in Malawi. J Infect Dis 187:1870–1875
45. Mita T, Kaneko A, Lum JK et al (2003) Recovery of chloroquine sensitivity and low prevalence of the *Plasmodium falciparum* chloroquine resistance transporter gene mutation K76T following the discontinuance of chloroquine use in Malawi. Am J Trop Med Hyg 68:413–415
46. Mwai L, Ochong E, Abdirahman A et al (2009) Chloroquine resistance before and after its withdrawal in Kenya. Malar J 8:106
47. Jullien V, Ogutu B, Juma E et al (2010) Population pharmacokinetics and pharmacodynamic considerations of amodiaquine and desethylamodiaquine in Kenyan adults with uncomplicated malaria receiving artesunate-amodiaquine combination therapy. Antimicrob Agents Chemother 54:2611–2617
48. Stepniewska K, Taylor W, Sirima SB et al (2009) Population pharmacokinetics of artesunate and amodiaquine in African children. Malar J 8:200
49. Basco LK, Ringwald P (2007) Molecular epidemiology of malaria in Cameroon. XXIV. Trends of *in vitro* antimalarial drug responses in Yaounde, Cameroon. Am J Trop Med Hyg 76:20–26
50. Tinto H, Rwagacondo C, Karema C et al (2006) *In vitro* susceptibility of *Plasmodium falciparum* to monodesethylamodiaquine, dihydroartemisinin and quinine in an area of high chloroquine resistance in Rwanda. Trans R Soc Trop Med Hyg 100:509–514
51. Mbacham WF, Evehe MS, Netongo PM et al (2010) Efficacy of amodiaquine, sulphadoxine-pyrimethamine and their combination for the treatment of uncomplicated *Plasmodium falciparum* malaria in children in Cameroon at the time of policy change to artemisinin-based combination therapy. Malar J 9:34
52. Echeverry DF, Holmgren G, Murillo C et al (2007) Short report: polymorphisms in the pfcrt and pfmdr1 genes of *Plasmodium falciparum* and in vitro susceptibility to amodiaquine and desethylamodiaquine. Am J Trop Med Hyg 77:1034–1038
53. Nsobya SL, Dokomajilar C, Joloba M et al (2007) Resistance-mediating *Plasmodium falciparum* pfcrt and pfmdr1 alleles after treatment with artesunate-amodiaquine in Uganda. Antimicrob Agents Chemother 51:3023–3025
54. Happi CT, Gbotosho GO, Folarin OA et al (2006) Association between mutations in *Plasmodium falciparum* chloroquine resistance transporter and *P. falciparum* multidrug resistance 1 genes and in vivo amodiaquine resistance in *P. falciparum* malaria-infected children in Nigeria. Am J Trop Med Hyg 75:155–161
55. Holmgren G, Gil JP, Ferreira PM et al (2006) Amodiaquine resistant *Plasmodium falciparum* malaria in vivo is associated with selection of pfcrt 76T and pfmdr1 86Y. Infect Genet Evol 6:309–314
56. Ochong EO, van den Broek IV, Keus K et al (2003) Short report: association between chloroquine and amodiaquine resistance and allelic variation in the *Plasmodium falciparum* multiple drug resistance 1 gene and the chloroquine resistance transporter gene in isolates from the upper Nile in southern Sudan. Am J Trop Med Hyg 69:184–187

57. Sa JM, Twu O, Hayton K et al (2009) Geographic patterns of *Plasmodium falciparum* drug resistance distinguished by differential responses to amodiaquine and chloroquine. Proc Natl Acad Sci USA 106:18883–18889
58. Alifrangis M, Dalgaard MB, Lusingu JP et al (2006) Occurrence of the Southeast Asian/South American SVMNT haplotype of the chloroquine-resistance transporter gene in *Plasmodium falciparum* in Tanzania. J Infect Dis 193:1738–1741
59. Sa JM, Twu O (2010) Protecting the malaria drug arsenal: halting the rise and spread of amodiaquine resistance by monitoring the PfCRT SVMNT type. Malar J 9:374
60. Sisowath C, Ferreira PE, Bustamante LY et al (2007) The role of pfmdr1 in *Plasmodium falciparum* tolerance to artemether-lumefantrine in Africa. Trop Med Int Health 12:736–742
61. Humphreys GS, Merinopoulos I, Ahmed J et al (2007) Amodiaquine and artemether-lumefantrine select distinct alleles of the *Plasmodium falciparum* mdr1 gene in Tanzanian children treated for uncomplicated malaria. Antimicrob Agents Chemother 51:991–997
62. Sisowath C, Stromberg J, Martensson A et al (2005) *In vivo* selection of *Plasmodium falciparum* pfmdr1 86N coding alleles by artemether-lumefantrine (Coartem™). J Infect Dis 191:1014–1017
63. Nsobya SL, Kiggundu M, Nanyunja S et al (2010) *In vitro* sensitivities of *Plasmodium falciparum* to different antimalarial drugs in Uganda. Antimicrob Agents Chemother 54:1200–1206
64. Price RN, Uhlemann AC, van Vugt M et al (2006) Molecular and pharmacological determinants of the therapeutic response to artemether-lumefantrine in multidrug-resistant *Plasmodium falciparum* malaria. Clin Infect Dis 42:1570–1577
65. Price RN, Uhlemann AC, Brockman A et al (2004) Mefloquine resistance in *Plasmodium falciparum* and increased pfmdr1 gene copy number. Lancet 364:438–447
66. Okombo J, Ohuma E, Picot S et al (2011) Update on genetic markers of quinine resistance in *Plasmodium falciparum*. Mol Biochem Parasitol 177:77–82
67. Praygod G, de Frey A, Eisenhut M (2008) Artemisinin derivatives versus quinine in treating severe malaria in children: a systematic review. Malar J 7:210
68. Jones KL, Donegan S, Lalloo DG (2007) Artesunate versus quinine for treating severe malaria. Cochrane Database Syst Rev CD005967
69. Myint HY, Tipmanee P, Nosten F et al (2004) A systematic overview of published antimalarial drug trials. Trans R Soc Trop Med Hyg 98:73–81
70. Achan J, Tibenderana JK, Kyabayinze D et al (2009) Effectiveness of quinine versus artemether-lumefantrine for treating uncomplicated falciparum malaria in Ugandan children: randomised trial. BMJ 339:b2763
71. Adam I, Salih I, Elbashir MI (2005) Quinine for the treatment of uncomplicated *Plasmodium falciparum* malaria in eastern Sudan. Trans R Soc Trop Med Hyg 99:736–738
72. Adegnika AA, Breitling LP, Agnandji ST et al (2005) Effectiveness of quinine monotherapy for the treatment of *Plasmodium falciparum* infection in pregnant women in Lambarene, Gabon. Am J Trop Med Hyg 73:263–266
73. de Vries PJ, Bich NN, Van Thien H et al (2000) Combinations of artemisinin and quinine for uncomplicated falciparum malaria: efficacy and pharmacodynamics. Antimicrob Agents Chemother 44:1302–1308
74. McGready R, Ashley EA, Moo E et al (2005) A randomized comparison of artesunate-atovaquone-proguanil versus quinine in treatment for uncomplicated falciparum malaria during pregnancy. J Infect Dis 192:846–853
75. Basco LK, Le Bras J (1993) *In vitro* activity of chloroquine and quinine in combination with desferrioxamine against *Plasmodium falciparum*. Am J Hematol 42:389–391
76. Brasseur P, Kouamouo J, Moyou RS et al (1992) Mefloquine resistant malaria in Cameroon and correlation with resistance to quinine. Mem Inst Oswaldo Cruz 87(Suppl 3):271–273
77. Brasseur P, Kouamouo J, Moyou-Somo R et al (1992) Multi-drug resistant falciparum malaria in Cameroon in 1987–1988. II. Mefloquine resistance confirmed in vivo and in vitro and its correlation with quinine resistance. Am J Trop Med Hyg 46:8–14

78. Simon F, Le Bras J, Charmot G et al (1986) Severe chloroquine-resistant falciparum malaria in Gabon with decreased sensitivity to quinine. Trans R Soc Trop Med Hyg 80:996–997
79. Warsame M, Wernsdorfer WH, Payne D et al (1991) Susceptibility of *Plasmodium falciparum in vitro* to chloroquine, mefloquine, quinine and sulfadoxine/pyrimethamine in Somalia: relationships between the responses to the different drugs. Trans R Soc Trop Med Hyg 85:565–569
80. Uhlemann AC, Krishna S (2005) Antimalarial multi-drug resistance in Asia: mechanisms and assessment. Curr Top Microbiol Immunol 295:39–53
81. Andriantsoanirina V, Menard D, Rabearimanana S et al (2010) Association of microsatellite variations of *Plasmodium falciparum* Na^+/H^+ exchanger (Pfnhe-1) gene with reduced *in vitro* susceptibility to quinine: lack of confirmation in clinical isolates from Africa. Am J Trop Med Hyg 82:782–787
82. Chaijaroenkul W, Wisedpanichkij R, Na-Bangchang K (2010) Monitoring of *in vitro* susceptibilities and molecular markers of resistance of *Plasmodium falciparum* isolates from Thai-Myanmar border to chloroquine, quinine, mefloquine and artesunate. Acta Trop 113:190–194
83. Lakshmanan V, Bray PG, Verdier-Pinard D et al (2005) A critical role for PfCRT K76T in *Plasmodium falciparum* verapamil-reversible chloroquine resistance. EMBO J 24:2294–2305
84. Sidhu AB, Verdier-Pinard D, Fidock DA (2002) Chloroquine resistance in *Plasmodium falciparum* malaria parasites conferred by pfcrt mutations. Science 298:210–213
85. Woodrow CJ, Krishna S (2006) Antimalarial drugs: recent advances in molecular determinants of resistance and their clinical significance. Cell Mol Life Sci 63:1586–1596
86. Ferdig MT, Cooper RA, Mu J et al (2004) Dissecting the loci of low-level quinine resistance in malaria parasites. Mol Microbiol 52:985–997
87. Reed MB, Saliba KJ, Caruana SR et al (2000) Pgh1 modulates sensitivity and resistance to multiple antimalarials in *Plasmodium falciparum*. Nature 403:906–909
88. Sidhu AB, Valderramos SG, Fidock DA (2005) Pfmdr1 mutations contribute to quinine resistance and enhance mefloquine and artemisinin sensitivity in *Plasmodium falciparum*. Mol Microbiol 57:913–926
89. Pickard AL, Wongsrichanalai C, Purfield A et al (2003) Resistance to antimalarials in Southeast Asia and genetic polymorphisms in pfmdr1. Antimicrob Agents Chemother 47:2418–2423
90. Anderson TJ, Nair S, Qin H et al (2005) Are transporter genes other than the chloroquine resistance locus (pfcrt) and multidrug resistance gene (pfmdr) associated with antimalarial drug resistance? Antimicrob Agents Chemother 49:2180–2188
91. Sidhu AB, Uhlemann AC, Valderramos SG et al (2006) Decreasing pfmdr1 copy number in *Plasmodium falciparum* malaria heightens susceptibility to mefloquine, lumefantrine, halofantrine, quinine, and artemisinin. J Infect Dis 194:528–535
92. Mu J, Ferdig MT, Feng X et al (2003) Multiple transporters associated with malaria parasite responses to chloroquine and quinine. Mol Microbiol 49:977–989
93. Yang Z, Li C, Miao M et al (2011) Multidrug-resistant genotypes of *Plasmodium falciparum*, Myanmar. Emerg Infect Dis 17:498–501
94. Baliraine FN, Nsobya SL, Achan J et al (2011) Limited ability of *Plasmodium falciparum* pfcrt, pfmdr1, and pfnhe1 polymorphisms to predict quinine *in vitro* sensitivity or clinical effectiveness in Uganda. Antimicrob Agents Chemother 55:615–622
95. Briolant S, Pelleau S, Bogreau H et al (2011) *In vitro* susceptibility to quinine and microsatellite variations of the *Plasmodium falciparum* Na^+/H^+ exchanger (Pfnhe-1) gene: the absence of association in clinical isolates from the Republic of Congo. Malar J 10:37
96. Henry M, Briolant S, Zettor A et al (2009) *Plasmodium falciparum* Na^+/H^+ exchanger 1 transporter is involved in reduced susceptibility to quinine. Antimicrob Agents Chemother 53:1926–1930
97. Meng H, Zhang R, Yang H et al (2010) *In vitro* sensitivity of *Plasmodium falciparum* clinical isolates from the China-Myanmar border area to quinine and association with polymorphism in the Na^+/H^+ exchanger. Antimicrob Agents Chemother 54:4306–4313

98. Nkrumah LJ, Riegelhaupt PM, Moura P et al (2009) Probing the multifactorial basis of *Plasmodium falciparum* quinine resistance: evidence for a strain-specific contribution of the sodium-proton exchanger PfNHE. Mol Biochem Parasitol 165:122–131
99. Bukirwa H, Orton L (2005) Artesunate plus mefloquine versus mefloquine for treating uncomplicated malaria. Cochrane Database Syst Rev CD004531
100. Croft AM, Garner P (2008) WITHDRAWN: mefloquine for preventing malaria in non-immune adult travellers. Cochrane Database Syst Rev CD000138
101. Schlagenhauf P, Adamcova M, Regep L et al (2010) The position of mefloquine as a 21st century malaria chemoprophylaxis. Malar J 9:357
102. Toovey S, Bustamante LY, Uhlemann AC et al (2008) Effect of artemisinins and amino alcohol partner antimalarials on mammalian sarcoendoplasmic reticulum calcium adenosine triphosphatase activity. Basic Clin Pharmacol Toxicol 103:209–213
103. Henry M, Diallo I, Bordes J et al (2006) Urban malaria in Dakar, Senegal: chemosusceptibility and genetic diversity of *Plasmodium falciparum* isolates. Am J Trop Med Hyg 75:146–151
104. Pradines B, Hovette P, Fusai T et al (2006) Prevalence of *in vitro* resistance to eleven standard or new antimalarial drugs among *Plasmodium falciparum* isolates from Pointe-Noire, Republic of the Congo. J Clin Microbiol 44:2404–2408
105. Ramharter M, Wernsdorfer WH, Kremsner PG (2004) *In vitro* activity of quinolines against *Plasmodium falciparum* in Gabon. Acta Trop 90:55–60
106. Randrianarivelojosia M, Randrianasolo L, Randremanana RV et al (2004) Susceptibility of *Plasmodium falciparum* to the drugs used to treat severe malaria (quinine) and to prevent malaria (mefloquine, cycloguanil) in Comoros Union and Madagascar. S Afr Med J 94:47–51
107. Ndong JM, Atteke C, Aubouy A et al (2003) *In vitro* activity of chloroquine, quinine, mefloquine and halofantrine against Gabonese isolates of *Plasmodium falciparum.* Trop Med Int Health 8:25–29
108. Yavo W, Bla KB, Djaman AJ et al (2010) *In vitro* susceptibility of *Plasmodium falciparum* to monodesethylamodiaquine, quinine, mefloquine and halofantrine in Abidjan (Cote d'Ivoire). Afr Health Sci 10:111–116
109. Adam I, Ali DA, Alwaseila A et al (2004) Mefloquine in the treatment of falciparum malaria during pregnancy in Eastern Sudan. Saudi Med J 25:1400–1402
110. Aubouy A, Fievet N, Bertin G et al (2007) Dramatically decreased therapeutic efficacy of chloroquine and sulfadoxine-pyrimethamine, but not mefloquine, in southern Benin. Trop Med Int Health 12:886–894
111. Bouyou-Akotet MK, Ramharter M, Ngoungou EB et al (2010) Efficacy and safety of a new pediatric artesunate-mefloquine drug formulation for the treatment of uncomplicated falciparum malaria in Gabon. Wien Klin Wochenschr 122:173–178
112. Faye B, Ndiaye JL, Tine R et al (2010) A randomized trial of artesunate mefloquine versus artemether lumefantrine for the treatment of uncomplicated *Plasmodium falciparum* malaria in Senegalese children. Am J Trop Med Hyg 82:140–144
113. Sowunmi A, Gbotosho GO, Happi C et al (2009) Therapeutic efficacy and effects of artesunate-mefloquine and mefloquine alone on malaria-associated anemia in children with uncomplicated *Plasmodium falciparum* malaria in southwest Nigeria. Am J Trop Med Hyg 81:979–986
114. Tietche F, Chelo D, Mina Ntoto NK et al (2010) Tolerability and efficacy of a pediatric granule formulation of artesunate-mefloquine in young children from Cameroon with uncomplicated falciparum malaria. Am J Trop Med Hyg 82:1034–1040
115. Fryauff DJ, Owusu-Agyei S, Utz G et al (2007) Mefloquine treatment for uncomplicated falciparum malaria in young children 6-24 months of age in northern Ghana. Am J Trop Med Hyg 76:224–231
116. Barends M, Jaidee A, Khaohirun N et al (2007) *In vitro* activity of ferroquine (SSR 97193) against *Plasmodium falciparum* isolates from the Thai-Burmese border. Malar J 6:81

117. Huttinger F, Satimai W, Wernsdorfer G et al (2010) Sensitivity to artemisinin, mefloquine and quinine of *Plasmodium falciparum* in northwestern Thailand. Wien Klin Wochenschr 122(Suppl 3):52–56
118. Noedl H, Krudsood S, Leowattana W et al (2007) *In vitro* antimalarial activity of azithromycin, artesunate, and quinine in combination and correlation with clinical outcome. Antimicrob Agents Chemother 51:651–656
119. Oduola AM, Milhous WK, Weatherly NF et al (1988) *Plasmodium falciparum*: induction of resistance to mefloquine in cloned strains by continuous drug exposure in vitro. Exp Parasitol 67:354–360
120. Wilson CM, Volkman SK, Thaithong S et al (1993) Amplification of pfmdr 1 associated with mefloquine and halofantrine resistance in *Plasmodium falciparum* from Thailand. Mol Biochem Parasitol 57:151–160
121. Cowman AF, Galatis D, Thompson JK (1994) Selection for mefloquine resistance in *Plasmodium falciparum* is linked to amplification of the pfmdr1 gene and cross-resistance to halofantrine and quinine. Proc Natl Acad Sci USA 91:1143–1147
122. Price RN, Cassar C, Brockman A et al (1999) The pfmdr1 gene is associated with a multidrug-resistant phenotype in *Plasmodium falciparum* from the western border of Thailand. Antimicrob Agents Chemother 43:2943–2949
123. Nothdurft HD, Clemens R, Bock HL et al (1993) Halofantrine: a new substance for treatment of multidrug-resistant malaria. Clin Investig 71:69–73
124. Bryson HM, Goa KL (1992) Halofantrine. A review of its antimalarial activity, pharmacokinetic properties and therapeutic potential. Drugs 43:236–258
125. Brasseur P, Bitsindou P, Moyou RS et al (1993) Fast emergence of *Plasmodium falciparum* resistance to halofantrine. Lancet 341:901–902
126. Basco LK (1993) Halofantrine resistance in African countries. Lancet 341:1283
127. Carme B, Gay F, Hayette MP et al (1993) Halofantrine resistance in African countries. Lancet 341:1282–1283
128. Chavchich M, Gerena L, Peters J et al (2010) Role of pfmdr1 amplification and expression in induction of resistance to artemisinin derivatives in *Plasmodium falciparum*. Antimicrob Agents Chemother 54:2455–2464
129. Kim HS, Okuda Y, Begum K et al (2001) Analysis of Pfmdr 1 gene in mefloquine-resistant *Plasmodium falciparum*. Nucleic Acids Res Suppl (1):231–232
130. Van Tyne D, Park DJ, Schaffner SF et al (2011) Identification and functional validation of the novel antimalarial resistance locus PF10_0355 in *Plasmodium falciparum*. PLoS Genet 7:e1001383
131. Beese SE, Negishi T, Levin DE (2009) Identification of positive regulators of the yeast fps1 glycerol channel. PLoS Genet 5:e1000738
132. Geisler M, Girin M, Brandt S et al (2004) *Arabidopsis* immunophilin-like TWD1 functionally interacts with vacuolar ABC transporters. Mol Biol Cell 15:3393–3405
133. Miletti-Gonzalez KE, Chen S, Muthukumaran N et al (2005) The CD44 receptor interacts with P-glycoprotein to promote cell migration and invasion in cancer. Cancer Res 65:6660–6667
134. Maude RJ, Pontavornpinyo W, Saralamba S et al (2009) The last man standing is the most resistant: eliminating artemisinin-resistant malaria in Cambodia. Malar J 8:31
135. Li J, Zhou B (2010) Biological actions of artemisinin: insights from medicinal chemistry studies. Molecules 15:1378–1397
136. Eastman RT, Fidock DA (2009) Artemisinin-based combination therapies: a vital tool in efforts to eliminate malaria. Nat Rev Microbiol 7:864–874
137. White NJ (2004) Antimalarial drug resistance. J Clin Investig 113:1084–1092
138. Anonymous (2011) WHO GPARC. http://www.who.int/malaria/publications/atoz/9789241500838/en/index.html
139. Dondorp AM, Yeung S, White L et al (2010) Artemisinin resistance: current status and scenarios for containment. Nat Rev Microbiol 8:272–280

140. Ding XC, Beck HP, Raso G (2011) *Plasmodium* sensitivity to artemisinins: magic bullets hit elusive targets. Trends Parasitol 27:73–81
141. Eckstein-Ludwig U, Webb RJ, Van Goethem ID et al (2003) Artemisinins target the SERCA of *Plasmodium falciparum*. Nature 424:957–961
142. Uhlemann AC, Cameron A, Eckstein-Ludwig U et al (2005) A single amino acid residue can determine the sensitivity of SERCAs to artemisinins. Nat Struct Mol Biol 12:628–629
143. Valderramos SG, Scanfeld D, Uhlemann AC et al (2010) Investigations into the role of the *Plasmodium falciparum* SERCA (PfATP6) L263E mutation in artemisinin action and resistance. Antimicrob Agents Chemother 54:3842–3852
144. Jambou R, Legrand E, Niang M et al (2005) Resistance of *Plasmodium falciparum* field isolates to *in-vitro* artemether and point mutations of the SERCA-type PfATPase6. Lancet 366:1960–1963
145. Ibrahim ML, Khim N, Adam HH et al (2009) Polymorphism of PfATPase in Niger: detection of three new point mutations. Malar J 8:28
146. Codd A, Teuscher F, Kyle DE et al (2011) Artemisinin-induced parasite dormancy: a plausible mechanism for treatment failure. Malar J 10:56
147. Teuscher F, Gatton ML, Chen N et al (2010) Artemisinin-induced dormancy in *Plasmodium falciparum*: duration, recovery rates, and implications in treatment failure. J Infect Dis 202:1362–1368
148. Witkowski B, Lelievre J, Barragan MJ et al (2010) Increased tolerance to artemisinin in *Plasmodium falciparum* is mediated by a quiescence mechanism. Antimicrob Agents Chemother 54:1872–1877
149. Nzila A (2006) The past, present and future of antifolates in the treatment of *Plasmodium falciparum* infection. J Antimicrob Chemother 57:1043–1054
150. Premji Z, Umeh RE, Owusu-Agyei S et al (2009) Chlorproguanil-dapsone-artesunate versus artemether-lumefantrine: a randomized, double-blind Phase III trial in African children and adolescents with uncomplicated *Plasmodium falciparum* malaria. PLoS One 4:e6682
151. Tiono AB, Dicko A, Ndububa DA et al (2009) Chlorproguanil-dapsone-artesunate versus chlorproguanil-dapsone: a randomized, double-blind, Phase III trial in African children, adolescents, and adults with uncomplicated *Plasmodium falciparum* malaria. Am J Trop Med Hyg 81:969–978
152. Nakato H, Vivancos R, Hunter PR (2007) A systematic review and meta-analysis of the effectiveness and safety of atovaquone proguanil (Malarone) for chemoprophylaxis against malaria. J Antimicrob Chemother 60:929–936
153. Looareesuwan S, Chulay JD, Canfield CJ et al (1999) Malarone (atovaquone and proguanil hydrochloride): a review of its clinical development for treatment of malaria .Malarone Clinical Trials Study Group. Am J Trop Med Hyg 60:533–541
154. Gregson A, Plowe CV (2005) Mechanisms of resistance of malaria parasites to antifolates. Pharmacol Rev 57:117–145
155. Sibley CH, Hyde JE, Sims PF et al (2001) Pyrimethamine-sulfadoxine resistance in *Plasmodium falciparum*: what next? Trends Parasitol 17:582–588
156. Cui L, Su XZ (2009) Discovery, mechanisms of action and combination therapy of artemisinin. Expert Rev Anti Infect Ther 7:999–1013
157. Wilson AL (2011) A systematic review and meta-analysis of the efficacy and safety of intermittent preventive treatment of malaria in children (IPTc). PLoS One 6:e16976
158. Gosling RD, Carneiro I, Chandramohan D (2009) Intermittent preventive treatment of malaria in infants: how does it work and where will it work? Trop Med Int Health 14:1003–1010
159. Briand V, Bottero J, Noel H et al (2009) Intermittent treatment for the prevention of malaria during pregnancy in Benin: a randomized, open-label equivalence trial comparing sulfadoxine-pyrimethamine with mefloquine. J Infect Dis 200:991–1001
160. Cisse B, Cairns M, Faye E et al (2009) Randomized trial of piperaquine with sulfadoxine-pyrimethamine or dihydroartemisinin for malaria intermittent preventive treatment in children. PLoS One 4:e7164

161. Hyde JE (2008) Antifolate resistance in Africa and the 164-dollar question. Trans R Soc Trop Med Hyg 102:301–303
162. Nzila A, Ochong E, Nduati E et al (2005) Why has the dihydrofolate reductase 164 mutation not consistently been found in Africa yet? Trans R Soc Trop Med Hyg 99:341–346
163. Sridaran S, McClintock SK, Syphard LM et al (2010) Anti-folate drug resistance in Africa: meta-analysis of reported dihydrofolate reductase (dhfr) and dihydropteroate synthase (dhps) mutant genotype frequencies in African *Plasmodium falciparum* parasite populations. Malar J 9:247
164. Nair S, Miller B, Barends M et al (2008) Adaptive copy number evolution in malaria parasites. PLoS Genet 4:e1000243
165. Sandefur CI, Wooden JM, Quaye IK et al (2007) Pyrimethamine-resistant dihydrofolate reductase enzymes of *Plasmodium falciparum* are not enzymatically compromised in vitro. Mol Biochem Parasitol 154:1–5
166. Sirawaraporn W, Sathitkul T, Sirawaraporn R et al (1997) Antifolate-resistant mutants of *Plasmodium falciparum* dihydrofolate reductase. Proc Natl Acad Sci USA 94:1124–1129
167. Wootton JC, Feng X, Ferdig MT et al (2002) Genetic diversity and chloroquine selective sweeps in *Plasmodium falciparum*. Nature 418:320–323
168. Maiga O, Djimde AA, Hubert V et al (2007) A shared Asian origin of the triple-mutant dhfr allele in *Plasmodium falciparum* from sites across Africa. J Infect Dis 196:165–172
169. McCollum AM, Basco LK, Tahar R et al (2008) Hitchhiking and selective sweeps of *Plasmodium falciparum* sulfadoxine and pyrimethamine resistance alleles in a population from central Africa. Antimicrob Agents Chemother 52:4089–4097
170. Mita T (2010) Origins and spread of pfdhfr mutant alleles in *Plasmodium falciparum*. Acta Trop 114:166–170
171. Mita T, Tanabe K, Kita K (2009) Spread and evolution of *Plasmodium falciparum* drug resistance. Parasitol Int 58:201–209
172. Roper C, Pearce R, Nair S et al (2004) Intercontinental spread of pyrimethamine-resistant malaria. Science 305:1124
173. Certain LK, Briceno M, Kiara SM et al (2008) Characteristics of *Plasmodium falciparum* dhfr haplotypes that confer pyrimethamine resistance, Kilifi, Kenya, 1987–2006. J Infect Dis 197:1743–1751
174. McCollum AM, Poe AC, Hamel M et al (2006) Antifolate resistance in *Plasmodium falciparum*: multiple origins and identification of novel dhfr alleles. J Infect Dis 194:189–197
175. Pearce RJ, Pota H, Evehe MS et al (2009) Multiple origins and regional dispersal of resistant dhps in African *Plasmodium falciparum* malaria. PLoS Med 6:e1000055

Chapter 14
Random and Rational Approaches to HIV Drug Discovery in Africa

R. Hewer, F.H. Kriel, and J. Coates

14.1 Drug Discovery and Development

The drug discovery and development process is an inherently lengthy endeavour, with the development of a compound from initial discovery to market approval averaging between 10 and >15 years. It is also widely acknowledged as a costly endeavour although the precise costs, and the methodology to calculate them, have been highly debated. An often-cited study places the average costs of bringing a new drug to market at $802 million (in 2000 dollars) [1, 2] while other studies have estimated these costs from $43.4 million [3] to $1.2 billion [4]. A high rate of attrition is observed in the process, and it estimated that only 1 in every 5,000–10,000 compounds entering the pipeline will achieve clinical approval. As evident in Fig. 14.1, the highest rate of attrition is in the drug discovery phase where a 'fail early, fail cheap' paradigm applies in order to reduce costly downstream failures. While risk is maintained throughout the pipeline, it reduces considerably as the research and development costs increase (Fig. 14.1).

14.1.1 HIV Drugs and Treatment Strategies

To date, 35 antiretroviral (ARV) drugs have been approved by the Food and Drug Administration (FDA) for use in the therapeutic treatment of human immuno-deficiency virus (HIV) infection [5]. These agents can be broadly categorised under five different drug classes, depending on which stage of the HIV life cycle they modulate, with an additional multi-class category (Table 14.1). The first drug

R. Hewer (✉) • F.H. Kriel • J. Coates
Advanced Materials Division, Mintek, Private Bag X3015, Randburg, 2125, Biomed, Johannesburg, South Africa
e-mail: raymondh@mintek.co.za

K. Chibale et al. (eds.), *Drug Discovery in Africa*,
DOI 10.1007/978-3-642-28175-4_14,

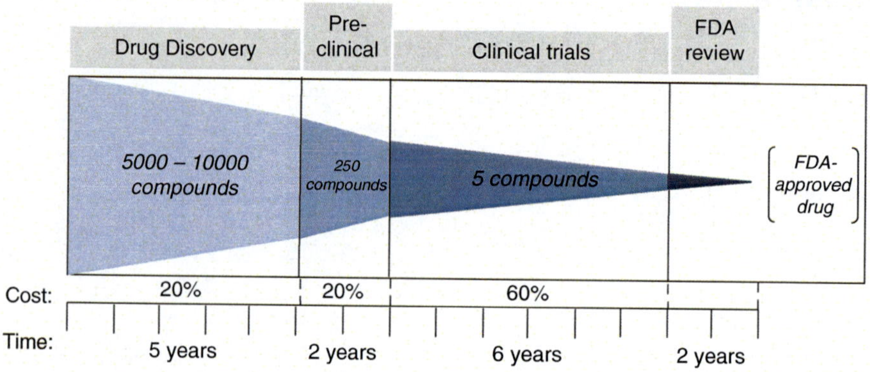

Fig. 14.1 Schematic representation of the drug discovery and development pipeline from initial discovery to market entry, illustrating the average success rates and estimated division of time and costs between the three broad phases that underpin the process

approved was zidovudine (AZT) in 1987 while the latest HIV drug approved by the FDA was Complera, a multi-class combination product developed by Gilead Sciences which entered the market in August 2011. Mostly, HIV drugs are used in combination with three or more drugs administered together, typically involving two reverse transcriptase inhibitors and an inhibitor from one of the other drug classes. This combination therapy, known as highly active antiretroviral therapy (HAART), has proven to be unequivocally successful in the treatment of HIV-positive/acquired immune deficiency syndrome (AIDS) patients. Since its inception in 1996, HAART has significantly reduced mortality and morbidity within the HIV-infected population and effectively transformed a life-threatening disease with short survival rate into a manageable condition. As a synergistic drug effect is induced through combination therapy, less of each drug is required for effective treatment, thus reducing viral resistance, toxicity and pill burden. Nonetheless, virologic failure (as characterised by a persistent viral load above 5,000 copies/ml) can occur primarily due to patient non-compliance and/or viral resistance and may necessitate a treatment regimen switch. Following failure of two or more ARV treatments, durable virologic suppression is unlikely, but continuation of HAART is recommended to prevent further immunologic deterioration.

14.1.2 HIV Drugs in Development

A substantial need exists for the ARV therapy described above. As of the end of 2009, an estimated 33 million people (of which 30 million are from low- and middle-income countries) were infected with HIV with 2 million new infections occurring annually [6]. An estimated 5.25 million HIV-positive patients from low- and middle-income countries are currently receiving treatment, but considering

Table 14.1 Complete list of drugs approved by the Food and Drug Administration (FDA) for the therapeutic treatment of HIV, categorised by class of drug and function

Class	Function	Brand name	Generic name	Manufacturer name	Approval date
Multi-class combination products	Combination of more than one drug from different drug classes	Atripla	Efavirenz, emtricitabine and tenofovir disoproxil fumarate	Bristol-Myers Squibb and Gilead Sciences	12-Jul-2006
		Complera	Emtricitabine, rilpivirine and tenofovir disoproxil fumarate	Gilead Sciences	10-Aug-2011
Reverse transcriptase inhibitors (RTIs) including nucleoside reverse transcriptase inhibitors (NRTIs), nucleotide reverse transcriptase inhibitors (NtRTIs) and non-nucleoside reverse transcriptase inhibitors (NNRTIs)	Suppress replication by inhibiting function of the HIV-1 reverse transcriptase enzyme. The nucleoside and nucleotide analogues cause premature termination of the proviral (viral precursor) DNA chain with all NRTIs requiring phosphorylation in the host cell prior to their incorporation into the viral DNA. The NNRTIs bind to a site distant to the active site and function through allosteric inhibition of the reverse transcriptase enzyme	Combivir	Lamivudine and zidovudine	GlaxoSmithKline	27-Sep-1997
		Emtriva	Emtricitabine/FTC	Gilead Sciences	02-Jul-2003
		Epivir	Lamivudine/3TC	GlaxoSmithKline	17-Nov-1995
		Epzicom	Abacavir and lamivudine	GlaxoSmithKline	02-Aug-2004
		Hivid	Zalcitabine/dideoxycytidine/ddC (no longer marketed)	Hoffmann-La Roche	19-Jun-1992
		Retrovir	Zidovudine/azidothymidine/AZT/ZDV	GlaxoSmithKline	19-Mar-1987
		Trizivir	Abacavir, zidovudine and lamivudine	GlaxoSmithKline	14-Nov-2000
		Truvada	Tenofovir disoproxil fumarate and emtricitabine	Gilead Sciences, Inc.	02-Aug-2004

(continued)

Table 14.1 (continued)

Class	Function	Brand name	Generic name	Manufacturer name	Approval date
		Videx EC	Enteric-coated didanosine/ddI EC	Bristol-Myers Squibb	31-Oct-2000
		Videx	Didanosine/ dideoxyinosine/ddI	Bristol-Myers Squibb	09-Oct-1991
		Viread	Tenofovir disoproxil fumarate/TDF	Gilead	26-Oct-2001
		Zerit	Stavudine/d4T	Bristol-Myers Squibb	24-Jun-1994
		Ziagen	Abacavir sulphate/ ABC	GlaxoSmithKline	17-Dec-1998
		Edurant	Rilpivirine	Tibotec Therapeutics	20-May-2011
		Intelence	Etravirine	Tibotec Therapeutics	18-Jan-2008
		Rescriptor	Delavirdine/DLV	Pfizer	04-Apr-1997
		Sustiva	Efavirenz/EFV	Bristol-Myers Squibb	17-Sep-1998
		Viramune	Nevirapine/NVP	Boehringer Ingelheim	21-Jun-1996
		Viramune XR (extended release)	Nevirapine/NVP	Boehringer Ingelheim	25-Mar-2011
Protease inhibitors (PIs)	Prevent HIV-1 protease from cleaving newly formed viral polyproteins and forming mature viral particles. The resulting viral particles are uninfectious	Agenerase	Amprenavir/APV	GlaxoSmithKline	15-Apr-1999
		Aptivus	Tipranavir/TPV	Boehringer Ingelheim	22-Jun-2005
		Crixivan	Indinavir/IDV	Merck	13-Mar-1996
		Fortovase	Saquinavir (no longer marketed)	Hoffmann-La Roche	07-Nov-1997

		Invirase	Saquinavir mesylate/SQV	Hoffmann-La Roche	06-Dec-1995
		Kaletra	Lopinavir/LPV and ritonavir/RTV	Abbott Laboratories	15-Sep-2000
		Lexiva	Fosamprenavir calcium/FOS-APV	GlaxoSmithKline	20-Oct-2003
		Norvir	Ritonavir/RTV	Abbott Laboratories	01-Mar-1996
		Prezista	Darunavir	Tibotec, Inc.	23-Jun-2006
		Reyataz	Atazanavir sulphate/ATV	Bristol-Myers Squibb	20-Jun-2003
		Viracept	Nelfinavir mesylate/NFV	Agouron Pharmaceuticals	14-Mar-1997
Fusion inhibitors (FI)	Disrupt fusion of the virus with the host cell by binding to HIV-1 gp41	Fuzeon	Enfuvirtide/T-20	Hoffmann-La Roche & Trimeris	13-Mar-2003
Entry inhibitors (EI)-CCR5 co-receptor antagonist	Bind to host cell CCR5 receptors preventing HIV gp120 from coupling with the receptor	Selzentry	Maraviroc	Pfizer	06-Aug-2007
Integrase strand-transfer inhibitors (INSTI)	Inhibit the strand-transfer function of HIV-1 integrase, therefore preventing integration of viral DNA into the host genome	Isentress	Raltegravir	Merck & Co., Inc.	12-Oct-2007

Table 14.2 Anti-HIV agents under investigation in either phase II or phase III clinical trials[a]

Drug	Class	Mechanism	Phase
Apricitabine	RT inhibitor	NRTI	III
BMS-663068	Fusion inhibitor	Binds gp120	II
Cenicriviroc	Entry inhibitor	CCR5 antagonist/anti-inflammatory	II
Dolutegravir	Integrase inhibitor	Strand-transfer inhibitor	II
Elvitegravir	Integrase inhibitor	Strand-transfer inhibitor	III
Elvucitabine	RT inhibitor	NRTI	II
Festinavir	RT inhibitor	NRTI	II
Ibalizumab	Entry inhibitor	CD4-specific monoclonal antibody	II
KP-1461	RT inhibitor	NRTI/viral decay accelerator	II
Lersivirine	RT inhibitor	NNRTI	II
PRO 140	Entry inhibitor	CCR5 monoclonal antibody	II
Quad	Multi-class	Fixed dose combination	III
Racivir	RT inhibitor	NRTI	II
Vivecon	Maturation inhibitor		II

[a]Information as obtained from [10, 11] and current as of November 2011

WHO guidance [7], which recommends the initiation of ARV therapy in all patients with a CD4 count ≤350 cells/mm^3 irrespective of clinical symptoms, a further 10 million patients are also in need of treatment [8].

At present, the global drug market for HIV is estimated at US $8 billion/year and expected to grow as HIV patients live longer, in the absence of a preventative vaccine and as access to antiviral treatment is extended to more patients. As a result, the market for HIV drugs is projected to reach US $15.5 million by 2015 [9]. Nonetheless, shortcomings of current ARV treatment have been highlighted, and a number of strategies have been put forward that may improve ARV regimens. Possibilities for ARV regimen improvement include higher efficacy approaching 100%, higher tolerability, increased barrier to resistance, lower associated costs, improved compatibility with tuberculosis (TB) and hepatitis medication, less dosing, safer drugs for pregnant women and in paediatric patients, and no cold-chain requirement [8]. Review of ongoing clinical trials indicates that 60 new compounds are currently in various stages of clinical investigation with three drugs (namely elvitegravir, apricitabine and Quad) in phase III clinical trials and 11 drugs in phase II clinical trials (Table 14.2 [10, 11]).

14.1.3 Approaches to HIV Drug Discovery

The need for improved ARV therapy and replenishment of the drug development pipeline necessitates the continual search for new antiviral agents. New chemical entities (NCE) are typically identified through the discovery phase of the drug development pipeline and are sought for both new targets and established/validated targets. As illustrated in Fig. 14.1, the discovery phase of the drug development pipeline consumes approximately one third of the total time and 20% of the cost

Fig. 14.2 An overview of the phases and some of the methods involved in early-stage drug discovery which lead to the preclinical development phase

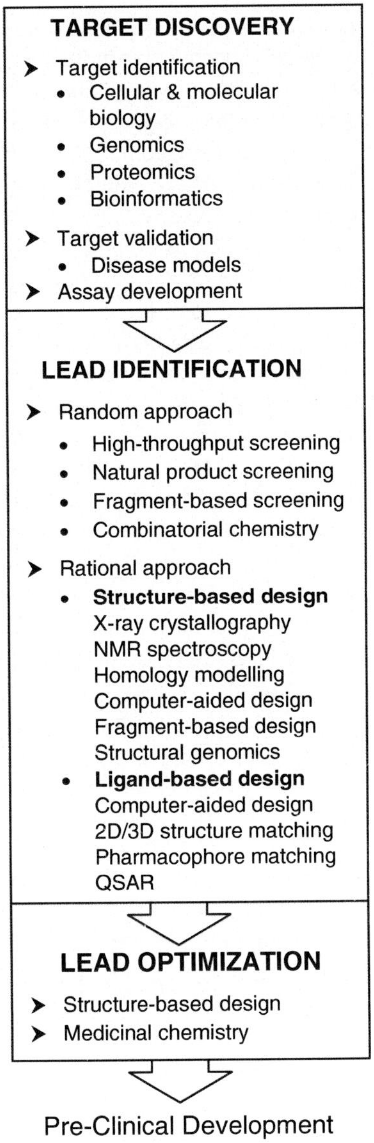

expended throughout the pipeline and carries the highest risk of failure. At this stage, the costs, time and risks associated with identifying inhibitors of new targets are higher than those associated with established targets.

A simplified overview illustrating a typical approach to drug discovery is illustrated in Fig. 14.2. Following target discovery and assay development, lead compounds are sought through either a random (screening) or rational (design) approach. Natural products and synthetic derivatives thereof have provided and continue to provide candidates for screening in this phase of drug discovery. Natural sources such as plants, microorganisms and the marine environment can provide low-molecular-mass natural products of vast and unique structural diversity from which lead compounds can be identified. In many cases, comprehensive libraries of natural products have been assembled for lead discovery. These natural product libraries as well as synthetic compound libraries (generated through techniques such as combinatorial chemistry and parallel synthesis) and fragment-based libraries can be tested for their ability to modulate the target through the use of automated high-throughput screening (HTS). The mean size of a compound library within a large pharmaceutical company is ±1.8 million while small pharmaceutical companies and biotech companies are estimated to have libraries comprising ±1.2 million compounds [12]. Current technological capabilities allow for automated testing of in-house compound libraries with a throughput of >10,000 compounds (HTS) to >100,000 compounds (ultra HTS, uHTS) per day with a recent survey revealing that HTS laboratories are testing an average of 505,530 compounds per screen in primary screening [13]. Consequently, this technology has expedited the speed at which drug discovery can be undertaken and increased the chemical space that has been explored.

However, even with HTS technology, exploring the entire chemical space may not be entirely realistic. At present, a chemical space approaching 10^{63} compounds with molecular weights below 500 Da is thought to exist while the largest compound libraries available do not exceed 10^{12} compounds. Techniques such as combinatorial chemistry may work towards filling the chemical space; however, it may also obscure the drug discovery phase rather than facilitate it through the addition of more non-relevant compounds and increased costs associated with the screening of larger libraries. This notwithstanding, the costs and resources required for HTS are prohibitive in many cases, preventing many research laboratories from implementing the HTS approach in the first place. To this end, rational drug design offers an alternative/complementary approach to screening for lead identification. Rational drug design can be broadly divided into structure-based design (or direct drug design) which relies on prior knowledge of the three-dimensional structure of the target (typically a protein) to identify new molecules and ligand-based design (or indirect drug design) which relies on prior knowledge of other active ligands that interact with and modulate the target's activity. A vast array of methods and technologies are required and can be utilised during rational drug design, some of which are denoted in Fig. 14.2.

For structure-based drug design, the prior knowledge of the structure of the biological target is generally determined via X-ray crystallography and/or NMR spectroscopy. In some cases in which both are missing, homology models have been generated based on similar and related proteins. Computer-aided drug design (CADD) has also been widely employed to predict and prepare 3-D models, model potential conformational changes in the target, identify in silico HIT compounds through virtual screen, model the interaction between ligands and the protein and optimise identified LEAD compounds. Fragment-based lead discovery can also be initiated and then later optimised through a rational drug design approach based on the 3D structure of the target of interest.

In ligand-based drug discovery, knowledge of active ligands/inhibitors of the target allows for the generation of a pharmacophore model which describes the characteristics and features required for the ligand to interact with the molecular target. Experimental means can then be employed to synthesise and evaluate novel analogues that fit the pharmacophore with the intention of discovering new inhibitors. As with structure-based drug design, computer-aided design can also facilitate ligand-based drug design through the in silico generation of a pharmacophore, database searching, library formation and quantitative structure–activity relationship (QSAR) development in which the experimentally determined activity of active compounds can be correlated with calculated properties of the molecules to discover new analogues.

Rational drug design has been used effectively over several decades to discover countless new chemical entities for further optimisation and development along the drug development pipeline. Ultimately, the use of rational drug design has also led to the development of several FDA-approved inhibitors. For HIV in particular, HIV protease inhibitors were initially discovered through the application of structure-based drug design [14]. In this case, knowledge of the structure and mode of action (MOA) of the active site of the HIV aspartyl protease led to the design of inhibitors that mimic the transition state of natural substrates of the enzyme. Hydroxyethylamine was the most promising transition-state analogue discovered in this manner, and it led to the development of saquinavir, the first HIV protease inhibitor approved by the FDA [5]. Thereafter, a range of HIV protease inhibitors (including ritonavir, indinavir, amprenavir, fosamprenavir, lopinavir, atazanavir and nelfinavir) have been developed according to the same principles and guided by structure-based drug design [14, 15]. The first FDA-approved HIV fusion inhibitor, enfuvirtide (approved in March 2003 and marketed by Roche under the name Fuzeon), has also been highlighted as a success of structure-based drug design [16]. Finally, a plethora of HIV nucleoside reverse transcriptase inhibitors (NRTIs) and non-nucleoside reverse transcriptase inhibitors (NNRTIs) stem from the application of ligand-based drug design. For example, several approved NNRTIs have been designed and developed to fit a 'butterfly' pharmacophore comprising a hydrophilic 'body' flanked by two hydrophobic aromatic 'wings' [17].

14.2 HIV Drug Discovery in Africa

Sub-Saharan Africa remains the region most affected by the HIV/AIDS epidemic with 67% of the world's HIV-infected people (approximately 22.5 million people) and with almost 72% of all AIDS-related deaths/year. Of the 11 countries that comprise southern Africa (Angola, Namibia, Zambia, Zimbabwe, Botswana, Malawi, Mozambique, South Africa, Lesotho, Swaziland, Madagascar), at least

Table 14.3 Summary of the HIV-related publications and HIV drug discovery publications from 1990 to present for Africa as a whole and select African countries with comparison to the United States of America (USA)

Country/continent	Number of HIV-related publications (1990–present) according to ISI Web of Knowledge	Number of HIV drug discovery publications
Africa	11,212	192
South Africa	3,921	57
Egypt	192	52
Morocco	91	37
Uganda	1,058	7
Nigeria	516	5
Kenya	869	4
Tanzania	601	4
Cameroon	312	4
Ivory Coast	261	4
Ethiopia	412	4
Algeria	594	4
Tunisia	471	3
USA	73,465	–

6 countries have infection rates exceeding 20%. According to the UNAIDS, South Africa is the country with the highest HIV-positive population in the world with an estimated five million HIV-infected people. In response, in 2006, South Africa launched the largest ARV rollout programme ever undertaken, and by 2010, over 500,000 people were receiving ARV therapy. Additionally, as the continent most affected by the disease, considerable HIV research has also been undertaken in Africa with work conducted and published in basic and clinical research, as well as the social, economic and political fields. Taken together, review of the ISI Web of Knowledge for the period from 1990 to present day reveals more than 11,000 HIV-related publications stemming from the African continent (using HIV as the search keyword with the condition of at least one African country in the affiliations) as reflected in Table 14.3.

In terms of HIV drug discovery research specifically, just more than 190 publications could be isolated from the ISI Web of Knowledge during the same period (1990–present). Reflecting the multidisciplinary nature of drug discovery, these publications cover a range of scientific fields including among others: chemistry, biochemistry, microbiology, computer science and physics. Of the 192 papers identified, 30% have South African authors, 27% have Egyptian authors, and 19% have Moroccan authors. The remainder of the publications are divided among a further 14 countries, each contributing no more than 4% of the total African publications. Only 5 of the 192 publications have more than one African country listed in the affiliations. Of the 54 sovereign countries on the African continent, 37 countries have not published within the field of HIV drug discovery according to the keywords and criteria utilised to search the ISI Web of Knowledge.

14.2.1 *HIV Drug Discovery in Africa: Random (Screening) Approach*

A number of HIV drug discovery publications originating from African laboratories demonstrate application of the random (screening) approach to drug discovery. Given the exceptional biodiversity present in Africa, and that over 60% of new chemical entities are derived from natural products or semi-synthetic derivatives of natural products [18–20], it is not surprising that natural product research dominates the published material. Through the use of bioprospecting, a large number of different plant species have been investigated for anti-HIV activity with over 150 different species reported in the literature (Table 14.4). In most cases, anecdotal evidence of activity obtained through use of the plant as traditional medicine has impelled its testing. For instance, Asres and co-workers [24] completed a comprehensive study of 21 plants which were ethnobotanically selected based on their use in Ethiopian traditional medicine for the treatment of various ailments including infectious diseases. In total, 71 extracts obtained from these plants were evaluated in cell-based antiviral assays for inhibitory activity towards the replication of HIV-1 (III_B) and HIV-2 (ROD) in MT-4 cells. Overall, the plants lacked significant antiviral activity, with the acetone extract from the leaf of *Combretum paniculatum* displaying the highest antiviral activity against both HIV-1 and HIV-2 with selective index (SI, calculated as CC_{50}/EC_{50}) values of 6.4 and 32 obtained, respectively [18].

A further example of an ethnobotanically selected plant is that of *Sutherlandia frutescens* which is widely used as a traditional medicine in South Africa and has gained international attention through its use as treatment for HIV/AIDS patients. Anecdotal evidence has suggested that HIV-infected people using *Sutherlandia* treatment have gained weight, have better appetites, improved CD4 counts and decreased viral loads [55, 56]. These claimed outcomes have been attributed to three bioactive compounds present in high quantities within the plant. L-Canavanine, present at 2.2 mg/g of dried leaf, is thought to confer antiviral and anti-inflammation activity, while pinitol has been suggested to be effective against wasting. The inhibitory neurotransmitter γ-amino butyric acid (GABA), isolated in levels up to 14 mg/g dry weight of plant material, is believed to contribute to the sense of general well-being observed in patients on *Sutherlandia* treatment. In direct enzymatic assays, plant material has been evaluated for its activity against specific viral targets such as reverse transcriptase [36, 40, 41], RNaseH [40] and protease [36]. Despite this, the exact mechanism of action by which *S. frutescens* confers its medicinal benefit still requires further delineation. A number of publications have also highlighted the need to exhibit caution in endorsing *Sutherlandia* (and other well-known herbal remedies such as *Hypoxis hemerocallidea*, commonly known as African potato) as a treatment for HIV until such time as its efficacy, toxicity and drug interactions have been fully determined through clinical trials (as reviewed in [57]). For this very purpose, the South African Department of Science and Technology (DST) has invested R10 million (US $1.4 million) towards

Table 14.4 Natural species tested for anti-HIV activity as reported in published literature from 1990 to present, indicating the biological targets investigated

Country	Species: plant	Biological assay[a]	Reference
Cameroon	*Treculia obovoidea, T. africana, T. acuminata*	RT	[21]
	Erythrina senegalensis	PR	[22]
	Enantia chlorantha	CB	[23]
Ethiopia	*Alcea rosea, Calpurnia aurea, Justitia schimperiana, Salvia leucantha, Solanecio gigas, Stephania abyssinica, Verbena officinalis, Ajuga integrifolia, Artemisia abyssinica, Artemisia afra, Coriandrum sativum, Melilotus elegans, Bersama abyssinica, Dovyalis abyssinica, Securidaca longepedunculata, Withania somnifera, Ximenia americana, Clerodendron discolor, Combretum paniculatum, Dodonaea angustifolia, Vernonia galamensis*	CB (HIV-1 and HIV-2)	[24, 25]
Ivory Coast	*Tieghemella heckelii*	Cell fusion	[26]
Kenya	*Maytenus buchananii, Maytenus senegalensis, Prunus africana, Acacia mellifera, Rhus natalensis, Vernonia jugalis, Melia azedarach*	RT	[27]
Morocco	*Euphorbia officinarum*	Gene transcription	[28]
Nigeria	*Aglaia* sp. *(Meliaceae)*	CB	[29]
	Combretum micranthum, Combretum dolchipeles, Richtiae capparoides, Cajanus cajan, Parmelia perlata, Ramalina farinacea.	CB	[30]
	Momordica balsamina	CB	[31]
South Africa	*Elaeodendron croceum*	CB,	[32, 33]
	Lippia javanica, Hoslundia opposita	RT	[34]
	Bulbine alooides, Crinum macowanii, Hypoxis sobolifera, Leonotis leonurus, Tulbaghia violacea	RT, PR	[35]
	Sutherlandia frutescens, Lobostemon trigonus	RT, PR	[36]
	Anredera cordifolia, Clerodendrum glabrum, Elaeodendron transvaalense, Polianthes tuberosa, Rauvolfia caffra, Rotheca myricoides, Senna occidentalis, Senna petersiana, Terminalia sericea, Zanthoxylum davyi	a-Glucosidase, RT, NF-kB, Tat	[37]
	Elaeodendron transvaalense, Faurea saligna, Parinari curatellifolia, Senna petersiana, Terminalia sericea, Xanthoxylum dayvi	CB	[38]
	Euclea natalensis	RT	[39]
	Bridelia micrantha, Combretum mollec, Elaeodendron transvaalense, Mucuna coriacea, Peltophorum africanum, Ricinus communis, Sutherlandia frutescens, Vernonia stipulacea, Ziziphus mucronata	RT, RNase H	[40]

(continued)

Table 14.4 (continued)

Country	Species: plant	Biological assay[a]	Reference
	Bridelia micrantha, Combretum molle, Elaendendron transvaalensis, Mucana coriacea, Vernonia stipulacea, Sutherlandia frutescens, Ricinus communis	RT, IN	[41]
	Commercial herbal mixtures, with various trade names: *Umzimba omubi, Umuthi wekukhwehlela ne zilonda, Mvusa ukunzi, Umpatisa inkosi, Imbiza ephuzwato, Vusa umzimba, Ingwe® muthi mixture, Ibhubezi™, Supreme one hundred™, Sejeso herbal mixture Ingwe®, Lion izifozonke Ingwe®, Stameta™ BODicare®, Ingwe® special muthi, African potato extract™*	RT	[42]
	Sutherlandia frutescens		[43]
Sudan	*Acanthospermum hispidum, Aristolochia bracteolate, Boscia senegalensis, Capparis deciduas, Cassia occidentalis, Cissus quadrangularis, Citrullus colocynthis, Combretum hartmannianum, Crinum amabile, Croton zambesicus, Erythrina abyssinica, Euphorbia thi, Gardenia lutea, Moringa oblongifolia, Moringa oleifera, Pamianthe peruviana, Tamarix nilotica, Ziziphus spina*	RT	[44]
	Acacia nilotica, Ambrosia maritime, Ammi visnaga, Aristolochia bracteolate, Azadirachta indica, Balanites aegyptiaca, Cassia obtusifolia, Croton zambesicus, Euphorbia granulate, Euphoria prostrate, Haplophyllum tuberculatum, Khaya senegalensis, Maytenus senegalensis, Nauclea latifolia, Salvadora persica, Solanostemma argel, Solanum dubium, Tribulus terrestris, Trichilia emetica	CB, PR	[45]
Tanzania	*Garcinia ferrea, G. edulis, G. bifasciculata, G. buchananni, G. semseii, G. volkensis, G. livingstoneii, G. kingaensis, G. huillensis*	PR	[46]
	Garcinia edulis	PR	[47]
	Elaeodendron schlechteranum	CB (HIV-1 and HIV-2)	[48]
		CB (HIV-1 and HIV-2)	[49]

(continued)

Table 14.4 (continued)

Country	Species: plant	Biological assay[a]	Reference
	Barleria eranthemoides, Zaleya pentandra, Lannea schweinfurthii, Ozoroa reticulate, Balanites aegyptiaca, Kigelia africana, Boscia angustifolia, Maerus edulis, Elaeodendron schlechteranum, Combretum adenogonium, Terminalia mollis, Emilia coccinea, Senecio discifolius, Tithonia diversifolia, Vernonia cinerascens, Euphorbia tirucalli, Crotalaria cf. caudata, Crotalaria retusa, Hoslundia opposite, Leonotis nepetaefolia, Acacia brevispica, Acacia tortilis, Ficus cycamorus, Boerhavia coccinea, Jasminum fluminense, Argemone mexicana, Aeschynomene indica, Dalbergia melanoxylon, Erythrina abyssinica, Indigofera colutea, Lonchocarpus eriocalyx, Ormocarpum kirkii, Rhynchosia sublobata, Sesbania sesban, Crossopteryx febrifuga, Cardiospermum halicacabum, Harrisonia abyssinica, Waltheria indica, Stachytarpheta jamaicaensis		

Country	Species: other	Tested	Reference
Egypt	*Ganoderma colossum* (Vietnamese mushroom)	PR	[50]
South Africa	*Alcyonium fauri* (South African soft coral)	CB	[51]
Uganda	Bacteria derived restriction modification nucleic enzymatic peptides		[52–54]

[a] *RT* reverse transcriptase, *PR* protease, *IN* integrase, *CB* cell-based antiviral assays; all cell-based assays were performed with HIV-1 unless otherwise indicated

a phase II trial of Sutherlandia to be conducted by the South African Herbal Science and Medicine Institute (SAHSMI). This trial, to be commenced in 2011 and involving 100 pre-ARV HIV patients, will build on a small pilot study (12 treatment patients and 13 placebo patients) completed in 2005 which demonstrated that 800 mg/day of *Sutherlandia* leaf powder for 3 months was tolerated by healthy adults [58]. Results of the phase II trial are expected to be released in 2013.

In contrast to international endeavours, minimal HIV research has been conducted in Africa into natural product sources other than plants with only five published reports identified through review of the ISI Web of Knowledge. These reports describe the investigation into bacteria [52–54], fungus [50] and sea coral [51] as potential sources for HIV inhibitors, respectively. In the latter-mentioned article, Hooper and Davies-Coleman describe a cell-based IC_{50} of 1.23 μM obtained for sesquiterpene hydroquinones isolated from *Alcyonium fauri*. In the study by EL Dine et al. [50], two farnesyl hydroquinones were isolated from the *Ganoderma colossum* mushroom and were experimentally shown to inhibit HIV-1 protease at 7.5 and 1.0 μg/ml.

As illustrated in the two previously discussed studies, isolation of active compounds from bio-fractions extracted from the natural source is commonplace and can greatly facilitate investigation and further evaluation. In many examples, active compounds are isolated through a variety of chromatographic techniques and characterised through methods such as mass spectrometry (MS) and NMR spectroscopy. For example, in a 2008 study by Hussein and co-workers [34], eleven compounds were isolated from *Lippia javanica* and *Hoslundia opposita* (aromatic herbs native to Mozambique) through separation on silica gel columns and Sephadex LH-20 columns and thereafter identified through ultraviolet (UV), NMR and MS before evaluation at a single dose in a direct reverse transcriptase (RT) inhibition assay. In this case, three of the eleven compounds inhibited direct RT enzymatic activity by more than 50%, although this was obtained with a relatively high test concentration of 100 μg/ml. Research of this manner has resulted in the isolation of a large number of natural compounds with demonstrated antiviral activity. As extensively described in a number of reviews [25, 59, 60], several natural products including quinones, saponins, alkaloids, sulphated polysaccharides, coumarins and various proteins have been regularly isolated as active ingredients and cited as promising antiviral agents. Expanding on this work, compounds with proven anti-HIV activity and analogues thereof have been synthetically produced by a number of research groups in Africa (Table 14.5). In particular, analogues of coumarins [62, 63], quinones [39, 64], alkaloids [23, 61] and others have been synthesised with the objective of delineating and improving on the anti-HIV activity of the originally identified natural compound.

Additionally, compounds that potentially act through non-specific inhibition, such as tannins [36] and flavonoids [65, 66], have also been regularly isolated and identified as the activity-conferring component [25, 59, 60]. As an approach to minimise the identification and effect of non-specific tannins, Harnett and co-workers [36] introduced 0.2% (w/v) bovine serum albumin (BSA) to assay buffers in order to absorb the tannins from the crude extracts. This inclusion resulted in a clear decrease in anti-RT and anti-glucohydrolase activity of several extracts from *Sutherlandia frutescens* and *Lobostemon trigonus*. As the extracts had demonstrated activity in the absence of BSA, the authors attributed the observed decrease in activity to the removal of tannins and emphasised the importance of controlling for tannin-based compounds in drug screening. Review of the published work also illustrates the dependence of direct enzymatic assays in testing of natural products (Table 14.4). While evaluation against a validated target for drug intervention such as RT, protease or integrase offers a reasonable initial approach to identifying inhibitors, it does not limit non-specific activity. Cross-screening of active compounds through additional assays and screening of active compounds in cell-based antiviral assays may aid in recognising compounds that act through non-specific inhibition. Additionally, testing of natural products within cell-based assays also offers the potential benefit of identifying compounds which would otherwise not have been identified through direct assays aimed at the inhibition of a particular target.

Table 14.5 Published reports from the period 1990 to present detailing compounds of natural origin and derivatives thereof, synthetically produced and investigated by African researchers

Country	Species	Family	Compounds	Reference
Cameroon	*Erythrina senegalensis*	Prenylated isoflavone	8-Prenylluteone, auriculatin, erysenegalensein O, erysenegalensein D, erysenegalensein N, derrone, alpinumisoflavone, 6,8-diprenylgenistein	[22]
	Enantia chlorantha	Alkaloid	7,8-Dihydro-8-hydroxypalmatine	[23]
Egypt	*Ganoderma colossum*	Hydroquinone	Ganomycin I, ganomycin B	[50]
Ivory Coast	*Tieghemella heckelii*	Saponin	Arganine C, tieghemelin, 16β-hydroxyprotobassic acid 3-O-α-D-glucopyranoside, 16β-hydroxyprotobassic acid 3-O-α-D-glucuronopyranoside	[26]
Morocco	*Euphorbia officinarum*	Ingol diterpene, spirotriterpene	7,8,12-Triacetate 3-phenylacetate, 7,8,12-triacetate 3-(4-methoxyphenyl)acetate, 8-methoxyingol 7,12-diacetate 3-phenylacetate, 3S,4S,5R,7S,9R,14R-3,7-dihydroxy-4,14-dimethyl-7 [8 → 9]-Abeo-cholestan-8-one	[28]
Nigeria	*Aglaia* sp. *(Meliaceae)*	Secodammarane triterpenoid	Dammarenolic acid	[29]
South Africa	*Elaeodendron croceum*	Glycoside	Digitoxigenin glucoside	[32, 33]
	Lippia javanica, Hoslundia opposita		4-Ethylnonacosane, (*E*)-2(3)-tagetenone epoxide, myrcenone, piperitenone, apigenin, cirsimaritin, 6-methoxyluteolin 4′-methyl ether, 6-methoxyluteolin, 3′,4′,7-trimethyl ether, 5,7-dimethoxy-6-methylflavone, hoslunddiol, euscaphic acid	[34]
	Euclea natalensis	Naphthoquinones	7-Methyljuglone (7-MJ), shinanolone, diospyrin, neodiospyrin, isodiospyrin, 6-methyljuglone (6 MJ), and 6 synthesised derivatives: (chloro-7-methyljuglone, bromo-7-methyljuglone, fluoro-7-methyljuglone, chloromethoxy-7-methyljuglone, chloro-ethoxy-7-methyljuglone, chloro-6-methyljuglone)	[39]

	Alcyonium fauri	Sesquiterpene hydroquinones	Rietone, 8′-acetoxyrietone, 8′-desoxyrietone	[51]
	Ancistrocladus korupensis	Alkaloid	Isochromane analogues of michellamines and korupensamines	[61]
		Coumarin, chromone, chromene, thriochromene		[62, 63]
		Quinolines		[64]
Tanzania	*Garcinia edulis*	Isoprenylated xanthone, pentacyclic triterpenoids	1,4,6-Trihydroxy-3-methoxy-2-(3-methyl-2-butenyl)-5-(1,1-dimethyl-prop-2-enyl)xanthone, forbexanthone, friedelin, lupeol, lupeol acetate	[47]
	Elaeodendron schlechteranum	Triterpene	3β,29-Dihydroxyglutin-5-ene, 22β-hydroxytingenone (tingenin B), cangoronine methyl ester, 4′-O-methylepigallocatechin, 4′-O-methylgallocatechin, 4′-O-methylgallocatechin-(4α → 8)-4′-O-methylepigallocatechin(4′,4‴-di-O-methyl-prodelphinidin B4)	[48]

As described earlier, high-throughput screening (HTS) technology has been widely implemented to facilitate the random drug discovery approach. Through review of the available literature, it seems apparent that HTS has not yet been utilised in Africa for the purpose of anti-HIV screening. This is almost certainly a consequence of the costs and resources required, although a HTS screen for *Mycobacterium tuberculosis* was developed and undertaken by a TDR-supported, multinational consortium which included African scientists from Cameroon, Ghana, South Africa, Zimbabwe and Kenya [67].

Finally, a number of collaborative efforts have been initiated to investigate natural products and affect the translation of indigenous knowledge to laboratory knowledge, for example, the Conserve Africa organisation, the Indigenous Knowledge Systems (IKS) Lead Programme of the South African Medical Research Council (MRC), The International Centre for Innovation Partnerships in Science (TICIPS), as well as the European Commission-funded MUTHI (Multidisciplinary University Traditional Health Initiative), which is a large network involving researchers from Mali, South Africa and Uganda.

14.2.2 HIV Drug Discovery in Africa: Rational Design Approach

While the majority of commercial drugs available today have been discovered through serendipitous discovery or through random screening, the need to increase the efficiency of drug discovery (both in terms of costs and time) has led to the rational design approach. Through an iterative process, inhibitors can be identified based on knowledge of the 3D structure of the molecular target (structure-based drug design) or based on extensive knowledge of other active ligands of a known target (ligand-based drug design). From review of the ISI Web of Knowledge, it is evident that the rational drug design approach has been employed in Africa more widely than the random screening approach. Tables 14.6–14.8 summarise published literature in which rational drug design was used within Africa for the purpose of HIV drug discovery.

As discussed above in Sect. 14.1.3, computer modelling is frequently used to aid rational drug design (both structure-based and ligand-based drug design), and several African research groups have made use of this technology (Table 14.6). The use of molecular modelling has been employed by Mintek and the Council for Scientific and Industrial Research (CSIR) (both South African Science Councils) to design and discover new inhibitors of HIV integrase and protease, respectively. In the former instance, Sybyl™ (Tripos Inc., St. Louis, Missouri, USA) and Discovery Studio™ (Accelrys Software Inc., San Diego, California, USA) were used to prepare models of HIV-1 integrase through which large commercial databases were screened in silico. Lead compounds identified in this manner, and families comprising synthetic derivatives of the lead compounds, have been synthesised and biologically screened through direct HIV-1 integrase inhibition assays. The CSIR, on the other hand, have utilised molecular modelling programmes for

Table 14.6 Reports detailing the use of computer-aided drug design (CADD) in the rational design of HIV inhibitors

Country	Chemical family	Computational method[a]	Target enzyme/ protein	Reference
Egypt	Nucleoside	QSAR	NRTI	[68]
	Peptide	QSAR	Protease	[69, 70]
Ethiopia	Piperazine; piperidine-carboxamide	QSAR	CCR5	[71, 72]
Morocco	TIBO and HEPT derivatives	QSAR	NNRTI	[73–78]
	HEPT	QSAR	NNRTI	[79, 80]
South Africa	Coumarin	Docking	PR; RT	[62, 63]
	Imidazo[1,2-α] pyridines	Docking	NNRTI	[81]
	Pentacycloundecane	Docking, QM, MD, MM	PR	[82–84]
	Metal complexes	Visualisation	RT, IN	[85]

[a]*QSAR* quantitative structure-activity relationship, *QM* quantum mechanical, *MD* molecular dynamics, *MM* molecular mechanics

Table 14.7 Publications detailing the rational drug design of RT inhibitors for HIV drug discovery in Africa

Country	Chemical family	References
Egypt	Nucleosides	[68]
	Nucleosides; peptide side chains	[86–91]
	Nucleosides; quinoxilane	[92–94]
	Imidazole nucleotides	[95]
	Purine	[96]
	Nucleosides	[97–99]
	Thiouracil	[100]
	Nucleosides	[101]
	Non-classic nucleosides	[102]
	Thieno[2,3-d]pyrimidine-2,4-dithiones	[103]
	Benzofuran-metal complexes	[104, 105]
Ivory Coast	Nucleosides	[106, 107]
Libya	Thiosugar nucleosides	[108]
Morocco	NNRTI	[73–78]
	NNRTI	[79, 80]
	Nucleoside	[109, 110]
	Pyridine	[111, 112]
	Nucleoside; glutaric acid ester carriers	[113–123]
South Africa	NRTI	[62, 63]
	NNRTI	[124–126]
	Metal complexes	[85, 127, 128]
	Nucleosides, double-drug inhibitors	[129–133]
	Imidazo[1,2-α]pyridines	[81]
	Nucleosides	[134]

Table 14.8 Publications employing rational drug design for HIV drug discovery in Africa for targets other than HIV-1 reverse transcriptase

Country	Family	Viral target	Reference
Algeria	Piperazine; piperidinecarboxamides	HIV-1	[135, 136]
	Cyclic sulfamides	PR	[137–139]
Egypt	Benzimidazole	HIV-1	[140, 141]
	Thiazoles	HIV-1	[142–144]
	Various	HIV-1	[145–147]
	Imidazoles	HIV-1	[148–150]
	Functionalised peptides	HIV-1/HIV-2	[151, 152]
	Benzoxazole; benzofuran; benzimidazole	HIV-1	[153, 154]
	Quinozoline	HIV	[155]
Gabon	Immunor 28 (IM28)	HIV-1	[156, 157]
Ivory Coast	TDS (triphenylene dimethyl succinimide)	TAT	[158, 159]
Mauritius	Metal complexes	HIV-1/HIV-2	[160, 161]
	Macrocyclic benzamides and dilactams	HIV-1/HIV-2	[162–164]
Morocco	O,N,O-Tridentate ligands	IN	[165–172]
	Spiroheterocycles	HIV-1	[173]
	Enol-lactones	PR	[174]
South Africa	Metal complexes	PR, IN	[85, 127, 128]
	Pentacycloundecane	PR	[82–84, 175]
	Coumarin	PR	[62]
	Quinoline	PR	[64]
Tunisia	Beta-aminosulfones	PR	[176]
	Gamma-butyrolactones	PR	[177]

pharmacophore searching, specifically to identify natural scaffolds that share structural similarity to existing HIV-1 protease inhibitors. Thereafter, computer modelling was again used, this time to optimise the docking of the identified lead compounds into the active site of protease prior to synthesis and evaluation of the compounds in a direct enzyme inhibition assay. Similarly, Pelly and co-workers from the University of Stellenbosch, South Africa, have utilised molecular modelling software to design indole-based NNRTIs which has led to identification of a novel, proof-of-concept compound.

Computer modelling can also be utilised for lead optimisation after the initial identification of an inhibitor. In a recent publication by Kaye and co-workers [62], Discovery Studio™ was utilised to visualise the interactions between a coumarin-AZT compound and the receptor binding sites of both HIV-RT and HIV-1 PR, providing direction on ligand modifications that may enhance docking to the receptor. Similarly, in an elegant study by Bode et al. [81], random screening of a small set of compounds led to the identification of an inhibitor reasonably active in a direct RT assay but poorly active in a whole cell-based assay. Based on this compound, a larger library comprising structurally similar compounds were synthesised and evaluated in both direct assays and cell-based assays, resulting in the identification of a number of useful inhibitors. In particular, one of the second-generation inhibitors yielded an IC_{50} of 0.18 μM and a corresponding SI value of

868. This compound was thereafter docked into the NNRTI binding site using CDOCKER (Accelrys) to delineate the interaction with the active site, rationalise the notable antiviral activity and suggest sites where further interaction may improve activity.

Computer modelling software has also been utilised to generate QSARs for the purpose of HIV drug discovery in Africa [68, 73]. For example, Cherqaoui and collaborators [73] applied a support vector machine (SVM) algorithm to generate a QSAR predication for the anti-HIV activity of tetrahydroimidazo [4,5,1-jk][1, 4] benzodiazepinone (TIBO) derivatives. In this study, a QSAR was successfully developed through the correlation of experimentally determined anti-HIV activity and ten molecular descriptors of 89 TIBO inhibitors, with hydrophobicity identified as the most pertinent descriptor.

As mentioned, several methods other than CADD have been used to facilitate the rational drug design approach. Despite the technological advancement offered by computer modelling, CADD has not yet fulfilled the initial expectations of its contribution to drug discovery. One particular reason for this is that CADD makes use of rigid protein structures in its preparation of the target negating (to some extent) the flexible nature of the protein. To this end, Montembault and co-workers developed a new approach to account for the dynamic behaviour of the receptor and aid the structure-based design of HIV-1 TAT inhibitors [158]. From a 2D NMR structure of TAT, compounds that bind to a hydrophobic pocket of the protein were designed, synthesised and found to allosterically inhibit TAT function by blocking essential structural changes.

It is also evident that a large body of rational drug design has focused on the design and discovery of RT inhibitors including NRTIs, NtRTIs and NNRTIs (Table 14.7). One particularly interesting approach is that of double-drug inhibitors developed in the Hunter laboratory at the University of Cape Town, South Africa [129–133]. The close proximity (10–15 Å) of the two reverse transcriptase inhibitor sites prompted the synthesis and development of bifunctional compound comprising NRTI (d4T or tenofovir analogous) and NNRTI inhibitors (HI-236 or DAPY TMC-125 analogues) tethered via a non-cleavable spacer (PEG or butynyl). Biological evaluation in both cell-based and direct RT assays demonstrated that the bifunctional compounds were more active than the NRTI alone; however, synergistic inhibition was not observed. Through further experimental analysis, the authors determined that the bifunctional compounds were actually acting as chain-extended NNRTIs without any antiviral contribution from the NRTI component of the double-drug, likely due to a lack of cellular phosphorylation as a result of kinase recognition failure.

Finally, a series of recent papers by the GGKM group and collaborators on pentacycloundecane (PCU)-bound protease substrates provide a leading example of drug discovery research conducted in Africa [82–84]. Through the effective use of a number of techniques including drug design, CADD, NMR spectroscopy and SAR generation, the authors determined the structural requirements of inhibitory activity and identified a lead compound with nanomolar activity against wild-type subtype C protease.

14.3 Challenges in the Field of HIV Drug Discovery Research in Africa

The work described above illustrates that a large body of basic research has been undertaken in the field of HIV drug discovery in Africa. To this end, progress has ensued and scientific contributions have been made to the field. Through review of the published literature and through discussion with authors of the discussed work, it is equally apparent that challenges exist that limit the likelihood of true scientific breakthroughs. An overriding challenge, consistently emphasised, is the apparent lack of sufficient funding for basic HIV drug discovery research. Drug discovery is an inherently costly endeavour with escalating costs arising with each technological advance. To conduct research into HIV drug discovery and to be genuinely competitive within the field requires substantial funding assured over a lengthy period of time. Co-ordinated research may aid in minimising the impact of suboptimal funding; however, review of the literature indicates that disparate research is undertaken, yielding minimal collaboration between research groups within the HIV drug discovery field. Notably, only 5 of the 192 publications detailed above result from collaborative efforts between two or more African countries [21, 32, 73, 75, 107]. It is feasible that HIV drug discovery in Africa could significantly benefit from a co-ordinated central body, such as the newly established African Network of Drug and Diagnostics Innovation (ANDI).

References

1. DiMasi J (2002) The value of improving the productivity of the drug development process: faster times and better decisions. Pharmacoeconomics 20(3):1–10
2. DiMasi J, Hansen R, Grabowski H (2003) The price of innovation: new estimates of drug development costs. J Health Econ 22(2):151–185
3. Light DW, Warburton R (2011) Demythologizing the high costs of pharmaceutical research. BioSocieties 6:34–50
4. Adams CP, Brantner VV (2010) Spending on new drug development. Health Econ 19(2):130–141
5. http://www.fda.gov/ForConsumers/byAudience/ForPatientAdvocates/HIVandAIDSActivities/ucm118915.htm. Accessed Nov 2011
6. UNAIDS (2008) Report on the global HIV/AIDS epidemic 2008: Annex 1, HIV and AIDS estimates and data, 2007 and 2001
7. World Health Organisation (2010) Antiretroviral therapy for HIV infection in adults and adolescents: recommendations for a public health approach – 2010 rev
8. Barnhart M, Shelton J (2011) A better state of ART improving antiviral regimens to increase global access to HIV treatment. J AIDS HIV Res 3:71–78
9. HIV drugs: a global strategic business report. 2010 Global Industry Analysts Inc., CA
10. http://www.avert.org/new-aids-drugs.htm. Accessed Nov 2011
11. i-BASE/Treatment Action Group (Second Edition, September 2011) '2011 Pipeline Report' HIV, hepatitis C virus (HCV), and tuberculosis drugs, diagnostics, vaccines, and preventive technologies in development

12. Comley J (2006) Tools and technology that facilitate automated screening. In: Huser J (ed) High throughput-screening in drug discovery. Wiley-VCH, Verlag GmbH & Co. KGaA, Weinheim, Germany
13. Downey W, Liu C, Hartigan J (2010) Compound profiling: size impact on primary screening libraries. Drug Discov World, Spring 2010
14. Davis AM, Teague SJ, Kleywegt GJ (2003) Application and limitations of X-ray crystallographic data in structure-based ligand and drug design. Angew Chem Int Ed 42:2718–2736
15. Wlodawer A (2002) Rational approach to AIDS drug design through structural biology. Annu Rev Med 53:595–614
16. Wang Y, Lu H, Zhu Q, Jiang S et al (2010) Structure-based design, synthesis and biological evaluation of new N-carboxyphenylpyrrole derivatives as HIV fusion inhibitors targeting gp41. Bioorg Med Chem Lett 20(1):189–192
17. Das K, Lewi PJ, Hughes SH et al (2005) Crystallography and the design of anti-AIDS drugs: conformational flexibility and positional adaptability are important in the design of non-nucleoside HIV-1 reverse transcriptase inhibitors. Prog Biophys Mol Biol 88(2):209–231
18. Chin Y-W, Balunas MJ, Chai HB et al (2006) Drug discovery from natural sources. AAPS J 8(2):E239–E253
19. Newman DJ, Cragg GM (2007) Natural products as sources of new drugs over the last 25 years. J Nat Prod 70:461–477
20. Molinari G (2009) Natural products in drug discovery: present status and perspectives. Adv Exp Med Biol 655:13–27
21. Kuete V, Metuno R, Keilah PL et al (2010) Evaluation of the genus *Treculia* for antimycobacterial, anti-reverse transcriptase, radical scavenging and antitumor activities. S Afr J Bot 76(3):530–535
22. Lee JS, Oh WK, Ahn JS et al (2009) Prenylisoflavonoids from *Erythrina senegalensis* as novel HIV-1 protease inhibitors. Planta Med 75:268–270
23. Wafo P, Nyasse B, Fontaine C (1999) A 7,8-dihydro-8-hydroxypalmatine from *Enantia chlorantha*. Phytochemistry 50:279–281
24. Asres K, Bucar F, Kartnig T et al (2001) Antiviral activity against human immunodeficiency virus type 1 (HIV-1) and type 2. Phytother Res 15:62–69
25. Asres K, Seyoum A, Veeresham C et al (2005) Antiviral activity against human immunodeficiency virus type 1 (HIV-1) and type 2 (HIV-2) of ethnobotanically selected Ethiopian medicinal plants. Phytother Res 19:557–581
26. Gosse B, Gnabre J, Bates RB et al (2002) Antiviral saponins from *Tieghemella heckelii*. J Nat Prod 65:1942–1944
27. Kanyara JN, Njagi ENM (2005) Anti-HIV-1 activities in extracts from some medicinal plants as assessed in an in vitro biochemical HIV-1 reverse transcriptase assay. Phytother Res 19:287–290
28. Daoubi M, Marquez N, Mazoir N et al (2007) Isolation of new phenylacetylingol derivatives that reactivate HIV-1 latency and a novel spirotriterpenoid from *Euphorbia officinarum latex*. Bioorg Med Chem 15:4577–4584
29. Esimone CO, Eck G, Nworu CS et al (2010) Dammarenolic acid, a secodammarane triterpenoid from *Aglaia* sp. shows potent anti-retroviral activity *in vitro*. Phytomedicine 17:540–547
30. Esimone CO, Grunwald T, Wildner O et al (2005) In vitro pharmacodynamic evaluation of antiviral medicinal plants using a vector-based assay technique. J Appl Microbiol 99:1346–1355
31. Bot YS, Mgbojikwe LO, Chika N et al (2007) Screening of the fruit pulp extract of Momordica balsamina for anti HIV property. Afr J Biotechnol 6(1):47–52
32. Prinsloo G, Meyer JJM, Hussein AA et al (2010) A cardiac glucoside with in vitro anti-HIV activity isolated from *Elaeodendron croceum*. Nat Prod Res 24(18):1743–1746
33. Prinsloo G, Meyer JJM, Hussein AA (2007) Anti-HIV activity of a cardiac glycoside isolated from *Elaeodendron croceum*. S Afr J Bot 73(2):308

34. Mujovo SF, Hussein AA, Meyer JJM et al (2008) Bioactive compounds from *Lippia javanica* and *Hoslundia opposite*. Nat Prod Res 22(12):1047–1054
35. Klos M, van de Venter M, Milne PJ et al (2009) In vitro anti-HIV activity of five selected South African medicinal plant extracts. J Ethnopharmacol 124:182–188
36. Harnett SM, Oosthuizen V, van de Venter M (2005) Anti-HIV activities of organic and aqueous extracts of *Sutherlandia frutescens* and *Lobostemon trigonus*. J Ethnopharmcol 96:113–119
37. Tshikalange TE, Meyer JJM, Lall N et al (2008) In vitro anti-HIV-1 properties of ethnobotanically selected South African plants used in the treatment of sexually transmitted diseases. J Ethnopharmacol 119:478–481
38. Tshikalange TE, Meyer JJM, Hattori T et al (2008) Anti-HIV screening of ethnobotanical selected SA plants. S Afr J Bot 74(2):391
39. Tshikalange TE, Lall N, Meyer JJM et al (2007) In vitro HIV-1 reverse transcriptase inhibitory activity of naphthoquinones and derivatives from *Euclea natalensis*. S Afr J Bot 73(2):308
40. Bessong PO, Obi CL, Andreola ML et al (2005) Evaluation of selected South African medicinal plants for inhibitory properties against human immunodeficiency virus type 1 reverse transcriptase and integrase. J Ethnopharmcol 99:83–91
41. Bessong PO, Rojas LB, Obi LC et al (2006) Further screening of Venda medicinal plants for activity against HIV type 1 reverse transcriptase and integrase. Afr J Biotechnol 5(6):526–528
42. Ndhlala AR, Finnie JF, Van Staden J (2010) *In vitro* antioxidant properties, HIV-1 reverse transcriptase and acetylcholinesterase inhibitory effects of traditional herbal. Preparations Sold in South Africa. Molecules 15:6888–6904
43. Van Wyk BE, Albrecht C (2008) A review of the taxonomy, ethnobotany, chemistry and pharmacology of *Sutherlandia frutescens* (Fabaceae). J Ethnopharmacol 119:620–629
44. Ali H, Konig GM, Khalid SA, Wright AD et al (2002) Evaluation of selected Sudanese medicinal plants for their in vitroactivity against hemoflagellates, selected bacteria, HIV-1-RT and tyrosine kinase inhibitory, and for cytotoxicity. J Ethnopharmacol 83:219–228
45. Hussein G, Miyashiro H, Nakamura N et al (1999) Inhibitory effects of Sudanese plant extracts on HIV-1 replication and HIV-1 protease. Phytother Res 13:31–36
46. Magadula JJ, Tewtrakul S (2010) Anti-HIV-1 protease activities of crude extracts of some *Garcinia* species growing in Tanzania. Afr J Biotechnol 9(12):1848–1852
47. Magadula JJ (2010) A bioactive isoprenylated xanthone and other constituents of *Garcinia edulis*. Fitoterapia 81:420–423
48. Maregesi SM, Hermans N, Dhooghe L et al (2010) Phytochemical and biological investigations of *Elaeodendron schlechteranum*. J Ethnopharmacol 129:319–326
49. Maregesi S, Van Miert S, Pannecouque C et al (2010) Screening of Tanzanian medicinal plants against *Plasmodium falciparum* and human immunodeficiency virus. Planta Med 76:195–201
50. El Dine RS, El Halawany AM, Ma CM et al (2009) Inhibition of the dimerization and active site of HIV-1 protease by secondary metabolites from the Vietnamese mushroom *Ganoderma colossum*. J Nat Prod 72:2019–2023
51. Hooper GJ, Davies-Coleman MT (1995) Sesquiterpene hydroquinones from the South African soft coral *Alcyonium fauri*. Tetrahedron Lett 36(18):3265–3268
52. Wayengera M, Byarugaba W, Kajjumbula H (2007) Frequency and site mapping of HIV 1/SIVcpz, HIV-2/SIVsmm and other SIV gene sequence cleavage by various bacteria restriction enzymes: precursors for a novel HIV inhibitory product. Afr J Biotechnol 6(10):1225–1232
53. Wayengera M (2008) Why bacteria derived R-M nucleic enzymatic peptides are likely efficient therapeutic molecules for use in the design and development of novel HIV inhibitory strategies. Afr J Biotechnol 7(12):1791–1796

54. Wayengera M (2007) A recombinant lactobacillus strain expressing genes coding for restriction enzymes cleaving the HIV genomes for use as a live microbicide strategy against heterosexual transmission of HIV. Afr J Biotechnol 6(15):1750–1756
55. Gericke N, Albrecht CF, Van Wyk B et al (2001) *Sutherlandia frutescens*. Aust J Med Herb 13:9–15
56. Chaffy N, Stokes T (2002) AIDS herbal therapy. Trends Plant Sci 7:57
57. Mills E, Cooper C, Seely D et al (2005) African herbal medicines in the treatment of HIV: *Hypoxis* and *Sutherlandia*. An overview of evidence and pharmacology. Nutr J 4:1–6
58. Johnson Q, Syce J, Nell H et al (2007) A randomized, double-blind, placebo-controlled trial of *Lessertia frutescens* in healthy adults. PLOS Clin Trials 2(4):e16
59. Singh IP, Bharate SP, Bhutani KK (2005) Anti-HIV natural products. Curr Sci 89(2):269–290
60. Chinsembu KC, Hedimbi M (2010) Ethnomedicinal plants and other natural products with anti-HIV active compounds and their putative models of action. Int J Biotechnol Mol Biol Res 1:74–91
61. De Koning CB, Michael JP, van Otterlo WAL (2000) Synthesis of isochromane analogues of the michellamines and korupensamines. J Chem Soc Perkin Trans 1:799–811
62. Olomola TO, Klein R, Lobb KA et al (2010) Towards the synthesis of coumarin derivatives as potential dual-action HIV-1 protease and reverse transcriptase inhibitors. Tetrahedron Lett 51:6325–6328
63. Kaye PT, Musa MA, Nchinda AT et al (2004) Novel heterocyclic analogues of the HIV-1 protease inhibitor, ritonavir. Synth Commun 34(14):2575–2589
64. Familoni OB, Klaas PJ, Lobb KA et al (2006) The Baylis–Hillman approach to quinoline derivatives. Org Biomol Chem 4:3960–3965
65. Webb MR, Ebeler SE (2004) Comparative analysis of topoisomerase IB inhibition and DNA intercalation by flavonoids and similar compounds: structural determinates of activity. Biochem J 384:527–541
66. Cowan MM (1999) Plant products as antimicrobial agents. Clin Microbiol Rev 12:564–582
67. http://apps.who.int/tdr/svc/research/lead-discovery-drugs/management. Accessed Nov 2011
68. Essa AH, Ibrahim M, Hameed AJ et al (2008) Theoretical investigation of 3'-subtituted-2'-3'-dideoxythymidines related to AZT. QSAR infrared and substituent electronic effect studies. ARKIVOC xiii:255–265
69. Sheha MM, El-Koussi NA, Farag HH (2003) Brain delivery of HIV protease inhibitors. Arch Pharm Pharm Med Chem 1:47–52
70. Sheha MM, Mahfouz NM, Hassan HY et al (2000) Synthesis of di- and tripeptide analogues containing a-ketoamide as a new core structure for inhibition of HIV-1 protease. Eur J Med Chem 35:887–894
71. Dessalew N (2009) Investigation of the structural requirement for blocking the human CCR5 chemokine receptor. An insight from quantitative structure activity relationships study. Lett Drug Des Discov 6:114–124
72. Dessalew N (2008) QSAR Study on piperidinecarboxamides as antiretroviral agents: an insight into the structural basis for HIV coreceptor antagonist activity. QSAR Comb Sci 27(7):901–912
73. Darnag R, Schmitzer A, Belmiloud Y et al (2010) Quantitative structure-activity relationship studies of TIBO derivatives using support vector machines. SAR QSAR Environ Res 21(3–4):231–246
74. Darnag R, Mazouz ELM, Schmitzer A et al (2010) Support vector machines: development of QSAR models for predicting anti-HIV-1 activity of TIBO derivatives. Eur J Med Chem 45:1590–1597
75. Darnag R, Schmitzer A, Belmiloud Y et al (2008) QSAR Studies of HEPT derivatives using support vector machines. QSAR Comb Sci 28(6–7):709–718
76. Douali L, Villemin D, Zyad A et al (2004) Artificial neural networks: non-linear QSAR studies of HEPT derivatives as HIV-1 reverse transcriptase inhibitors. Mol Divers 8:1–8

77. Douali L, Villemin D, Cherqaoui D (2004) Exploring QSAR of non-nucleoside reverse transcriptase inhibitors by neural networks: TIBO derivatives. Int J Mol Sci 5:48–55
78. Douali L, Villemin D, Cherqaoui D (2003) Neural networks: accurate nonlinear QSAR model for HEPT derivatives. J Chem Inf Comput Sci 43:1200–1207
79. Zahouily M, Rakik J, Lazar M et al (2007) Exploring QSAR of non-nucleoside reverse transcriptase inhibitors by artificial neural networks: HEPT derivatives. ARKIVOC xiv:245–256
80. Bazoui H, Zahouily M, Sebti S et al (2002) Structure-cytotoxicity relationships for a series of HEPT derivatives. J Mol Model 8:1–7
81. Bode ML, Gravestock D, Moleele SS et al (2011) Imidazo[1,2-a]pyridin-3-amines as potential HIV-1 non-nucleoside reverse transcriptase inhibitors. Bioorg Med Chem 19:4227–4237
82. Makatini MM, Petzold K, Sriharsha SN et al (2011) Pentacycloundecane-based inhibitors of wild-type C-South African HIV-protease. Bioorg Med Chem Lett 21:2274–2277
83. Makatini MM, Petzold K, Sriharsha SN et al (2011) Synthesis and structural studies of pentacycloundecane-based HIV-1 PR inhibitors: a hybrid 2D NMR and docking/QM/MM/MD approach. Eur J Med Chem 46(9):3976–3985
84. Karpoormath R, Sayed Y, Govender P et al (2012) Pentacycloundecance derived hydroxyl acid peptides: a new class of irreversible non-scissile ether bridged isoster as potential HIV-1 wild type C-SA protease inhibitors. Bioorg Chem 40(1):19–29
85. Mphahlele M, Papathanasopoulos M, Cinellu MA et al (2012) Modification of HIV-1 reverse transcriptase and integrase activity by gold(III) complexes in direct biochemical assays. Bioorg Med Chem 20(1):401–407
86. El-Sayed WA, El-Essawy FA, Ali OM et al (2010) Synthesis and antiviral evaluation of new 2,5-disubstituted 1,3,4-oxadiazole derivatives and their acyclic nucleoside analogues. Monatsh Chem 141:1021–1028
87. Abdel-Rahman AAH, El-Etrawy AASH, Abdel-Megied AES et al (2008) Synthesis and antiviral evaluation of novel 2,3-dihydroxypropyl nucleosides from 2- and 4-thiouracils. Nucleosides Nucleotides Nucleic Acids 27:1257–1271
88. Abdel-Rahman AAH (2008) Amino acid derivatives, VI [1]: synthesis, antiviral, and antimicrobial evaluation of a-amino acid esters bearing a 1,2,3-triazolo[4,5-d]pyrimidinedione side chain. Monatsh Chem 139:61–68
89. Abdel-Rahman AAH (2008) Amino acid derivatives, VII [1]: synthesis and antiviral evaluation of a-amino acid esters bearing an indazole side chain. Monatsh Chem 139:289–297
90. Abdel-Rahman AAH, Abdel-Megied AES, Goda AES et al (2003) Synthesis and anti-HBV activity of thiouracils linked via S and N-1 to the 5-position of methyl b-D-ribofuranoside. Nucleosides Nucleotides Nucleic Acids 22(11):2027–2038
91. Abdel-Rahman AAH (2003) Synthesis and anti-hepatitis B virus activity of glucosylated 2-O-ethyluracils. Monatsh Chem 134:929–939
92. Ali IAI, Al-Masoudi IA, Hassan HG et al (2007) Synthesis and anti-HIV activity of new homo acyclic nucleosides, 1-(pent-4-enyl)quinoxalin-2-ones and 2-(pent-4-enyloxy) quinoxalines. Chem Heterocycl Compd 43(8):1052
93. Al-Masoudi NA, Al-Soud YA, Ali IAI et al (2007) New AZT analogues having 5-alkylsulfonyl groups: synthesis and anti-HIV activity. Nucleosides Nucleotides Nucleic Acids 26:223–230
94. Ali IAI, Al-Masoudi IA, Aziz NM et al (2008) New acyclic quinoxaline nucleosides. Synthesis and anti-HIV activity. Nucleosides Nucleotides Nucleic Acids 27:146–156
95. Al-Masoudi IA, Khodair AI, Al-Soud YA et al (2003) Synthesis of N-substituted 1-amino-2,3-dihydro- 1 H-imidazole-2-thione-N-nucleosides and S-glycosylated derivatives. Nucleosides Nucleotides Nucleic Acids 22(3):299–307
96. Rida SM, Ashour FA, El-Hawash SAM et al (2007) Synthesis of some novel substituted purine derivatives as potential anticancer, anti-HIV-1 and antimicrobial agents. Arch Pharm Chem Life Sci 340:185–194
97. Attia AM, Sallam MA, Almehdi AA et al (1999) Synthesis and biological activity of modified thiopyrimidine nucleosides. Nucleosides Nucleotides 18:2307–2315

98. Attia AM, Elgemeie GH, Alnaimi IS (1998) Synthesis of 1-(β-D-glycopyranosyl)-3-deazapyrimidines from 2-hydroxy and 2-mercaptopyridines. Nucleosides Nucleotides 17(8):1355–1363
99. Elgemeie GEH, Attia AME, Hussain BAW (1998) A synthetic strategy to a new class of cycloalkane ring-fused pyridine nucleosides as potential anti HIV agents. Nucleosides Nucleotides 17(5):855–868
100. El-Emam AA, Nasr MNA, Pedersen EB et al (2001) Synthesis of certain 6-(arylthio)uracils as potential antiviral agents. Phosphorus Sulfur Silicon Relat Elem 174:25–35
101. Elshehry MF, Balzarini J, Meier C (2009) Synthesis of new cyclic and acyclic 5-halouridine derivatives as potential antiviral agents new 5-halouridine derivatives as potential antiviral agents. Synthesis 5:841–847
102. Elgemeie GEH, Mansour OA, Metwally NH (1999) Synthesis and anti-HIV activity of different novel nonclassical nucleosides. Nucleosides Nucleotides 18(1):113–123
103. Hafez HN, Hussein HAR, El-Gazzar ARBA (2010) Synthesis of substituted thieno[2,3-d] pyrimidine-2,4-dithiones and their S-glycoside analogues as potential antiviral and antibacterial agents. Eur J Med Chem 45:4026–4034
104. Galal SA, Abd El-All ASA, Hegab KH et al (2010) Novel antiviral benzofuran-transition metal complexes. Eur J Med Chem 45:3035–3046
105. Galal SA, Abd El-All ASA, Abdallah MM et al (2009) Synthesis of potent antitumor and antiviral benzofuran derivatives. Bioorg Med Chem Lett 19:2420–2428
106. Prestat G, Dubreui D, Adjou A et al (2000) Synthesis of 3′-O2-(azaheterocycle)- thymidines nucleosides. Nucleosides Nucleotides Nucleic Acids 19(4):735–748
107. Ané A, Prestat G, Manh GT et al (2001) Synthesis of nucleoside analogs and new Tat protein inhibitors. Pure Appl Chem 73(7):1189–1196
108. Al-Masoudi NA, Al-Soud YA, Al-Masoudi WA (2004) Thiosugar nucleosides. Synthesis and biological activity of 1,3,4-thiadiazole, thiazoline and thiourea derivatives of 5-thio-d-glucose. Nucleosides Nucleotides Nucleic Acids 23(11):1739–1749
109. Len C, Selouane A, Weiling A et al (2003) Asymmetric synthesis of (3S) 3-benzoyloxymethylisobenzofuranone and its 3R enantiomer as analogues of α, β-butenolides. Tetrahedron Lett 44:663–666
110. Selouane A, Vaccher C, Villa P et al (2002) Enantiomeric d4T analogues and their structural determination. Tetrahedron Asymmetry 13:407–413
111. Chaouni-Benabdallah A, Galtier C, Allouchi H et al (2001) A 3-benzamido, ureido and thioureidoimidazo[1,2-a]pyridine derivatives as potential antiviral agents. Chem Pharm Bull 49(12):1631–1635
112. Chaouni-Benabdallah A, Galtier C, Allouchi H et al (2001) Synthesis of 3-nitrosoimidazo [1,2-a]pyridine derivatives as potential antiretroviral agents. Arch Pharm Pharm Med Chem 334:224–228
113. Lazrek HB, Vasseur JJ, Secrist J et al (2007) A glutaric acid ester as carrier system for sustained delivery of lamuvidine (3tc) dimers. Nucleosides Nucleotides Nucleic Acids 26:1103–1106
114. Taourirte M, Lazrek HB, Rochdi A et al (2005) Homo and heterodimers of ddi, d4t and azt: influence of (5′-5′) thiolcabonate-carbamate linkage on anti-HIV activity. Nucleosides Nucleotides Nucleic Acids 24(5–7):523–525
115. Tourirte M, Oulih T, Lazrek HB et al (2003) Synthesis of 3O-deoxy-3O-[4-(pyrimidin-1-yl) methyl-1,2,3-triazol-1-yl]thymidine via 1,3-dipolar cycloaddition. Nucleosides Nucleotides Nucleic Acids 22(11):1985–1993
116. Ait Mohamed L, Taourirte M, Rochdi A et al (2003) Synthesis of new homo and heterodimers of 2O,3O-dideoxyinosine (ddI) using ester linkages. Nucleosides Nucleotides Nucleic Acids 22(5–8):829–831
117. Moukha-chafiq O, Taha ML, Lazrek HB et al (2002) Synthesis and biological evaluation of some acyclic a-(1H-pyrazolo- [3,4-d]pyrimidin-4-yl)thioalkylamide nucleosides. Nucleosides Nucleotides Nucleic Acids 21(2):165–176

118. Lazrek HB, Taourirte M, Oulih T et al (2001) Synthesis and anti-HIV activity of new modified 1,2,3-triazole acyclonucleosides. Nucleosides Nucleotides Nucleic Acids 20(12):1949–1960
119. Moukha-chafiq O, Taha ML, Lazrek HB et al (2001) Synthesis and biological activity of 4-substituted 1-[1-(2-hydroxyethoxy)- methyl-1,2,3-triazol-(4 & 5)-ylmethyl]-1h-pyrazolo[3,4-d]pyrimidines. Nucleosides Nucleotides Nucleic Acids 20(10–11):1797–1810
120. Moukha-chafiq O, Taha ML, Lazrek HB et al (2001) Synthesis and biological evaluation of some 4-substituted 1-[1-(4-hydroxybutyl)-1,2,3-triazol- (4 & 5)-ylmethyl]-1h-pyrazolo-[3,4-d]pyrimidines. Nucleosides Nucleotides Nucleic Acids 20(10–11):1811–1821
121. Taourirte M, Lazrek HB, Vasseur JJ et al (2001) Synthesis of new homo and heterodinucleosides containing the 2′,3′-dideoxynucleosides AZT and D4T. Nucleosides Nucleotides Nucleic Acids 20(4–7):959–962
122. Lazrek HB, Engels JW, Pfleidere W (1998) Synthesis of novel branched nucleoside dimers containing a 1,2,3-triazolyl linkage. Nucleosides Nucleotides 17(9–11):1851–1856
123. Lazrek HB, Rochdi A, Khaider H et al (1998) Synthesis of (Z) and (E) alpha – alkenyl phosphonic acid derivatives of purines and pyrimidines. Tetrahedron 54:3807–3816
124. Stieger N, Caira MR, Liebenberg W et al (2010) Influence of the composition of water/ethanol mixtures on the solubility and recrystallization of nevirapine. Cryst Growth Des 10:9
125. Stieger N, Liebenberg W, Wessels JC et al (2010) Channel inclusion of primary alcohols in isostructural solvates of the antiretroviral nevirapine: an X-ray and thermal analysis study. Struct Chem 21:771–777
126. Caira MR, Stieger N, Liebenberg W et al (2008) Solvent inclusion by the anti-HIV drug nevirapine: X-ray structures and thermal decomposition of representative solvates. Cryst Growth Des 8:1
127. Fonteh PN, Keter FK, Meyer D et al (2009) Tetra-chloro-(bis-(3,5-dimethylpyrazolyl) methane)gold(III) chloride: an HIV-1 reverse transcriptase and protease inhibitor. J Inorg Biochem 103:190–194
128. Fonteh PN, Keter FK, Meyer D (2010) HIV therapeutic possibilities of gold compounds. Biometals 23:185–196
129. Younis Y, Hunter R, Muhanji CI et al (2010) [d4U]-Spacer-[HI-236] double-drug inhibitors of HIV-1 reverse-transcriptase. Bioorg Med Chem 18:4661–4673
130. Hunter R, Younis Y, Muhanji CI et al (2008) C-2-Aryl O-substituted HI-236 derivatives as non-nucleoside HIV-1 reverse-transcriptase inhibitors. Bioorg Med Chem 16:10270–10280
131. Hunter R, Muhanji CI, Hale I et al (2007) [d4U]-butyne-[HI-236] as a non-cleavable, bifunctional NRTI/NNRTI HIV-1 reverse-transcriptase inhibitor. Bioorg Med Chem Lett 17:2614–2617
132. Arnott G, Hunter R, Mbeki L et al (2005) New methodology for 2-alkylation of 3-furoic acids: application to the synthesis of tethered UC-781/d4T bifunctional HIV reverse-transcriptase inhibitors. Tetrahedron Lett 46:4023–4026
133. Muhanji CI, Hunter R (2007) Current developments in the synthesis and biological activity of HIV-1 double-drug inhibitors. Curr Med Chem 14:1207–1220
134. N'Da DD, Breytenbach JC, Legoabe LJ et al (2009) Synthesis and in vitro transdermal penetration of methoxypoly(ethylene glycol) carbonate derivatives of stavudine. Med Chem 5:497–506
135. Serradji N, Martin M, Bensaid O et al (2004) Structure-activity relationships in platelet-activating factor. 12. Synthesis and biological evaluation of platelet-activating factor antagonists with anti-HIV-1 activity. J Med Chem 47:6410–6419
136. Serradji N, Bensaid O, Martin M et al (2006) Structure–activity relationships in platelet-activating factor. Part 13: synthesis and biological evaluation of piperazine derivatives with dual anti-PAF and anti-HIV-1 or pure antiretroviral activity. Bioorg Med Chem 14:8109–8125
137. Bendjeddou A, Djebbar H, Berredjem M et al (2006) Cyclosulfamides as constraint dipeptides: the synthesis and structure of Chiral substituted 1,2,5-thiadiazolidine 1,1-dioxides: evaluation of the toxicity. Phosphorus Sulfur Silicon Relat Elem 181:1351–1362

138. Bendjeddou A, Djeribi R, Regainia Z et al (2005) N, N′-substituted 1,2,5 thiadiazolidine 1,1-dioxides: synthesis, selected chemical and spectral proprieties and antimicrobial evaluation. Molecules 10:1387–1398
139. Regaɪnia Z, Winum JY, Smaine FZ et al (2003) General synthesis of n-membered cyclic sulfamides. Tetrahedron 59:6051–6056
140. Badawey EAM, Kappe T (1997) Synthesis and in vitro anti-HIV activity of certain 2-(1H-benzimidazol-2-ylamino)pyrimidin-4(3H)-ones and related derivatives. Arch Pharm Pharm Med Chem 330:59
141. Badawey E, Kappe T (1995) Benzimidazole condensed ring system. IX. Potential antineoplastics. New synthesis of some pyrido[l,2-a]benzimidazoles and related derivatives. Eur J Med Chem 30:327–332
142. Habib NS, Rida SM, Badawey EAM et al (1996) Condensed thiazoles, i: synthesis of 5,7-disubstituted thiazolo [4,5-d]pyrimidines as possible anti-HIV, anticancer, and antimicrobial agents. Monatsh Chem 127:1203–1207
143. Habib NS, Rida SM, Badawey EAM et al (1996) Condensed thiazoles, ii: synthesis of 7-substituted thiazolo[4,5-d]pyrimidines as possible anti-HIV, anticancer, and antimicrobial agents. Monatsh Chem 127:1209–1214
144. Rida SM, Habib NS, Badawey EAM et al (1995) Synthesis and biological investigations of some new thiazolylbenzimidazoles and benzirnidazolylthiazo10[3,2-a]pyridines. Arch Pharm 328:325–328
145. El-Barbary AA, Abou El-Ezz AZA, Sharaf AM (2007) Studies on 2,4-dithioxo and 2-thioxoimidazolidene derivatives. Phosphorus Sulfur Silicon Relat Elem 182:1621–1632
146. El-Barbary AA, Abou El-Ezz AZA, Sharaf AM et al (2006) The synthesis of some new quinazolone derivatives of potential biological activity. Phosphorus Sulfur Silicon Relat Elem 181:1895–1912
147. El-Barbary AA, Abou-El-Ezz AZA, Abdel-Kader AA et al (2004) Synthesis of some new 4-amino-1,2,4-triazole derivatives as potential anti-HIV and anti-HBV. Phosphorus Sulfur Silicon Relat Elem 179:1497–1508
148. Loksha YM, Pedersen EB, Loddo R et al (2009) Synthesis and anti-HIV-1 activity of 1-substituted 6-(3-cyanobenzoyl) and [(3-cyanophenyl)fluoromethyl]-5-ethyluracils. Arch Pharm Chem Life Sci 342:501–506
149. Loksha YM, El-Barbary AA, El-Badawi MA et al (2005) Synthesis of 2-(aminocarbonyl-methylthio)-1 H-imidazoles as novel Capravirine analogues. Bioorg Med Chem 13:4209–4220
150. Loksha YM, El-Badawi MA, El-Barbary AA et al (2003) Synthesis of 2-methylsulfanyl-1 h-imidazoles as novel non-nucleoside reverse transcriptase inhibitors (NNRTIs). Arch Pharm Pharm Med Chem 336:175–180
151. Al-Masoudi NA, Al-Masoudi IA, Ali IAI et al (2005) Amino acid derivatives, part 3: new peptide and glycopeptide derivatives conjugated naphthalene. synthesis, antitumor, anti-HIV, and BVDV evaluation. Heteroatom Chem 16(7):576
152. Ali IAI, Al-Masoudi IA, Saeed B et al (2005) Amino acid derivatives, part 2: synthesis, antiviral, and antitumor activity of simple protected amino acids functionalized at n-terminus with naphthalene side chain. Heteroatom Chem 16(2):148
153. Rida SM, Ashour FA, El-Hawash SAM et al (2005) Synthesis of some novel benzoxazole derivatives as anticancer, anti-HIV-1 and antimicrobial agents. Eur J Med Chem 40:949–959
154. Rida SM, Ashour FA, El-Hawash SAM et al (2006) Synthesis of novel benzofuran and related benzimidazole derivatives for evaluation of *in vitro* anti-HIV-1, anticancer and antimicrobial activities. Arch Pharm Res 29(10):826–833
155. El-Hamid A, Ismail AA, Attia AME (2003) Synthesis of some new quinazoline derivatives analogues to MKC-442 and TNK 561. Phosphorus Sulfur Silicon Relat Elem 178:1231–1240
156. Diallo K, Loemb H, Oliveir M et al (2000) Inhibition of human immunodeficiency virus type-1 (HIV-1) replication by immunor (Im28), a new analog of dehydroepiandrosterone. Nucleosides Nucleotides Nucleic Acids 19(10–12):2019–2024

157. Mavoungou D, Poaty-Mavoungou V, Akoume MY et al (2005) Inhibition of human immunodeficiency virus type-1 (HIV-1) glycoprotein-mediated cell-cell fusion by immunor (IM28). Virol J 2:9
158. Montembault M, Vo-Thanh G, Deyine A et al (2004) A possible improvement for structure-based drug design illustrated by the discovery of a Tat HIV-1 inhibitor. Bioorg Med Chem Lett 14:1543–1546
159. Ané A, Prestat G, Manh GT et al (2001) Synthesis of nucleoside analogs and new Tat protein inhibitors. Pure Appl Chem 73(7):1189–1196
160. Bhowon MG, Laulloo BSJ (2004) Synthesis and anti-HIV activity of metal complexes of SRR-SB3. Indian J Chem 43(5):1131–1133
161. Bhowon MG (2000) Synthesis, catalytic and biological activity of ruthenium(II) complexes. Indian J Chem 39(11):1207–1209
162. Jhaumeer-Laulloo BS (2000) Synthesis and anti-HIV activity of novel macrocyclic disulphide compounds with thioureylene group. Asian J Chem 12(3):775–780
163. Jhaumeer-Laulloo S, Witvrouw M (2000) Synthesis and anti-HIV activity of novel macrocyclic benzamides with a disulphide bridge. Indian J Chem 39(11):842–846
164. Jhaumeer-Laulloo BS, Ramadas SR (1999) Synthesis and anti-HIV activity of macrocyclic dilactams containing disulphide bridge. Indian J Heterocycl Chem 9(1):1–6
165. Meskini I, Toupet L, Daoudi M et al (2010) Structure of 2-[(phenyl)-(3,5-dimethyl-pyrazol-1-yl)-methyl]- malonic acid diethyl ester. J Chem Crystallogr 40:812–815
166. Meskini I, Daoudi M, Daran JC et al (2010) Poly[[bis{13-2-[(3,5-dimethyl-1H-pyrazol- 1-yl) (phenyl)methyl]propanedioato} tetrasodium(I)] 7.5-hydrate]. Acta Crystallogr Sect E Struct Rep Online E66:m1009–m1010
167. Meskini I, Daoudi M, Daran JC et al (2010) Diethyl 2-[(3,5-dimethyl-1 H-pyrazol-1- yl) (4-methoxyphenyl)methyl]propanedioate. Acta Crystallogr Sect E Struct Rep Online 66:o1965
168. Meskini I, Toupet L, Daoudi M et al (2010) An efficient protocol for accessing b-amino dicarbonyl compounds through aza-Michael reaction. J Braz Chem Soc 21(6):1129–1135
169. Meskini I, Daoudi M, Daran JC et al (2010) Diethyl 2-[phenyl(pyrazol-1-yl)methyl]-propanedioate. Acta Crystallogr Sect E Struct Rep Online 66:o1014
170. Meskini I, Daoudi M, Daran JC et al (2010) Diethyl 2-[(N-benzyl-N-methylamino)- (phenyl) methyl]propanedioate. Acta Crystallogr Sect E Struct Rep Online 66:o746
171. Meskini I, Daoudi M, Daran JC et al (2010) Diethyl 2-{(dibenzylamino)[4-(trifluoromethyl) phenyl]methyl}malonate. Acta Crystallogr Sect E Struct Rep Online E66:o961–o962
172. Meskini I, Toupet L, Akkurt M et al (2010) Crystal structure of diethyl[(4chlorophenyl) (dibenzylamino)methyl]propanedioate. J Chem Crystallogr 40:391–395
173. Bennani B, Kerbal A, Daoudi M et al (2007) Combined drug design of potential *Mycobacterium tuberculosis* and HIV-1 inhibitors: 3′,4′-di-substituted -4′H-spiro [isothiochromene-3,5′-isoxazol]-4(1H)-one. ARKIVOC xvi:19–40
174. Ibrahimi S, Sauvé G, Yelle J et al (2005) Synthèse racémique et énantiosélective d'énol-lactones et leur évaluation comme inhibiteurs de la protéase du VIH-1. C R Chim 8:75–83
175. Onajole OK, Makatini MM, Govender P et al (2010) Synthesis and NMR assignment of pentacycloundecane precursors of potential pharmaceutical agents. Magn Reson Chem 48:249–255
176. Touati R, Hassine BB (2008) Asymmetric synthesis of beta-aminosulfones via the enantioselective hydrogenation of the corresponding beta-ketosulfones. Lett Org Chem 5:240–243
177. Samarat A, Amria H, Landaisb Y (2004) Enantioselective synthesis of functionalized g-butyrolactones. Tetrahedron 60:8949–8956

Chapter 15
Structure-Based Design of Domain-Selective Angiotensin-Converting Enzyme Inhibitors

Ross G. Douglas and Edward D. Sturrock

Abbreviations

ACE	Angiotensin-converting enzyme
AcSDKP	*N*-acetyl-serylaspartyllysylproline
ADE	Adverse drug event
AngI	Angiotensin I
AngII	Angiotensin II
Ang1-7	Angiotensin 1-7
AT_1R	Angiotensin type 1 receptor
AT_2R	Angiotensin type 2 receptor
BK	Bradykinin
BPF	Bradykinin-potentiating factor
CPA	Carboxypeptidase A
CVD	Cardiovascular disease
kAF	keto-ACE Phe inhibitor
kAW	keto-ACE Trp
LisW	Lisinopril Trp
PDB	Protein Data Bank (http://www.rcsb.org/pdb)
RAAS	Renin-angiotensin-aldosterone system
tACE	Testis angiotensin-converting enzyme
WHO	World Health Organisation
ZMRG	Zinc Metalloprotease Research Group

R.G. Douglas • E.D. Sturrock (✉)
Division of Medical Biochemistry, Institute of Infectious Disease and Molecular Medicine, University of Cape Town, Observatory, Cape Town, South Africa 7935
e-mail: edward.sturrock@uct.ac.za

K. Chibale et al. (eds.), *Drug Discovery in Africa*,
DOI 10.1007/978-3-642-28175-4_15, © Springer-Verlag Berlin Heidelberg 2012

15.1 Introduction

Cardiovascular disease (CVD) accounts for a large number of deaths and is a major cause for morbidity and disability worldwide. While the occurrence of CVD is global, statistics indicate that an alarming 80% of cases occur in developing nations [1]. The prevalence of CVD in such countries warrants a combined effort in the establishment of hygiene education, primary health-care facilities, improved monitoring systems and the development of improved and cost-effective treatments for those afflicted with CVD.

Hypertension is a major risk factor in the development of CVD, as illustrated by a decrease in incidence of adverse cardiovascular events (including stoke and myocardial infarction) when blood pressure is controlled. As already stated, hypertension affects predominantly developing nations, with Africa being no exception. In 1998, approximately 21% of the South African adult population was estimated to have hypertension [2], with general sub-Saharan data estimates at approximately 27% of the adult population [1]. Ten years later, even with the increased knowledge of vasoaction, CVD and antihypertensive drug design, the World Health Organisation (WHO) estimates that approximately 37% of adult South Africans have raised blood pressure (SBP $\geq$ 140 or DBP $\geq$ 90) [3].

Given the prevalence of hypertension in sub-Saharan Africa, and given the diversity of populations present, African researchers have a distinct responsibility in the development of suitable technologies and appropriate strategies to combat this dangerous risk factor. Certainly, the employment of early warning diagnosis, education and reputable primary health-care facilities are indispensable to prevention and early intervention; however, the use of necessary therapies is vital in the continuing efforts to minimise this threat. The use of angiotensin-converting enzyme (ACE) inhibitors has made a substantial contribution in the control of blood pressure [4]. However, the prevalence of adverse drug events (ADEs) associated with current ACE inhibitor treatment has necessitated the need to develop second-generation ACE inhibitors. Interestingly, for factors currently not fully understood, individuals of African origin can tend to have increased risk of occurrence of these drug events including life-threatening angioedema and persistent cough [5, 6]. Also, second-generation ACE inhibitors could be efficacious in the treatment of conditions not limited to blood pressure control [7]. Thus, the Zinc Metalloprotease Research Group (ZMRG) at the University of Cape Town, South Africa, has been investigating a single piece of the cardiovascular puzzle: the structure–function relationship of ACE with emphasis on the design of domain-selective ACE inhibitor drug candidates.

15.2 The Renin-Angiotensin-Aldosterone System: A Key System in Blood Pressure Regulation

The renin-angiotensin-aldosterone system (RAAS) is one of the central pathways in the regulation of blood pressure [8]. Angiotensinogen is a 55-kDa protein expressed in the liver, and an N-terminal decapeptide of this protein is released through

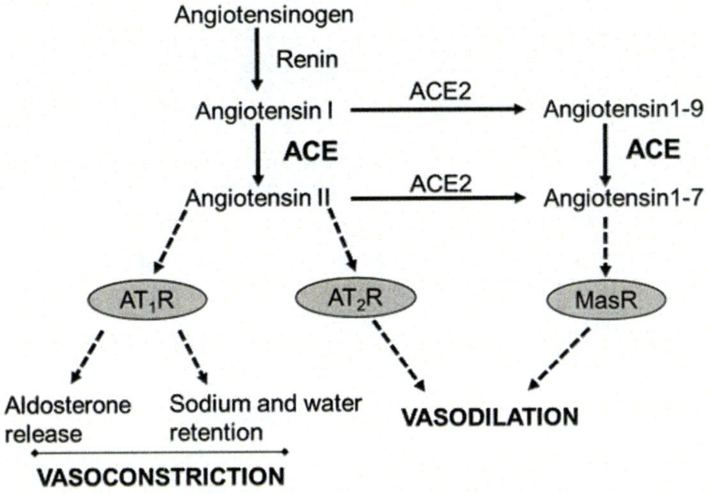

Fig. 15.1 A schematic representation of the renin-angiotensin-aldosterone system (RAAS). Angiotensin I is produced through the proteolytic action of the enzyme renin, and angiotensin I, in the classic pathway, is subsequently converted to angiotensin II (AngII) by ACE (indicated in *bold font*). AngII mediates its vasconstrictive effects through the AT type 1 receptor (AT_1R) due to the predominant tissue distribution of this receptor over the type 2 isoform. Recently added elements of the RAAS include the production of angiotensin1–9 and angiotensin1–7 (Ang1–7) by the action of ACE homologue ACE2. Ang1–7 mediates vasodilatory effects through the Mas receptor and thus serves as a counter-regulatory arm of this system

the action of the highly specific aspartyl protease renin [9]. This N-terminal decapeptide, referred to as angiotensin I (AngI), is an inactive form of the angiotensin hormone and is cleaved at the penultimate C-terminal peptide bond by ACE to yield the vasoactive octapeptide angiotensin II (AngII) [10, 11]. AngII binds and mediates its vasoconstrictive action predominantly through the angiotensin type 1 receptor (AT_1R). Downstream signalling from this receptor interaction promotes aldosterone release and subsequent water and sodium retention, thus resulting in vasoconstrictive events [4, 12]. While a type 2 receptor also exists that mediates vasodilatory events, the predominant tissue expression and distribution of AT_1R in the adult cardiovascular system accounts for a net increase of systemic blood pressure upon AngII production [13] (Fig. 15.1).

In addition to the above classical description of the RAAS, other angiotensin peptides have been discovered to also play a regulatory role within this system. Of perhaps greatest relevance is Ang1–7, a peptide produced either by the carboxypeptidase action of ACE homologue ACE2 on AngII or by the initial cleavage of AngI by ACE2 with subsequent cleavage by ACE [14] (Fig. 15.1). Ang1–7 mediates its effects through the Mas receptor [15] and promotes vasodilatory actions, thereby serving as a counter-regulatory arm of the RAAS [16].

The use of gene knockout technology in the production of mice lacking certain aspects of the RAAS has provided vital information regarding the importance of

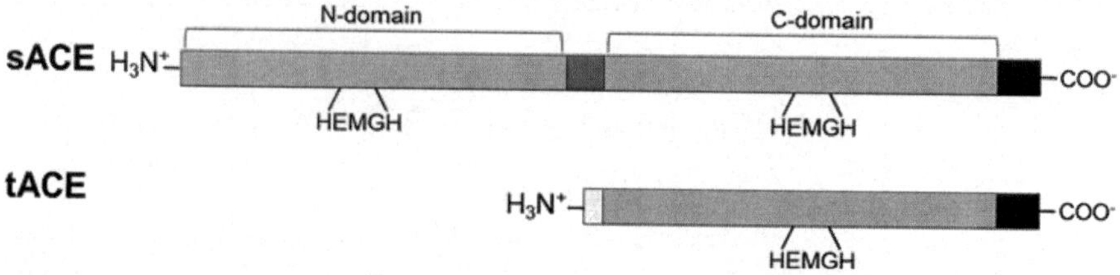

Fig. 15.2 The domain assembly of angiotensin-converting enzyme (ACE) isoforms. Each domain is presented in *light grey* with the zinc coordinating motif crucial for catalytic function (HEMGH) indicated. In somatic ACE (sACE), the linker region between the two domains is shown in *dark grey* and the C-terminal transmembrane and cytoplasmic tail in *black*. Testis ACE (tACE) is identical to the C-domain of sACE with the exception of a unique 36 amino acid sequence on the N-terminus (*white*)

the RAAS elements in the regulation of blood pressure and general physiological function. Mice containing increasing numbers of angiotensinogen gene copies tended to display increased blood pressure [17]. Mice homozygous for renin knockout showed significantly reduced blood pressure compared to controls [18], as did mice lacking AT_1R expression [19]. Similarly, homozygous ACE knockout mice displayed significantly reduced blood pressure, renal defects and fertility compared to wild-type mice [20, 21]. Thus, while there are other systems contributing to blood pressure regulation, such models emphasise the central importance of the RAAS in the systemic regulation of blood pressure.

15.3 Angiotensin-Converting Enzyme: A Central Figure in the RAAS

15.3.1 General Background

ACE belongs to the gluzincin family (MA clan) of metallopeptidases and contains a characteristic HEXXH zinc-binding motif critical for the catalytic mechanism of substrate cleavage [22]. Furthermore, it is a zinc dipeptidyl carboxypeptidase and, while best known and named for its cleavage of AngI, possesses the ability to cleave a diverse set of substrates and even displays endopeptidase activity *in vitro* [23–26]. ACE is heavily glycosylated and such a posttranslational modification is important for correct intracellular folding prior to export, thermal stability and homodimerisation [27–30]. Somatic ACE is 1,277 amino acids in length and expressed in many human tissues, especially vascular endothelial cells [31]. Interestingly, somatic ACE possesses two homologous domains (designated N- and C-domains based on their location in the polypeptide chain), each of which contains a functionally active catalytic site (Fig. 15.2) [31, 32]. Comparison of overall sequence identity between the N- and C-domains reveals approximately 60% similarity. In the domain active sites regions, the sequence identity is even more conserved with 90% sequence similarity. With this being said and perhaps surprisingly, despite high-sequence identity and similar structural topology between the domains, differences are noticed in domain-specific substrate processing and inhibitor-binding profiles (see Sects. 15.3.2, 15.6.2 and 15.6.4). An appreciation of the differential functioning of each domain has prompted efforts to develop domain-selective inhibitors that could be useful in distinct disease states.

15.3.2 Pertinent Substrates of ACE

15.3.2.1 Vasoactive Peptides: Angiotensin I and Bradykinin

ACE was first isolated and characterised on the basis of the enzyme's ability to convert AngI (DRVYIHPFHL) to produce the potent vasoconstrictive octapeptide AngII (DRVYIHPF) [10, 11]. While vasoconstriction is one response from AngII production, the downstream effects of AT_1R activation also result in increased cell proliferation, hypertrophy and fibrosis, thus making ACE a promising drug target for the treatment of disorders relating to vascular function [33].

Bradykinin (RPPGFSPFR, BK) is a nonapeptide member of the kallikrein-kinin system and is currently the best physiological substrate of ACE, with a tenfold higher catalytic efficiency than AngI conversion [34]. Contrasting AngII action, BK mediates nitric oxide release and prostaglandin synthesis that promotes vasodilation [35]. ACE is able to abolish the vasodilatory ability of this peptide through its sequential dipeptidase activity. Thus, to put it simply, ACE activity serves to result in a net increase in blood pressure through the introduction of vasoconstrictive and removal of vasodilatory peptides.

Gene modification technologies that allow for the generation of mice that contain only one functional domain of the somatic ACE molecule have provided important in vivo understanding of the domain contribution in substrate processing. BK has been shown to be cleaved with approximately similar catalytic efficiencies by both domains [36, 37]. In contrast, the C-domain has been shown to be the primary site for the conversion of AngI [37].

15.3.2.2 *N*-acetyl-Ser-Asp-Lys-Pro: Another Important ACE Substrate

While AngI and BK are certainly the most celebrated substrates of ACE, others exist that could be important in disease progression or prevention thereof. These include gonadotropin-releasing hormone and amyloid β-peptide, although the *in vivo* role of ACE cleavage of these peptides remains controversial [24, 38]. *N*-acetyl-Ser-Asp-Lys-Pro (AcSDKP) is a negative regulator of haematopoiesis and a potent anti-fibrotic agent. ACE is the principal enzyme involved in the cleavage of AcSDKP [39]. Numerous mouse models of hypertension have provided strong support for the role of AcSDKP in minimising collagen deposition in the heart and kidney through peptide infusion or ACE inhibition and thus suggest an important role of this peptide in tissue injury [40–47].

Cleavage of AcSDKP is highly N-domain specific both *in vitro* and *in vivo* [48, 49]. Mice lacking a functional N-domain catalytic site displayed increased plasma concentration of AcSDKP, as did mice treated with an N-domain-selective inhibitor [49, 50]. N-domain active site knockout mice also showed increased protection against bleomycin-induced lung fibrosis, with this protection being linked to increased AcSDKP levels [51].

With an understanding of the centrality of ACE substrate cleavage in the regulation of vascular function, it is not surprising that ACE was identified as a promising antihypertensive drug target.

15.4 First-Generation ACE Inhibitors

The discovery of the first orally active ACE inhibitors in the late 1970s is one of the classic examples of structure-based rational drug design that appears in most medicinal chemistry textbooks. However, remarkably, these inhibitors were successfully developed with no knowledge of the enzyme's three-dimensional structure and instead were based on an assumed mechanistic homology to carboxypeptidase A (CPA). There are several excellent reviews of the discovery and development of the first-generation ACE inhibitors captopril, lisinopril and enalaprilat that have been published over the last 30 years (Fig. 15.3) [4, 52–57]. Thus, the intention of this chapter is not to provide a comprehensive review of this fascinating story, but to highlight some of the key events and insights in the context of the structural data that has been elucidated more recently in our laboratory and those of our collaborators.

The key discovery of Kevin Ng and John Vane that AngI is converted to AngII in the lung [58] and the speculation that the enzyme responsible was a carboxypeptidase sparked a research programme at the pharmaceutical company Squibb to find an inhibitor of the candidate enzyme with potential application in hypertension. Ferreira et al. convincingly showed that snake venom contained inhibitors of the degradation of both bradykinin and AngI [59]. At Squibb, Miguel Ondetti and David Cushman built on these findings and were the first to synthesise the potent snake venom inhibitor SQ20881—the same as the nonapeptide bradykinin-potentiating factor (BPF) teprotide. This peptide was rapidly progressed into clinical trials and shown to decrease blood pressure in hypertensive patients [60]. A key insight derived from work on the BPFs was that the C-terminal sequence Phe-Ala-Pro and its acylated derivatives provided optimal enzyme inhibition. The next challenge was to synthesise orally active, non-peptide analogues of BPF, and the programme at Squibb was temporarily halted in 1973 due to failure to achieve this objective.

One of the key breakthroughs in the development of the first antihypertensive drug captopril was the hypothesis that ACE is structurally and mechanistically related to CPA. This led Ondetti and Cushman to propose a model of the ACE

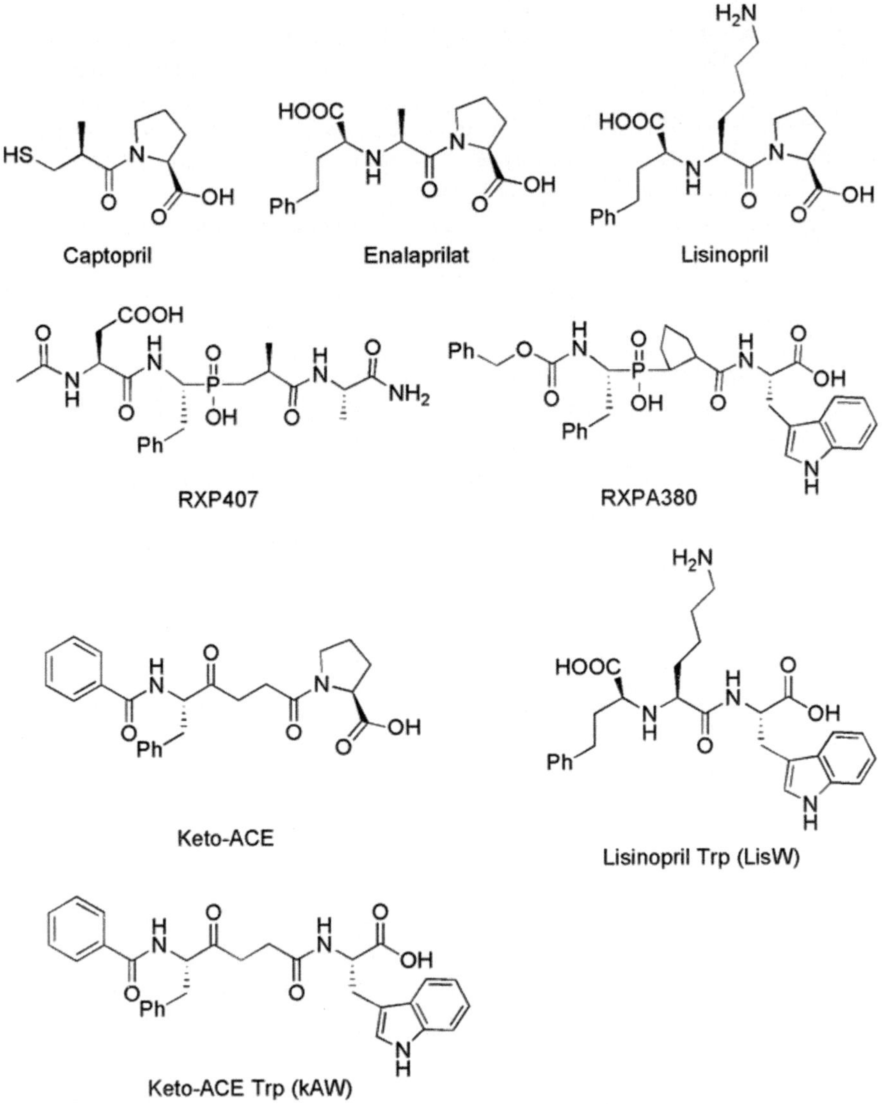

Fig. 15.3 Chemical structures of a selection of ACE inhibitors

active site with the critical zinc ion positioned to activate the scissile carbonyl of peptidic substrates. This key concept, together with work by Byers and Wolfenden showing that D-2-benzylsuccinic acid was a potent by-product inhibitor of CPA [61], prompted the design of similar succinyl amino acid derivatives with analogous structures to the dipeptide product of ACE hydrolysis. On the basis of the BPF Phe-Ala-Pro sequence, Ondetti synthesised methylsuccinyl Pro which was a specific inhibitor of ACE with an IC_{50} in the low micromolar range. By replacing the carboxyl zinc-binding group with a sulfhydryl, Cushman and Ondetti were able to improve the potency of the inhibitor and thus the competitive inhibitor captopril with a K_i of 1.7 nM was launched (Fig. 15.3) [62, 63]. Its proposed contacts with the active site S_1' and S_2' pockets and the interaction of the prolyl carboxylate with a positively charged residue are remarkably similar to those observed in the captopril-ACE crystal structure almost 30 years later [64]. Captopril was approved for clinical use in 1981 and widely used for the treatment of hypertension and congestive heart failure. However, various ADEs caused by the sulfhydryl group led to renewed efforts by Patchett and associates at Merck to design tripeptide mimetics with a carboxyl zinc-binding group [65]. Like captopril, their efforts were based on the work of Byers and Wolfenden, and they reasoned that it should be possible to raise the potency of their early lead carboxylate compounds by making them more similar to the C-terminal peptides of BPF. To achieve this, an amine group was introduced into the backbone of the molecule affording the *N*-carboxymethyl derivative of Ala-Pro. Disappointingly, this compound had a similar inhibition constant to its CH_2 analogue. The realisation that compensating hydrophobic effects likely influenced these results prompted the addition of a P_1 methyl followed by a P_1 Phe that had vastly improved IC_{50} values of 92 nM and 3.8 nM, respectively [65]. It was thought that these residues made contact with a hydrophobic surface in the S_1 pocket, and once maximal interactions had been made, changes to affect lipophilicity are tolerated because these groups do not interact with the active site. The Phe-Ala-Pro compound (enalaprilat, Fig. 15.3) had poor oral activity [66]; however, conversion of the zinc-binding group to an ethyl ester (enalapril) altered the lipophilicity of the compound and dramatically improved oral availability. The approach of synthesising a prodrug that required hydrolysis of the ester in the liver to yield the active free acid was subsequently used with a number of ACE inhibitors that are presently in clinical use. Substitution of a P_1' Lys for the Ala in enalaprilat yielding the Phe-Lys-Pro analogue (lisinopril, Fig. 15.3) also improved the oral activity resulting in longer duration of action and no requirement for metabolic activation.

With the benefit of the three-dimensional crystal structures of enalaprilat-ACE and lisinopril-ACE [64, 67], it is now clear that the postulated hydrophobic interactions of the amino-terminal Phe of the inhibitors, made by the Merck group, are validated by contacts with Phe^{512} and Val^{518} in the S_1 subsite. In the lisinopril-ACE complex, the P_1' lysyl amine, which occupies the deep S_1' pocket, interacts ionically with Glu^{162} and with Asp^{377} via water-mediated interactions. Interestingly, Glu^{162} is replaced by an Asp in the N-domain, and this amino acid substitution is likely responsible for lisinopril's modestly increased affinity for the C-domain active site [68] (see Sect. 15.6.3 on C-selective inhibitors). Enalaprilat and captopril, on the other hand, have a methyl in the P_1' position which does not make significant contact with any of the S_1' residues. The interaction of the C-terminal carboxylate with an arginine in the case of carboxypeptidase was shown to be with Lys^{511} in the C-domain, consistent with the notion that this interaction was important for the correct positioning of the scissile bond carbonyl with respect to the catalytic zinc. Finally, the hypothesis that large conformational changes of the Ala-Pro in captopril and enalaprilat occurred on binding to the active site of ACE led to the design and synthesis of numerous conformationally restricted analogues [69]. These compounds were created by bridging a carbon chain between the P_1' methyl and the prolyl C-5. Medium-sized lactams were predicted to be good mimics of the low-energy conformations of Ala-Pro. These bulky conformationally restricted analogues were well accommodated by the prime side of the ACE active site, and this is in agreement with the architecture of the cavernous S_1'/S_2' subsite revealed in the crystal structure.

The successful development of these first-generation ACE inhibitors was achieved prior to the discovery of ACE containing two catalytically active and differentially functioning domains [31, 32, 37, 49]. These inhibitors, while they interact slightly differently with each domain, lack any significant domain selectivity [68]. The high-affinity binding of inhibitors to both domain active sites is thought to be a contributor of the observed ADEs associated with treatment. While the mechanisms of ACE inhibitor associated ADEs are complex, the elevation of BK due to dual domain blockade has been implicated as a key contributor to these events [70–72]. As presented above, BK is cleaved at approximately the same rate by both domains. Thus, the selective inhibition of a single domain would allow the other active site to be involved in BK clearance and potentially lower ADE incidence. This would allow for inhibitors that could have efficacious effects in treatment of hypertension (C-domain-selective inhibitors) or pulmonary fibrosis (without affecting blood pressure, N-domain-selective inhibitors) with reduced side effect profiles [4, 7]. The development of such inhibitors has been a strong research focus in our laboratory and stands on a strong tradition of ACE research in South Africa.

15.5 Historical African Contributions

One of the first South African studies on ACE started at almost the same time as the first clinical ACE inhibitors became available. The high incidence of tuberculosis and sarcoidosis in the Western Cape prompted the evaluation of serum ACE levels in disease activity in relation to remission and steroid therapy [73]. The serum ACE assay was shown to provide a good monitor of disease activity and found to be useful in the management of patients with sarcoidosis. Serum ACE is today still used as a marker for sarcoidosis; however, ACE levels probably reflect the total granulomatous burden rather than the degree of lung involvement by sarcoidosis [74].

This work was followed by a detailed biochemical investigation of the somatic form of ACE which was also carried out in the Department of Medicine at the University of Cape Town. The paradigm of collaborating with the Department of Chemistry in an effort to solve ACE-related biological chemistry problems was initiated by Mario Ehlers who synthesised a novel N-carboxyalkyl dipeptide CA-Gly-Gly that was used for the rapid affinity purification of human lung and kidney ACE [75]. The weak competitive ACE inhibitor CA-Gly-Gly had a terminal amino group available for the covalent attachment to Sepharose. This approach, together with the use of CA-Phe-Gly [76], revolutionised the previously tedious and time-consuming purification of ACE. In addition, the CA-Gly-Gly ligand could be prepared by a relatively simple and efficient 4-step synthetic procedure [77]. CA-Gly-Gly was one of the weakest inhibitors in the N-carboxyalkyl dipeptide class, and thus ACE could be easily eluted from a CA-Gly-Gly affinity resin under mild conditions without denaturing the enzyme. Today, this purification method has been adapted and the N-carboxyalkyl dipeptide replaced with the highly specific ACE inhibitor lisinopril which is coupled to Sepharose via its lysyl amine [78, 79]. Ehlers went on to kinetically characterise the unique chloride activation of human ACE [80]. The effect of chloride on ACE-catalysed hydrolysis of AngI was shown to be complex and dependent on both chloride concentration and pH. In addition, at high chloride concentrations, the enzyme was inhibited by chloride by a complex mechanism which may involve more than one low-affinity inhibitory binding site on free enzyme, enzyme–substrate complex and/or enzyme–product complex. This *in vitro* study suggested that ACE is probably maximally activated in the vascular endothelium and renal proximal tubular epithelium, since at these sites, pH and chloride concentration are optimal and do not fluctuate. However, in the intestinal microvilli that contain large amounts of ACE, the enzyme's activity could be further controlled by ion fluxes.

This solid biochemical foundation paved the way for the cloning studies of ACE carried out by Ehlers in collaboration with James Riordan at Harvard. A race between the Harvard group and Pierre Corvol's group in Paris to clone the enzyme ensued. The cloning of somatic ACE revealed the striking feature that the large ectoprotein region was comprised of two homologous domains that had likely resulted from a gene duplication [31]. Moreover, each domain contained short sequences of the critical residues identical to those of the active sites of other metalloproteases, such as thermolysin and neutral endopeptidase, suggesting that each contained a putative active site. Molecular cloning of the human testis ACE (tACE) isoform showed that it was virtually identical to the C-terminal half of somatic ACE also bearing a functionally active catalytic site (Fig. 15.2) [81]. At that stage, scientists could only speculate on the significance of the curious finding that there was a unique testis isoform localised only in sperm cells and that this isozyme was a simpler version of the ubiquitously expressed somatic ACE. These early studies provided important biochemical and biophysical insights into the mechanism of action of ACE and paved the way for the engineering of soluble constructs that would be suitable for three-dimensional structure determination.

15.6 Current Activities: The Structure-Based Design of Second-Generation ACE Inhibitors

15.6.1 Crystal Structures of ACE

The three-dimensional structure of drug targets has played a pivotal role in drug design and development. However, the crystal structure of ACE has eluded scientists for almost half a century since its discovery, and this has been largely due to the "glycosylation problem", namely, the heterogeneity of the surface glycans prevent crystal lattice formation and the growth of diffraction-quality crystals. Herculean efforts by a number of groups in the late 1980s and early 1990s to express large quantities of somatic ACE and to crystallise the enzyme were unsuccessful due to the high level of surface glycosylation and the flexibility of the linker region joining the two domains. However, the complex problem of somatic ACE crystallisation has been solved in part by expressing the N- and C-domains separately and solving their crystal structures (Fig. 15.4). In the mid-1990s, we started investigating the glycan occupancy of tACE (the equivalent of the C-domain) and the effect of glycosylation on the functional integrity of the enzyme [82]. This study was followed by mutagenesis of the Asn-Xxx-Ser/Thr glycosylation sites of tACE in order to produce minimally glycosylated forms of the enzyme [83]. Two of the seven glycans at the N-terminus of tACE were found to be necessary and sufficient for the expression of correctly folded and enzymatically active enzyme. Because of the homology between the two domains, we assumed that the N-terminal preference for a minimally glycosylated C-domain would be similar for the N-domain. Surprisingly, this was not the case, and the N-domain showed an additional requirement for C-terminal glycosylation [28]. Thus, a construct with glycans in positions Asn^{45}, Asn^{416} and Asn^{480} was fully active and gave reproducible crystallisation results yielding diffraction-quality crystals. These studies blazed the trail for the high-resolution crystal structures of the individual N- and C-domains.

A truncated form of tACE that lacked the N-terminal glycosylated region and the transmembrane domain was expressed in the presence of a glucosidase inhibitor N-butyldeoxynojirimycin, and the N-linked glycans were removed with endo-glycosidase-H (which leaves just a single *N*-acetylglucosamine residue attached to the Asn). This deglycosylated construct formed diffractable crystals and led to the solution of the first X-ray crystal structure for human tACE [67]. The hypoglycosylated form of tACE that had only two intact glycosylation sites at Asn^{72} and Asn^{109} (tACE numbering) was also used to solve the structure of native tACE and tACE in complex with different inhibitors [84–86]. The overall structures of tACE [67] and the N-domain [87] are very similar. Both molecules are ellipsoid in shape and comprised mainly of α-helices with a few short regions of β-strand. They are divided by a deep channel in which the active site zinc ion is located, and this channel is shuttered by the N-terminal two α-helices or lid helices. Normal mode analysis of tACE suggests that a hinge movement opens the active site to allow substrate and inhibitor access [84]. The lid helices α1 and α2 swivel to one side, and the channel opens up more on one end than the other with a similar observation having been seen in a N-domain structure [28]. Helices α13 (tACE) and α 15 (N-domain) contain the HEXXH zinc-binding motif with its two zinc-coordinating histidines. A glutamate on α14 (α17 for the N-domain) and an acetate ion (from the crystallisation medium) or a water molecule complete the tetrahedral coordination of the zinc ion. The N- and C-domain crystal structures revealed differences in chloride binding: two buried chloride ions were observed in tACE

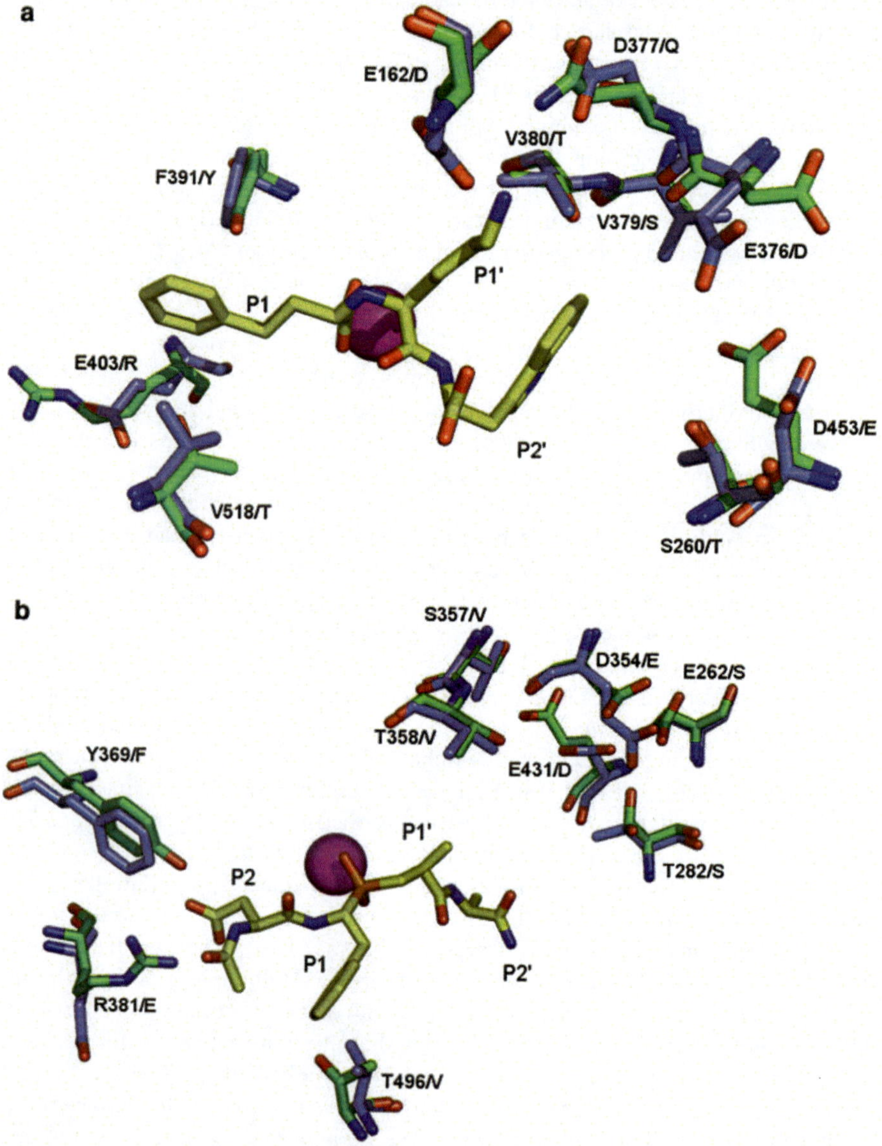

Fig. 15.4 Three-dimensional crystal structures stick representation of ACE with inhibitors. (**a**) Stick representation of lisinopril-Trp (lisW-S, *yellow*) within the tACE (C-domain) active site (PDB accession code 3L3N). Unique residues of interest in the C- and N-domains are shown in *blue* and *green sticks* respectively and residues labelled with the C-domain residue number followed by the corresponding N-domain counterpart. The catalytic zinc ion is shown in *magenta*. (**b**) Stick representation of RXP407 (*yellow*) within the N-domain active site (PDB accession code 3NXQ). Unique residues of interest in the C- and N-domains are shown in *blue* and *green sticks*, respectively, and residues labelled with the N-domain residue number followed by the corresponding C-domain counterpart. Domains were aligned using the programme *ALIGN* [103] and figures generated with PyMOL software (v 0.99, DeLano Scientific)

(chloride I and II), whereas in the N-domain structure, there was a single chloride ion corresponding to chloride II, located distally to the zinc ion.

When comparing the superimposed N- and C-domain structures, the most clearly observable difference is the extra length of the N-domain at the N- and C-termini, the latter of which includes the linker region. The well-defined nature of this region in the electron density map allowed us to propose a model for the two-domain somatic ACE. The loop between helices 19 and 20 (residues 409–417) that was not visible in the tACE structure is well defined in the N-domain. Furthermore, three other flexible loops, between helices 3 and 4, strands 1 and 2 and strand 6 and helix 23, show small differences between the two domains. Despite the structural homology between the two domains, there are some important differences between the N- and C-active sites that are responsible for the domain-selectivity of certain substrates and inhibitors. These residue substitutions in the obligatory binding site have been instrumental in our design of novel domain-selective ACE inhibitors. These active site differences will be discussed in more detail under the sections below (see Sects. 15.6.3 and 15.6.5).

15.6.2 C-Domain-Selective ACE Inhibitors

The first highly selective C-domain inhibitor, the phosphinic peptide RXPA380, effectively blocked the hydrolysis of AngI in a mouse model [36]. In contrast, bradykinin protection from ACE cleavage required the inhibition of both N- and C-active sites. A subsequent study of several RXPA380 analogues investigating the contribution of each residue towards its selectivity revealed that the P_1' pseudoproline and the P_2' tryptophan of RXPA380 were the key determinants of its C-domain specificity [88].

In our design and preparation of C-domain inhibitors, initial synthetic efforts focussed on analogues of the moderately C-selective ketomethylene derivative 5-S-5-Benzamido-4-oxo-6-phenylhexanoyl-L-proline (keto-ACE) [89]. The keto-methylene scaffold has been used extensively in the design of protease inhibitors, and it facilitated the design of non-peptidic small molecules. Keto-ACE has a proline in the P_2' position and inspection of the tACE-lisinopril structure [67] revealed that the P_2' proline of lisinopril did not make contact with any S_2' residues that are unique to the C-domain. Thus, the objective was to introduce bulky hydrophobic P_2' residues that were more likely to interact with the Val^{379} and Val^{380} that are replaced by a Ser and Thr in the N-domain, respectively. A simple five-step synthetic approach yielded a carboxylic acid which could be coupled with different *O*-protected amino acids to provide the corresponding esters in excellent yields [90]. These were hydrolysed at room temperature to afford the desired compounds in quantitative yields. Compounds with a P_2' Trp (kAW) or Phe (kAF) were 1,300-fold and at least 600-fold more selective for the C-domain using the substrate Abz-FRK(Dnp)P-OH and had K_i values of 0.83 μM and 0.68 μM, respectively [85]. Moreover, the co-crystal structures of these compounds with the C-domain confirmed our hypothesis that the nonpolar P_2' moieties of the keto-ACE derivatives would make close contacts with residues in the cavern-like S_1'/S_2' pocket, e.g. Val^{379}, Val^{380}, His^{387}, Phe^{457}, Phe^{512}, Val^{518} and Phe^{527} and contribute favourably to the binding entropy by the hydrophobic effect [85].

In an effort to increase the binding strength of the C-selective keto-ACE inhibitors and improve the drug-like characteristics of the molecule, we used the insights from the inhibitor RXPA380 [88] and kAW to design a lisinopril analogue with a carboxylic acid zinc-binding group and a P_2' tryptophan (lisW) [91]. Synthesis of lisW was achieved by reductive amination of ethyl 2-oxo-4-phenyl butyrate 1 and N-ε-(*tert*-butoxycarbonyl)-L-lysine to give a key intermediate that was coupled with the L-tryptophan methyl ester affording the Boc-protected ester as a mixture of diastereoisomers. Saponification of the ester with LiOH followed by semi-preparative HPLC afforded the two lisW diastereomers. The K_i value of the S enantiomer of lisW is 6.6 nM, 600-fold lower than the keto-ACE equivalent [86]. However, the R configuration bound considerably more weakly with a K_i three orders of magnitude greater than that of the lisW-S diastereomer. This decreased affinity is probably due to a steric clash between the β-carbon of the P_1 lysine and the hydroxyl group of the conserved Tyr^{523} if the R diastereomer is modelled into the active site. In the lisW-S-tACE crystal structure, there are nine amino acids that differ between the N- and C-domains and are thus likely to contribute to its C-selective binding (Fig. 15.4). In the S_1' pocket, Glu^{162} is substituted with Asp^{140} in the N-domain and Asp^{377} with Gln^{355}. Both these C-domain residues are close enough to the amine of the P_1' lysine to make ionic interactions or weak hydrogen bonds. LisW-S makes interactions with unique C-domain residues Glu^{376} and Val^{380} in the S_2' pocket and Val^{518} in the S_1 subsite, similar to the key contacts in the crystal structures of kAW and RXPA380. Interestingly, the P_2' tryptophan of lisW-S takes a different conformation from that observed in the co-crystal structures of tACE with kAW [85] and RXPA380 [92]. This conformation is likely favoured because of the hydrophobic interactions between the P_2' tryptophan, the P_1' lysine and Val^{380} that is positioned between the two. The two different conformations of the P_2' tryptophan emphasises the large volume of the S_2' pocket, which allows different orientations of this side chain during binding.

Surprisingly, trandopril which is also a potent ACE inhibitor lacks the domain selectivity seen in lisW-S and RXPA380 despite the fact that all three compounds have bulky hydrophobic P_2' groups. A possible explanation for the lack of selectivity of trandolapril is that its P_2' tryptophan lacks an indole nitrogen, which would prevent it from hydrogen bonding with Asp^{415}, as seen in tACE- lisW-S structure [86].

15.6.3 Site-Directed Mutagenesis: An Assessment of C-Domain Residue Contribution

The structural modification of selective inhibitors is an important and useful strategy in the assessment of functional group contribution to selective binding.

However, further information regarding the relative contribution of unique amino acids in the domain active sites provides vital information in the drug discovery process, allowing for a further basis for improved inhibitor contacts. In our laboratory, site-directed mutagenesis was carried out, whereby C-domain residues implicated for selectivity in several crystal structures were converted to their corresponding N-domain counterparts. Binding affinities of these mutants with C-selective inhibitors were then assessed and revealed that contributions of amino acids across several subsites were important for the observed C-selectivity.

Assessment of RXPA380-binding affinity with these mutants suggested an important hydrophobic interaction between Phe^{391} of the C-domain and the carboxyphenyl moiety of the inhibitor [93]. In addition, S_2' residues Thr^{282}, Glu^{376}, Val^{380} and Asp^{453} have been shown to contribute to C-domain inhibitor selectivity [85, 86, 93]. Variations in the magnitude of the effect of the mutations on different inhibitors can be attributed both to the different degrees to which individual interactions contribute to the overall affinity of each inhibitor and to the flexibility of the P_2' side chain. Interestingly, the mutation of Val^{380} to Thr had little effect on lisW-S affinity in contrast to the decrease in affinity observed for kAW, kAF and RXPA380. This emphasises the unpredictable effects of flexibility in this position. Such an approach was useful in fully clarifying the contribution of novel C-selective ACE inhibitors that were based on nominally selective inhibitor templates keto-ACE and lisinopril. Observations from co-crystal structures with different domain-selective inhibitors such as kAW and lisW-S (see above) are important as this information has allowed for identification of a consistent set of unique binding residues (Fig. 15.5 and Sect. 15.6.2). This mutagenic approach has thus allowed for a further refinement of our understanding of the active site residue contributions and provides a solid foundation for the development of next-generation inhibitors. In addition, inhibitors kAW and lisW-S provide important leads in the drug discovery process that are currently under investigation in animal studies.

15.6.4 N-Domain-Selective Inhibitors

Original efforts focussed on the development of selective inhibitors that could be used in the treatment of CVD by reduced AngII production (C-selective ACE inhibitors). However, the increasing number of publications supporting the prominent role of AcSDKP in fibrotic tissue injury (presented in Sect. 15.3.2.2) has identified the N-domain of ACE as another important drug target for the treatment of diseases relating to collagen deposition. This includes the use of an N-selective inhibitor as a co-treatment with bleomycin [51], a potent anti-neoplastic agent that results in discontinuation of treatment in many patients due to excessive lung fibrosis. Moreover, N-domain inhibitors could also be used for the treatment of idiopathic pulmonary fibrosis, for which there are no currently no approved effective drugs [94]. Thus, N-selective inhibitors could be used as a treatment in these diseases without affecting blood pressure.

RXP407, like RXPA380, is a phosphinic peptidomimetic inhibitor that has provided an important starting point in the design of N-selective ACE inhibitors (Fig. 15.3) [95]. A combinatorial chemistry approach was employed to synthesise such an inhibitor and provides important guidance in functional group contribution to N-selectivity. Keeping the P_1 and P_1' moieties in a consistent R-Pheψ-(PO_2-CH_2)-Ala-R′ form, amino acids were modified in the P_2 and P_2' positions and such products were assessed for their N-selective inhibitory ability. This resulted in the inhibitor Ac-Asp-Pheψ-(PO_2-CH_2)-Ala-Ala-NH_2, a compound with two unusual structural characteristics compared with previous ACE inhibitors: a C-terminal amide (traditionally a free carboxylate) and a protected P_2 group. Synthesis of structural analogues lacking these groups indicated the importance of these unique features in allowing for N-selective binding [95].

Several models have been proposed previously to account for the unique contacts of RXP407 in the N-domain active site, with differences in precise inhibitor functional group positioning [87, 96–98]. The recently resolved crystal structure of the N-domain in complex with RXP407 provides helpful information and is a very important step in the design of novel N-selective ACE inhibitors (Fig. 15.4) [28]. In this structure, RXP407 makes 12 direct hydrogen bonds with the N-domain active site. Of these interactions, Tyr^{369} and Arg^{381}, residues present in the S_2 subsite, are the only residues that differ to their corresponding C-domain counterparts and appear to have a prominent interaction with the P_2 Asp of RXP407. Thus, the crystal structure provides a structural basis for the importance of the Asp in N-selectivity. Interestingly, the C-terminal amide has no unique contacts with the S_2' subsite. Further mechanistic insight as to the residue contribution in the dynamic active site, as stated in a previous section, provides information that is useful in the actual contribution of these residues.

15.6.5 Site-Directed Mutagenesis: An Assessment of N-Domain Residue Contribution

Owing to the implicated importance of the S_2 and the S_2' subsites in providing interactions with RXP407 and the importance of assessing residue contribution in the drug development process, our laboratory sought to generate mutations in these pockets and analyse the effects of these amino acid substitutions on inhibitor binding. Consistent with that observed in the crystal structure, conversions to C-domain residues in the S_2' subsite had little effect on the binding ability of RXP407 [93]. Individual mutations in the S_2 subsite, namely, Y369F and R381E, had modest effects on RXP407 inhibitor binding. However, production of a mutant active site with both Y369F and R381E mutations resulted in a decrease in inhibitor-binding affinity of more than 100-fold compared to the wild-type

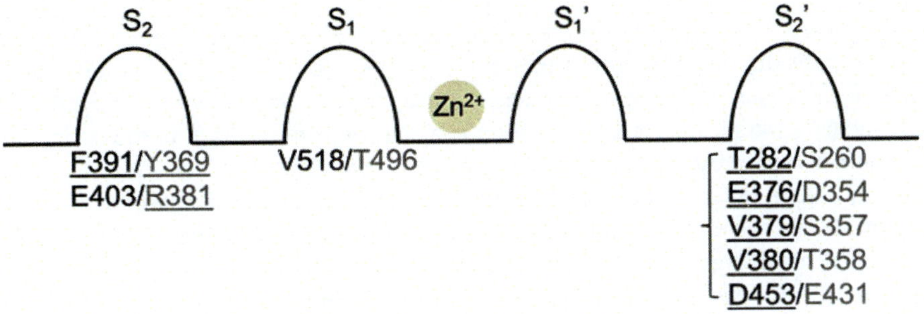

Fig. 15.5 Schechter and Berger representation [104] of subsites indicating a selection of unique residues in the N- (*grey*) and C-domain (*black*) active sites. Residues that were shown to be important contributors to selective binding are underlined. The *bracket* indicates the combined role of all the C-domain S_2' residues to facilitate C-selective binding

N-domain, suggesting crucial interactions with RXP407. Thus, the mutational analysis is in agreement with the crystal structure while the exact role of the C-terminal amide in contributing to RXP407 N-selectivity remains unclear (Fig. 15.5). In further support of the lack of importance of the C-terminal amide, it has been noted that keto-ACE analogues containing a C-terminal amide are poor inhibitors and not significantly more N-selective than the keto-ACE parent compound (unpublished observations).

Such information provides a solid basis for the design of novel N-selective ACE inhibitors. Current work involves the addition of P_2 functionalities that would have optimal interactions with residues Tyr^{369} and Arg^{381} in a keto-ACE template. This will provide an important "proof of concept molecule" that could be a promising lead in the development of N-selective inhibitors with clinical relevance.

15.7 Conclusions and Future Directions

Over the past 15 years, great strides have been made in the development of second-generation domain-selective ACE inhibitors that possess similar efficacy to that of original inhibitors but with reduced side effect profiles. In addition, increased understanding of the *in vivo* role of each domain has allowed for the initiation of the design of inhibitors that perhaps 10 years ago would not have been identified as potentially useful (e.g. AcSDKP and its role in lung fibrotic injury).

Several technologies have been indispensable in the process of ACE drug design and refinement. Firstly, the resolution of the crystal structures of each ACE domain has been invaluable in providing a structural basis in the rational design of inhibitors. This has further allowed for an identification and implication of unique residues present in the active sites that could play a role in selective binding. Another technology that has been important in allowing for a functional assessment of residue contribution towards binding has been the invaluable technique of site-directed mutagenesis for studying protein structure–function relationships. This has allowed for the prioritisation of particular "contact points" with the inhibitor and therefore assisted in the design process. These, coupled with appropriate synthetic chemistry strategies, have allowed for additional steps to be taken in the design of domain-selective ACE inhibitors. Future drug discovery work certainly involves the incorporation of other new technologies. For example, isothermal titration calorimetry is a rapid technique that will provide detailed binding analyses that will fuel the design process. Such a technique has been used for ACE inhibitors previously [99] and is currently being optimised for use in our research group.

The novel C-selective compounds LisW-S and kAW have displayed good *in vitro* selectivity (presented above). The next logical step in the drug discovery pipeline has been the investigation of the *in vivo* potential of such compounds. The efficacy of these inhibitors in hypertensive rat models is currently under investigation and will provide further indications of the druggability of these candidates.

The development of N-selective ACE inhibitors is a more recent and exciting addition to the possible therapeutic benefits of domain-selective inhibition. The identification of the important contribution of the S_2 residues provides a good starting point in terms of inhibitor design. Such compounds are the focus of current work and could provide important early lead compounds. Following this, compound refinement in terms of ADMET properties will probably be required in the move towards an appropriate drug candidate.

An interesting development in the field of CVD treatment has been the development of vasopeptidase inhibitors: compounds that are capable of inhibiting two or more enzymes involved in vasoaction (reviewed in [100]). Most work has indicated that these inhibitors tended to be associated with increased risk for ADEs, as was the case for dual ACE/neprilysin inhibitor omapatrilat [100, 101]. In an effort to lower risks associated with this form of treatment, interest has now developed in trying to produce molecules that are selective within enzymes while still inhibiting two or more enzymes. Such molecules include compound 8 F_2, an inhibitor that is C-selective for ACE (to free the N-domain for BK clearance) and also inhibits endothelin-converting enzyme (another enzyme that produces a peptide in vasoaction), which has proved efficacious in a rat model [102]. With increasing knowledge of the structure–function relationships of these important enzymes, the possibility of achieving success in these developments seems possible.

In closing, our research has employed technologies that have provided a strong impetus in the drug development process of ACE inhibitor design. While there remains a significant amount of work to be done prior to the possibility of approval, the information generated during the last 15 years could contribute to such inhibitors being produced. With the need of better outcomes of CVD in the African context and the need for treatment of lung fibrosis, it is hoped that our and others' pursuit of the development of novel domain-selective ACE inhibitors will be of lasting benefit to the people of Africa and beyond.

References

1. Kearney PM, Whelton M, Reynolds K et al (2005) Global burden of hypertension: analysis of worldwide data. Lancet 365:217–223
2. Steyn K, Gaziano TA, Bradshaw D et al (2001) Hypertension in South African adults: results from the Demographic and Health Survey, 1998. J Hypertens 19:1717–1725
3. World Health Organisation Online Data Bank Repository (2008) Accessible through http://www.who.int
4. Acharya KR, Sturrock ED, Riordan JF et al (2003) ACE revisited: a new target for structure-based drug design. Nat Rev Drug Discov 2:891–902
5. Elliott WJ (1996) Higher incidence of discontinuation of angiotensin converting enzyme inhibitors due to cough in black subjects. Clin Pharmacol Ther 60:582–588
6. Gibbs CR, Lip GY, Beevers DG (1999) Angioedema due to ACE inhibitors: increased risk in patients of African origin. Br J Clin Pharmacol 48:861–865
7. Ehlers MR (2006) Safety issues associated with the use of angiotensin-converting enzyme inhibitors. Expert Opin Drug Saf 5:739–740
8. Fyhrquist F, Saijonmaa O (2008) Renin-angiotensin system revisited. J Intern Med 264:224–236
9. Page IH, Helmer OM (1940) A crystalline pressor substance (angiotonin) resulting from the interaction between renin and renin activator. J Exp Med 71:29–42
10. Skeggs LT, Marsh WH, Kahn JR et al (1954) The purification of hypertensin I. J Exp Med 100:363–370
11. Skeggs LT, Kahn JR, Shumway NP (1956) The preparation and function of the hypertensin-converting enzyme. J Exp Med 103:295–299
12. Timmermans PB, Wong PC, Chiu AT et al (1993) Angiotensin II receptors and angiotensin II receptor antagonists. Pharmacol Rev 45:205–251
13. Lemarie CA, Schiffrin EL (2010) The angiotensin II type 2 receptor in cardiovascular disease. J Renin Angiotensin Aldosterone Syst 11:19–31
14. Rice GI, Thomas DA, Grant PJ et al (2004) Evaluation of angiotensin-converting enzyme (ACE), its homologue ACE2 and neprilysin in angiotensin peptide metabolism. Biochem J 383:45–51
15. Santos RA, Simoes e Silva AC, Maric C et al (2003) Angiotensin-(1-7) is an endogenous ligand for the G protein-coupled receptor Mas. Proc Natl Acad Sci USA 100:8258–8263
16. Roks AJ, van Geel PP, Pinto YM et al (1999) Angiotensin-(1-7) is a modulator of the human renin-angiotensin system. Hypertension 34:296–301
17. Kim HS, Krege JH, Kluckman KD et al (1995) Genetic control of blood pressure and the angiotensinogen locus. Proc Natl Acad Sci USA 92:2735–2739
18. Yanai K, Saito T, Kakinuma Y et al (2000) Renin-dependent cardiovascular functions and renin-independent blood-brain barrier functions revealed by renin-deficient mice. J Biol Chem 275:5–8
19. Ito M, Oliverio MI, Mannon PJ et al (1995) Regulation of blood pressure by the type 1A angiotensin II receptor gene. Proc Natl Acad Sci USA 92:3521–3525
20. Krege JH, John SW, Langenbach LL et al (1995) Male-female differences in fertility and blood pressure in ACE-deficient mice. Nature 375:146–148
21. Esther CR, Howard TE, Marino EM et al (1996) Mice lacking angiotensin-converting enzyme have low blood pressure, renal pathology, and reduced male fertility. Lab Invest 74:953–965
22. Rawlings ND, Barrett AJ, Bateman A (2010) MEROPS: the peptidase database. Nucleic Acids Res 38:D227–D233
23. Skidgel RA, Engelbrecht S, Johnson AR et al (1984) Hydrolysis of substance p and neurotensin by converting enzyme and neutral endopeptidase. Peptides 5:769–776

24. Skidgel RA, Erdos EG (1985) Novel activity of human angiotensin I converting enzyme: release of the NH2- and COOH-terminal tripeptides from the luteinizing hormone-releasing hormone. Proc Natl Acad Sci USA 82:1025–1029
25. Hu J, Igarashi A, Kamata M et al (2001) Angiotensin-converting enzyme degrades Alzheimer amyloid beta-peptide (A beta); retards A beta aggregation, deposition, fibril formation; and inhibits cytotoxicity. J Biol Chem 276:47863–47868
26. Sun X, Becker M, Pankow K et al (2008) Catabolic attacks of membrane-bound angiotensin-converting enzyme on the N-terminal part of species-specific amyloid-beta peptides. Eur J Pharmacol 588:18–25
27. O'Neill HG, Redelinghuys P, Schwager SL et al (2008) The role of glycosylation and domain interactions in the thermal stability of human angiotensin-converting enzyme. Biol Chem 389:1153–1161
28. Anthony CS, Corradi HR, Schwager SL et al (2010) The N domain of human angiotensin-I-converting enzyme: the role of N-glycosylation and the crystal structure in complex with an N domain-specific phosphinic inhibitor, RXP407. J Biol Chem 285:35685–35693
29. Kost OA, Bovin NV, Chemodanova EE et al (2000) New feature of angiotensin-converting enzyme: carbohydrate-recognizing domain. J Mol Recognit 13:360–369
30. Kohlstedt K, Gershome C, Friedrich M et al (2006) Angiotensin-converting enzyme (ACE) dimerization is the initial step in the ACE inhibitor-induced ACE signaling cascade in endothelial cells. Mol Pharmacol 69:1725–1732
31. Soubrier F, Alhenc-Gelas F, Hubert C et al (1988) Two putative active centers in human angiotensin I-converting enzyme revealed by molecular cloning. Proc Natl Acad Sci USA 85:9386–9390
32. Wei L, Alhenc-Gelas F, Corvol P et al (1991) The two homologous domains of human angiotensin I-converting enzyme are both catalytically active. J Biol Chem 266:9002–9008
33. Turner AJ, Hooper NM (2002) The angiotensin-converting enzyme gene family: genomics and pharmacology. Trends Pharmacol Sci 23:177–183
34. Jaspard E, Wei L, Alhenc-Gelas F (1993) Differences in the properties and enzymatic specificities of the two active sites of angiotensin I-converting enzyme (kininase II). Studies with bradykinin and other natural peptides. J Biol Chem 268:9496–9503
35. Kakoki M, Smithies O (2009) The kallikrein-kinin system in health and in diseases of the kidney. Kidney Int 75:1019–1030
36. Georgiadis D, Beau F, Czarny B et al (2003) Roles of the two active sites of somatic angiotensin-converting enzyme in the cleavage of angiotensin I and bradykinin: insights from selective inhibitors. Circ Res 93:148–154
37. Fuchs S, Xiao HD, Hubert C et al (2008) Angiotensin-converting enzyme C-terminal catalytic domain is the main site of angiotensin I cleavage *in vivo*. Hypertension 51:267–274
38. Kehoe PG (2009) Angiotensins and Alzheimer's disease: a bench to bedside overview. Alzheimers Res Ther 1:3–10
39. Rieger KJ, Saez-Servent N, Papet MP et al (1993) Involvement of human plasma angiotensin I-converting enzyme in the degradation of the haemoregulatory peptide N-acetyl-seryl-aspartyl-lysyl-proline. Biochem J 296(Pt 2):373–378
40. Peng H, Carretero OA, Raij L et al (2001) Antifibrotic effects of N-acetyl-seryl-aspartyl-Lysyl-proline on the heart and kidney in aldosterone-salt hypertensive rats. Hypertension 37:794–800
41. Peng H, Carretero OA, Brigstock DR et al (2003) Ac-SDKP reverses cardiac fibrosis in rats with renovascular hypertension. Hypertension 42:1164–1170
42. Peng H, Carretero OA, Vuljaj N et al (2005) Angiotensin-converting enzyme inhibitors: a new mechanism of action. Circulation 112:2436–2445
43. Peng H, Carretero OA, Liao TD et al (2007) Role of N-acetyl-seryl-aspartyl-lysyl-proline in the antifibrotic and anti-inflammatory effects of the angiotensin-converting enzyme inhibitor captopril in hypertension. Hypertension 49:695–703
44. Rasoul S, Carretero OA, Peng H et al (2004) Antifibrotic effect of Ac-SDKP and angiotensin-converting enzyme inhibition in hypertension. J Hypertens 22:593–603

45. Lin CX, Rhaleb NE, Yang XP et al (2008) Prevention of aortic fibrosis by N-acetyl-seryl-aspartyl-lysyl-proline in angiotensin II-induced hypertension. Am J Physiol Heart Circ Physiol 295:H1253–H1261
46. Liao TD, Yang XP, D'Ambrosio M et al (2010) N-acetyl-seryl-aspartyl-lysyl-proline attenuates renal injury and dysfunction in hypertensive rats with reduced renal mass: council for high blood pressure research. Hypertension 55:459–467
47. Wang M, Liu R, Jia X et al (2010) N-acetyl-seryl-aspartyl-lysyl-proline attenuates renal inflammation and tubulointerstitial fibrosis in rats. Int J Mol Med 26:795–801
48. Rousseau A, Michaud A, Chauvet MT et al (1995) The hemoregulatory peptide N-acetyl-Ser-Asp-Lys-Pro is a natural and specific substrate of the N-terminal active site of human angiotensin-converting enzyme. J Biol Chem 270:3656–3661
49. Fuchs S, Xiao HD, Cole JM et al (2004) Role of the N-terminal catalytic domain of angiotensin-converting enzyme investigated by targeted inactivation in mice. J Biol Chem 279:15946–15953
50. Junot C, Gonzales MF, Ezan E et al (2001) RXP 407, a selective inhibitor of the N-domain of angiotensin I-converting enzyme, blocks in vivo the degradation of hemoregulatory peptide acetyl-Ser-Asp-Lys-Pro with no effect on angiotensin I hydrolysis. J Pharmacol Exp Ther 297:606–611
51. Li P, Xiao HD, Xu J et al (2010) Angiotensin-converting enzyme N-terminal inactivation alleviates bleomycin-induced lung injury. Am J Pathol 177:1113–1121
52. Ondetti MA, Cushman DW (1982) Enzymes of the renin-angiotensin system and their inhibitors. Annu Rev Biochem 51:283–308
53. Patchett AA, Cordes EH (1985) The design and properties of N-carboxyalkyldipeptide inhibitors of angiotensin-converting enzyme. Adv Enzymol Relat Areas Mol Biol 57:1–84
54. Cushman DW, Ondetti MA (1999) Design of angiotensin converting enzyme inhibitors. Nat Med 5:1110–1113
55. Menard J, Patchett AA (2001) Angiotensin-converting enzyme inhibitors. Adv Protein Chem 56:13–75
56. Patchett AA (2002) 2002 Alfred Burger Award Address in Medicinal Chemistry. Natural products and design: interrelated approaches in drug discovery. J Med Chem 45:5609–5616
57. Smith CG, Vane JR (2003) The discovery of captopril. FASEB J 17:788–789
58. Ng KK, Vane JR (1968) Fate of angiotensin I in the circulation. Nature 218:144–150
59. Ferreira SH, Bartelt DC, Greene LJ (1970) Isolation of bradykinin-potentiating peptides from Bothrops jararaca venom. Biochemistry 9:2583–2593
60. Gavras H, Brunner HR, Laragh JH et al (1974) An angiotensin converting-enzyme inhibitor to identify and treat vasoconstrictor and volume factors in hypertensive patients. N Engl J Med 291:817–821
61. Byers LD, Wolfenden R (1973) Binding of the by-product analog benzylsuccinic acid by carboxypeptidase A. Biochemistry 12:2070–2078
62. Ondetti MA, Rubin B, Cushman DW (1977) Design of specific inhibitors of angiotensin-converting enzyme: new class of orally active antihypertensive agents. Science 196:441–444
63. Cushman DW, Cheung HS, Sabo EF et al (1977) Design of potent competitive inhibitors of angiotensin-converting enzyme. Carboxyalkanoyl and mercaptoalkanoyl amino acids. Biochemistry 16:5484–5491
64. Natesh R, Schwager SL, Evans HR et al (2004) Structural details on the binding of antihypertensive drugs captopril and enalaprilat to human testicular angiotensin I-converting enzyme. Biochemistry 43:8718–8724
65. Patchett AA, Harris E, Tristram EW et al (1980) A new class of angiotensin-converting enzyme inhibitors. Nature 288:280–283
66. Biollaz J, Burnier M, Turini GA et al (1981) Three new long-acting converting-enzyme inhibitors: relationship between plasma converting-enzyme activity and response to angiotensin I. Clin Pharmacol Ther 29:665–670
67. Natesh R, Schwager SL, Sturrock ED et al (2003) Crystal structure of the human angiotensin-converting enzyme-lisinopril complex. Nature 421:551–554

68. Wei L, Clauser E, Alhenc-Gelas F et al (1992) The two homologous domains of human angiotensin I-converting enzyme interact differently with competitive inhibitors. J Biol Chem 267:13398–13405
69. Wyvratt MJ, Patchett AA (1985) Recent developments in the design of angiotensin-converting enzyme inhibitors. Med Res Rev 5:483–531
70. Nussberger J, Cugno M, Amstutz C et al (1998) Plasma bradykinin in angio-oedema. Lancet 351:1693–1697
71. Emanueli C, Grady EF, Madeddu P et al (1998) Acute ACE inhibition causes plasma extravasation in mice that is mediated by bradykinin and substance P. Hypertension 31:1299–1304
72. Adam A, Cugno M, Molinaro G et al (2002) Aminopeptidase P in individuals with a history of angiooedema on ACE inhibitors. Lancet 359:2088–2089
73. Ainslie GM, Benatar SR (1985) Serum angiotensin converting enzyme in sarcoidosis: sensitivity and specificity in diagnosis: correlations with disease activity, duration, extra-thoracic involvement, radiographic type and therapy. Q J Med 55:253–270
74. Westall GP (2003) Interstitial lung disease. BC Decker, London
75. Ehlers MR, Maeder DL, Kirsch RE (1986) Rapid affinity chromatographic purification of human lung and kidney angiotensin-converting enzyme with the novel N-carboxyalkyl dipeptide inhibitor N-[1(S)-carboxy-5-aminopentyl]glycylglycine. Biochim Biophys Acta 883:361–372
76. Pantoliano MW, Holmquist B, Riordan JF (1984) Affinity chromatographic purification of angiotensin converting enzyme. Biochemistry 23:1037–1042
77. Ehlers MR, Kirsch RE, Giles RGF, Yorke SC (1986) Synthesis of (1 S)-N-(1-carboxy-5-aminopentyl)-glycylglycine: A prospective competitive inhibitor for angiotensin-converting enzyme. S Afr J Chem 39:134–136
78. El Dorry HA, Bull HG, Iwata K et al (1982) Molecular and catalytic properties of rabbit testicular dipeptidyl carboxypeptidase. J Biol Chem 257:14128–14133
79. Bull HG, Thornberry NA, Cordes EH (1985) Purification of angiotensin-converting enzyme from rabbit lung and human plasma by affinity chromatography. J Biol Chem 260:2963–2972
80. Ehlers MRW, Kirsch RE (1988) Catalysis of angiotensin I hydrolysis by human angiotensin-converting enzyme: effect of chloride and pH. Biochemistry 27:5538–5544
81. Ehlers MR, Fox EA, Strydom DJ et al (1989) Molecular cloning of human testicular angiotensin-converting enzyme: the testis isozyme is identical to the C-terminal half of endothelial angiotensin-converting enzyme. Proc Natl Acad Sci USA 86:7741–7745
82. Yu XC, Sturrock ED, Wu Z et al (1997) Identification of N-linked glycosylation sites in human testis angiotensin-converting enzyme and expression of an active deglycosylated form. J Biol Chem 272:3511–3519
83. Gordon K, Redelinghuys P, Schwager SL et al (2003) Deglycosylation, processing and crystallization of human testis angiotensin-converting enzyme. Biochem J 371:437–442
84. Watermeyer JM, Sewell BT, Schwager SL et al (2006) Structure of testis ACE glycosylation mutants and evidence for conserved domain movement. Biochemistry 45:12654–12663
85. Watermeyer JM, Kroger WL, O'Neill HG et al (2008) Probing the basis of domain-dependent inhibition using novel ketone inhibitors of angiotensin-converting enzyme. Biochemistry 47:5942–5950
86. Watermeyer JM, Kroger WL, O'Neill HG et al (2010) Characterization of domain-selective inhibitor binding in angiotensin-converting enzyme using a novel derivative of lisinopril. Biochem J 428:67–74
87. Corradi HR, Schwager SL, Nchinda AT et al (2006) Crystal structure of the N domain of human somatic angiotensin I-converting enzyme provides a structural basis for domain-specific inhibitor design. J Mol Biol 357:964–974
88. Georgiadis D, Cuniasse P, Cotton J et al (2004) Structural determinants of RXPA380, a potent and highly selective inhibitor of the angiotensin-converting enzyme C-domain. Biochemistry 43:8048–8054

89. Deddish PA, Marcic B, Jackman HL et al (1998) N-domain-specific substrate and C-domain inhibitors of angiotensin-converting enzyme: angiotensin-(1-7) and keto-ACE. Hypertension 31:912–917
90. Nchinda AT, Chibale K, Redelinghuys P et al (2006) Synthesis of novel keto-ACE analogues as domain-selective angiotensin I-converting enzyme inhibitors. Bioorg Med Chem Lett 16:4612–4615
91. Nchinda AT, Chibale K, Redelinghuys P et al (2006) Synthesis and molecular modeling of a lisinopril-tryptophan analogue inhibitor of angiotensin I-converting enzyme. Bioorg Med Chem Lett 16:4616–4619
92. Corradi HR, Chitapi I, Sewell BT et al (2007) The structure of testis angiotensin-converting enzyme in complex with the C domain-specific inhibitor RXPA380. Biochemistry 46:5473–5478
93. Kroger WL, Douglas RG, O'Neill HG et al (2009) Investigating the domain specificity of phosphinic inhibitors RXPA380 and RXP407 in angiotensin-converting enzyme. Biochemistry 48:8405–8412
94. du Bois RM (2010) Strategies for treating idiopathic pulmonary fibrosis. Nat Rev Drug Discov 9:129–140
95. Dive V, Cotton J, Yiotakis A et al (1999) RXP 407, a phosphinic peptide, is a potent inhibitor of angiotensin I converting enzyme able to differentiate between its two active sites. Proc Natl Acad Sci USA 96:4330–4335
96. Tzakos AG, Gerothanassis IP (2005) Domain-selective ligand-binding modes and atomic level pharmacophore refinement in angiotensin I converting enzyme (ACE) inhibitors. Chembiochem 6:1089–1103
97. Jullien ND, Cuniasse P, Georgiadis D et al (2006) Combined use of selective inhibitors and fluorogenic substrates to study the specificity of somatic wild-type angiotensin-converting enzyme. FEBS J 273:1772–1781
98. Akif M, Georgiadis D, Mahajan A et al (2010) High-resolution crystal structures of *Drosophila melanogaster* angiotensin-converting enzyme in complex with novel inhibitors and antihypertensive drugs. J Mol Biol 400:502–517
99. Andujar-Sanchez M, Camara-Artigas A, Jara-Perez V (2004) A calorimetric study of the binding of lisinopril, enalaprilat and captopril to angiotensin-converting enzyme. Biophys Chem 111:183–189
100. Douglas RG, Ehlers MRW, Sturrock ED (2011) Vasopeptidase inhibition – solving the cardiovascular puzzle? Drug Future 36:33–43
101. Kostis JB, Packer M, Black HR et al (2004) Omapatrilat and enalapril in patients with hypertension: the Omapatrilat Cardiovascular Treatment vs. Enalapril (OCTAVE) trial. Am J Hypertens 17:103–111
102. Jullien N, Makritis A, Georgiadis D et al (2010) Phosphinic tripeptides as dual angiotensin-converting enzyme C-domain and endothelin-converting enzyme-1 inhibitors. J Med Chem 53:208–220
103. Cohen GH (1997) ALIGN: a program to superimpose protein coordinates, accounting for insertions and deletions. J Appl Cryst 30:1160–1161
104. Schechter I, Berger A (1967) On the size of the active site in proteases. I. Papain. Biochem Biophys Res Comm 27:157–162

Chapter 16
Natural Products and Antimalarial Drugs: Will Africa Provide the Next Major Breakthrough?

Ivan Addae-Mensah and Dorcas Osei-Safo

16.1 Introduction

The two epoch-making events that have really made a lasting impact on malaria chemotherapy have emanated from natural products. The discovery of quinine and quinidine from the *Cinchona* bark was the first of these, and compounds that emerged from this discovery were the mainstay of malaria chemotherapy for decades. Chloroquine was until recently the cheapest, safest and most effective antimalarial drug, and it is a direct descendant of the *Cinchona* tree story. The *Cinchona* tree is native to South America.

The other epoch-making discovery was that of the artemisinin-based antimalarials which are now the mainstay of first-line malaria treatment but are now also being threatened with possible drug resistance problems. Artemisinin is from the Chinese plant *Artemisia annua*.

Several African medicinal plants are alleged to be efficacious for treatment of fevers including malaria. In a recent major revision of ethnobotanical and floristic studies of Ghanaian medicinal plants undertaken with support from the African Union Scientific, Technical and Research Commission (OAU-STRC), the authors of the resultant book examined the traditional medicine and ethnobotanical uses of 581 plant species from 107 genera. Of these, 37 were listed specifically for the treatment of malaria, 55 for fevers in general, of which some could be for malaria, and another 27 for the treatment of headaches [1]. In his seminal book *"Woody Plants of Ghana"*, Irvine deals comprehensively with 1,833 monocotyledonous and dicotyledonous woody plant species covering 707 genera from 108 families.

I. Addae-Mensah (✉) • D. Osei-Safo
Department of Chemistry, University of Ghana, P. O. Box LG 56, Legon, Ghana
e-mail: a-mensah@ug.edu.gh

K. Chibale et al. (eds.), *Drug Discovery in Africa*,
DOI 10.1007/978-3-642-28175-4_16, © Springer-Verlag Berlin Heidelberg 2012

Of these, 754 species are listed as having various medicinal uses while 421 species are classified as poisons and antidotes; 146 species listed as being poisonous to humans; 47 species poisonous to domestic animals, game and wildlife; another 71 as fish poisons; 50 species as insecticides and insectifuges; and 107 species as vermifuges and anthelmintics [2]. Kokwaro in his book *Medicinal Plants of East Africa* discusses the ethnomedicinal uses of 1301 species from 140 woody and herbaceous plant families. Fifty seven of these species are specifically listed for the treatment of malaria while another 190 species are generally classified for treating headaches due to colds, unspecified fevers and influenza [3]. Most of the plant species documented in these three books have not been investigated phytochemically or pharmacologically even though a few of them have been developed as phytopharmaceuticals. In Ghana, Nigeria, Kenya, Zimbabwe and some other African countries where there are legal and regulatory frameworks for herbal and other traditional medical practices, some of these phytopharmaceuticals have been standardised and licensed with the various countries' drug regulatory authorities for sale to the general public through designated pharmacies and licensed chemical sellers.

Relatively few of those plants used for treating malaria have been evaluated comprehensively for antimalarial activity. Such studies have been limited to crude extracts tested *in vitro*. Toxicological studies are even rarer. Phytochemical investigations of some of these plants have yielded many structurally unique compound classes. These phytochemical studies have however far outnumbered biological and pharmacological assessments. In a few cases, the compounds have been assessed for various activities in animals such as mice, rats, rabbits or dogs. Even though crude plant extracts have been used for human treatment of malaria, virtually none of the pure active compounds isolated from these plants have as yet made it into human trials or clinical use. A few of them can however be considered as potential lead compounds or structural scaffolds for further investigation.

Within the context of current efforts in drug development in Africa, we would like to pose the following questions:

- Will the next major breakthrough in malaria chemotherapy also come from natural products?
- Will Africa be the source of such a drug?
- What has Africa got to offer from its vast biodiversity resource in answer to what Asia and South America have given to the world?
- What challenges do African researchers face in their quest to play a major or relevant role in the search for such a drug?

Traditional medicine has always played and will continue to play a major role in health-care delivery in Africa and the world. A 1991 survey by the USAID has shown that in many African countries, traditional medical practitioners outnumber allopathic practitioners by 100 to 1 [4]. In Ghana, the ratio was approximately 10 to 1 [5]. Currently, there are approximately 20,770 registered traditional medical practitioners to about 2,010 allopathic doctors for a population of 24 million, giving a ratio of 1 traditional medical practitioner to about 1,200 persons and 1 allopathic medical practitioner to about 12,000 people with some regions having as high a figure as 1 to over 65,000 persons [6].

Studies by the WHO roll back malaria programme in 1998 showed that in Ghana, Mali, Nigeria and Zambia, more than 60% of children with high fever are treated at home with herbal medicines, because of ready accessibility and perceived lower cost. The cost of the herbal treatment is only about 6% of the cost of the clinical or hospital treatment. Whereas it costs about US $1.6 to treat a child in a conventional clinic or hospital, self-treatment with drugs bought from a pharmacy costs about US $0.35 while home treatment with herbs costs US $0.10 [7]. This survey does not however indicate how effective the herbal treatments are on the children. But even if it is not as effective as the conventional but more expensive hospital treatment, in a situation where it may be the only available treatment within a radius of more than 30 km, it will be far better than no treatment, in circumstances where no treatment at all may mean certain death.

16.2 Natural Products as Potential Sources of New Drugs

Drug discovery and development is a very expensive venture, and the success rate for the development of a new drug for any disease type is only about 1:6,000 to 1:10,000. A higher ratio of hits is normally encountered in natural products since many naturally occurring compounds would have undergone some level of preselection and optimisation in the course of evolution or biogenesis. Only about 5,000 of the almost 300,000 existing species of higher plants have been chemically and biologically investigated. It is estimated that there may also be another 100,000 species waiting to be discovered, while some of the already known species are also facing extinction due to human activities, even before they have been scientifically investigated. This still makes the plant kingdom one of the richest possible sources of potential lead compounds for all types of diseases including malaria. The marine flora has even been less well investigated. Notwithstanding these potential rich sources of medicines, true authentic naturally occurring compounds that have been introduced into human use without any further modifications or improvements have been relatively rare. Ethnomedicines or phytopharmaceuticals developed on the basis of ethnobotanical properties may therefore be common, but development of single-compound drugs for human use proves to be very difficult. Most ethnomedicines probably owe their efficacy and relative lack of toxicity to the presence of relatively small quantities of the same compound types in a given plant extract, acting synergistically. This may well be the reason why crude extracts of plants regularly show marked activity without significant toxicity, while isolation of the pure chemical constituents may lead to no or weak activity and significant toxicity. This presents a major challenge to African researchers in their efforts at developing usable phytomedicines while at the same time searching for active drugable lead compounds from these plant sources.

In this chapter, we will discuss previous and current efforts at developing African plants both as usable phytopharmaceuticals and as a source of potential lead compounds or structural scaffolds for antimalarial drug development. We contend that there is the need for a paradigm shift in our research efforts. While investigating as yet un-investigated plant species, it might be worthwhile re-examining some of the species that have already been investigated either for malaria or for other diseases including the neglected tropical diseases. We shall however select six plant genera containing mainly alkaloids and critically review them. Some specific compound types from these plant species that have been tested for antimalarial activity will be reviewed. Arguments in favour of revisiting some of the already investigated medicinal plants and re-evaluating their potential as sources of antimalarial drugs will be advanced. This chapter will conclude with prospects for the future and what is to be done. The selected genera are *Ancistrocladus*, *Triphyophyllum*, *Tabernaemontana*, *Zanthoxylum*, *Azadirachta* and *Cryptolepis*.

16.2.1 The Ancistrocladaceae and Dioncophyllaceae as Potential Sources of Antimalarial Drugs

These two families of tropical lianas contain a unique class of natural products, the naphthylisoquinolines. These compounds consist of a naphthalene and isoquinoline moiety linked together by a biaryl axis. Examples are dioncophylline A and B from *Triphyophyllum peltatum* (Dioncophyllaceae) and ancistrocladine from *Ancistrocladus heyneanus* and the West African species *Ancistrocladus abbreviatus* (Ancistrocladaceae). A whole series of over 50 of these alkaloids have been isolated and fully characterised from these two plant genera. These include ancistrobrevines A, B, C and D, dioncopeltine A, dioncolactone A and hamatine, an atropisomer of ancistrocladine A [8–13]. Many of the naphthylisoquinoline alkaloids from the genera *Ancistrocladus* and *Triphyophyllum* have been found to have significant fungicidal, antitumour, anti-HIV, spasmolytic, antifeedant and growth-retarding activities [14, 15] but even more importantly very pronounced *in vitro* and *in vivo* antimalarial activity against *Plasmodium falciparum* and *P. berghei* [16]. The botany, chemistry, agronomy, biogenesis and other significant aspects of these plant genera and their alkaloid constituents have been comprehensively reviewed [17–20]. Many of the alkaloids display atropisomerism and axial chirality due to the bulky ortho-substituents adjacent to the biaryl axis.

ANCISTROCLADUS AND DIONCOPHYLLUM ALKALOIDS

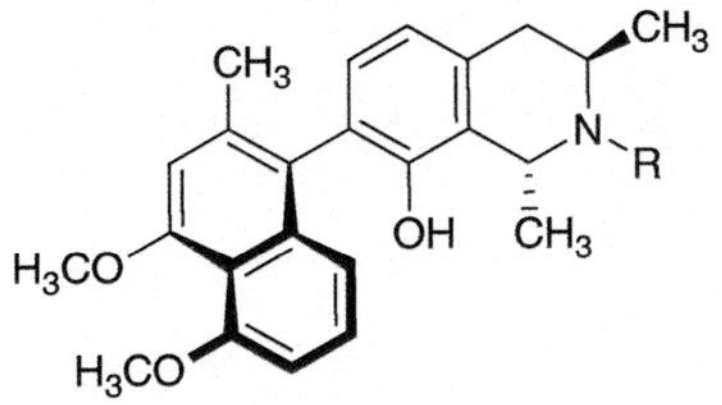

Dioncophylline A : R = H
N-Methyl dioncophylline A : R = CH_3

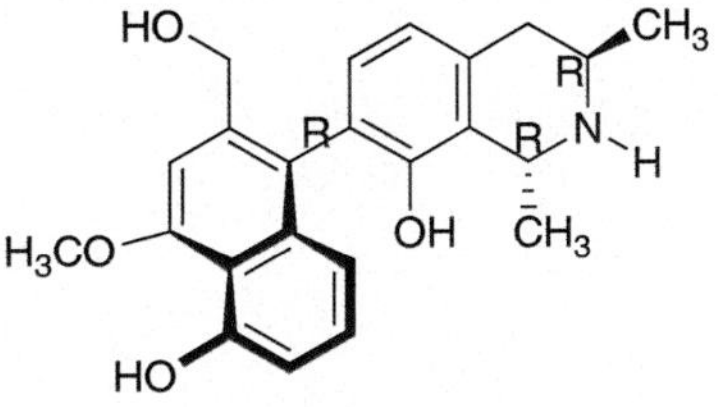

Dioncopeltine A

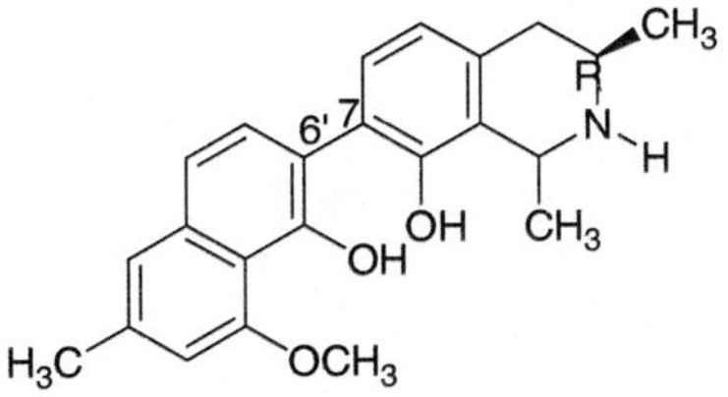

Dioncophylline B

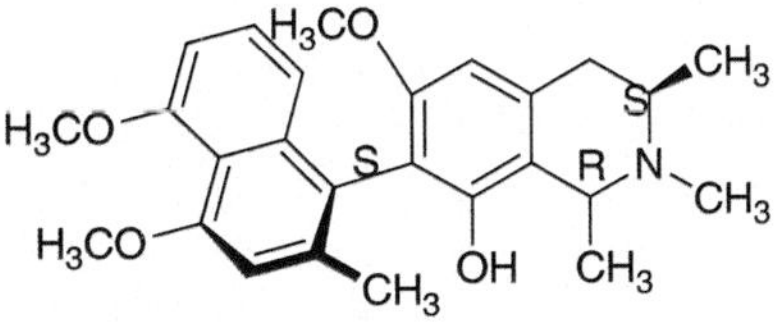

Ancistrobrevine D

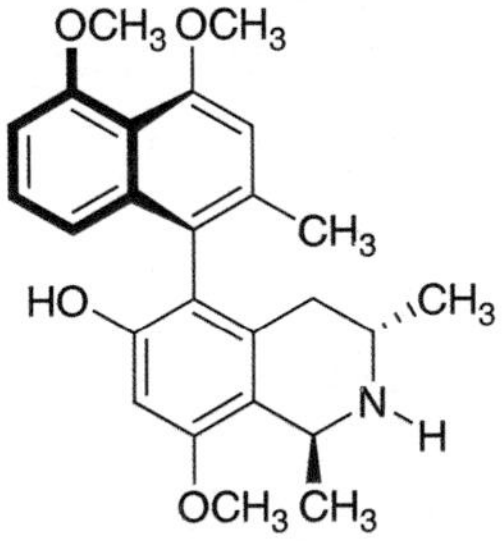

Ancistrocladine

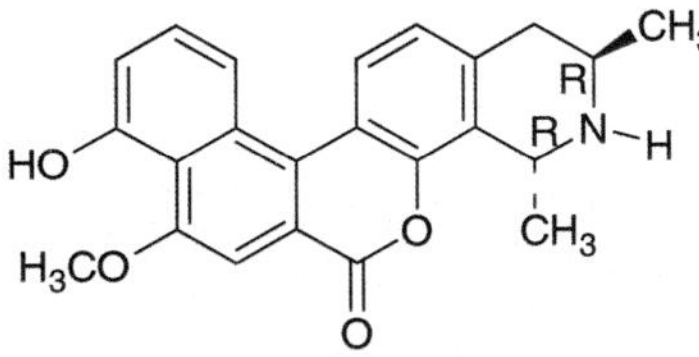

Dioncolactone A

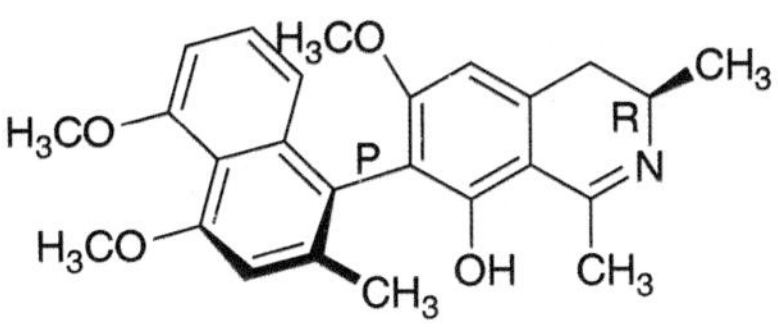

Ancistrobrevine C

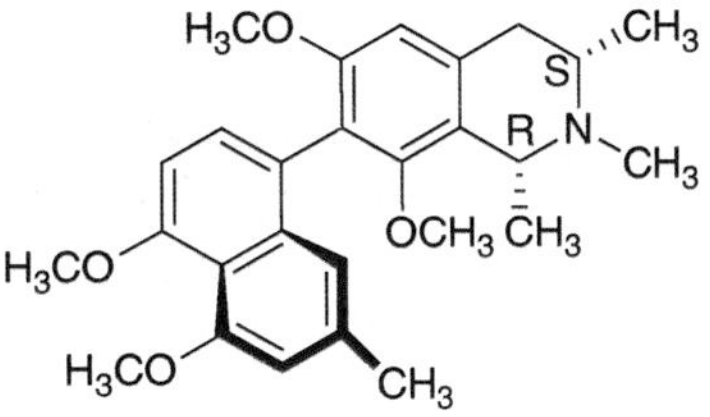

Ancistrobrevine A

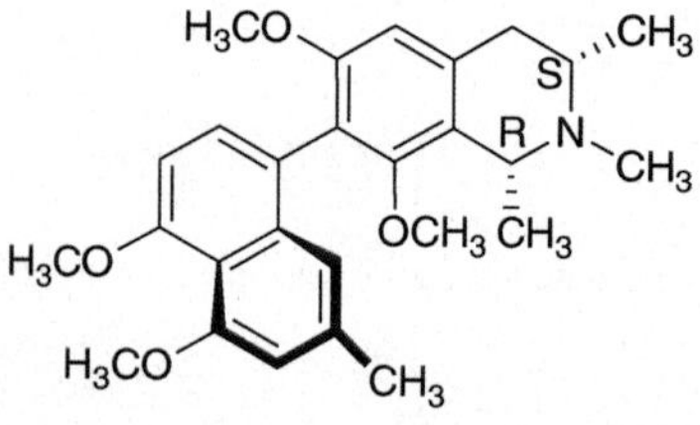

Ancistrobrevine B

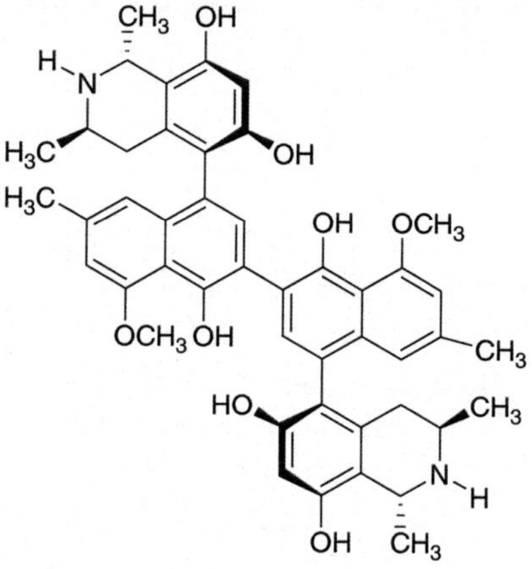

Michellamine A

Table 16.1 Antiplasmodial activity of crude extracts and pure alkaloids of *Triphyophyllum peltatum, Ancistrocladus abbreviatus* and *A. barteri* [16]

Extract/pure alkaloid	IC_{50} (KI strain) (μg/ml)	IC_{50} (NF 54/64) (μg/ml)
T. peltatum root (CH_2Cl_2)	0.017	0.053
T. peltatum shoot (CH_2Cl_2)	Not tested	0.076
A. abbreviatus root (CH_2Cl_2)	0.153	0.105
A. abbreviatus shoot (CH_2Cl_2–NH_3)	Not tested	0.031
A. barteri root (CH_2Cl_2–NH_3)	0.553	0.648
Dioncophylline B	0.063	0.224
Dioncopeltine A	0.330	0.021
Dioncophylline A	0.860	1.443
N-methyldioncophylline	5.749	13.637
Ancistrobrevine	10.440	12.222
Ancistrocladine	25.350	18.353
Chloroquine	0.068	0.002

Extracts of *T. peltatum, A. abbreviatus* and *A. barteri,* and some of the pure alkaloids have been evaluated *in vitro* against *P. falciparum* [16]. The results are shown in Table 16.1. Structure–activity relationship studies suggest that an R-configuration at C-3 associated with the lack of an oxygen substituent at C-6 and the absence of N-methylation are essential for antiplasmodial activity [16].

In vivo tests on dioncophylline C and dioncophylline A showed that oral administration of dioncophylline C (50 mg/kg/day) cured mice infected with *P. berghei* without obvious toxic effects, while dioncophylline A almost suppressed parasitaemia. Administration of both alkaloids by implanted mini-osmotic pumps resulted in total cure [21].

After a relative lull in research on the antimalarial activity of these naphthylisoquinoline alkaloids, interest has recently been aroused. It has been reported that dioncophylline A is metabolised by rat liver microsomes to 5′-O-methyl-dioncophylline A. Oral bioavailability in rats is about 25%, and the plasma half-life is 2.5 h [22]. Dioncophylline C forms a complex with haem very similar to the haem–drug complexes of the quinoline antimalarials [23]. Several other related alkaloids from the Ancistrocladaceae have more recently been assessed and found to be very active *in vitro* against chloroquine-resistant and chloroquine-sensitive strains of *P. falciparum* [24], but no *in vivo* tests are reported on them. Attempts have also been made to functionalise dioncophylline A to improve its mode of action, including bioavailability [25]. Simplified analogues of dioncopeltine A have also been synthesised for antimalarial activity assessment [25]. It is clear that a lot more work still remains to be completed on the many alkaloids isolated from these plant genera, particularly *in vivo* studies on the more promising compounds as well as the crude extracts, if the crude plant extracts are to be developed into usable phytopharmaceuticals.

Following the pioneering work by Bringmann and his co-workers, and in a search for potential anti-HIV compounds from natural sources, another new class of alkaloids, the bis-benzylisoquinolines, were found to be major constituents of these plant species. The most significant of these compounds were the michellamines, first isolated from a hitherto unknown species of *Ancistrocladus*, *Ancistrocladus korupensis*, native to Cameroun. The michellamines, which are dimers of the ancistrocladine- and dioncophylline-type alkaloids, created a lot of excitement in the search for potential anti-HIV drugs and were at one time regarded as potential candidates for first-in-human trials. The same types of compounds have since been isolated from other *Ancistrocladus* species including the more common West African species *A. abbreviatus*. It is significant to stress that in addition to *A. abbreviatus*, two more new and much rarer species, *A. robertsoniorum* and *A. korupensis,* have been discovered in Kenya and Cameroun, respectively [26–28]. In addition to the anti-HIV michellamines [29, 30], these two species have also yielded some novel alkaloids including the korupensamines that have significant antimalarial activities [31, 32].

Thus, more intensive and extensive examination of these plant genera for antimalarial drugs is warranted. The peculiar stereochemical orientation of the isoquinoline ring system that confers some stereochemical rigidity and chirality, as well as the unique substitution patterns in many of these alkaloids and their facile conversion into semi-synthetic analogues, makes this class of alkaloids attractive for further examination including structure–activity relationship studies for antimalarial activity. What is left therefore is an equally thorough investigation of the medicinal chemistry of these alkaloids and their analogues, including *in vivo* antimalarial assessment and *in silico* ADME studies, which studies have seriously lagged behind the chemistry. The *in silico* studies will also require identification of an appropriate molecular target.

One very important objective has however been already achieved. A number of the species of these plants have been successfully cultivated under greenhouse conditions and in the field [17, 33]. This constitutes a very important solution to possible threat of extinction and sustainability, especially of the rarer *A. korupensis* and *A. robertsoniorum*. Other African *Ancistrocladus* species that are yet to receive any attention are *A. congolensis, A. ealaensis, A. guineensis, A. letestui, A. likoko, A. pachyrrhadis* and *A. uncinatus* [17].

16.2.2 The Genus Tabernaemontana *(Family Apocynaceae)*

This genus is a very good example of medicinal plants whose phytochemical examination has been far ahead of any bioactivity studies. It is a good example of why we strongly suggest in this chapter the need for re-examination of constituents of certain medicinal plants with activities other than antimalarial, for potential antimalarial activities.

Various *Tabernaemontana* species occur in Central and South America, as well as West and East Africa. It has been reported that the principal constituent of the Brazilian species *Tabernaemontana fuchsiafolia* is voacamine [34]. Voacamine has been found to be active against chloroquine-resistant *Plasmodium falciparum* with IC_{50} of 411 nM and a selectivity index of 47. The compound suppressed parasitaemia in mice by 43% when given orally at 10 mg kg^{-1} in Peters' 4-day suppressive test [34].

Several years ago, Achenbach et al. carried out extensive phytochemical examination of three South American and two West African *Tabernaemontana* species and isolated several novel indole and bis-indole alkaloids, including voacamine from *T. accedens*. Some of these were tested for their antihypertensive and antibiotic activities. Structure–activity relationship studies on the antibiotic activities of some of the alkaloids were investigated. Table 16.2 shows some of the alkaloids isolated from these *T.* species [35–40]. For example, tabernulosine from *T. glandulosa* was found to lower blood pressure by about 25 mmHg (about 15%) at a single dose of 18 mg/kg, and the effect was maintained for up to 6 h [38, 41]. 19-Hydroxycoronaridine was found to be a very active antibiotic, effective at extremely low concentrations against a number of gram-negative and gram-positive bacteria. Table 16.3 shows the antibiotic activity of this compound against various gram-negative bacteria. Structure–activity relationship studies showed that removal of the carbomethoxy group at position 18 of 19-hydroxycoronaridine to give 19-hydroxyibogamine resulted in a considerable increase in the antibiotic activity [41]. Whereas coronaridine and voacangine were found to have no antibiotic activity, converting them to their lactam derivatives resulted in very weak antibiotic activity [41].

Table 16.2 Some alkaloids of *Tabernaemontana* species

Plant species	Compound	Status at publication	Reference
Tabernaemontana accedens	Accedine	New	[35–37]
	Accedinine	New	
	Accedinisine	New	
	N-demethyl-16-epi-accedine	New	
	Voacamine	Known	
	Voacamidine	Known	
	Voacamidine N-oxide	Known	
	N-demethyl voacamine	New	
	Nα-methyl-epi-affinin	New	
T. glandulosa	Tabernulosine	New	[38]
	19-Hydroxycoronaridine	New	
	19-Ethoxycoronaridine	New	
	12-Demethoxytabernulosine	New	
	Plus 15 other minor alkaloids		
T. olivaceae	Condylocarpine	Known	[39]
	Condylocarpine N-oxide	New	
T. flavicans	Ibophyllidine	New	[39]
	Ibophyllidine N-oxide	New	
T. psorocarpa	12-Methoxy-17,18-dehydrovincamin	Known	[40]
	16-Epi-isositsirikin	Known	

Table 16.3 Antibiotic activity of 19-hydroxy coronaridine against selected gram-negative bacteria [41]

Organism	MIC (μg/ml)	Organism	MIC (μg/ml)
Achromobacter geminiani	10	*E. coli ton A von Aro B*	10
Aerobacter aerogenes	>100	*Proteus vulgaris*	100
Agrobacterium tumefaciens	0.01	*Pseudomonas aeruginosa*	0.01
Chromobacterium violaceum	>100	*P. fluorescens*	>100
Escherichia coli ATCC 8739	100	*Salmonella typhimurium*	>100
E. coli wild-type aro B	10		

Taking a cue from the antimalarial activity of voacamine, there is no reason why there should not be further investigation of the indole and bis-indole alkaloids listed in Table 16.2 and their many semi-synthetic analogues obtained in good yield through facile chemical modifications, for antimalarial activity.

TABERNAEMONTANA ALKALOIDS

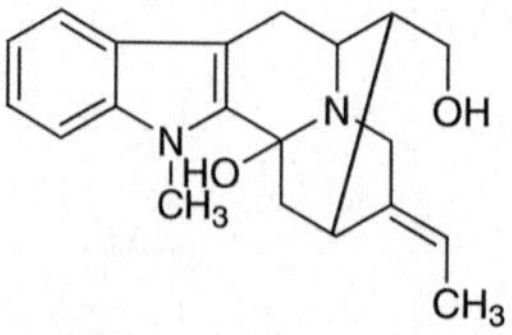

Accedine

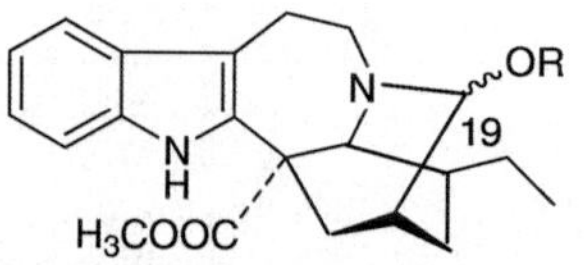

19-Hydroxycoronaridine R=OH
19-Ethoxycoronaridine R=OC_2H_5

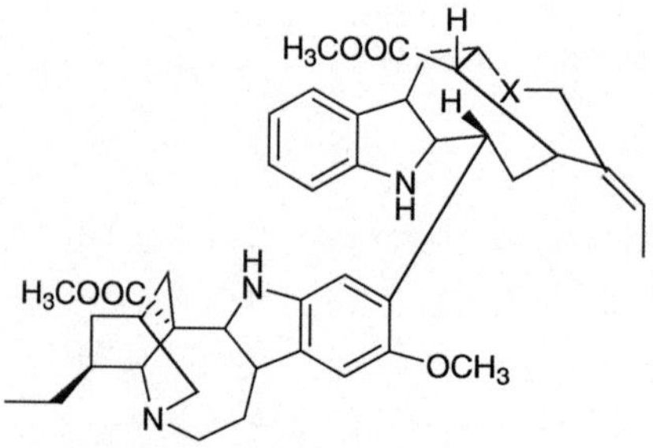

Voacamine X = NCH_3
Voacamine N_β-oxide X = NCH_3 → O

N-Demethylvoacamine X = NH

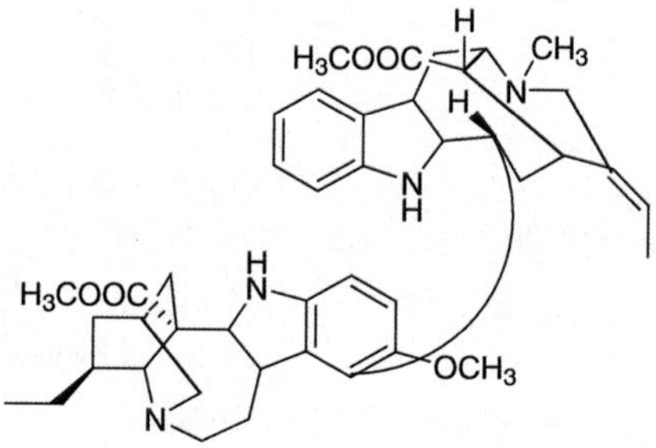

Voacamidine

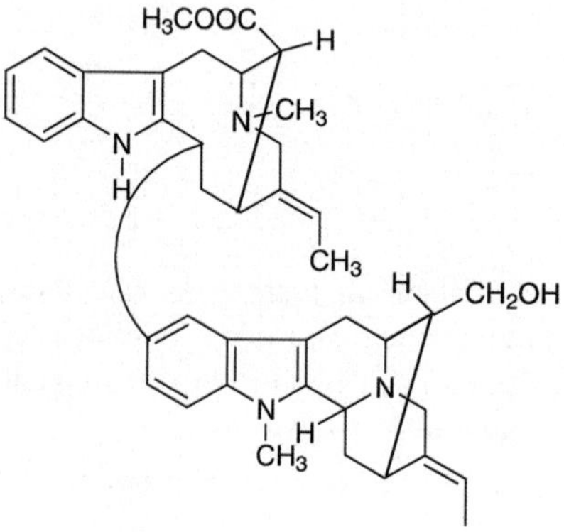

Accedinisine

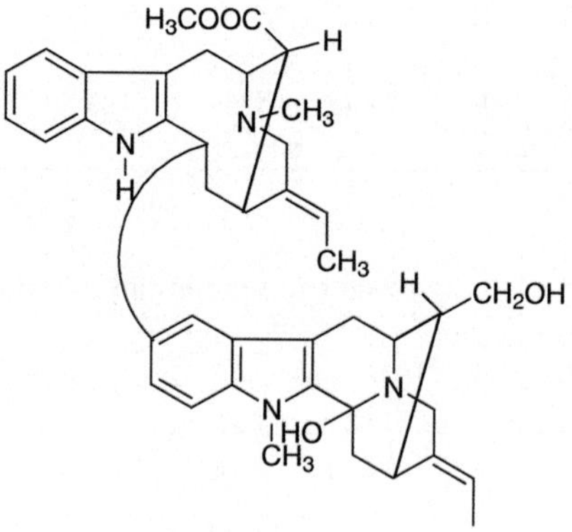

Accedinine

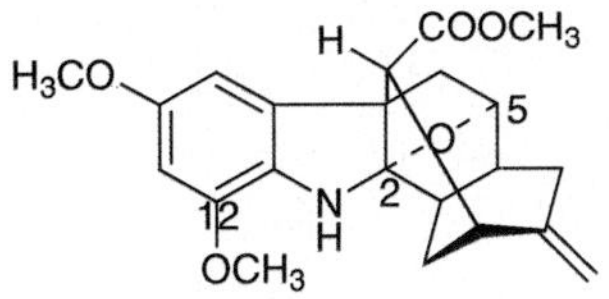

Tabernulosine

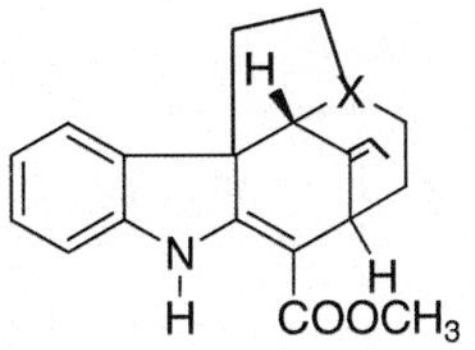

Condylocarpine X = N
Condylocarpine N-oxide X = N→O

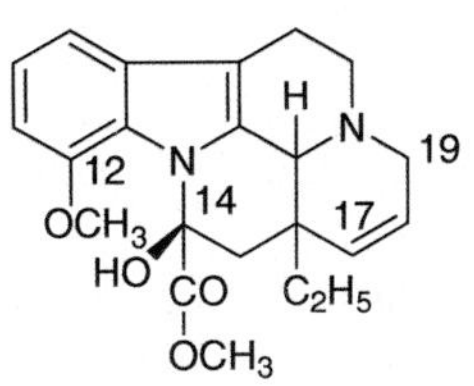

12-Methoxy-17,18-dehydrovincamine

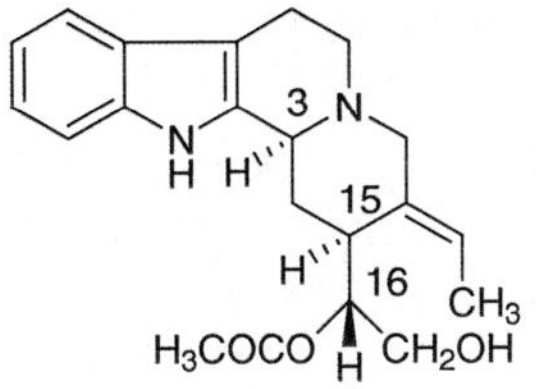

16-epi-Isositsirikin

There may be some correlation between the antimicrobial and antimalarial activities of the monomeric and dimeric forms of these alkaloids. If the indoloquinoline alkaloids of *Cryptolepis sanguinolenta* which have been extensively investigated for antimalarial activity (see later) are now proving to possess other significant biological activities such as antidiabetic and anticancer properties, then there is no reason why the indole and bis-indole alkaloids of *Tabernaemontana*, some of which have shown good antibiotic and antihypertensive activities, cannot also be further investigated as potential antimalarial lead compounds.

Even though voacamine has a relatively low selectivity index, it is quite possible that other bis-indole alkaloids or their monomeric analogues like tabernulosine and 19-hydroxycoronaridine, which have shown other significant biological activities, might also possess better antimalarial activity with improved selectivity. The substitution patterns and the stereochemistry of these alkaloids and their facile chemical modifications to other semi-synthetic products, as well as total synthesis of some of them, indicate that they could lend themselves to considerable variation for bioactivity and structure–activity relationship studies. *In vitro*, *in vivo* and *in silico* studies of these alkaloids could be useful in determining their potential as lead compounds, provided a suitable molecular target is found.

Other indole alkaloids that have been found to possess significant antimalarial activities include strychnobrasiline and malagashanine from *Strychnos myrtoides*. These compounds possess relatively weak antiplasmodial activity, but in combination with chloroquine, they reverse the chloroquine resistance of *P. falciparum* in vitro. Although malagashanine has weak activity *in vitro*, it is active *in vivo*. *S. usambarensis* contains dihydrousambarensine which is very active against chloroquine-resistant strain W-2 with good selectivity (IC_{50} 23nm; SI 1474) but inactive against chloroquine-sensitive strains [42]. Several African *Strychnos* species are yet to be investigated.

16.2.3 Zanthoxylum *and* Toddalia *Benzophenanthridine Alkaloids*

Another class of alkaloids worth revisiting as potential sources of lead compounds for malaria chemotherapy are the benzophenanthridine alkaloids, mainly from the genera *Zanthoxylum (Fagara)* and *Toddalia*. This class of alkaloids has been extensively investigated since the 1960s. The chemistry, including total synthesis of a number of them, is well documented. Bioactivity studies have been centred mainly on their anticancer, antimicrobial, cardiovascular and lately their anti-HIV properties [43–50]. In addition to their reverse transcriptase inhibition properties [50], nitidine and fagaronine have also been found to inhibit proliferation of *Pneumocystis carinii* of rat origin in short-term cultures, thus giving credence to their potential as drugs for managing some of the known opportunistic infections associated with HIV. It is regarded as an excellent candidate for further study [51]. It is only recently that attention has focused on their potential antimalarial activity. In this regard, nitidine chloride seems to have received the greatest attention.

The *Zanthoxylum* alkaloids include fagaronine and fagaridine from *Z. xanthoxyloides (Fagara xanthoxyloides)*; chelerythrine, its 9-methoxy and 9-ethoxy derivatives from the same species and from *Z. chalybeum (Fagara chalybea)*; dihydrochelerythrine and skimmianine from *Z. xanthoxyloides*; and dihydrochelerythrine and dihydronitidine from *Z. gillettii (F. macrophylla)*. The presence of oxynitidine is believed to arise from disproportionation of nitidine chloride to give the dihydro derivative and oxynitidine [52–61]. Also present in many of these African Z. species are a number of amide alkaloids such as pellitorine [62], fagaramide [56] and 2E, 4E-N-isobutyl-2,4-dodecadienamide [63] with various biological activities including their organoleptic properties and insecticidal activity. All the biological activities so far observed for these Z. alkaloids seem to provide a very strong rational scientific basis for the ethnomedical uses of the plants. However, even though almost all the *Zanthoxylum* and *Toddalia* species occurring in Africa are reputed to be useful for treatment of various fevers, their potential antimalarial activities have only recently received requisite attention.

BENZO[c]PHENANTHRIDINE ALKALOIDS

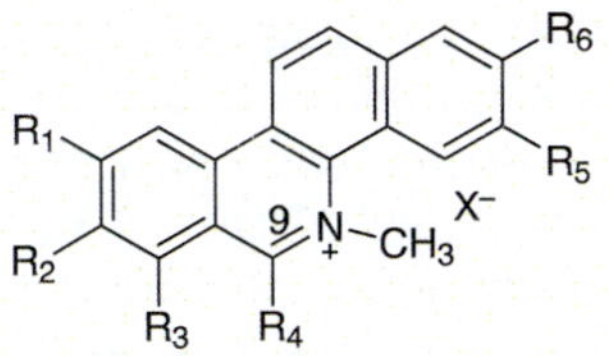

Nitidine Chloride: $R_1 = R_2 = OCH_3$: $R_3 = R_4 = H$; R_5-R_6 = -O-CH_2-O- ; $X^- = Cl^-$
Fagaridine: $R_1 = R_4 = H$; R_2 = OCH3 or OH; R_3 = OH or OCH_3; R_5-R_6 = -O-CH_2-O- ; X = OH
Fagaronine: $R_1 = R_4 = H$; $R_2 = R_3$ = OCH3; $R_5 = OCH_3$; R_6 = OH
Chelerythrine: $R_1 = R_4 = H$; $R_2 = R_3 = OCH_3$; R_5-R_6 = -O-CH_2-O- ; X = OH

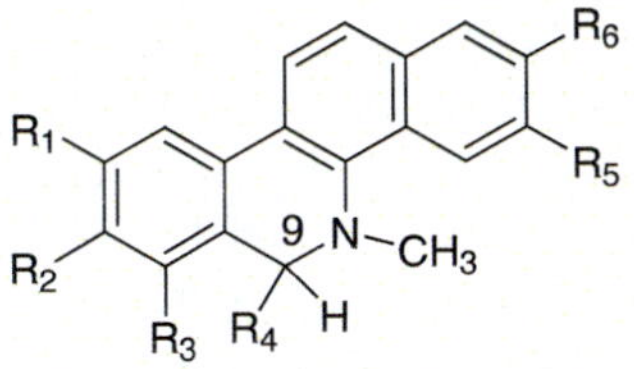

9-Methoxychelerythrine: R_1 = H $R_4 = OCH_3$; $R_2 = R_3 = OCH_3$; R_5-R_6 = -O-CH_2-O-
9-Ethoxychelerythrine: R_1 = H $R_4 = OC_2H_5$; $R_2 = R_3 = OCH_3$; R_5-R_6 = -O-CH_2-O-
Dihydrochelerythrine: $R_1 = R_4$ = OH; $R_2 = R_3 = OCH_3$; R_5-R_6 = -O-CH_2-O-

Nitidine chloride was first isolated from *Zanthoxylum nitidum* in 1958 by Arthur et al. [54] and in 1970 by Torto and Addae-Mensah [57] from *Z. gillettii* (*Fagara macrophylla*). Wall et al. (1971) and Fish and Waterman (1972) later reported the isolation of the same compound from the same species in 1971 [58, 59]. It is also present as the main alkaloid in *Z. chalybeum* [51] and *Z. tessmannii* [61] *(F. chalybea and F. tessmannii)*. It has also been reported from *Toddalia asiatica* [64].

Gakunju et al. reported that nitidine chloride from *T. asiatica* was active against various chloroquine-sensitive and chloroquine-resistant strains of *P. falciparum, in vitro*. Nitidine was active against the chloroquine-sensitive strain K39 at 45.1 ng/ml while dihydronitidine was active at 1.03 μg/ml, compared with 4.4 ng/ml for chloroquine. Against the chloroquine-resistant strains, nitidine had an IC_{50} of 42.0 ng/ml compared with 65.9 ng/ml for chloroquine. In all, nitidine and chloroquine were tested against four chloroquine-sensitive and three chloroquine-resistant strains [64]. In investigating the anticancer activities of nitidine and other benzophenanthridine alkaloids, it has been found that both nitidine and fagaronine are very strong inhibitors of topoisomerase I enzyme at 0.15–0.3 μM with significant selectivity and relatively low inhibition of topoisomerase II [65, 66], suggesting that both the anticancer and antimalarial activities might be due to enzyme inhibition. However, fagaronine appears to have no antiplasmodial activity, implying that the anticancer mechanism of these alkaloids might be different from the antimalarial activity [66].

Synthetic analogues of nitidine have been prepared and found to be very potent *in vitro* against both chloroquine-resistant and chloroquine-sensitive *P. falciparum* [66]. In this study, whereas the natural fagaronine was inactive with IC_{50} in the range of 0.8–1.0 μg/ml, synthetic O-methyl fagaronine was 6 to 30 times more potent than the parent compound. It had an IC_{50} of 33 ng/ml against chloroquine-sensitive K39 strains while the chloroquine-resistant strain V1/S was susceptible at an IC_{50} of 140 ng/ml. Chloroquine had values of 4 and 57ng/ml, respectively, for these two strains. O-methyl fagaronine showed some cross-resistance with chloroquine, unlike nitidine chloride. However, the fully synthetic analogue was equally active against both the chloroquine-sensitive and chloroquine-resistant strains at 60 ng/ml [66]. Structure–activity relationship studies show that the activity of these analogues is very sensitive to small changes in structure [66].

Chelerythrine has been found to be a potent and specific inhibitor of protein kinase C with an IC_{50} of 0.66 μM [67]. The protein kinases are considered as legitimate targets for antimalarial activity. It is therefore possible that chelerythrine and its congeners might be good candidates for antimalarial studies.

One of the problems encountered in studying nitidine and other benzophenanthridine alkaloids *in vivo* for cancer chemotherapy and other diseases is their poor absorption and hence bioavailability since they normally occur as quaternary salts. In attempts to find possible drug delivery systems to enhance their cancer chemotherapeutic profiles, the pharmacokinetics and pharmacodynamics of nitidine chloride on various preparations of solid lipid nanoparticles were studied in rat models and found to considerably improve the drug delivery to various organs and tissues of the rats [68]. Nanoparticle delivery systems may therefore also be worth some attention in *in vivo* antimalarial activity studies aimed at improving bioavailability of these alkaloids.

The various observations discussed above further support our hypothesis and contention that it is worth revisiting these *Zanthoxylum* and *Toddalia* benzophenanthridine alkaloids as potential lead compounds for malaria systems to improve the chemotherapy.

*16.2.4 The Neem (*Azadirachta indica*)*

No discussion of herbal treatment of malaria anywhere in Africa can be complete without even a brief mention of the neem tree (*Azadirachta indica*). Although the neem tree was introduced to Africa from Asia about 110 years ago, it has become very much a part of the daily lives of many Africans. It is widely used both in agriculture and for health purposes.

The best known and most popular medicinal use of the neem tree is the treatment of malaria and other feverish conditions. Many Africans swear to its efficacy and even claim it to be more effective than chloroquine, the artemisinins and other antimalarial drugs. Both the leaves and stem bark are used for this purpose. A hot decoction of the leaves is used as a steam bath to "sweat out" the fever, and a decoction of either the leaves or the stem is then drunk.

Notwithstanding the vast amount of research conducted and the wealth of scientific information available in the literature, no commercially exploitable single-compound antimalarial drug has to date been developed from it. This could, among other reasons, be attributed to the fact that earlier researchers were looking mainly for alkaloids and other nitrogen-containing compounds. It is only since the discovery of artemisinin that researchers have started looking more seriously at the terpenoids and other constituents of the neem for their potential antimalarial activity.

16.2.4.1 Chemical Constituents

Hundreds of chemical constituents have been found in various parts of the neem tree. The main constituents are limonoids, also known as meliacins. These are tetranortriterpenoids related to meliacin, a cyclopentanoperhydrophenanthrene derivative with an α-methyl group at C-13 and an α-furyl substituent at C-17. Several such compounds have been isolated from the neem. These include azadiron, azadiradion, epoxyazadiradion, gedunin, nimbin, salannin, azadirachtin, 1,3-diacetyl vilasinin, 1-tiglyl-3-acetyl vilasinin, 3-deacetyl salannin, salannol, salannol acetate, nimbandiol, 6-acetyl nimbandiol, nimbinen, 6-deacetyl nimbinen, meliantriol and nimbidin. Other constituents reported and demonstrated to have some bearing on the medicinal and other uses of the plant are quercetin-3-rhamnoside (quercitrin) and quercetin-3-rutinoside (rutin). Neem constituents are comprehensively covered in various books and journals [68–71]. We shall therefore discuss the constituents and antimalarial activity of this plant very briefly in view of their tremendous ethnopharmaceutical importance and interest in the whole of Africa.

16.2.4.2 Antimalarials

Recent studies continue to strengthen the evidence that neem extracts contain antimalarial principles. Warhurst [72] showed that neem extracts exhibited schizontocidal activity. Subsequent work however did not seem to corroborate Warhurst's findings until Ekanem reported that there was a fall in parasite count in chloroquine-sensitive strains of *Plasmodium berghei*-infected mice when the mice were treated subcutaneously with a decoction of the leaves [73]. Etkin [74], Okpanyi [75] and Okpanyi and Ezeukwu [76] reported that neem extracts had antimalarial, antipyretic, analgesic and anti-inflammatory properties. The antimalarial activity was believed to be due to quercetin-3-rhamnoside and quercetin-3-

Table 16.4 Comparison of the antimalarial activities of gedunin-containing plant extracts with standard antimalarial drugs [83]

Species or standard drug	Extract	IC_{50} (μg/mL) Clone D-6	IC_{50} (μg/mL) Clone W-2
Azadirachta indica	Leaf in ethanol	2.50	2.48
Cedrela odorata	Wood in ethanol	3.88	3.26
Cedrela odorata	Wood in toluene	9.29	2.77
Chloroquine		3.4×10^{-3}	0.101
Quinine		9.5×10^{-3}	38.2×10^{-3}
Mefloquine		2.7×10^{-3}	0.9×10^{-3}
Artemisinin		$4.5 \text{ x } 10^{-3}$	2.2×10^{-3}

rutinoside. The antimalarial activity of these flavonoids has not been corroborated by any further studies. The compounds however are well known to have antiviral and antidiabetic properties, which account for other known folkloric uses of the plant [77–79]. Ade-Serrano [80] reported the growth-inhibitory effect of leaf extracts of neem on *P. falciparum* culture. The antimalarial effect was found to be about half of the therapeutic dose of chloroquine sulphate on dry weight basis. Khalid showed that neem extracts and some neem-derived principles had *in vivo* antimalarial activity [81, 82]. Recent studies reported by McKinnon et al. (1996) and cited by Gbeassor et al. (1996) have confirmed these findings showing that neem extracts have an IC_{50} below the threshold for promising extracts (20 μg/g) for both chloroquine-sensitive (D-6) and chloroquine-resistant (W-2) clones of *P. falciparum.* Extracts of neem from Togo proved to have the greatest activity. HPLC measurements showed that the neem extracts contained about 0.1% of gedunin, which has been demonstrated to be the active antimalarial principle. Table 16.4 shows the results obtained compared with extracts of *Cedrela odorata,* another gedunin-containing plant, and known antimalarial compounds [83].

GEDUNIN AND ANALOGUES

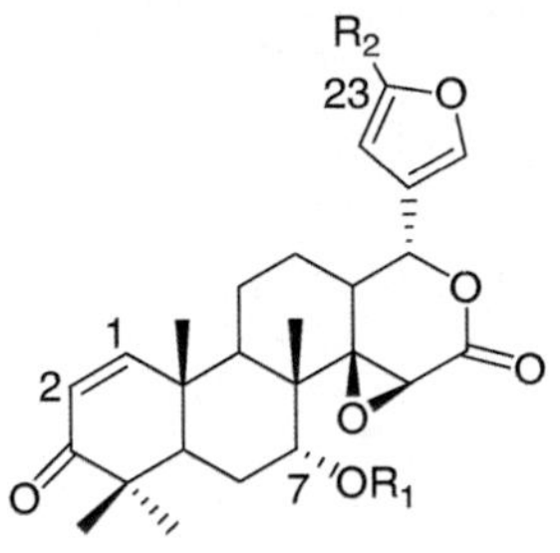

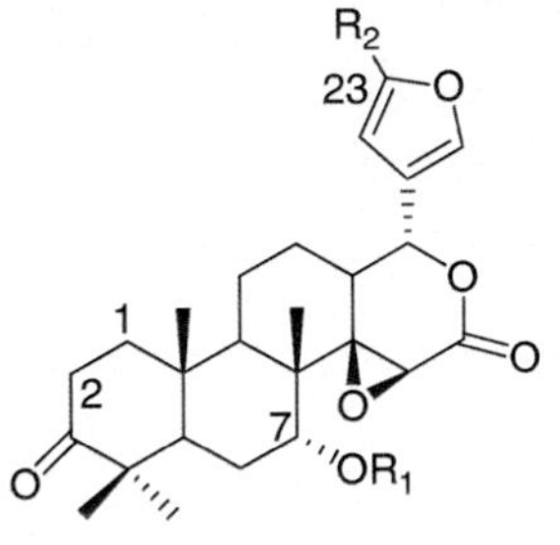

Gedunin: R_1 = Ac, R2 = H
23-Acetylgedunin: R_1 = Ac, R_2 = OAc
7-deacetylgedunin: R_1 = H, R_2 = H

1,2-Dihydrogedunin: : R_1 = Ac, R2 = H

In further studies by Gbeassor et al. (1996) on gedunin as a potential marker for standardisation of neem extracts, the antimalarial activity of pure gedunin was compared with other standard antimalarials. The results are as shown in Table 16.5 [83]. These results show clear structure–activity relationships between gedunin and its derivatives. It shows that an α,β-unsaturated carbonyl group in ring A and an unsubstituted furan ring at position 18 of ring D are both essential for activity. Although the therapeutic index of chloroquine was obviously much better than that of gedunin in sensitive *falciparum* strains(D-6), gedunin appears to be superior to both chloroquine and quinine in resistant *falciparum* strains (W-2). Artemisinin seems to have the best therapeutic index for both the sensitive and resistant strains of *P. falciparum* [83].

Even though no commercially exploitable single-compound antimalarial drug has been developed from neem natural products, the plant is still widely used for malaria chemotherapy all over Africa. Several standardised phytopharmaceuticals are available on various African markets. In Ghana, Nigeria and Kenya, it is known that some of these phytopharmaceuticals are officially registered for sale and use from accredited pharmacies and shops by the countries' drug regulatory authorities. So should the population stop using it because no specific compound has as yet been developed from it as a drug?

16.2.5 Cryptolepis sanguinolenta

Cryptolepis sanguinolenta (family Periplocaceae) is a less widely distributed plant. This plant, also known as Ghana quinine, is officially approved for the treatment of malaria at the clinic of the Centre for Scientific Research into Plant Medicine (CSRPM) at Akwapim Mampong in Ghana under the trade name *Nibima*. Other herbalists have various approved preparations which they sell under different brand

Table 16.5 Comparison of antimalarial activities of selected neem constituents with known antimalarial drugs [83]

Compound	Clone D-6 IC_{50} (ng/mL)	Clone W-2 IC_{50} (ng/mL)
Gedunin	39 (100)[a]	20 (100)[a]
1,2-dihydrogedunin	>10,000 (0.39)	>840 (2.38)
23-acetylgedunin	832(4.69)	156 (12.8)
Obacunone	>10,000 (<0.39)	>10,000 (<5 × 10^{-4})
Limonin	>10,000 (<0.39)	>10,0 00 (<5 × 10^{-4})
Hirtin	173 (22.6)	96 (20.8)
Chloroquine	1.3 (3,000)	29.5 (67.8)
Quinine	14.8 (264)	34.9 (57.3)
Mefloquine	7.5 (521)	1.4 (1,429)
Artemisinin	1.8 (2,170)	0.5 (4,000)

[a]Figures in parenthesis indicate toxicity relative to potency of gedunin (100%)

names such as *Phyto-Laria*. The aqueous root extract of *C. sanguinolenta* is a popular antimalarial in West African ethnomedicine perhaps second only to neem. Various studies have confirmed the antimalarial activity of the crude extract. Boye and Ampofo were the first to demonstrate clinically that the preparation used at the CSRPM cleared over 60% of parasitaemia in two days and completely cleared parasitaemia (100%) within five days without recrudescence after 28 days [84]. Other subsequent studies have confirmed these observations to varying degrees [85, Sittie AA, Asiedu-Larbi J (2002) CSRPM technical report (Unpublished)]. However, use of the crude extract for the treatment of malaria is inefficient, unreliable and cumbersome. Patients have been known to withdraw from treatment due to the bitterness and volume of the liquid dosage form, which sometimes induces nausea and vomiting [Sittie AA, Asiedu-Larbi J (2002) CSRPM technical report (Unpublished)]. Moreover, the use of the roots also threatens the plant with possible non-sustainable use or even extinction unless ways of cultivation and maximisation of the active constituents in the cultivars are developed. Research in this regard is currently being undertaken as a joint multidisciplinary project by the Department of Crop Science, the Department of Chemistry and the Noguchi Memorial Institute for Medical Research (NMIMR), University of Ghana, and the CSRPM at Mampong in Ghana.

The main alkaloid constituent of *C. sanguinolenta* is the indoloquinoline alkaloid cryptolepine. Paradoxically, this alkaloid happens to have been synthesised some 20 years earlier (1907) by Fichter as a synthetic chemistry curiosity [86–88] before it was first isolated as a natural product in 1929 by Clinquart from *C. triangularis* [89]. In 1951, Gellert et al. isolated it from a Nigerian species of *C. sanguinolenta* [90]. The alkaloid did not receive much attention until about thirty years later, when, after introduction of the plant to researchers at the Kwame Nkrumah University of Science and Technology by Dr. Oku Ampofo of the CSRPM in 1974, the same compound and its analogue quindoline (norcryptolepine) were isolated from the Ghanaian species [91]. From then on, investigation of the compound and its natural and synthetic analogues for antimalarial activity in particular and other biological activities were intensified [85, 92–97]. A few of the results of these studies that are pertinent to the present discussion will be highlighted here.

SOME CRYPTOLEPIS ALKALOIDS

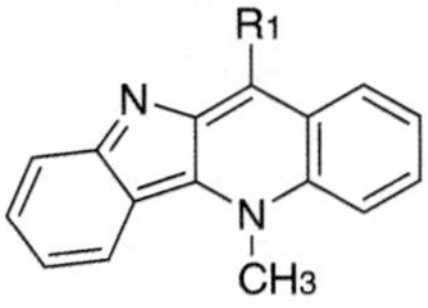

R1 = H = Cryptolepine

R1 = OH = 11-Hydroxycryptolepine

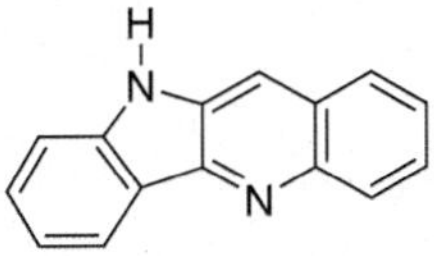

Quindoline

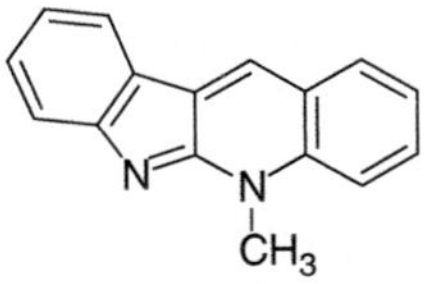

Cryptotakieine

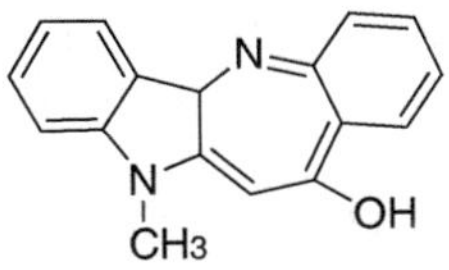

Cryptoheptine

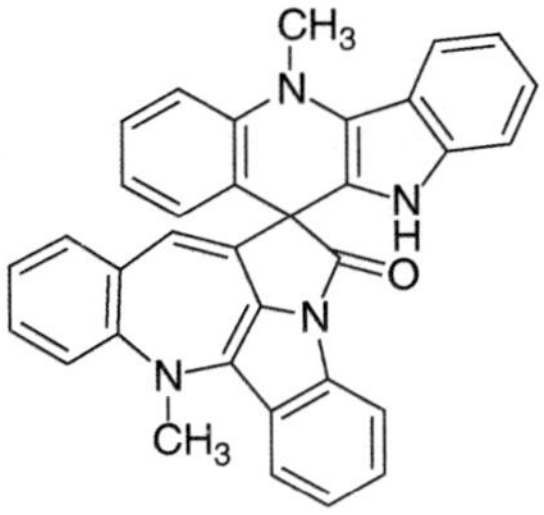

Cryptospirolepine

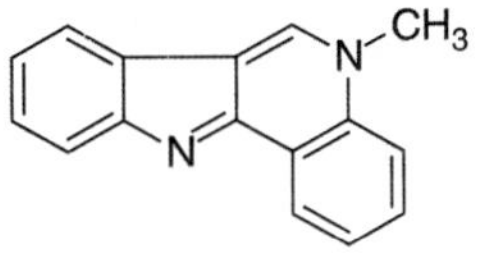

Isocryptolepine

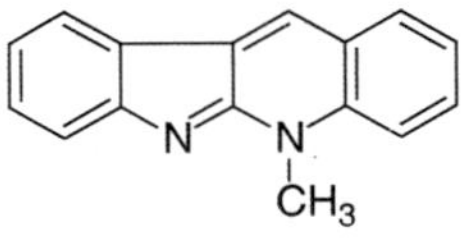

Neocryptolepine

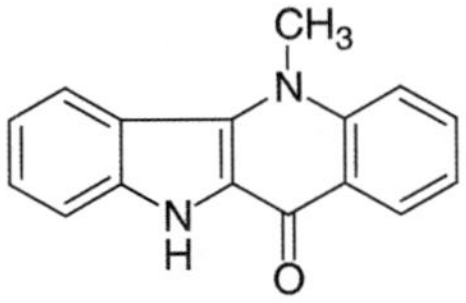

Cryptolepinone

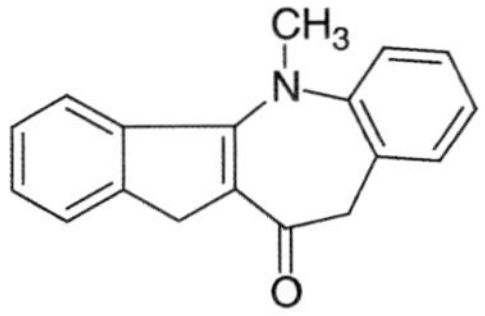

Homocryptolepinone

Cryptolepine has potent *in vitro* antiplasmodial activity. Kirby et al. have reported that the compound is as active against the K1 chloroquine-resistant strain (IC_{50} 0.11 μM) as chloroquine (IC_{50} 0.20 μM) [97]. But it is also a DNA intercalator [98] with cytotoxic properties [99]. It inhibits DNA synthesis in B16 melanoma cells and interferes with topoisomerase II [100]. However, its antimalarial mechanism of action is believed to be due, at least in part, to a chloroquine-like action that does not depend on intercalation into DNA. Various substituted analogues have been found to possess potent activities against both chloroquine-sensitive and chloroquine-resistant strains of *Plasmodium falciparum* and also inhibit *beta*-hematin formation in a cell-free system [101]. Several analogues are also active against *P. berghei* in mice, the most potent being 2,7-dibromocryptolepine which suppressed parasitaemia by 89% as compared to untreated infected controls at a dose of 12.5 mg/kg/day ip [101].

Some of the novel analogues of cryptolepine have also been found to be useful for the treatment of insulin-dependent diabetes mellitus (IDDM or type I) and non-insulin dependent diabetes mellitus (NIDDM or type II) [102]. The crude extract and its major alkaloid have also been demonstrated to be potently cytotoxic to a number of organ-specific human cancer cells. The broad-spectrum activity on a variety of cancer cell lines, apoptotic cell death and apparently low genotoxicity suggest that these agents may have potential as candidates for cancer chemotherapy [103]. The alkaloid is also active against various mycobacteria with MIC ranging between 2 and 32 μg/ml [104].

The minor alkaloid cryptospirolepine has also been found to be even more potent as a potential antimalarial agent that can be administered as a combination drug with other known antimalarials [105, 106]. But unlike the parent alkaloid cryptolepine, this alkaloid has so far received very little attention even though the compound has been patented [106].

Cryptospirolepine is an indoloquinoline–indolobenzazepine spiro-nonacyclic dimeric alkaloid.

The HCl salt, administered daily in combination with, prior to, concurrent with or subsequent to the administration of a therapeutically effective dose of chloroquine, hydroxychloroquine or mefloquine, effectively clears *P. falciparum* from test animals at ng ml^{-1} concentrations and is comparable in activity to chloroquine, quinine and mefloquine. It can also be formulated as the acetate, aspartate, benzoate, benzenesulfonate, bisulfate, butyrate, citrate, dihydrogen phosphate, dodecylsulfate, ethanesulfonate, hydrochloride, hydrobromide, hydroiodide, lactate, methanesulfonate, succinate, sulphate, tartrate, thiocyanate, maleate or fumarate salt, but the preferred formulation is the hydrochloride salt [106].

The IC_{50} of the drug on two independent clones of *Plasmodium falciparum*, D-6 and W-2, compared with quinine, mefloquine and chloroquine are as shown below.

Parasite clone	Cryptospirolepine (ng/ml)	Quinine (ng/ml)	Chloroquine (ng/ml)	Mefloquine (ng/ml)
W-2 clone	14.84	29.85	29.81	0.57
D6 clone	32.2	10.0	1.17	3.52

Even though just like the neem none of the active alkaloids of cryptolepine have as yet found its way into malaria and other chemotherapy, the crude extract is still widely used for treating malaria. The same question that was posed for the neem extract has to be posed here too. Does the totality of scientific data on the plant extract and the pure alkaloids support the continued use of the phytopharmaceutical, or are there any grounds to discontinue its use? Should the people of Africa who use this preparation with reasonable success wait until a pure totally non-toxic natural or synthetic cryptolepine analogue has been developed into a drug?

16.3 Future Considerations

The examples discussed above suggest that there is still considerable potential for investigating plants as potential sources of antimalarial lead compounds. If Africa is to succeed in making the next major impact on malaria chemotherapy from natural products research, then there ought to be a major paradigm shift in our approach to research in this area. There is the need to adopt a completely different strategy in the development of traditional medicine. Assessment of plants for single active compounds will have to go hand in hand with development of usable and well-standardised phytopharmaceuticals. This will make sense in this age of malaria combination therapy. We believe that some of the questions that need to be asked and answered are:

- Do the remedies work?
- If they do, are they acutely and/or chronically toxic?
- Can herbalists continue to use their own crude formulations once non-toxicity has been determined while further investigations are ongoing?
- What are the active constituents and how do they function?
- Can dosage forms be compounded and standardised according to clearly defined norms such as those of the WHO?
- Will the constituents, if isolated, have the same activity as the crude drug?
- Can scientists, through such investigations, rationalise scientifically the traditional healers' claims and help to improve upon the practice?
- What are the intellectual property rights issues that need to be addressed?
- Are funding agencies prepared to fund such research which may not necessarily put a new single-compound drug on the shelf in the conventional sense but which will provide something useful and usable to the vast majority of the population that may not have access to or be able to afford conventional hospital or clinical treatment?

As has already been mentioned, for a very long time, there has been a lag between conventional chemical research and the search for biologically active compounds. This has resulted in the isolation and characterisation of a vast number of potentially active natural products from many African medicinal plants but which have never been tested for any biological activity, or even if they have been, these tests have not been carried beyond preliminary laboratory assessments. This serious lag between chemical investigations and attempts to discover natural product lead compounds for further development should be narrowed, and this can only be achieved if researchers work as teams on the same project.

While investigating the vast number of yet to be investigated medicinal plants, there is also the urgent need to go back to certain plants which have already been chemically and biologically evaluated for other diseases and re-evaluate them specifically for antimalarial activity. While re-evaluating the already investigated plants, efforts should also be intensified to look at some of the rarer species. Only about 10% of the over 5,000 African plant species have to date been investigated. Who knows what new structural feature could be lurking undiscovered somewhere that could be Africa's answer to the two major breakthroughs from plants indigenous to South America and China? Our recent discovery of a new class of triterpenoids, the dichapetalins, is a good example. They include dichapetalin A which has very potent *in vitro* antitumour activity against L1210 murine leukaemia cells (EC_{90} <0.0001 μg/ml), with cytotoxicity several times greater than podophyllotoxin, and dichapetalin M which is even more potent than dichapetalin A. They are being assessed for potential activity and selectivity against other neglected tropical disease pathogens, including malaria [107–110].

The search for antimalarial lead compounds should go hand in hand with research into novel tools for vector management in order to achieve a more comprehensive malaria control. Natural products have a role to play in this. Apart from the neem, there are several other plants such as *Piper guineense, Clausena anisata* and several essential oil-bearing plants, for example, *Cymbopogon* (citronella and lemon grass) and *Lippia*, whose constituents have been demonstrated to have significant activity against the insect vectors of malaria and other neglected tropical diseases [68, 111, 112]. However, while investigating these plants or exploiting them for medicinal purposes, the need for conservation measures should not be ignored. Many plant species which are yet to be investigated are either virtually extinct or are seriously threatened by uncontrolled human activities including urban development as well as overexploitation for ethnomedicinal use.

African institutions need to be strengthened. Synthetic chemistry is not a strong point in many African research institutions due to lack of infrastructure. Many African chemists who may have studied overseas are likely to have been trained as synthetic chemists, only to come back home and be forced by circumstances to revert to other areas of research. But if this vicious circle is to be broken, then attempts should also be made to improve infrastructural capacity including equipment, to be able to carry research into natural products beyond its present status of isolation, structural elucidation and preliminary biological activity screening *in vitro* and *in vivo*. Lack of the type of funding necessary to go beyond this level and the non-existence of major pharmaceutical industries that can afford the sort of funding required for such drug development are major challenges.

Governments should seriously address issues of scientific research and Science and Technology (S&T) governance structures. Our governments should realise that our science will never make any meaningful global impact if spending on S&T research continues to depend primarily on donor-funded programmes. Governments must take the lead in research funding and devote a substantial portion of GDP to research. Local industries are usually not strong enough for the development of strong industry-research links to be forged. Most of our pharmaceutical industries are into the manufacture of generic dosage forms from imported active pharmaceutical ingredients (APIs). An insignificant number of these industries have WHO accreditation for production of APIs. Few of them have developed drugs that they can call their own flagship invention or innovation.

As a first step, our local industries should at least move to fill the gap between product development and production of pharmaceutical dosage forms. They should aim at local production of APIs. That will be the first step in driving local research, because there will then be the incentive for a local industry to aim at producing at least one brand that will be its own flagship product and not the generic form of somebody else's innovation whose patent period is past. If such an industry existed anywhere in Africa, we are sure cryptolepine and its analogues and the ancistrocladines and dioncophyllines, as well as the indole alkaloids of the genus *Tabernaemontana* and the benzophenathridines of *Zanthoxylum* and *Toddalia*, might by now have gone far beyond their present stage of development for malaria chemotherapy. We believe an improved analogue of gedunin might have been developed by now. We are sure that if a consortium of local industries in West Africa had come together to fund the development of the constituents of *Cryptolepis*, there would have been a strong chance of putting this compound or its analogue on to the market by now, 83 years after its first isolation by Clinquart.

African scientists need to come together to promote south–south and north–south cooperation in drug development if Africa is to achieve what China and South America have given to the world in malaria chemotherapy. Many of the compounds from African medicinal plants that have been assessed for antimalarial activity are at a stage similar to where artemisinin was about 30 years ago. We are all aware of the difficulties China faced in making the rest of the world take note of their developments in this field. A non-alkaloidal antimalarial with an unfamiliar mechanism of action was strange to the world pharmaceutical industry. Reasons of potential toxicity, recrudescence and all the usual reasons were given. It took a bold decision by China to go it alone and develop the drug. The difficult problem of chloroquine resistance forced the rest of the world to sit up and take notice of what was going on in China. Will Africa be bold enough to take such a decision and support its scientists to pursue a similar goal? This is where we believe that the recent establishment of the African Network for Drug and Diagnostics Innovation (ANDI) is a step in the right direction. ANDI has a major mobilisation and advocacy role to play in this and deserves the support and encouragement of all African governments, the world scientific community and the major funding agencies.

References

1. Mshana NR, Abbiw DK, Addae-Mensah I et al (2000) Traditional medicine and pharmacopoeia. Contribution to the revision of ethno-botanical and floristic studies in Ghana. Organization of African Unity/Scientific and Technical Research Commission (OAU/STRC). Lagos, Nigeria. ISBN 978-2453-66-2 (920 pages)
2. Irvine FR (1961) Woody plants of Ghana. Oxford University Press, London (868 pages)
3. Kokwaro JO (1976) Medicinal plants of East Africa. East African Literature Bureau, Kampala, Nairobi, Dar es Salaam (384 pages)
4. World Health Organisation (2002) WHO traditional medicine strategy 2002–2005. World Health Organisation, Geneva, p 12
5. See various relevant reports of the Ghana Statistical Service 2000 Ghana Population & Housing Census and subsequent updates
6. Twum-Baah K, Addae-Mensah I, Kumekpor T (eds) (2005) 2000 population and housing census. Analysis of district data and implications for planning. Regional reports (10 volumes) Ghana Statistical Service. Asante and Hittscher Press
7. World Health Organisation (2002) WHO traditional medicine strategy 2002–2005. World Health Organisation, Geneva, p 13
8. Bringmann G, Rubenacker M, Jansen JR et al (1990) On the structure of the Dioncophyllaceae alkaloids dioncophylline A and O-Methyl triphyophylline. Tetrahedron Lett 31:639–642
9. Bringmann G, Zagst R, Ake Assi L (1991) Ancistrobrevine B: a naphthylisoquinioline alkaloid with a novel coupling type from *Ancistrocladus abbreviatus*. Planta Med 57 (2):96–97
10. Bringmann G, Pokorny F, Stablein M et al (1993) Ancistrobrevine C from *Acistrocladus abbreviatus*, the first mixed Ancistrocladaceae/dioncophyllaceae-type naphthylisoquinoline alkaloid. Phytochemistry 33:1511–1515; 34:1663
11. Bringmann G, Kinzinger L, Ortmann T et al (1994) Isoancistrocladine from *Ancistrocladus heyneanus*: the first naturally occurring N-unsusbstituted cis-configurated naphthylisoquinoline alkaloid. Phytochemistry 35:259–261
12. Bringmann G, Rubenacker M, Vogt P et al (1991) Dioncopeltine and dioncolactone A: alkaloids from *Triphyophyllum peltatum*. Phytochemistry 30(5):1691–1696
13. Bringmann G, Rubenacker M, Geuder T et al (1991) Dioncophylline B, a naphthylisoquinoline alkaloid with a new coupling type from *Triphyophyllum peltatum*. Phytochemistry 30:3845–3847
14. Sharma SC, Shukla YN, Tandon JS (1975) Alkaloids and terpenoids of *Ancistrocladus heyneanus, Sagittaria sagittifolia, Lyonia formosa* and *Heydichium spicatum*. Phytochemistry 14:578
15. Bringmann G, Gramatski S, Grimm C et al (1992) Feeding deterrency and growth retarding activity of the naphthylisoquinoline alkaloid dioncophylline A against *Spodoptera littoralis*. Phytochemistry 31(11):3821–3825
16. François G, Bringmann G, Phillipson JD et al (1994) Activity of extracts and naphthylisoquinoline alkaloids from *Triphyophyllum peltatum, Ancistrocladus abbreviatus* and *A. barteri* against *Plasmodium falciparum in vitro*. Phytochemistry 35(6):1461–1464
17. Bringmann G, Pokorny F, Zinsmeister HD (1991) *Ancistrocladus*, eine botanisch und chemisch bemerkenswerte Gattung. Der Palmengarten 55(3):13–18
18. Bringmann G (1986) The naphthylisoquinoline alkaloids. In: Brossi A (ed) The alkaloids, vol 29. Academic, New York, pp 141–184
19. Bringmann G, Pokorny F (1995) The naphthylisoquinoline alkaloids. In: Cordell G (ed) The alkaloids, vol 46. Academic, New York, pp 127–271, and references therein
20. Bringmann G (1995) Michellamines and their monomeric halves: anti-HIV, resp. antimalarial agents from African Plants. In: Asomaning WA, Gadzekpo VPY (eds) Chemistry in health, industry and the environment. Proceedings of the 6th international chemistry conference in Africa, Accra, Ghana. School of Communication Studies Press, University of Ghana, Legon, Accra, 31 July–4 August

21. Francois G, Timperman G, Eling W et al (1997) Naphthylisoquinoline alkaloids against malaria: evaluation of the curative potentials of dioncophylline C and dioncopeltine A against *Plasmodium berghei* in vivo. Antimicrob Agents Chemother 41:2533–2539
22. Sieber M, Dekant W, Faber JH et al (2006) Biotransformation and pharmacokinetics of the antiplasmodial naphthylisoquinoline alkaloid dioncophylline A. Xenobiotica 36:750–762
23. Schwedhelm KF, Hortsmann M, Faber JH et al (2007) The novel antimalarial compound dioncophylline C forms a complex with heme in solution. ChemMedChem 2:541–548
24. Kaur K, Jain M, Kaur T et al (2009) Antimalarials from Nature. Bioorg Med Chem 17:3229–3256
25. Kumar V, Mahajan A, Chibale K (2009) Synthetic medicinal chemistry of selected antimalarial natural products. Bioorg Med Chem 17:2235–2275
26. Léornard J, (1984) *Ancistrocladus robertsoniorum* J Léonard (Ancistrocladaceae), Espece nouvelle du Kenya Bull. Jard Bot Nat Belg. (Bull Nat. Planterium Belg) 54: 465-470
27. Thomas DW, Gereu RE (1993) *Ancistrocadus korupensis* (Ancistrocladaceae) a new species of liana from Cameroon. Novon 3:494–498
28. Bringmann G, Haller RD, Bär S et al (1994) *Ancistrocladus robertsoniorum* J Léonard*:* eine erst spät entdeckte *Ancistrocladus* Art. Der Palmengarten 58:148–153
29. Manfredi KP, Blunt JW, Cardelina JH II et al (1991) J Med Chem 34:3402–3405
30. Boyd MR, Hallock YF, Cardelina JH II et al (1994) Anti-HIV Michellamines from *Ancistrocladus korupensis*. J Med Chem 37:1740–1745
31. Hallock YF, Manfredi KP, Blunt JW et al (1994) Korupensamines A-D, novel antimalarial alkaloids from *Ancistrocladus korupensis*. J Org Chem 59(21):6349–6355
32. Boyd MR, François G, Bringmann G et al (1994) Antimalarial korupensamines and pharmaceutical compositions and medical uses thereof. US Patent Application No 08/195,260 (14.02.94)
33. Bringmann G, Schneider Ch, Pokorny F et al (1993) The cultivation of tropical lianas of the genus *Ancistrocladus*. Planta Medica 59(suppl):623–624
34. Ramanitrahasimbola D, Rasoanivo P, Ratsimamanga-Uverg S et al (2001) Biological activities of the plant-derived bisindole voacamine with reference to malaria. Phytother Res 15:30–33
35. Achenbach H, Schaller E (1975) Accedin und Nα-Methyl-epi-affinin, zwei neue alkaloide aus *Tabernaemontana accedens*. Chem Ber 108:3842–3854
36. Achenbach H, Schaller E (1976) N-Demethyl-16-epi-accedin, ein neues alkaloid aus *Tabernaemontana accedens*. Tetrahedron lett 5:351–352
37. Achenbach H, Schaller E (1976) Über einige bisindolalkaloide aus *Tabernaemontana accedens*. Chem Ber 109:3527–3536
38. Achenbach H, Raffelsberger B, Addae-Mensah I (1982) Tabernulosin und 12-Demethoxytabernulosin, zwei neue alkaloide vom Picrinin-typ aus *Tabernaemontana glandulosa*. Liebigs Ann Chem 5:830–844
39. Achenbach H (1983) Chemical investigation on *Tabernaemontana* species and on West African medicinal plants. Rev Latinoamer Quim 14:6–16
40. Achembach H, Renner C, Addae-Mensah I (1982) Alkaloide in *Tabernaemontana* Arten; XVI(1); 12-Methoxy-17,18-dehydro-vincamin und 16-epi-isositsirikin, Alkaloide aus *Tabernaemontana psorocarpa*. Planta Med 46:88–90
41. Achenbach H (1986) Investigation on West African medicinal plants. Pure Appl Chem 58 (8):653–662
42. Wright CW (2010) Recent developments in research on terrestrial plants for the treatment of malaria. Nat Prod Rep 27:961–968, and references therein
43. Messmer WM, Tin-Wa M, Fong HMS et al (1972) Fagaronine, a new tumor inhibitor isolated from *Fagara zanthoxyloides* Lam (Rutaceae). J Pharm Sci 6:1858
44. Zee-Cheng KY, Cheng CC (1975) Preparation and antileukaemic activity of some alkoxybenzo[c]phenanthridinium salts and corresponding dihydro derivatives. J Med Chem 18:66–71
45. Stermitz FR, Larson KA, Kim DK (1973) Some structural relationships among cytotoxic and antitumour benzophenanthridine alkaloid derivatives. J Med Chem 16:939–940

46. Wall ME, Wani MC, Taylor H (1987) Plant antitumor agents 27. Isolation, structure and structure activity relationship of alkaloids from *Fagara macrophylla*. J Nat Prod 50 (6):1095–1099
47. Mwasaki H, Okabe T, Takara K et al (2010) Tumor –selective cytotoxicity of Benzo[c] phenanthridine derivatives from *Toddalia asiatica* Lam. Cancer Chemother Pharmacol 65 (4):719–726
48. Odebiyi OO, Sofowora EA (1973) Antimicrobial alkaloids from a Nigerian chewing stick (*Fagara xanthoxyloides*). Planta Med 36:204–207
49. Addae-Mensah I, Njonge E (1988) 9-methoxychelerythrine as a true natural product – its antimicrobial and cardiovascular effects. Planta Medica 60 GA conf. Abstract KI-8 pp 4–5 (Abstracts of the 36 annual congress on medicinal plants, Freiburg)
50. Paris R, Moyse-Mignon H (1951) African Fagaras III *Fagara macrophylla*. Annales de Pharmacie de France 9:479–493
51. Addae-Mensah I, Munenge R, Guantai A (1989) Comparative examination of two *Zanthoxylum* benzophenanthridine alkaloids for cardiovascular effects in rabbits. Phytother Res 3(5):165–169
52. Tan GT, Pezzuto JM, Kingham AD (1990) Evaluation of natural products as inhibitors of human immunodeficiency virus (HIV) reverse transcriptase. Abstract 0_3 7 of the international joint symposium of the Society of Medicinal Plant Research (GA), American Society of Pharmacognosy, Association pour l'enseignement et la Recherche en Pharmacognosie and the Phytochemical Society of Europe (BONN-BACANS) Bonn, Germany, p 13
53. Shaw MM, Queener SF, Smith JW et al (1997) Nitidine inhibits proliferation of *Pneumocystis carinii* of rat origin in short-term cultures. Abstract. General meeting of American Society for Microbiology. 4–8 May; 97: 5 (Abstract No A24)
54. Arthur HR, Hui WH, Ng YL (1959) An examination of the Rutaceae of Hong Kong II. The alkaloids nitidine and oxynitidine from *Z. nitidum*. J Chem Soc 1840–1845. See also (1958) the same authors in Chem Ind (London 1514)
55. Torto FG, Sefcovic P, Dadson BA (1966) Medicinal plants of Ghana: identity of alkaloid from *Fagara Xanthoxyloides*. Tetrahedron lett 2:181–183
56. Torto FG, Sefcovic P, Dadson BA, Addae Mensah I (1969) Alkaloids from Fagara species. Ghana J Sci 9:3–7
57. Torto FG, Addae Mensah I (1970) Alkaloids of *Fagara macrophylla*. Phytochemistry 9:911–914
58. Wall ME, Wani MC, Taylor H (1971) Plant antitumour agents. VIII. Isolation and structure of antitumor alkaloids from *Fagara macrophylla*. Presented at the 62nd American Chemical Society national meeting. Washington, DC, MEDI-34
59. Fish F, Waterman PG (1972) Methanol-soluble quaternary alkaloids from African *Fagara* species. Phytochemistry 11:3007–3014
60. Torto FG, Addae-Mensah I, Baxter I (1973) Fagaridine: a phenolic Benzophenanthridine alkaloid from *Fagara macrophylla*. Phytochemistry 12:2315–2317
61. Addae-Mensah I, Sofowora A (1979) Constituents of *Fagara tessmannii*. Planta Med 35:94–95
62. Bowden K, Ross WJ (1963) The local anaesthetic in *Fagara zanthoxyloides*. J Chem Soc 3503–3505
63. Adesina SK (2005) The Nigerian *Zanthoxylum*: chemical and biological values (review article). Afr J Trad. CAM. 282-301 and references therein. ISSN 0189-6016
64. Gakunju DMN, Mberu EK, Dossaji SF et al (1995) Potent antimalarial activity of the alkaloid nitidine, isolated from a Kenyan herbal remedy. Antimicrob Agents Chemother 39 (12):2606–2609
65. Wang LK, Johnson RK, Hecht SM (1993) Inhibition of topoisomerase I function by nitidine and fagaronine. Chem Res Toxicol 6(6):613–618
66. Nyangulu JM, Hargreaves SL, Sharples SL et al (2005) Antimalarial benzo(c) phenanthridines. Bioorg Med Chem Lett 15:2007–2010

67. Herbert JM, Augereu JM, Gleye J, Maffrand JP (1990) Chelerythrine is a potent and specific inhibitor of protein kinase C. Biochem Biophys Res Commun 172(3):993–999
68. Schmutterer H (ed) In close cooperation with Ascher KRS, Isman MB, Jacobson M et al (1995) The neem tree, source of unique natural products for integrated pest management, medicine, industry and other purposes. VCH Verlagsgesellschaft MBH, Weinheim, Germany
69. Kraus W (1984) Biologically active compounds from Meliaceae. In: Szantay CS (ed). Chemistry and biotechnology of biologically active natural products. 2nd international conference. Budapest, Hungary (1983). Akademiai Kiado, Budapest
70. Addae-Mensah I (1998) The potential of the neem tree in Ghana. In: Proceedings of a seminar held in Dodowa, Ghana. Deutsche Gesellschaft for Technische Zusammenarbeit (GTZ), Eschborn, Germany, pp 11–19
71. Wilcox M, Chamberlain J (2004) Neem (*Azadirachta indica).* In: Willcox M, Boedecker G, Rasoanaivo P (eds) Traditional medicinal plants and malaria. CRC, Boca Raton, FL
72. Warhurst DD (1966) Bioassay of *Plasmodium berghei*. Trans R Soc Trop Med Hyg 60:6
73. Ekanem OJ (1978) Has *Azadirachta indica* (dogonyaro) any antimalarial activity? Niger Med J 8:8–10
74. Etkin NL (1981) A Hausa herbal pharmacopoeia: biomedical evaluation of commonly used plant medicines. J Ethnopharmacol 4:75–98
75. Okpanyi SN (1977) In perspectives in medicinal plant research today. Drug Research Unit (DRU), University of Ife, Nigeria, p 89
76. Okpanyi SN, Ezeukwu GC (1981) Anti-inflammatory and antipyretic activities of *Azadirachta indica*. Planta Med 41:34–39
77. Addae-Mensah I, Achenbach H (1985) Terpenoids and flavonoids of *Bridelia ferruginea*. Phytochemistry 24(8):1817–1819
78. Addae-Mensah I, Munenge RW (1989) Quercetin -3-neohesperidoside (Rutin) and other flavonoids as the active hypoglycaemic agents of *Bridelia ferruginea*. Fitoterapia IX (4):359–362
79. Addy ME, Mills-Robertson FC, Addae-Mensah I (1995) A mixture of flavonoids extracted from *Bridelia ferruginea* is antihyperglycaemic in both insulin dependent and non-insulin dependent models of *Diabetes mellitus* in mice. Ghana J Biochem Biotech Mol Biol 3:15–25
80. Ade-Serrano OL (1982) Growth-inhibitory and lymphocytotoxic effect of *Azadirachta indica*. J. Afr. Med. Plants 5:137–139
81. Khalid SA, Farouk A, Geary TG et al (1986) Potential antimalarial candidates from African plants. J Ethnopharmacol 15:201–209
82. Khalid SA, Dudeck H, Gonzale-Sierra M (1989) Isolation and characterisation of an antimalarial agent from the neem tree. J Nat Prod 52:922–927
83. Gbeassor M, Koumaglo HK, Awang DVC et al (1996) Development of ethical phytomedicines for Togo, West Africa. In: Hostettmann K, Chinyanganya F, Maillard M et al (eds) Chemistry, biological and pharmacological properties of African medicinal plants. Proceedings of the first international IOCD symposium. Victoria Falls, Zimbabwe, Feb 25–28 and references therein
84. Boye GL, Ampofo O (1983) Clinical uses of *Cryptolepis sanguinolenta* (Ascelpiadaceae). In: Boakye-Yiadom K, Bamgbose SOA (eds) Proceedings of the first international symposium on Cryptolepine, University of Science and Technology, Kumasi, Ghana, pp 37–40
85. Boye GL (1989) Studies on the antimalarial action of *Cryptolepis sanguinolenta* extract. In: Proceedings of the international symposium on East-West medicine, Seul Korea, pp 242–251
86. Fichter F, Boehringer R (1906) Ueber Chindolin. Chem Ber 39:3932
87. Fichter F, Probst H (1907) Zur kenntnis des methylchindolinols. Chem Ber 40:3498
88. Fichter F, Rohmer F (1910) Über einige derivate des chindolins. Chem Ber 43:3489
89. Clinquart E (1929) The chemical composition of *Cryptolepis triangularis*, a plant from the Belgian Congo. Bull Acad R Med Belg 9:627
90. Gellert E, Raymond-Hamet CR, Schlitterler W (1951) Constitution of the alkaloid cryptolepine. Helv Chim Acta 31:642–654, and references therein

91. Dwuma-Badu D, Ayim JSK, Fiagbe NIY et al (1978) Constituents of West African Medicinal Plants XX. Quindoline from *Cryptolepis sanguinolenta*. J Pharm Sci 67:433–434
92. Noamesi BK, Bamgbose SOA (1983) Cryptolepine – a pharmacological review. Clinical uses of *Cryptolepis sanguinolenta* (Ascelpiadaceae). In: Boakye-Yiadom K, Bamgbose SOA (eds) Proceedings of the first international symposium on Cryptolepine, University of Science and Technology, Kumasi, Ghana, pp 41–50. See also references therein for other studies
93. Boakye-Yiadom K, Heman-Ackah SM (1979) Cryptolepine hydrochloride effect on *Staphylococcus aureus*. J Pharm Sci 12:1510–1514
94. Noamesi BK, Bangbose SOA (1984) Studies on cryptolepine III: effect of cryptolepine on the tone and prostaglandin production in isolated rabbit duodenum. Planta Med 1:98–101
95. Boakye-Yiadom K (1979) Antimicrobial properties of some West African medicinal plants II. Antimicrobial activity of aqueous extracts of *Cryptolepis sanguinolenta* (Lindl) Schlechter. Q J Crude Drug Res 17:78–80
96. Noamesi BK, Larson BS, Laryea DL et al (1991) Whole body autoradiographic study on the distribution of 3H-cryptolepine in mice. Arch Int Pharmacodyn Ther 313:5–14
97. Kirby Gl, Noamesi BK, Paine A et al (1995) *In vitro* and *in vivo* antimalarial activity of cryptolepine, a plant-derived indoloquinoline. Phytother Res 9:359–363
98. Lisgarten JN, Coll M, Portugal J et al (2002) The antimalarial and cytotoxic drug cryptolepine intercalates into DNA at cytosine-cytosine sites. Nat Struct Biol 9:57
99. Dassonville L, Lansiaux A, Wattelet A et al (2000) Cytotoxicity and cell cycle effects of the plant alkaloids cryptolepine and neocryptolepine: relation to drug-induced apoptosis. Eur J Pharmacol 409:9–18
100. Bonjean K, De Pauw-Gillet MC, Defresne MP et al (1998) The DNA-intercalating alkaloid cryptolepine interferes with topoisomerase II and inhibits primary DNA synthesis in B16 melanoma cells. Biochemistry 37:5136–5146
101. Wright CW, Addae-Kyeremeh J, Breen AG et al (2001) Synthesis and evaluation of cryptolepine analogues for their potential as new antimalarial agents. J Med Chem 44:3187–3194
102. Bierer DE (1996) Cryptolepine analogs with hypoglycaemic activity. World Intellectual Property Organization (WIPO). Patent No WO/1996/010015
103. Ansah C, Gooderham NJ (2002) The popular herbal antimalarial extract of *Cryptolepis sanguinolenta* is potently cytotoxic. Toxicol Sci 70:245–251
104. Gibbons S, Fatemeh F, Wright C (2002) Cryptolepine hydrochloride: a potent antimycobacterial alkaloid derived from *Cryptolepis sanguinolenta*. Phytother Res 17(4):434–436
105. Tackie AN, Boye GL, Sharaf MHM et al (1993) Cryptospirolepine, a unique spiro-nonacyclic alkaloid isolated from *Cryptolepis sanguinolenta*. J Nat Prod 56(5):653–670
106. Tackie AN, Schiff PL (1994) Compound and method of treatment for *falciparum* malaria. US Patent 5362726
107. Achenbach H, Asunka SA, Waibel R et al (1995) Dichapetalin A, a novel plant constituent from *Dichapetalum madagascariense* with potential antineoplastic activity. Nat Prod Lett 7:93–100
108. Addae-Mensah I, Waibel R, Asunka SA et al (1996) The Dichapetalins – a new class of triterpenoids. Phytochemistry 43(3):649–656
109. Fang F, Ito A, Chai H et al (2006) Cytotoxic constituents from the stem bark of *Dichapetalum gelonoides* collected in the Philippines. J Nat Prod 69:332–337
110. Osei-Safo D, Anti-Chama M, Addae-Mensah I et al (2008) Dichapetalin M from *Dichapetalum madagascariensis*. Phytochem Lett 1:147–150
111. Addae-Mensah I, Achieng G (1986) Mosquito larvicidal effects of six amide alkaloids from *Piper guineense*. Planta Med 52: 432. Abstracts of the 34th annual congress of the society for medicinal plant research (Gezellschaft fur Arzneipflanzenforschung), Hamburg, Germany
112. Osei-Safo D, Addae-Mensah I, Asomaning WA et al (2002) Insecticidal properties of *Clausena anisata* (Willd), Hook F. ex Benth. Actes du Colloque de Sainte Foy (Quebec) 5e Colloque Produit Naturel d'Origine Vegetale Laseve, Universite Du Quebec A Chicoutimi, Canada, pp 179–186

Chapter 17
Nanomedicine in the Development of Drugs for Poverty-Related Diseases

Rose Hayeshi, Boitumelo Semete, Lonji Kalombo, Lebogang Katata, Yolandy Lemmer, Paula Melariri, Belle Nyamboli, and Hulda Swai

Abbreviations

ACTs	Artemisinin-based combination therapies
ADME	Absorption, distribution, metabolism and excretion
ARV	Antiretroviral
AUC	Area under the curve
C_{max}	Maximum plasma concentration
CYP	Cytochrome P450
ESE	Emulsion-solvent-evaporation
ESSE	Emulsion-solvent-surfactant-evaporation
ETB	Ethambutol
HIV	Human immunodeficiency virus
INH	Isoniazid
IV	Intravenous
MIC	Minimum inhibitory concentration
NTDs	Neglected tropical diseases
PBCA	Poly(butyl-2-cyanoacrylate)
PCL	Polycaprolactone
PEG	Polyethylene glycol
PK	Pharmacokinetics
PLGA	Poly(D,L-lactic-co-glycolic acid)
PRDs	Poverty-related diseases
PZA	Pyrazinamide
RES	Reticuloendothelial system

R. Hayeshi (✉) • B. Semete • L. Kalombo • L. Katata • Y. Lemmer • P. Melariri • B. Nyamboli • H. Swai
Council for Scientific and Industrial Research, Polymers and Composites, P O Box 395, Pretoria 0001, South Africa
e-mail: RHayeshi@csir.co.za

K. Chibale et al. (eds.), *Drug Discovery in Africa*,
DOI 10.1007/978-3-642-28175-4_17, © Springer-Verlag Berlin Heidelberg 2012

RECG	Reverse-emulsion-cationic-gelification
RESCG	Reverse-emulsion-surfactant-cationic-gelification
RIF	Rifampicin
R&D	Research and development
TB	Tuberculosis

17.1 Introduction

Nanotechnology is a multidisciplinary field covering the design, manipulation, characterisation, production and application of structures, devices and systems at nanometer scale (1–500-nm-size range) which, at this size range, presents with unique or superior physicochemical properties. This scale represents the size of atoms, molecules and macromolecules [1]. Nanomedicine is the application of nanotechnology in medical sciences for imaging, diagnosis, drug delivery (nanocarriers) and therapeutics used for treating and preventing disease.

Nanomedicine has gained ground over the past several years as can be observed from the increase in the number of nanopharmaceutical patents to over 1,000 by the year 2008 [2]. Nanomedicine-based drug delivery systems offer a tool for expanding current drug markets as they can facilitate reformulation of classical drugs and failed leads resulting in improved half-life, controlled release over short or long durations and highly specific site-targeted delivery of therapeutic compounds. Examples of nanocarriers utilised in nanomedicine include nanocapsules, liposomes, dendrimers, gold nanoparticles, polymeric micelles, nanogels and solid lipid nanoparticles, among others. This technology has successfully revolutionised therapies for diseases like cancer with a number of nanomedicine products for cancer, such as Doxil® (liposome) and Abraxane® (albumin-bound nanoparticles), already on the market [3]. The current growth in this field is mainly due to the advances in nanoscience in better approaches of molecular assembly and the design of more controlled and efficient nanomaterial.

The field of drug development experiences very low success rates with regard to drugs that enter the market. These shortfalls are due to factors such as toxicity of the therapeutic compounds, poor solubility leading to lowered bioavailability and thus reduced efficacy. These challenges are even more pronounced in poverty-related diseases (PRDs), such as tuberculosis (TB), malaria and human immunodeficiency virus (HIV). The annual global death toll of HIV/AIDS, malaria and TB approaches 6 million people. According to the World Health Organisation (WHO) 2010 Global TB report, one third of the world's population is currently infected with *Mycobacterium tuberculosis (M.tb)* and an estimated 1.7 million people died from TB in 2009 with the highest number of deaths occurring in Africa [4]. It has been reported that malaria remains one of the world's most prevalent infectious diseases. Forty percent of the world's population is at risk of infection, and in 2009, there were an estimated 225 million cases of malaria reported worldwide and an estimated 781,000 deaths [5]. Sub-Saharan Africa still bears a large share of the global HIV burden with the highest number of people living with HIV, new HIV infections,

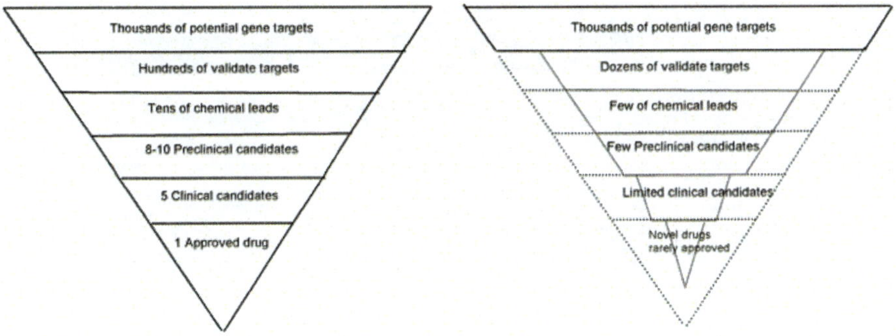

Fig. 17.1 Funding for PRD drug development does not span the whole drug development process in comparison to funding for drug development in the developed world

AIDS-related deaths and the highest adult HIV prevalence [6]. In addition, due to the weakening of the immune system by HIV/AIDS, coinfection with other diseases such as TB, malaria and leishmaniasis is beginning to gain attention. Apart from HIV, malaria and TB, neglected tropical diseases (NTDs) such as leishmaniasis also affect more than one billion people, primarily low-income populations living in tropical and subtropical climates. Visceral leishmaniasis is usually fatal in the absence of treatment [7], and there are an estimated 500,000 new cases of visceral leishmaniasis annually affecting mostly South East Asia and East Africa.

Although effective therapeutic regimens against these diseases are available, treatment failure due to poor adherence (which in turn leads to the emergence of drug-resistant strains) remains a challenge. Many of the drugs require high doses and high-dose frequency due to poor bioavailability, hence the long treatment durations and associated negative side effects. These in turn lead to poorer treatment outcomes and increased cost of treatment. In addition to these drug-related challenges, drug discovery and development research in these PRDs is not at a scale that corresponds with the impact of these diseases in the developing world [8].

The field of drug development for PRDs could benefit greatly from nanomedicine in terms of addressing the aforementioned shortfalls such as poor solubility and limited bioavailability. However, nanomedicine has not been widely applied to transform therapies for PRDs with only a few groups in Africa [9], including the authors of this chapter (DST/CSIR Nanomedicine Platform) [10–12], exploring the application of the technology for PRDs. The CSIR group as well as a group at the University of the Witwatersrand, South Africa, is investigating sustained-release nanodrug delivery systems that will enable anti-TB drugs to be administered at lower doses [9, 12].

Although statistics indicate an urgent need for the development of novel or better drugs, the investment in the research and development (R&D) of these drugs is not significant (Fig. 17.1). Pharmaceutical companies have lagged in the discovery of drugs for the diseases of the developing world due to the cost of the R&D, the risk involved and the time-consuming nature of this field. This is exemplified by a simple comparison of the global TB drug pipeline and the Novartis cancer drug pipeline (Fig. 17.2) where there are only 2 compounds in phase III for TB [13] and

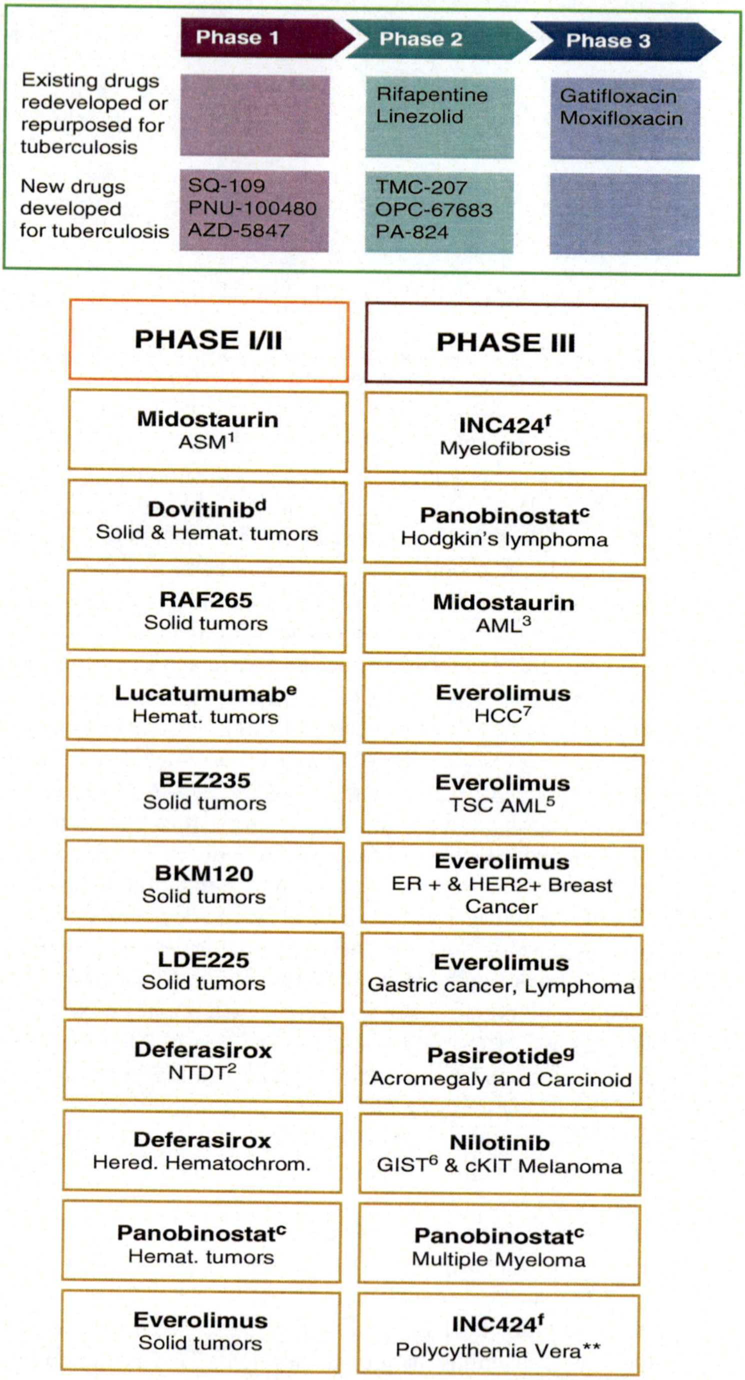

Fig. 17.2 The global TB drug development (**a**) pipeline is less promising than the Novartis oncology pipeline (**b**)

11 for cancer [14]. In the case of NTDs which, unlike HIV, malaria and TB, do not spread widely to high-income countries, there is even less incentive to industry to invest in developing new or better products for a market with low returns. Thus, for drug discovery and development for PRDs, where minimal returns if any can be expected, new approaches such as nanotechnology have to be explored.

To address the challenges in the treatment of PRDs, the investigation into nanomedicine by African researchers has revealed promising approaches for improving treatment of TB. Basic research in nanomedicine for malaria, leishmaniasis, HIV/AIDS and schistosomiasis is also being carried out, but no one is seriously developing a product in this regard.

17.2 Pharmacokinetics in Drug Development and Benefits of Nanomedicine

Pharmacokinetics (PK) is the science that describes the processes of bodily absorption, distribution, metabolism and excretion (ADME) of compounds and medicines. In drug development, PK parameters are required to determine route of administration and dose regimen.

Absorption describes the movement of molecules from the site of administration to the systemic circulation. Distribution is the movement from systemic circulation to extravascular sites. Metabolism is the enzymatic biotransformation of the molecules, and excretion is the passive or active transport of molecules into, e.g. bile and urine [15].

The oral route of drug administration is preferred due to its convenience and cost-effectiveness. However, to be absorbed into the systemic circulation and reach its target site, a drug must be able to cross cell membranes. In fact, each of the ADME processes involves passage of compounds across cell membranes. Several routes may be utilised depending on the physicochemical properties of the compound. Generally, lipophilic compounds are rapidly absorbed because they distribute into the cell membranes of epithelia via the passive transcellular route. Hydrophilic compounds are absorbed more slowly due to their poor distribution into cell membranes. Such compounds are, therefore, more likely to be transported by carrier-mediated pathways.

The bioavailability is the fraction of an administered dose of drug that reaches the systemic circulation. When administered intravenously, the bioavailability is 100%. When administered by other routes such as orally, the drug must first be absorbed in the intestine, which may be limited by efflux transporters such as P-glycoprotein in the intestinal epithelium. As the drug passes through the liver and intestine, metabolism mainly by the cytochrome P450 (CYP) family of enzymes (first-pass metabolism) and further excretion may take place thus reducing bioavailability.

Nanomedicine offers an alternative to address PK-related shortfalls in drug development, and the following sections will discuss the properties that make them advantageous as emerging therapies.

17.2.1 Factors Affecting Drug Development for PRDs

Poor PK is a major cause of PRD treatment failure due to the inability to achieve effective drug levels (poor solubility and intestinal permeability leading to poor bioavailability for orally administered drugs), production of toxic effects (poor elimination or levels above therapeutic levels) and drug interactions. For example, zalcitabine, an antiretroviral (ARV) drug, was discontinued due to adverse side effects and drug interactions [16]. The ultimate result is poor patient compliance which in turn leads to emergence of resistance. The small number of current drugs for PRDs is inadequate to address these treatment challenges, and development of new drugs is high on the agenda.

Drug discovery and development are long and complex, more so for PRDs which in addition to being pharmacologically active must meet the following criteria: oral administration with good bioavailability, well tolerated with minimal side effects and short treatment course [17]. A look at the PRD drug development pipeline reveals that there are too few compounds in clinical development with 10 for TB [13] and 17 for malaria [18] and even fewer for NTDs [19]. It is well known that the majority of compounds entering clinical testing do not make it to market due to poor PK, poor efficacy, side effects and toxicity [20]. The clinical success rate for infectious diseases has been estimated at 15% with a failure rate of about 60% at phase II [20]. Therefore, the need to strengthen the pipeline for PRDs to ensure that new products emerge requires a range of solutions. Strategies to increase the development of new treatments include re-optimising the use of current drugs, repurposing drugs used to treat other diseases, exploring natural resources and modifying existing drugs [18]. This chapter will endeavour to show the advantage of including nanomedicine in drug development programmes. The modification of existing drugs using nanomedicine has revolutionised treatment of diseases such as cancer but has not been extensively applied to PRDs. Doxil® and Abraxane® are two of several nanomedicine-based cancer therapies already on the market. Doxil® is a liposomal formulation of the anthracycline drug doxorubicin. It is used to treat cancer in AIDS-related Kaposi sarcoma and multiple myeloma. Its advantages over free doxorubicin are greater efficacy and lower cardiotoxicity due to altered PK [3]. Abraxane® consists of the anticancer drug paclitaxel bound to human albumin nanoparticles which confers it with a longer circulation half-life [3].

17.2.2 Pharmacokinetics of Nanomedicines

Nanotechnology-based therapies can lead to improved half-life, controlled release over short or long durations and highly specific site-targeted delivery of therapeutic compounds. This section will explain how nanomedicine can attain these improvements.

Nanopharmacokinetics [21] is distinct from pharmacokinetics of small molecules. The latter depends mainly on diffusion and transport (through blood) or metabolism as outlined in Sect. 17.2.1. However, nanopharmacokinetics is defined by physiological processes undergone by nanomaterials such as cellular recognition, opsonisation, adhesion, lymphatic transport and uptake processes such as phagocytosis [21]. The reduction in blood concentrations of nanomaterials might be related to movement into tissue from which further excretion does not occur. Indeed, many nanomaterials tend to accumulate in the liver and to be sequestered in the reticulo-endothelial system (RES) or bound to tissue proteins. In addition, nanomaterials may be transported through lymphatic pathways which must be taken into account in pharmacokinetic analysis based on blood sampling. However, this altered pharmacokinetics at the nanoscale means that nanomedicines present pharmaceutic improvement as drug delivery systems as they can:

- Improve drug stability *ex vivo* (long shelf life) and *in vivo* (protection from first-pass metabolism) [22, 23]
- Have a high carrying capacity (ability to encapsulate large quantities of drug molecules) [23]
- Incorporate hydrophilic and hydrophobic substances [23]
- Increase drug dissolution rate, leading to enhanced absorption and bioavailability [24]
- Target to specific tissues due to selective uptake by those tissues [3]
- Reduce clearance to increase drug half-life for a prolonged pharmacological effect [3]
- Present the capacity to be formulated for the purpose of controlled release [25], therefore posing the possibility to reduce dose frequency and subsequent dose-related side effects [26]
- Be actively targeted to a specific site by functionalising the nanoparticle surface with specific molecules or ligands such as monoclonal antibodies, RNA/DNA aptamers or peptides to enhance binding and interactions with specific receptors which are expressed by the cell populations at the diseased site [27] and thus reduce toxicity

The protection from first-pass metabolism is an important factor in enhancing systemic bioavailability. However, in terms of intracellular PK, targeting with specific ligands further enhances the intracellular bioavailability due to enhanced drug delivery directly into target cells [24].

17.2.2.1 Physicochemical Factors Influencing PK of Nanocarriers

When material is at a nanometre size range, it acquires unique physical and chemical properties. Specifically, the physicochemical properties attributed to the effectiveness of nanocarriers include the nano-sized range, surface properties and relative hydrophobicity.

Size

The sub-micron size of nanoparticles offers a number of distinct advantages, e.g. the ability to reach virtually all tissues in the body, particularly for particles less than 100 nm in size [28]. Desai et al. (1997) demonstrated that 100-nm-size nanoparticles showed 2.5-fold greater uptake compared to 1 μm and sixfold higher uptake compared to 10 μm microparticles in Caco-2 cell line [29]. This aspect of intracellular uptake is more so critical for intracellular pathogens such as infectious diseases, where the drug needs to act intracellularly. Thus, by nanoencapsulating the drug, one can attain intracellular delivery of drugs. Furthermore, these particles can cross barriers that in general make it difficult for conventional therapeutic compounds to reach the target. Reports on nanoparticles crossing the blood-brain barrier (BBB), the stomach epithelium and even the skin have been presented [30]. In addition, orally administered nanoparticles can enter the lymphatic system through intestinal Peyer's patches, followed by uptake via M cells.

Surface Properties

The surface charge in nanoparticles reflects the electrical potential of particles and is influenced by the chemical composition of the particle and the medium in which it is dispersed. A positive surface charge which can be attained by attaching positively charged polymers such as chitosan on the surface of nanoparticles enhances attachment to the negatively charged cellular membrane, thus improving cellular uptake. Chitosan-based or chitosan-coated particles have been reported to efficiently be taken up by cells and also cross cellular barriers such as the BBB. This is as function of chitosan opening the tight junctions between cells and thus facilitates transcellular particle transport [31]. The surface charge in nanoparticles reflects the electrical potential of particles and is influenced by the chemical composition of the particle and the medium in which it is dispersed. In the case of drug delivery, opsonisation, a process that involves the adsorption of proteins particularly of the complement system, to any foreign material, is also influenced by zeta potential. These proteins make the particle more susceptible to phagocytosis thus leading to their clearance from the body. To circumvent this effect, various groups have coated the particles with hydrophilic polymers, such as polyethylene glycol (PEG), Pluronics etc., thus affecting both the surface charge and hydrophobicity of the particles and therefore increasing the circulation time of the particles in the blood and in turn prolonging the release of the drugs from the particles [32, 33]. Thus, minimising opsonisation via changing the surface charge is important for controlled-release formulations. In addition, by coating the polymeric particles with hydrophilic polymers, the half-life of the drugs can be improved and thus their efficacy. This approach can reduce the dose and dose frequency of many effective but poorly soluble drugs and thus in turn minimise the adverse side effect since lower doses will be administered. Furthermore, nano-sized particles have a larger surface area due to the fact that a decrease in particle size results in an increase in surface-to-volume ratio and that size is inversely proportional to specific surface area. This larger surface area allows for a higher loading of the drug, thus leading to a reduction in the dose administered [34].

Hydrophobicity

Aqueous solubility, gastrointestinal permeability and low first-pass metabolism are important for high oral bioavailability. Nano-based drug delivery systems can increase drug dissolution rate, leading to enhanced absorption and bioavailability [24]. A combination of both particle surface charge and increased hydrophobicity of the material has been reported to improve gastrointestinal uptake in case of oral delivery. Hydrophobicity also plays a role in the drug release profile by impacting the kinetics of the degradation of the polymeric shell. Mittal et al. (2007) reported that by changing the hydrophobicity of a nanocarrier, the structure/composition of the polymer/copolymer or the molecular weight, the polymer degradation and thus the drug release mechanism and/or duration are impacted [35]. Nanoparticles have the advantage of improving the solubility of drugs, particularly for the very hydrophilic or poorly soluble drugs which in most cases are not easy to formulate and have poor bioavailability. By encapsulating these drugs into polymeric particles, which are coated with hydrophilic polymers, the solubility of the drugs can be greatly enhanced, in turn improving the bioavailability of the drug. Kondo et al. (1993) documented an increase in bioavailability as a result of a 10-fold reduction in particle size, which is a result of an increase in surface area and consequently an increase in dissolution rate [34].

17.2.3 Functional Nanocarriers Used in Drug Delivery

A drug delivery system is defined as a formulation or a device that enables the introduction of a therapeutic substance in the body and improves its efficacy and safety by controlling the rate, time and location of release of drugs in the body. Nanotechnology has been increasingly used in drug delivery for nanoencapsulation of medicinal drugs (nanomedicine) [36]. Several nanocarrier devices (Table 17.1, Fig. 17.3) have been used for nanodrug delivery applications. The nanocarriers may be further modified for active disease targeting by functionalizing the surface with

Table 17.1 Nanotechnology-based drug delivery systems

Nanocarrier	Characteristics
Liposomes	Self-assembling spherical, closed colloidal structures composed of phospholipid bilayers that surround a central aqueous space [3]
Polymeric micelles	Supramolecular assembly of amphiphilic block copolymers or polymer-lipid based conjugates [37–39]
Dendrimers	Globular repeatedly branched macromolecules exhibiting controlled patterns of branching with multiple arms extending from central core [40]
Solid lipid nanoparticles	Particulate systems made from lipids where melted lipids are dispersed in an aqueous surfactant by high pressure homogenization or emulsification [41]
Polymeric nanoparticles	Solid colloidal particles existing as nanospheres (matrix structure) or nanocapsules (polymeric shell and inner liquid core). Engineered from synthetic or natural polymers. The former are essentially polyesters and poly-acids including polylactic acid (PLA), poly(D,L-lactic-co-glycolic acid) (PLGA), polycaprolactone (PCL) and poly(butyl-2-cyanoacrylate) (PBCA). The latter include oligomers that are abundant in nature such as chitosan, alginate and starch [42, 43]

ligands such as antibodies, aptamers, peptides or small molecules that recognise disease-specific antigens (Fig. 17.4). In this way, the nanoparticles become "multiple nanocarriers". For example, a nanoparticle may be functionalised with aptamers to recognise macrophages infected with TB.

Some nanomedicine products currently on the market are summarised in Table 17.2 from which it can be noted that very little progress in the area of PRDs has been made. There is currently no nanomedicine-based product on PRDs. However, African research institutes are now initiating research in the application of nanomedicine to improve PRD therapies which shall be discussed in Sect. 17.3.

17.3 Nanomedicine Research for PRDs in Africa

The field of nanotechnology is relatively new in Africa and is not well exploited in terms of its application to the improvement of PRD therapies. The most significant progress has been made by research groups mainly in South Africa due to the expensive infrastructure the nanotechnology requires. The government of South Africa has taken nanotechnology very seriously providing all the support required as outlined in the following section. In the rest of sub-Saharan Africa, nanotechnology activities are minimal.

17.3.1 Nanomedicine Research for PRDs in South Africa

In South Africa, the national Science and Technology Ministry (the Department of Science and Technology, DST) has been the principal agency guiding

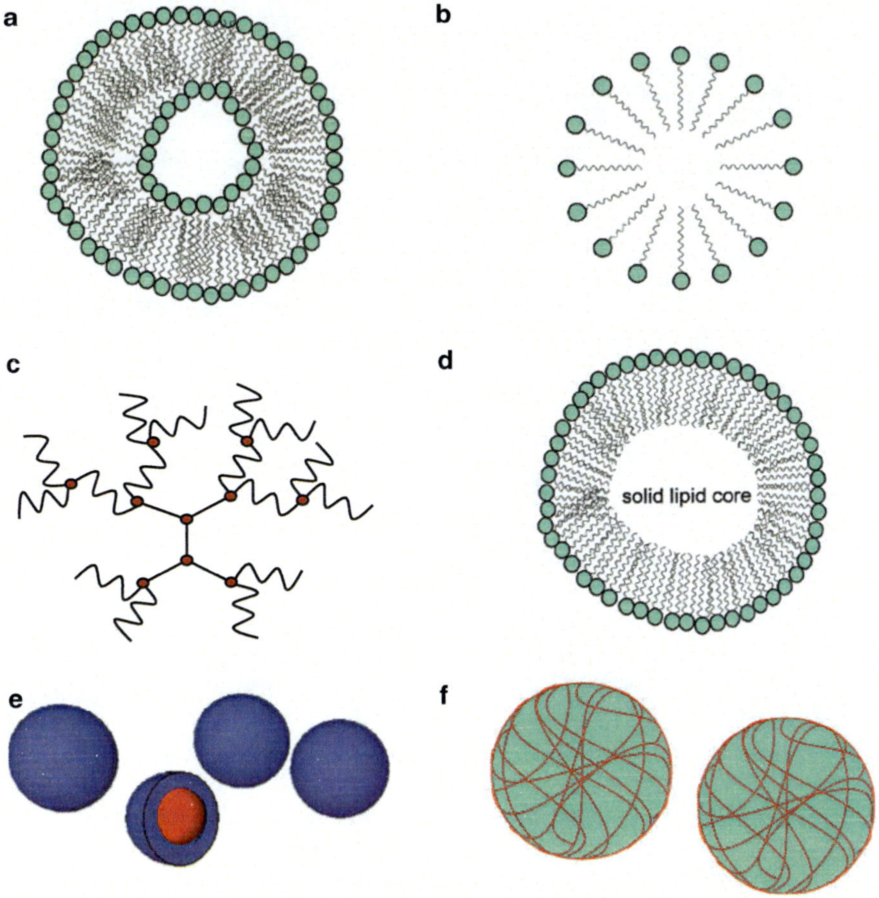

Fig. 17.3 Schematic illustration of nanotechnology-based drug delivery systems, (**a**) liposome, (**b**) polymeric micelles, (**c**) dendrimer, (**d**) solid lipid nanoparticle, (**e**) nanocapsules, (**f**) nanospheres

nanotechnology research direction and policy. In 2007, the DST launched a national nanotechnology strategy with six focus areas of high priority for the country. One of the focus areas is health with the aim of using nanomedicine to improve drug delivery systems, including traditional medicine through packaging medicine for ailments such as TB, HIV/AIDS and malaria in nanocapsules. In this regard, a nanotechnology flagship project (DST/CSIR Nanomedicine Platform) led by the authors of this chapter is being used to develop a drug delivery system for the existing TB drugs, to enhance their efficacy and to reduce dosage and dose frequency. This flagship project has now grown into a nanomedicine centre of excellence for poverty-related diseases for Africa. The centre is one of the recognised African Network for Drug Diagnostics and Innovation (ANDI) centres of excellence.

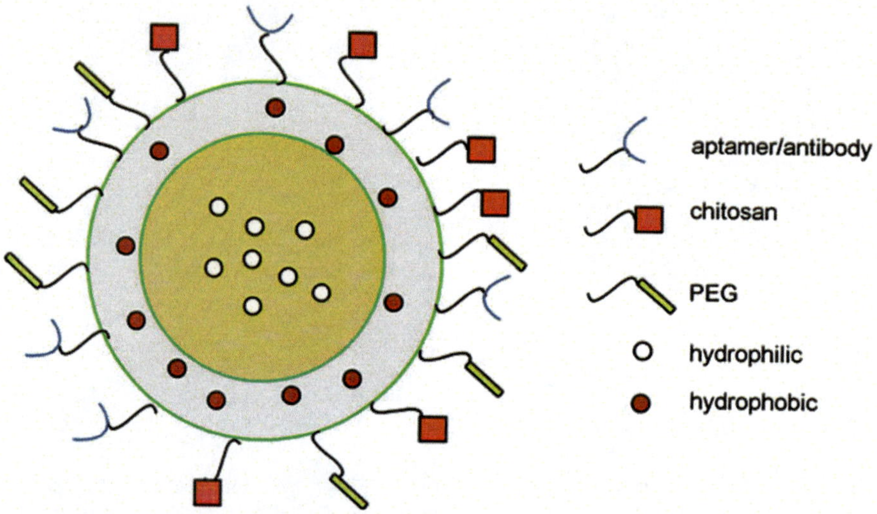

Fig. 17.4 Schematic illustration of a multifunctional nanocarrier

Other South African institutions carrying out nanomedicine research for PRDs include the University of the Witwatersrand (Wits) and North-West University. The group at Wits has also been recognised as a centre of excellence in drug delivery by ANDI.

17.3.1.1 CSIR ANDI Centre of Excellence in Nanomedicine Research

The authors of this chapter are applying nanomedicine to enhance efficacy, half-life, safety, structure and function of TB, malaria and HIV drugs. In addition, we have been spearheading several nanomedicine sensitization activities on the continent, e.g. establishing nanomedicine research programmes in Kenya, hosting international nanomedicine workshops, summer schools and lab exchange programmes.

Research in Progress for Improving TB Treatment Through Nanomedicine

We have encapsulated anti-TB drugs using a novel spray-drying technique as well as a freeze-drying technology. We will illustrate how we have managed to modify physiochemical properties of the particles and attain sustained drug release over a period of days, both *in vitro* and *in vivo*. We further indicate that our particles are taken up by cells and also that the activity of the drugs against *Mycobacterium tuberculosis* is still maintained in the process of encapsulation.

Nanoencapsulation of Anti-TB Drugs in PLGA Nanoparticles
Poly(D,L-lactic-co-glycolic acid) (PLGA) 50:50 (Mw: 45,000–75,000) nanoparticles were loaded with anti-tuberculosis drug prepared using a patented

Table 17.2 Nanomedicine-based products currently on the market

Product	Drug	Formulation	Route of administration	Application	Company
Abraxane	Paclitaxel	Albumin-bound nanoparticles	IV injection	Metastatic breast cancer	American Biosciences (Blauvelt, NY)
Amphocil	Amphotericin B	Lipocomplex	IV infusion	Serious fungal infections	Sequus Pharmaceuticals
Ambisome	Amphotericin B	Liposome	IV infusion	Serious fungal infections	NeXstar Pharmaceutical (Boulder, Colorado)
Abelcet	Amphotericin B	Lipid complex	IV infusion	Serious fungal infections	The Liposome Company (Princeton, NJ)
DaunoXome	Daunorubicin citrate	Liposome	IV	Kaposi sarcoma in AIDS	NeXstar Pharmaceutical (Boulder, Colorado)
Doxil	Doxorubicin	Liposome	IV injection	Kaposi sarcoma in AIDS	Sequus Pharmaceuticals
Elestrin	Estradiol	Calcium-phosphate-based nanoparticles	Transdermal	Moderated to severe vasomotor symptoms (hot flashes) in menopausal women	BioSante (Lincolnshire, Illinois)
Emend	Aprepitant, MK869	Nanocrystal particles	Oral	To delay nausea and vomiting	Merck/Elan(Whitehouse Sation, NJ)
Megace ES	Megaestrol acetate	Nanocrystal particles	Oral	Anorexia, cacheixa or unexplained significant weight loss	PAR Pharmaceutical (WoodCliff Lake, NJ)
Rapamune	Sirolimus	Nanocrystal particles	Oral	Immunosuppressant in kidney transplant patients	Wyeth/Elan (Madison, NJ)
Tricor	Fenofibrate	Nanocrystal particles	Oral	Primary hypercholestrolemiamixed lipidemia, hypertriglyceridemia	Abbott(Abbot Park Illinois)

IV intravenous, *NY* New York, *NJ* New Jersey

Table 17.3 Characterisation of nanoparticles ($n = 3$)

Drug	Type of drying	Ave size ± SD (nm)	Zeta potential (mV)
INH	Freeze-dried	210 ± 13	−14 ± 2
INH	Spray-dried	321 ± 33	+19 ± 1
RIF	Freeze-dried	280 ± 23	−10 ± 4
RIF	Spray-dried	297 ± 22	+16 ± 2

Adapted from [44]
SD standard deviation

multiple emulsion-solvent-evaporation technique followed by freeze-drying or spray-drying. Polyvinyl alcohol (PVA) was included as a stabiliser, polyethylene glycol (PEG) to increase bloodstream residence time and chitosan as a mucoadhesive, positively charged polymer to enhance gastrointestinal uptake. Using this technique, we have successfully encapsulated all four first-line anti-TB drugs, i.e. RIF, INH, ETB and PZA, in PLGA nanoparticles for oral delivery, with an encapsulation efficiency of 50–65% for INH and RIF, 84% for PZA and 60% for ETH [12], in particles of 250–350 nm [44]. A PCT patent application has been filed (WO 2009/105792) and has already proceeded to the national phase, with the European patent granted recently.

All samples made via freeze-drying showed a negative zeta potential. The addition of chitosan to provide positive surface charge resulted in microparticles. This problem was overcome by spray-drying the double emulsion containing chitosan and PEG in the formulation as shown in Table 17.3 for INH and RIF.

The particles were relatively uniform with an average polydispersity index of 0.2, and analysis of surface morphology revealed a smooth spherical surface achieved by the addition of lactose to the formulation (Fig. 17.5) [44]. Spherical particles offer maximum volume for drug penetration, and it has been reported that spherical particles possess the right curvature allowing attachment onto the cell [45] giving rise to enhanced efficiency of cell internalisation.

In Vitro and In Vivo Characterisation of PLGA Nanoparticles

The PLGA nanoparticles used to encapsulate anti-TB drugs were evaluated *in vitro* and *in vivo* with respect to cellular uptake and biodistribution. To investigate intracellular uptake, Caco-2 cells were exposed to rhodamine-labelled PLGA nanoparticles prepared in the same manner as the anti-TB drug nanoparticles. The labelled particles were taken up by Caco-2 cells and appeared to co-localise with lysosomes (Fig. 17.6) [44]. This indicates the feasibility of intracellular uptake by intestinal enterocytes in patients. *In vivo*, the PLGA nanoparticles were taken up by macrophages of the peritoneum when administered orally and peritoneally to female Balb/C mice [11].

The PLGA nanoparticles displayed no toxicity towards Caco-2 and HeLa cells as determined via the WST assay [10]. Subsequent to oral administration to mice, the particles remained detectable in the brain, heart, kidney, liver, lungs and spleen after 7 days, with the liver being the major organ of accumulation (Fig. 17.7). However, no pathological lesions were detected in any of the organs [10].

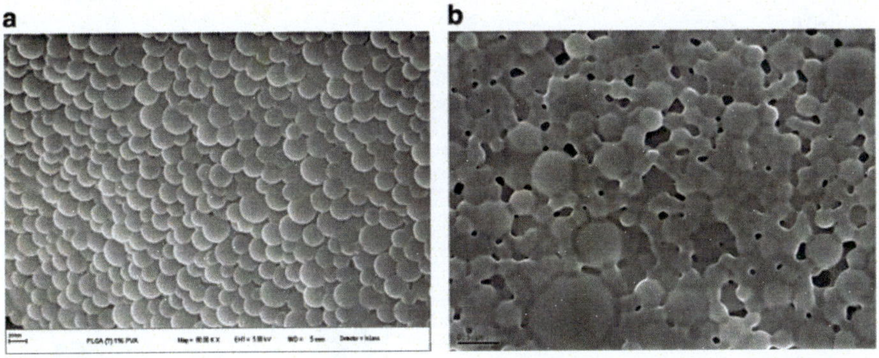

Fig. 17.5 (**a**) SEM image of spray-dried particles with lactose. (**b**) Spray-dried particles without 5% w/v lactose

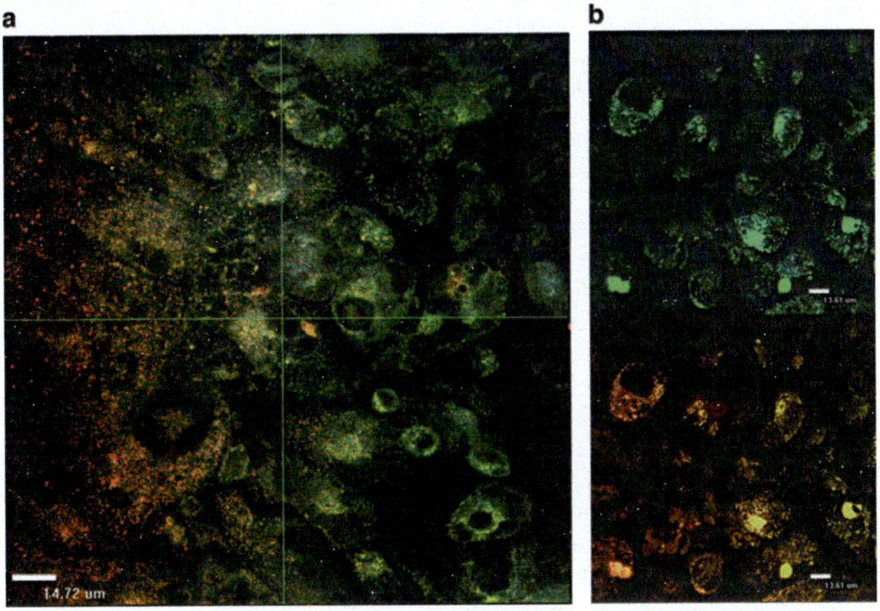

Fig. 17.6 (**a**) Indicates a Z-stack of 30 min incubation and (**b**) depicts a 60 min incubation period. Rhodamine-loaded nanoparticles co-localised with the lysosomes, as indicated by the lighter shaded areas

In Vitro and In Vivo Characterisation of Nanoencapsulated Anti-TB Drugs
The nanoparticles containing anti-TB drugs were evaluated with respect to release of the drugs from the nanoparticles as well as efficacy.

In vitro release assays in phosphate-buffered saline (PBS) showed that the drugs were released in a slow manner over a period of several days preceded by an initial

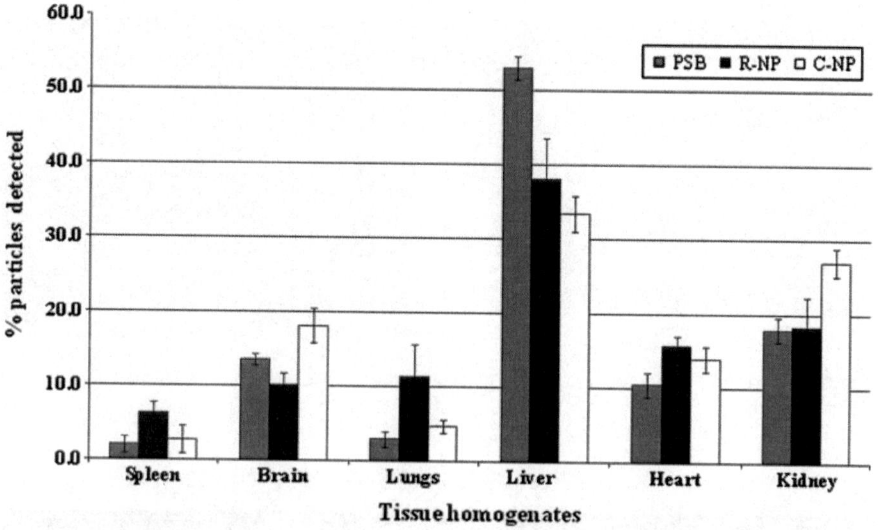

Fig. 17.7 Tissue distribution of nanoparticles after 7 days graphically represented as a measure of percentage of particles detected of the total particles. The data represent three repeats of $n = 6$; *error bars* indicate SEM. *PSB* polystyrene beads, *R-NP* rhodamine nanoparticles, *C-NP* coumarin nanoparticles

burst release. Since hydrolytic enzymes were not included in the PBS, the slow rate of nanoparticle degradation could be attributed to this factor. Faster release rates should be observed in the biological milieu with hydrolytic enzymes present.

The *in vitro* potency of encapsulated INH and RIF with free INH and RIF was compared using the Bactec 460 assay. The Bactec 460 assay is generally conducted to analyse the susceptibility of *M.tb* to test drugs. The efficacy of the encapsulated anti-TB drugs against $H_{37}R_V$ was comparable to the free drugs (Fig. 17.8) [44]. Therefore, the multiple emulsion spray-drying technique does not have any effect on the potency of the drugs.

When orally administered to mice, nanoparticles containing INH and RIF maintained a sustained-release profile (Fig. 17.9) over a period of at least 5 days when compared to free drugs which reached levels below the minimum inhibitory concentration (MIC) within 16 h. With the encapsulated drugs, drug concentration in plasma above the MIC level of RIF and INH was sustained for the 5 days [44].

An efficacy study in which equal doses of free anti-TB drugs were administered to TB-challenged mice once every day and encapsulated drug once every 7 days indicated comparable efficacy (unpublished data).

These are important results because they confirm the feasibility of slow release and reduced dose frequency.

Targeting of Nanoencapsulated Anti-TB Drugs

The PLGA nanoparticles containing anti-TB drugs were further functionalised with mycolic acids (MAs) or nucleic acid aptamers for active targeting of *Mycobacterium tuberculosis*-infected macrophages. MA (a lipid molecule on the cell wall of

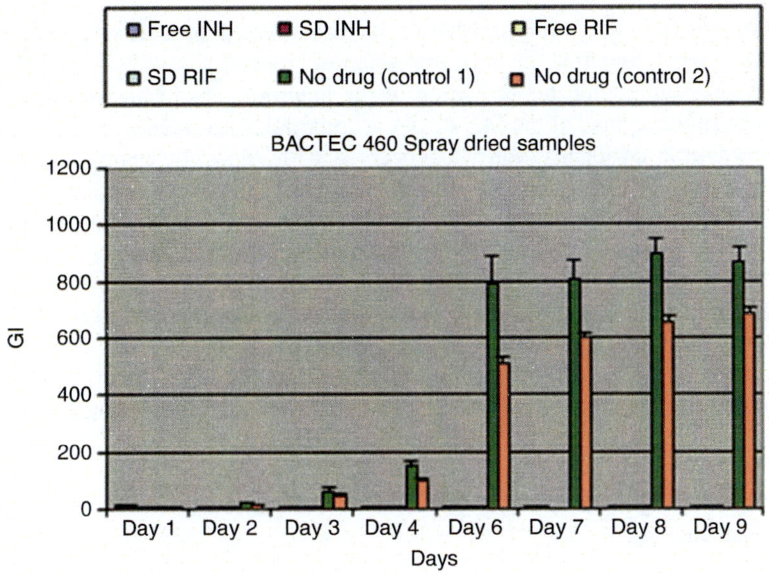

Fig. 17.8 BACTEC 460 data indicating bacterial growth index of $H_{37}R_V$ treated with encapsulated RIF and INH and unencapsulated drugs. *SD* spray-dried

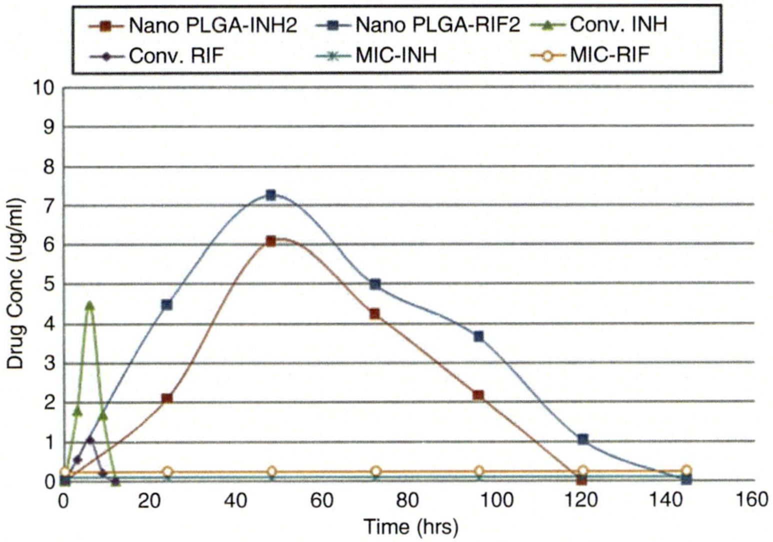

Fig. 17.9 In vivo release of free drugs versus spray-dried nanoparticles encapsulating RIF and INH and PZA. *Conv.* Conventional, *MIC* minimum inhibitory concentration

M. tuberculosis) was explored due to its cholesteroid properties [46], and the aptamers were prepared against the mannose receptor, which is significantly over-expressed during the activation of the macrophages in the presence of *M. tb*. Intracellular uptake of the MA PLGA nanoparticles was achieved in U937 cells. However, little co-localization was observed with endocytic markers, indicating that they could be localised in the cytosol. Vesicles bearing these particles were also observed in the cell membrane of the cells [47]. Uptake of the aptamers into THP-1 cells was also observed, illustrating the feasibility of using the nucleic acid species for active targeted delivery of the encapsulated anti-TB drugs [47]. A provisional patent application titled "High Affinity Nucleic Acid Ligands to the Mannose Receptor" has already been filed on the method. The success of these two approaches of anti-TB drug targeting will greatly address the challenges of poor bioavailability, reduced efficacy and adverse side effects for diseases such as TB.

Research in Progress for Improving HIV and Malaria Treatment Through Nanomedicine

Based on the successes and experiences obtained through the research work on nanomedicine for TB, the authors have begun on nanoencapsulation of antiretroviral and antimalarial drugs. To date, efavirenz and lamivudine have been encapsulated in PCL nanoparticles with an average size of 230 nm (unpublished data). For malaria, nanocarriers are being designed to target parasites in the liver (pre-erythrocytic) and the red blood cell (erythrocytic) stage of the parasites transmission cycle. Prophylactic and curative measures of the chemical agents will be investigated before and after the application of drug delivery systems.

Research Strategy for Improving NTDs Using Nanomedicine

The parasites causing NTDs such as leishmaniasis and trypanosomiasis often disseminate throughout the RES, e.g. leishmaniasis in the lymph nodes [48] and schistosomiasis in the spleen [49]. Therefore, the strategy for nanomedicine for these diseases is to take advantage of the selective uptake of nanocarriers by the RES which may be further enhanced by actively targeting the nanocarriers to the parasites in the, e.g. lymphatic system.

Activities to Build Nanomedicine Research Capacity in Africa

Towards advancing nanomedicine and the benefits of the technology in Africa, the authors organised the first international sensitisation workshop on nanomedicine for infectious diseases of poverty, in South Africa on March 2011. Officially opened by the minister of the Department of Science and Technology, this workshop brought together about 90 delegates from over 20 different countries and included representatives from academia, the pharmaceutical industry, regulatory authorities, donor agencies, international organisations and policymakers, all interested in supporting the advancement of nanomedicine in Africa. The workshop comprised a panel of highly accomplished experts in various aspects of nanomedicine and drug delivery as well as experts in drug development for poverty-related diseases. Oral and poster presentations encompassed basic science through to translational efforts and addressed topics on various initiatives and funding. The 4-day workshop featured plenary lectures, invited talks and round table discussions focusing on specific tenets of nanomedicine and drug development. The fourth day was dedicated to discuss intellectual property rights and technology transfer, an aspect which must be kept in mind when developing new technologies.

Following the workshop, the authors presented a series of nanomedicine sensitisation seminars (road shows) to students and young researchers at a total of 18 institutions in Kenya, Nigeria and Ethiopia with more seminars planned for Cameroon and other African countries such as Uganda, Sudan and Tanzania. These nanomedicine road shows highlight the urgent need for more in-depth training in nanomedicine for PRDs. Accordingly, the authors are planning the first Pan-African summer school in nanomedicine for PRDs, in collaboration with leading nanomedicine experts that have nanomedicines on the market and also have experience in operating such nanomedicine schools and conferences in Europe and the USA annually, as well as African PRD experts. The school aims to bridge the gap between the sciences, health and development in Africa, by educating young African scientists on the potential of applying nanomedicine in PRD drug development research. To achieve this, the school will focus on crucial areas to build capacity in nanomedicine. Furthermore, the school will assist in establishing networks and collaborations among trainees, to ensure that every trainee can confidently enhance knowledge dissemination and skills acquisition. The school will also encourage the young scientists to bring with them any compound which has failed to reach the market due to the above-mentioned shortfalls. In this workshop, they will have the opportunity to apply different nanocarriers to address the shortfalls.

17.3.1.2 University of the Witwatersrand (Wits)

The Wits Advanced Drug Delivery Platform (WAADP) is focused on advancements in polymeric science, formulation stability and drug delivery design including nanomedicine for infectious diseases such as TB. In a recent publication, the group evaluated sustained release of INH and RIF from polymeric nanoparticles synthesised via four emulsion-based processing strategies, namely emulsion-solvent-surfactant-evaporation (ESSE) and emulsion-solvent-evaporation (ESE) approaches for PLGA nanoparticles and reverse-emulsion-cationic-gelification (RECG) and reverse-emulsion-surfactant-cationic-gelification (RESCG) approaches for alginate hydrogel nanoparticles [9]. Encapsulation efficiencies were in the range of 73–82%. The ESSE and RESCG approaches which included sorbitan monooleate as a stabiliser yielded smaller sizes of nanoparticles in the range of 200–290 nm for INH and RIF and displayed sustained release over 8 h with zero-order kinetics in vitro.

Another group at Wits, the Antiviral Gene Therapy Research Unit (AGTRU), is using nanocarriers [50] to deliver nucleic acids that are capable of silencing gene expression of viruses that are responsible for infections of serious public health importance to South Africa such as HIV infection [51].

17.3.1.3 North-West University (NWU)

The Unit for Drug Research and Development at the NWU is conducting research aimed at optimising the delivery of anti-TB and antimalarial drugs using Pheroid™ technology. Pheroid™ technology is a drug delivery system patented by the NWU which can be described as a colloidal system that contains stable, submicron- and micron-sized active pharmaceutical ingredient dispensing vehicles. Recently, entrapment of the new artemisinin derivative, artemisone, in Pheroid™ vesicles has been shown to significantly enhance the absorption of the drug. The C_{max} was improved by 90%, and the $T_{1/2}$ increased three times after oral administration in a mouse model [52]. In addition, a Pheroid™ formulation for TB drugs is currently undergoing phase I clinical trials. The CSIR and NWU research groups are now collaborating on entrapping PLGA nanoparticles in Pheroids to further improve bioavailability and achieve controlled release for TB drugs.

17.3.2 Nanomedicine Research for PRDs in the Rest of Africa

In the rest of sub-Saharan Africa where PRDs are endemic, there is little advancement in nanomedicine research for the treatment of these diseases. A few groups exist carrying out basic research into nanomedicine-based therapies, with only two identified thus far at the University of Mauritius (UOM) and American University in Cairo, focusing on PRDs.

At the 4th ANDI Conference in October 2011, the Centre for Biomedical and Biomolecular Research at the UOM presented its unpublished work focusing on engineering novel block copolymer nanomicelles for the delivery of anti-TB drugs. The group has engineered amphiphilic block copolymers based on poly(ester-ether)s, polyLysine-b-caprolactone and oligoagarose-g-polycaprolactone. They reported loading of rifampicin up to 70% and sustained drug release over 72 h. The group in Cairo is investigating nanomedicine for schistosomiasis and filariasis but has not published any data as yet.

In terms of non-PRD nanomedicine-based therapies, Prof. Wole Soboyejo at the African University of Science and Technology (Abuja, Nigeria) is working on nanoparticles for cancer detection and treatment in collaboration with Princeton University, USA (Personal communication). In Ghana, Dr. Ofori-Kwakye and

Dr. Stanley Moffat are conducting basic research in pharmaceutical nanotechnology. Dr. Moffat was recently appointed the African coordinator for USEACANI (US-Europe-Asia Pacific-Caribbean Nanotechnology Initiative).

17.4 Conclusions

The number of discovery programmes for PRDs is too low to ensure a steady stream of treatments on to the market [53]. This is mainly due to the lack of activity from the pharmaceutical industry because refinancing the high development costs will not be profitable. Only 1.3 products are expected to reach the market out of 100 entering the screening phase of drug discovery [53]. These figures indicate that there is an urgent need for new strategies, such as nanomedicine, in drug development programmes for PRDs. Nanomedicine has been successfully applied for treatment of cancer with several products already on the market. Critical properties of nanomedicine systems include protection of instable drugs, cell-adhesion properties, intracellular delivery of drugs and the ability to be surface-modified by conjugation of specific ligands, enabling targeted delivery and controlled release. Thus, nanodrug delivery systems seem to be a promising and viable strategy for improving treatment of PRDs. However, in Africa, there is minimal application of this technology for the treatment of PRDs with only a few groups in South Africa making significant progress. Therefore, serious efforts need to be focused on the exploitation of the potential of applying nanomedicine in drug development for PRDs. We believe this is one way of taking failed leads through commercialisation and ultimately bridging the 90/10 gap. To this end, the DST/CSIR nanomedicine platform is sensitising African researchers and building capacity to include nanomedicine in drug development programmes in Africa.

References

1. Bawa R, Bawa SR, Maebius SB et al (2005) Protecting new ideas and inventions in nanomedicine with patents. Nanomedicine 1:150–158
2. Mishra B, Patel BB, Tiwari S (2010) Colloidal nanocarriers: a review on formulation technology, types and applications toward targeted drug delivery. Nanomedicine 6:9–24
3. Malam Y, Loizidou M, Seifalian AM (2009) Liposomes and nanoparticles: nanosized vehicles for drug delivery in cancer. Trends Pharmacol Sci 30:592–599
4. WHO (2010) Global tuberculosis control: WHO report 2010. World Health Organisation, Geneva
5. WHO (2010) World malaria report 2010. World Health Organisation, Geneva
6. UNAIDS (2010) UNAIDS report on the global AIDS epidemic
7. Davidson RN (2005) Leishmaniasis. Medicine 33:43–46
8. Anwabani GM (2002) Drug development: a perspective from Africa. Paediatr Perinat Drug Ther 5:4–11

9. Choonara YE, Pillay V, Ndesendo VMK et al (2011) Polymeric emulsion and crosslink-mediated synthesis of super-stable nanoparticles as sustained-release anti-tuberculosis drug carriers. Colloids Surf B Biointerfaces 87:243–254
10. Semete B, Booysen L, Lemmer Y et al (2010) *In vivo* evaluation of the biodistribution and safety of PLGA nanoparticles as drug delivery systems. Nanomedicine 6:662–671
11. Semete B, Booysen LI, Kalombo L et al (2010) *In vivo* uptake and acute immune response to orally administered chitosan and PEG coated PLGA nanoparticles. Toxicol Appl Pharmacol 249:158–165
12. Swai H, Semete B, Kalombo L et al (2008) Potential of treating tuberculosis with a polymeric nano-drug delivery system. J Control Release 132:e48
13. Ma Z, Lienhardt C, McIlleron H et al (2010) Global tuberculosis drug development pipeline: the need and the reality. Lancet 375:2100–2109
14. http://www.novartisoncology.com/research-innovation/pipeline.jsp. Accessed 22 July 2011
15. Jang GR, Harris RZ, Lau DT (2001) Pharmacokinetics and its role in small molecule drug discovery research. Med Res Rev 21:382–396
16. http://hivinsite.ucsf.edu/InSite?page=ar-01-03. Accessed 23 June 2011
17. Nzila A, Chilengi R (2010) Modulators of the efficacy and toxicity of drugs in malaria treatment. Trends Pharmacol Sci 31:277–283
18. Grimberg BT, Mehlotra RK (2011) Expanding the antimalarial drug arsenal-now, but how? Pharmaceuticals (Basel) 4:681–712
19. Chatelain E, Ioset JR (2011) Drug discovery and development for neglected diseases: the DNDi model. Drug Des Devel Ther 5:175–181
20. Kola I, Landis J (2004) Can the pharmaceutical industry reduce attrition rates? Nat Rev Drug Discov 3:711–715
21. Riviere JE (2009) Pharmacokinetics of nanomaterials: an overview of carbon nanotubes, fullerenes and quantum dots. Wiley Interdiscip Rev Nanomed Nanobiotechnol 1:26–34
22. Couvreur P, Vauthier C (2006) Nanotechnology: intelligent design to treat complex disease. Pharm Res 23:1417–1450
23. Gelperina S, Kisich K, Iseman MD et al (2005) The potential advantages of nanoparticle drug delivery systems in chemotherapy of tuberculosis. Am J Respir Crit Care Med 172:1487–1490
24. Li SD, Huang L (2008) Pharmacokinetics and biodistribution of nanoparticles. Mol Pharm 5:496–504
25. Pandey R, Ahmad Z, Sharma S et al (2005) Nano-encapsulation of azole antifungals: potential applications to improve oral drug delivery. Int J Pharm 301:268–276
26. Medina C, Santos-Martinez MJ, Radomski A et al (2007) Nanoparticles: pharmacological and toxicological significance. Br J Pharmacol 150:552–558
27. Kingsley JD, Dou H, Morehead J et al (2006) Nanotechnology: a focus on nanoparticles as a drug delivery system. J Neuroimmune Pharmacol 1:340–350
28. McNeil SE (2005) Nanotechnology for the biologist. J Leukoc Biol 78:585–594
29. Desai MP, Labhasetwar V, Walter E et al (1997) The mechanism of uptake of biodegradable microparticles in Caco-2 cells is size dependent. Pharm Res 14:1568–1573
30. Koziara JM, Lockman PR, Allen DD et al (2003) *In situ* blood-brain barrier transport of nanoparticles. Pharm Res 20:1772–1778
31. Park JH, Saravanakumar G, Kim K et al (2010) Targeted delivery of low molecular drugs using chitosan and its derivatives. Adv Drug Deliv Rev 62:28–41
32. Freiberg S, Zhu XX (2004) Polymer microspheres for controlled drug release. Int J Pharm 282:1–18
33. Mohanraj VJ, Chen Y (2006) Nanoparticles – a review. Trop J Pharm Res 5:561–573
34. Kondo N, Iwao T, Kikuchi M et al (1993) Pharmacokinetics of a micronized, poorly water-soluble drug, HO-221, in experimental animals. Biol Pharm Bull 16:796–800
35. Mittal G, Sahana DK, Bhardwaj V et al (2007) Estradiol loaded PLGA nanoparticles for oral administration: effect of polymer molecular weight and copolymer composition on release behavior *in vitro* and *in vivo*. J Control Release 119:77–85

36. Kumari A, Yadav SK, Yadav SC (2010) Biodegradable polymeric nanoparticles based drug delivery systems. Colloids Surf B Biointerfaces 75:1–18
37. Bae Y, Kataoka K (2009) Intelligent polymeric micelles from functional poly(ethylene glycol)-poly(amino acid) block copolymers. Adv Drug Deliv Rev 61:768–784
38. Gaucher G, Dufresne MH, Sant VP et al (2005) Block copolymer micelles: preparation, characterization and application in drug delivery. J Control Release 109:169–188
39. Jones M, Leroux J (1999) Polymeric micelles – a new generation of colloidal drug carriers. Eur J Pharm Biopharm 48:101–111
40. Svenson S, Tomalia DA (2005) Dendrimers in biomedical applications–reflections on the field. Adv Drug Deliv Rev 57:2106–2129
41. Muller RH, Mader K, Gohla S (2000) Solid lipid nanoparticles (SLN) for controlled drug delivery – a review of the state of the art. Eur J Pharm Biopharm 50:161–177
42. Couvreur P, Barratt G, Fattal E et al (2002) Nanocapsule technology: a review. Crit Rev Ther Drug Carrier Syst 19:99–134
43. Sosnik A, Carcaboso AM, Glisoni RJ et al (2010) New old challenges in tuberculosis: potentially effective nanotechnologies in drug delivery. Adv Drug Deliv Rev 62:547–559
44. Semete B, Kalombo L, Katata L et al. (2011) Potential of improving the treatment of tuberculosis through nanomedicine. Mol Cryst Liq Cryst 556:317–330
45. Trewyn BG, Nieweg JA, Zhao Y et al (2008) Biocompatible mesoporous silica nanoparticles with different morphologies for animal cell membrane penetration. Chem Eng J 137:23–29
46. Benadie Y, Deysel M, Siko DG et al (2008) Cholesteroid nature of free mycolic acids from *M. tuberculosis*. Chem Phys Lipids 152:95–103
47. Lemmer Y, Semete B, Booysen L et al (2008) Targeted nanodrug delivery systems for the treatment of tuberculosis. Drug Discov Today 15:1098
48. Murray HW, Berman JD, Davies CR et al (2005) Advances in leishmaniasis. Lancet 366:1561–1577
49. Abdulla M-H, Lim K-C, Sajid M et al (2007) *Schistosomiasis mansoni*: Novel chemotherapy using a cysteine protease inhibitor. PLoS Med 4:130–138
50. Islam RU, Hean J, van Otterlo WAL et al (2009) Efficient nucleic acid transduction with lipoplexes containing novel piperazine- and polyamine-conjugated cholesterol derivatives. Bioorg Med Chem Lett 19:100–103
51. Arbuthnot P (2009) Applying nanotechnology to gene therapy for treatment of serious viral infections. Nano News, South Africa. http://www.sani.org.za/pdf/NanoNovember09.pdf. Accessed 24 Oct 2011
52. Steyn JD, Wiesner L, du Plessis LH et al (2011) Absorption of the novel artemisinin derivatives artemisone and artemiside: potential application of Pheroid technology. Int J Pharm 414:260–266
53. Lowell JE, Earl CD (2009) Leveraging biotech's drug discovery expertise for neglected diseases. Nat Biotechnol 27:323–329

Index

K. Chibale et al. (eds.), *Drug Discovery in Africa*,
DOI 10.1007/978-3-642-28175-4, © Springer-Verlag Berlin Heidelberg 2012